PRINCIPLES AND PRACTICE OF CLINICAL RESEARCH

PRINCIPLES AND PRACTICE OF CLINICAL RESEARCH

Second Edition

JOHN I. GALLIN

FREDERICK P. OGNIBENE

Amsterdam • Boston • Heidelberg • London
New York • Oxford • Paris • San Diego
San Francisco • Singapore • Sydney • Tokyo

ELSEVIER

Academic Press is an imprint of Elsevier
30 Corporate Drive, Suite 400, Burlington, MA 01803, USA

This book is printed on acid-free paper. ∞

Library of Congress Cataloging-in-Publication Data

Principles and practice of clinical research / [edited by] John I. Gallin, Frederick P. Ognibene.—2nd ed.
    p. ; cm.
  Includes bibliographical references and index.
  ISBN-13: 978-0-12-369440-9 (hard cover : alk. paper)
  ISBN-10: 0-12-369440-X (hard cover : alk. paper)   1. Clinical medicine Research.   I. Gallin, John I.
II. Ognibene, Frederick P.
  [DNLM: I. Research–methods.   W 20.5 P957 2007]

R850.G35   2007
616.0072—dc22

2007004577

British Library Cataloguing-in-Publication Data
A catalogue record for this book is available from the British Library.

ISBN 13: 978-0-12-369440-9
ISBN 10: 0-12-369440-X

For information on all Academic Press publications
visit our Web site at www.books.elsevier.com

Printed in the United States of America
      11 12 13   9 8 7 6

Table of Contents

Preface

The positive reception of the first edition of *Principles and Practice of Clinical Research* prompted the second edition, which was written in the context of continued growth and scope of clinical research as a discipline since the publication of the first edition in 2002. The course at the National Institutes of Health (NIH) Clinical Center, which led to the production of the first edition, has been in existence for ten years and is now taught to nearly 1,000 students annually at the NIH Clinical Center and at multiple long-distance learning sites, including both domestic and international partners.

This second edition includes new chapters on clinical research from a patient's perspective, managing conflicts of interest in clinical research, the clinical researcher and the media, clinical research from an industry perspective, data management in clinical research, how to evaluate a protocol budget, and the role of the human genome project and genomics in clinical research. All other chapters have been updated with extensive changes in the chapters on technology transfer and how to successfully navigate the NIH peer review process for grants.

We hope that this book provides the reader with an expanded awareness of the broad scope of clinical research and the tools to conduct such research safely and effectively. Our goals as investigators should be to strive to improve the well being of patients in general while ensuring the safety of our research subjects enrolled in investigational protocols.

John I. Gallin, M.D.
Frederick P. Ognibene, M.D.
National Institutes of Health Clinical Center
Bethesda, Maryland

Acknowledgements

The editors extend special thanks to Benita Bazemore for her tremendous support in coordinating the many activities associated with the development of the second edition of this textbook, to Patricia Piringer for her assistance with the design of the cover of the textbook and for assistance in securing a number of images used on the cover and in the text, to Rona Buchbinder for excellent editorial assistance, and Tari Broderick, Renske van Dijk and Philip Bugeau at Elsevier for their help in bringing the undertaking to fruition. Thanks also to all of the authors who contributed outstanding, up-to-date chapters to this second edition.

Contributors

Numbers in parentheses indicate the pages on which each author's contribution begins.

Paul S. Albert (p. 165), Biometric Research Branch, Division of Cancer Treatment and Diagnosis, National Cancer Institute, National Institutes of Health, Bethesda, Maryland

Steven Banks (p. 265), Critical Care Medicine Department, National Institutes of Health Clinical Center, Bethesda, Maryland

Olivia T. Bartlett (p. 359), Research Programs Review Branch, National Cancer Institute, National Institutes of Health, Bethesda, Maryland

Angela Bates (p. 129), Office of Research on Women's Health, Office of the Director, National Institutes of Health, Bethesda, Maryland

Craig B. Borkowf (p. 165), Centers for Disease Control and Prevention, Atlanta, Georgia

John Burklow (p. 155), Office of Communications and Public Liaison, National Institutes of Health, Bethesda, Maryland

Susan Lowell Butler (p. 143), DC Cancer Consortium, Washington, DC

Robert M. Califf (p. 237), Duke Clinical Research Institute, Durham, North Carolina

Ezekiel J. Emanuel (p. 27), Department of Clinical Bioethics, National Institutes of Health Clinical Center, Bethesda, Maryland

Bradley D. Freeman (p. 265), Washington University School of Medicine, St. Louis, Missouri

Lawrence M. Friedman (p. 59), Formerly, National Institutes of Health, Bethesda, Maryland

John I. Gallin (p. 1), National Institutes of Health Clinical Center, Bethesda, Maryland

Lynn H. Gerber (p. 283), Center for Study of Chronic Illness and Disability, College of Health and Human Services, George Mason University, Fairfax, Virginia

Bruce Goldstein (p. 291), Office of Technology Transfer, National Institutes of Health, Rockville, Maryland

Michael M. Gottesman (p. 121), Office of the Director, National Institutes of Health, Bethesda, Maryland

Christine Grady (p. 15), Section on Human Subjects Research, Department of Clinical Bioethics, National Institutes of Health Clinical Center, Bethesda, Maryland

Jack M. Guralnik (p. 197), Laboratory of Epidemiology, Demography, and Biometry, National Institute on Aging, National Institutes of Health, Bethesda, Maryland

Laura Lee Johnson (p. 165, 273), Office of Clinical and Regulatory Affairs, National Center for Complementary and Alternative Medicine, National Institutes of Health, Bethesda, Maryland

Miriam Kelty (p. 129), National Institute on Aging, National Institutes of Health, Bethesda, Maryland

Bruce R. Korf (p. 405), Department of Genetics, University of Alabama, Birmingham, Alabama

Patricia A. Kvochak (p. 109), NIH Legal Advisor's Office, Office of the General Counsel, U.S. Department of Health and Human Services, Bethesda, Maryland

Helen N. Lyon (p. 405), Division of Genetics, Program in Genomics, Children's Hospital Boston, Boston, Massachusetts

Teri A. Manolio (p. 197), National Human Genome Research Institute, National Institutes of Health, Bethesda, Maryland

Margaret A. Matula (p. 341), Clinical Research Management Branch, National Institute of Allergy and Infectious Diseases, National Institutes of Health, Bethesda, Maryland

Mitchell B. Max (p. 219), Pain and Neurosensory Mechanism Program, National Institute of Dental and Craniofacial Research, National Institutes of Health, Bethesda, Maryland

Charles Natanson (p. 265), Critical Care Medicine Department, National Institutes of Health Clinical Center, Bethesda, Maryland

Robert B. Nussenblatt (p. 121, 335), Laboratory of Immunology, National Eye Institute and Office of Protocol Services, National Institutes of Health Clinical Center, Bethesda, Maryland

Vivian W. Pinn (p. 129), Office of Research on Women's Health, Office of the Director, National Institutes of Health, Bethesda, Maryland

Elliott Postow (p. 359), Division of Biologic Basis of Disease, Center for Scientific Review, National Institutes of Health, Bethesda, Maryland

Denise T. Resnik (p. 391), Medical Research Consulting Services, Yonkers, New York

Stephen Rosenfeld (p. 351), MaineHealth, Portland, Maine

Joan P. Schwartz (p. 39), Office of Intramural Research, Office of the Director, National Institutes of Health, Bethesda, Maryland

Joanna H. Shih (p. 273), Biometric Research Branch, Division of Cancer Treatment and Diagnosis, National Cancer Institute, National Institutes of Health, Bethesda, Maryland

Jack Spiegel (p. 315), Office of Technology Transfer, Office of the Director, National Institutes of Health, Rockville, Maryland

Stephen E. Straus (p. 77), Laboratory of Clinical Infectious Disease, National Institute of Allergy and Infectious Diseases and Office of the Director, National Center for Alternative and Complementary Medicine, National Institutes of Health, Bethesda, Maryland

Anne Tompkins (p. 67), Division of Cancer Prevention, National Cancer Institute, National Institutes of Health, Bethesda, Maryland

Alison Wichman (p. 47), Office of Human Subjects Research, Intramural Research Program, National Institutes of Health, Bethesda, Maryland

Robert A. Yetter (p. 97), Center for Biologics Evaluation and Research, Food and Drug Administration, Rockville, Maryland

Kathryn C. Zoon (p. 97), Division of Intramural Research, National Institute of Allergy and Infectious Diseases, National Institutes of Health, Bethesda, Maryland

1

A Historical Perspective on Clinical Research

JOHN I. GALLIN

National Institutes of Health Clinical Center, Bethesda, Maryland

If I have seen a little further it is by standing on the shoulders of giants.
　　　　　　　　　　　　　　—Sir Isaac Newton, 1676

The successful translation of a basic or clinical observation into a new treatment of disease is rare in an investigator's professional life, but when it occurs, the personal thrill is exhilarating and the impact on society may be substantial. The following historical highlights provide a perspective of the continuum of the clinical research endeavor. These events also emphasize the contribution that clinical research has made to advances in medicine and public health.

In this chapter, and throughout this book, a broad definition of clinical research of the Association of American Medical Colleges Task Force on Clinical Research is used.[1] That task force defined clinical research as

> a component of medical and health research intended to produce knowledge essential for understanding human disease, preventing and treating illness, and promoting health. Clinical research embraces a continuum of studies involving interaction with patients, diagnostic clinical materials or data, or populations, in any of these categories: disease mechanisms; translational research; clinical knowledge; detection; diagnosis and natural history of disease; therapeutic interventions including clinical trials; prevention and health promotion; behavioral research; health services research; epidemiology; and community-based and managed care-based research.

1. THE EARLIEST CLINICAL RESEARCH

Medical practice and clinical research are grounded in the beginnings of civilization. Egyptian medicine was dominant from approximately 2850 BC to 525 BC. The Egyptian Imhotep, whose name means "he who gives contentment," lived slightly after 3000 BC and was the first physician figure to rise out of antiquity.[2] Imhotep was a known scribe, priest, architect, astronomer, and magician (medicine and magic were used together), and he performed surgery, practiced some dentistry,[1] extracted medicine from plants, and knew the position and function of the vital organs.

There is also evidence that ancient Chinese medicine included clinical studies. For example, in 2737 BC Shen Nung, the putative father of Chinese medicine, experimented with poisons and classified medical plants,[3] and I. Yin (1176–1123 BC), a famous prime minister of the Shang dynasty, described the extraction of medicines from boiling plants.[4]

Documents from early Judeo-Christian and Eastern civilizations provide examples of a scientific approach to medicine and the origin of clinical research. In the Old Testament, written from the 15th century BC to approximately the 4th century BC,[5] a passage in the first chapter of the Book of Daniel describes a comparative "protocol" of diet and health. Daniel demonstrated the preferred diet of legumes and water made for healthier youths than the king's rich food and wine:

> Then Daniel said to the steward . . .
>
> "Test your servants for ten days; let us be given vegetables to eat and water to drink. Then let your appearance and the appearance of the youths who eat the king's rich food be observed by you, and according to what you see deal with your servants:
>
> So he harkened to them in this matter; and tested them for ten days.
>
> At the end of ten days it was seen that they were better in appearance and fatter in flesh than all the youths who ate the king's rich food. So the steward took away their rich food and the wine they were to drink, and gave them vegetables."
>
> Daniel 1:11–16

The ancient Hindus also excelled in early medicine, especially in surgery, and there is evidence of Indian hospitals in Ceylon in 437 and 137 bc.[4]

2. THE GREEK AND ROMAN INFLUENCE

Although early examples of clinical research predate the Greeks, Hippocrates (460–370 bc) is considered the father of modern medicine, and he exhibited the strict discipline required of a clinical investigator.

His emphasis on the art of clinical inspection, observation, and documentation established the science of medicine. In addition, as graduating physicians are reminded when they take the Hippocratic oath, he provided physicians with high moral standards. Hippocrates' meticulous clinical records were maintained in 42 case records representing the first known recorded clinical observations of disease.[6] These case studies describe, among other maladies, malarial fevers, diarrhea, dysentery, melancholia, mania, and pulmonary edema with remarkable clinical acumen.

On pulmonary edema, he wrote the following:

> Water accumulates; the patient has fever and cough; the respiration is fast; the feet become edematous; the nails appear curved and the patient suffers as if he has pus inside, only less severe and more protracted. One can recognize that it is not pus but water . . . if you put your ear against the chest you can hear it seethe inside like sour wine.[7]

Hippocrates also described the importance of cleanliness in the management of wounds. He wrote, "If water was used for irrigation, it had to be very pure or boiled, and the hands and nails of the operator were to be cleansed."[8] Hippocrates' teachings remained dominant and unchallenged until Galen of Pergamum (ca. 130–200 AD), the physician to the Roman Emperor Marcus Aurelius.[9] Galen was one of the first individuals to utilize animal studies to understand human disease. By experimenting on animals, he was able to describe the effects of transection of the spinal cord at different levels. According to Galen, health and disease were the balance of four humors (blood, phlegm, black bile, and yellow bile), and veins contained blood and the humors, together with some spirit.[9]

3. MIDDLE AGES AND RENAISSANCE

In the Middle Ages, improvements in medicine became evident, and the infrastructure for clinical research began to develop. Hospitals and nursing, with origins in the teachings of Christ,[10] became defined institutions (although the forerunner of hospitals can be traced to the ancient Babylonian custom of bringing the sick into the marketplace for consultation, and the Greeks and Romans had military hospitals). By the 1100s and 1200s, hospitals were being built in England, Scotland, France, and Germany.

Early progress in pharmacology can be linked to the Crusades and the development of commerce. Drug trade became enormously profitable during the Middle Ages. Drugs were recognized as the lightest, most compact, and most lucrative of all cargoes. The influences of Arabic pharmacy and the contact of the Crusaders with their Moslem foes spread the knowledge of Arabic pharmaceuticals and greatly enhanced the value of drugs from the Far East. The records of the customhouse at the port of Acre (1191–1291) show a lively traffic in aloes, benzoin, camphor, nutmegs, and opium.[11]

Documentation through case records is an essential feature of clinical research. Pre-Renaissance medicine of the 14th and 15th centuries saw the birth of "Consilia" or medical-case books, consisting of clinical records from the practice of well-known physicians.[12] Hippocrates' approach of case studies developed 1700 years earlier was reborn, particularly in the Bolognese and Paduan regions of Italy. Universities became important places of medicine in Paris, Bologna, and Padua.

Clinical research remained mostly descriptive, resembling today's natural history and disease pathogenesis protocols. In 1348, Gentile da Foligno, a Paduan professor, described gallstones.[12] Bartolommeo Montagnana (1470), an anatomist, described strangulated hernia, operated on lachrymal fistula, and extracted decayed teeth.[12] There was also evidence of the beginning of a statistical approach to medical issues during this period. For example, a 14th-century letter from Petrach to Boccaccio states that

> I once heard a physician of great renown among us express himself in the following terms: . . . I solemnly affirm and believe, if a hundred or a thousand of men of the same age, same temperament and habits, together with the same

surroundings, were attacked at the same time by the same disease, that if the one half followed the prescriptions of the doctors of the variety of those practicing at the present day, and that the other half took no medicine but relied on Nature's instincts, I have no doubt as to which half would escape.[13]

The Renaissance (1453–1600) represented the revival of learning and transition from medieval to modern conditions; many great clinicians and scientists prospered. At this time, many of the ancient Greek dictums of medicine, such as Galen's four humors, were discarded. Perhaps the most important anatomist of this period was Leonardo da Vinci (1453–1519) (Fig. 1-1).[14] Da Vinci created more than 750 detailed anatomic drawings (Fig. 1-2).

4. SEVENTEENTH CENTURY

Studies of blood began in the 17th century. William Harvey (1578–1657) convincingly described the circu-

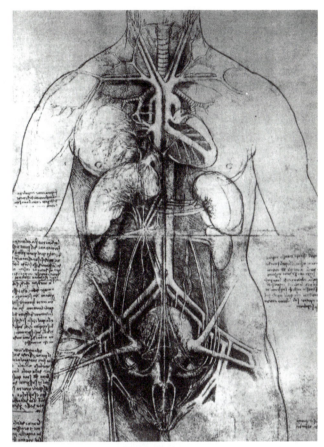

FIGURE 1-2 Example of anatomic drawing by Leonardo da Vinci. Trunk of female human body, with internal organs seen as though ventral side were transparent. From reference 14, p. 369.

FIGURE 1-1 Leonardo da Vinci self-portrait (red chalk); Turin, Royal Library. From reference 14, Figure 1.

lation of blood from the heart through the lungs and back to the heart and then into the arteries and back through the veins.[16] Harvey emphasized that the arteries and veins carried only one substance, the blood, ending Galen's proposal that veins carried a blend of multiple humors. (Of course, today we know that blood contains multiple cellular and humoral elements, so to some extent Galen was correct.) The famous architect Sir Christopher Wren (1632–1723), originally known as an astronomer and anatomist (Fig. 1-3), in 1656 assembled quills and silver tubes as cannulas and used animal bladders to inject opium into the veins of dogs.[17] The first well-documented transfusions of blood into humans were done in 1667 by Richard Lower and Edmund King in London[18] and mentioned in Pepys' diary.[19]

The 17th century also brought the first vital statistics, which were presented in Graunt's book, *Natural and Political Observations Mentioned in a Following Index, and Made Upon the Bills of Mortality*.[20] In this book of comparative statistics, populations and mortality statistics were compared for different countries, ages, and sex for rural and urban areas. The importance of using mortality among groups would have major importance in future clinical studies.

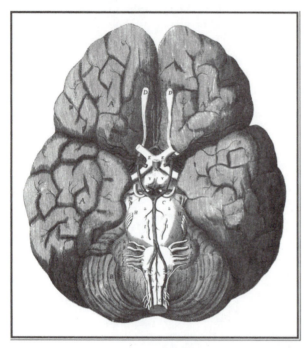

FIGURE 1-3 Christopher Wren's drawing of the brain shows blood vessels discovered by Thomas Willis.[15]

FIGURE 1-4 Antony van Leeuwenhoek. From reference 21.

5. EIGHTEENTH CENTURY

The 18th century brought extraordinary advances in the biological sciences and medicine. At the end of the 17th century, Antony van Leeuwenhoek of Delft (1632–1723) invented the microscope. Although he is best known for using his microscope to provide the first descriptions of protozoa and bacteria, Leeuwenhoek also provided the first description of striated voluntary muscle, the crystalline structure of the lens, red blood cells, and spermatozoa (Figs. 1-4 and 1-5).[21]

Modern clinical trials can be recognized in the 1700s. Scurvy was a major health problem for the British Navy. William Harvey earlier had recommended lemons to treat scurvy but argued that the therapeutic effect was a result of the acid in the fruit. James Lind (Fig. 1-6), a native of Scotland and a Royal Navy surgeon, conducted a clinical trial in 1747 to assess this hypothesis comparing three therapies for scurvy (Table 1-1).[22] Twelve sailors with classic scurvy were divided

0*0

PHILOSOPHICAL

TRANSACTIONS.

For the Months of *Jng*Ji* and *Stftemttr*.

Stftemt. 21. 1674,

The CONTENTS.

llxnfatictl Qtftrtfttutt firna «#r. Leeuwcnboeck, «fotf **Blood, Milk, Bones,** *tkt* **Brain, Spitle, Coticula, Sweat, Fact, Teares 5** *unmtmutei t» t*** **Letttrs to the PaUfitr.** *An AtuHit tf* ntttkU Ctfe if** **Dropfy,** *mfoktn fir Grt-vibtk* at+yt**g W*mt* ^ imftrttAh* Lttnttd Ptyfoi* MI** **Holland.** *jfa4tt*at,f three E»hr LDE SE-CRET IQKE JN-IM.ALI Ctgtutt, 4»tb.* **Gall. Co'e, At D. IF.** *Ertfmi BtrthiM SELECT A GEO eX£ fR \CA.* III. *LOG ICJ, fat An CritoUi; f * Gtllit* i» iMnnu* Strnutem vtrfc SUM A*t*u&o*rfait *p* tke* Latin *Verfon, mult by* C. S. *of the* **Pbil.TranUflions*/'** *J. 1665.1666.1667.*

l OtftnnttoKfr** e^C **Leeowrahoeck,** *u#tr»-*7* **Blood, Milk, Bones,** */fc* **Bnio,SpitIe,36^ Cuticula,^.** *MmmmtiUAk tkt JUt Obfervtr* **i»** *tin fMjbtr in ** **Letttr, lUteljUK r. 1674,**

Sir,

Y   Ours of *14** of *Mrit\a&* was *my* welcome to roe j Wheocelnnd«ftoodwitfj great cooteotfneot, that *my* Microfcopical Cooxnookatiom badnot been nnaocepoble co yoaaod yoor Philofophkal Frkndr; wUcb hath encouraged   R   **

FIGURE 1-5 Title page from Leeuwenhoeck's paper on *Microscopical Observations.* From reference 16.

FIGURE 1-6 James Lind.

TABLE 1-1   Treatment of Scurvy by James Lind

Treatment Arm	Cured	p Value[a]
Sulfuric acid	0/2	>0.05
Vinegar	0/2	>0.05
Seawater	0/2	>0.05
Cider	0/2	>0.05
Physicians	0/2	>0.05
Citrus fruit	2/2	>0.05

[a]Compared to patients in the five areas of the trial; no placebo group.

into six groups of two each, all given identical diets, and the various groups supplemented with vinegar, dilute sulfuric acid, cider, seawater, a nutmeg, garlic, and horseradish mixture, and with two oranges and one lemon daily.

Sulfuric acid, vinegar, seawater, cider, and physician's remedy had no benefit. Two sailors receiving citrus fruit avoided scurvy. Although not significant because of sample size, this early clinical study formed the basis for successfully avoiding scurvy with citrus fruit. The studies with sulfuric acid, vinegar, and cider excluded acid as a likely explanation for the beneficial effect of citrus fruit.

The 18th century saw great progress in the area of surgery. A remarkable succession of teachers and their students led these studies. Percival Pott of St. Bartholomew's Hospital described tuberculosis of the spine or "Pott's disease."[23] John Hunter, Pott's pupil, was the founder of experimental and surgical pathology as well as a pioneer in comparative physiology and experimental morphology. Hunter described shock, phlebitis, pyremia, and intussusception and made major findings of inflammation, gunshot wounds, and the surgical diseases of the vascular system.[23] Hunter's student, Edward Jenner (1749–1823),[23] introduced vaccination as a tool to prevent infectious diseases (Fig. 1-7).[24] Jenner was aware that dairymaids who had contacted cowpox through milking did not get smallpox. In 1798, Jenner conceived of applying the observation on a grand scale to prevent smallpox.[25]

Jenner was not the first to conceive of the idea of inoculation for smallpox. For example, the Chinese had thought of this earlier and Sir Hans Sloane had done small studies in 1717 using variolation (inoculating healthy people with pus from blisters obtained from patients with smallpox).[26] In addition, James Jurin published several articles between 1723 and 1727 comparing death from natural smallpox in people who had not been inoculated with those who had been inoculated. Jurin showed that death occurred in 5 of 6 subjects in the first group compared to 1 in 60 in the latter,[27] providing one of the first studies using mortality as a critical clinical end point. In 1734, Voltaire wrote, "The Cirassians [a Middle Eastern people] perceived that of a thousand persons hardly one was attacked twice by full blown smallpox; that in truth one sees three or four mild cases but never two that are serious and dangerous; that in a word one never truly has that illness twice in life."[28] Thus, Voltaire recognized natural immunity to smallpox, which was an important concept for future vaccinology. In 1721, Cotton Mather demonstrated that variolation protected citizens of the American colonies in Massachusettes,[29] and in 1777 George Washington used variolation against smallpox to inoculate the Continental Army, the first massive immunization of a military.[30] Jenner was the first to try vaccination on a large scale using scabs from cow pox to protect against human smallpox and the first to use experimental approaches to establish the scientific basis for vaccination. Jenner transformed a local country tradition into a viable prophylactic principle. Jenner's vaccine was adopted quickly in Germany and then in Holland, Denmark, the rest of Europe, and the United States.

FIGURE 1-7 Edward Jenner (painting by Sir Thomas Lawrence). From reference 3, p. 373.

The 1700s were also when the first known blinded clinical studies were performed. In 1784, a commission of inquiry was appointed by King Louis XVI of France to investigate medical claims of "animal magnetism" or "mesmerism." The commission, headed by Benjamin Franklin and consisting of such distinguished members as Antoine Lavoisier, Jean-Sylvain Bailly, and Joseph-Ignace Guillotin, had as a goal to assess whether the reported effects of this new healing method were due to "real" force or due to "illness of the mind." Among the many tests performed, blindfolded people were told that they were either receiving or not receiving magnetism when in fact, at times, the reverse was happening. The results showed that study subjects felt effects of magnetism only when they were told they received magnetism and felt no effects when they were not told, whether or not they were receiving the treatment.[31] This was the beginning of the use of blinded studies in clinical research.

The 18th century also provided the first legal example that physicians must obtain informed consent from patients before a procedure. In an English lawsuit, *Slater v. Baker & Stapleton*, two surgeons were found liable for disuniting a partially healed fracture without the patient's consent.[32] This case set the important precedent described by the court: "Indeed it is reasonable that a patient should be told what is about to be done to him that he may take courage and put himself in such a situation as to enable him to undergo the operation."

6. NINETEENTH CENTURY

In the first days of the 19th century, Benjamin Waterhouse, a Harvard professor of medicine, brought Jenner's vaccine to the United States, and by 1802 the first vaccine institute was established by James Smith in Baltimore, Maryland. This led to a national vaccine agency, which was established by the Congress of the United States under the direction of James Smith in 1813.[33]

Jenner's vaccination for smallpox was followed by other historic studies in the pathogenesis of infectious diseases. The French physician Pierre Charles Alexandre Louis (1787–1872) realized that clinical observations on large numbers of patients were essential for meaningful clinical research. He published classical studies on typhoid fever and tuberculosis, and his research in 1835 on the effects of bloodletting demonstrated that the benefits claimed for this popular mode of treatment were unsubstantiated.[34] On February 13, 1843, one of Louis' students, Oliver Wendell Holmes (1809–1894), the father of the great Justice Holmes, read his article, *On the Contagiousness of Puerperal Fever*,[35] to the Boston Society for Medical Improvement (Fig. 1-8). Holmes stated that women in childbed should never be attended by physicians who have been conducting postmortem sections on cases of puerperal fever; that the disease may be conveyed in this manner from patient to patient, even from a case of erysipelas; and that washing the hands in calcium chloride and changing the clothes after leaving a puerperal fever case was likely to be a preventive measure. Holmes' essay stirred up violent opposition by obstetricians. However, he continued to reiterate his views, and in 1855 in a monograph, *Puerperal Fever as a Private Pestilence*, Holmes noted that Semmelweis, working in Vienna and Budapest, had lessened the mortality of puerperal fever by disinfecting the hands with chloride of lime and the nail brush.[36]

Ignaz Philipp Semmelweis (1818–1865) performed the most sophisticated preventive clinical trial of the 19th century that established the importance of hand washing to prevent the spread of infection (Fig. 1-9).[37] Semmelweis, a Hungarian pupil, became an assistant in the first obstetric ward of the Allgemeines Kranken-

FIGURE 1-8 Oliver Wendell Holmes. From reference 3, p. 435.

FIGURE 1-9 Ignaz Philipp Semmelweis. From reference 4, p. 436.

haus in Vienna in 1846. Semmelweis was troubled by the death rate associated with puerperal or "childbed" fever. From 1841 to 1846, the maternal death rate from puerperal sepsis averaged approximately 10%, and in some periods as high as 50%, in the First Maternity Division of the Vienna General Hospital. In contrast, the rate was only 2 or 3% in the Second Division, which was attended by midwives rather than physicians. The public knew the disparity, and women feared being assigned to the First Division. Semmelweis became frustrated by this mystery and began to study cadavers of fever victims. In 1847, his friend and fellow physician, Jakob Kolletschka, died after receiving a small cut on the finger during an autopsy. The risk of minor cuts during autopsies was well-known, but Semmelweis made the further observation that Kolletschka's death was characteristic of death from puerperal fever. He reasoned that puerperal fever was "caused by conveyance to the pregnant women of putrid particles derived from living organisms, through the agency of the examining fingers." In particular, he identified the cadaveric matter from the autopsy room, with which the midwives had no contact, as the source of the infection.

In 1847, Semmelweis insisted that all students and physicians scrub their hands with chlorinated lime before entering the maternity ward, and during 1848 the mortality rate on his division dropped from 9.92%

to 1.27%. Despite his convincing data, his colleagues rejected his findings and accused him of insubordination. The dominant medical thinking at the time was that the high mortality in the charity hospital related to the poor health of the impoverished women, despite the difference between the control (no chlorinated lime hand washing) and experimental (washing with chlorinated lime) divisions. Without any opportunity for advancement in Vienna, Semmelweis returned to his home in Budapest and repeated his studies with the same results. In 1861, he finally published *The Etiology, Concept, and Prophylaxis of Childhood Fever*.[37] Although Holmes' work antedated Semmelweis by 5 years, the superiority of Semmelweis' observation lies not only in his experimental data but also in his recognition that puerperal fever was a blood poisoning. The observations of Holmes and Semmelweis were a critical step for medicine and surgery.

In addition to discovering the importance of hand washing, the first well-documented use of ether for surgery (1846) by William Thomas Green Morton with

Dr. John Collins Warren as the surgeon at the Massachusetts General Hospital also occurred during the 19th century.[38] Oliver Wendell Holmes is credited with proposing the words *anesthetic* and *anesthesia*.[38] Recognition of the importance of hand washing and the discovery of anesthetics were essential findings of the 19th century that were critical for the development of modern surgery.

The work of Holmes and Semmelweis on the importance of hand washing also opened the door for Pasteur's work on the germ basis of infectious diseases. Louis Pasteur (1822–1895) was perhaps the most outstanding clinical investigator of the 19th century (Fig. 1-10). He was trained in chemistry. His fundamental work in chemistry led to the discovery of levo and dextro isomers. He then studied the ferments of microorganisms, which eventually led him to study the detrimental causes of three major industries in France: wine, silk, and wool. Pasteur discovered the germ basis of fermentation, which formed the basis of the germ theory of disease.[39] He discovered *Staphylococcus pyogenes* as a cause of boils and the role of *Streptococcus pyogenes* in puerperal septicemia. In other studies, he carried forward Jenner's work on vaccination and developed approaches to vaccine development using attenuation of a virus for hydrophobia (rabies) and inactivation of a bacterium for anthrax.

The work of Pasteur was complemented by the studies of Robert Koch (1843–1910), who made critical technical advances in bacteriology. Koch was the first to use agar as a culture media and he introduced the petri dish, pour plates, and blood agar to make bacterial culture and identification easy and widely available. Koch cultured the tubercle bacillus and identified the etiologic agent for anthrax, which was later used by Pasteur to develop a vaccine, and he established "Koch's postulates" to prove that an infectious agent causes disease (Fig. 1-11).[39]

FIGURE 1-10 Louis Pasteur. One of the remarkable facts about Pasteur was his triumph over a great physical handicap. In 1868 at age 46, just after completing his studies on wine, he had a cerebral hemorrhage. Although his mind was not affected, he was left with partial paralysis of his left side, which persisted for the remainder of his life. This photograph, taken after he was awarded the Grand Cross of the Legion of Honor in 1881, gives no hint of his infirmity. From reference 23, p. 116.

FIGURE 1-11 Robert Koch. His career in research began in 1872 when his wife gave him a microscope as a birthday present. He was then 28 years old, performing general practice in a small town in Silesia. This was an agricultural region where anthrax was common among sheep and cattle, and it was in the microscopic study of this disease in rabbits that Koch made his first great discovery of the role of anthrax bacilli in disease. From reference 23, p. 132.

The studies of Pasteur and Koch were performed during the same period as the work of the Norwegian Gerhard Armauer Hansen (1841–1912). In 1874, based on epidemiological studies in Norway, Hansen concluded that *Mycobacterium leprae* was the microorganism responsible for leprosy. Hansen's claim was not well received, and in 1880, in an attempt to prove his point, he inoculated live leprosy bacilli into humans, including nurses and patients, without first obtaining permission. One of the patients brought legal action against Hansen. The court, in one of the early cases demonstrating the importance of informed consent in clinical research, removed Hansen from his position as director of Leprosarium No. 1, where the experiments had taken place. However, Hansen retained his position as chief medical officer for leprosy[40] and later in his life received worldwide recognition for his life's work on leprosy.

In the same era, Emil von Behring (1854–1917) demonstrated in 1890 that inoculation with attenuated diphtheria toxins in one animal resulted in production of a therapeutic serum factor (antitoxin) that could be delivered to another, thus discovering antibodies and establishing a role for passive immunization. On Christmas eve of 1891, the first successful clinical use of diphtheria antitoxin occurred.[39] By 1894,

diphtheria antiserum became commercially available as a result of Paul Ehrlich's work establishing methods for producing high-titer antisera. Behring's discovery of antitoxin was the beginning of humoral immunity, and in 1901 Behring received the first Nobel prize. Koch received the prize in 1905 (Fig. 1-12).

The Russian scientist Elie Metchnikoff (1845–1916) discovered the importance of phagocytosis in host defense against infection and emphasized the importance of the cellular components of host defense against infection.[41] Paul Ehrlich (1854–1915) discovered the complement system and asserted the importance of the humoral components of host defense. In 1908, Metchnikoff and Ehrlich shared the Nobel prize (Figs. 1-13 and 1-14).

At the end of the 19th century, studies of yellow fever increased the awareness of the importance of the informed consent process in clinical research. In 1897, Italian bacteriologist Giuseppe Sanarelli announced that he had discovered the bacillus for yellow fever by

FIGURE 1-12   Emil von Behring. From reference 39, p. 7.

FIGURE 1-13 Elie Metchnikoff in his forties. Reprinted frontispiece of E. Metchnikoff, *The Nature of Man: Studies in Optimistic Philosophy.* New York, Putnam, 1903. From reference 40, Figure 5.

FIGURE 1-14   Paul Ehrlich. From Reference 39, p. 9.

FIGURE 1-15   Marie Curie (1867–1934).

injecting the organism into five people. William Osler was present at an 1898 meeting at which the work by Sanarelli was discussed, and Osler said, "To deliberately inject a poison of known high degree of virulency into a human being, unless you obtain that man's sanction . . . is criminal."[42] This commentary by Osler had substantial influence on Walter Reed, who demonstrated in human volunteers that the mosquito is the vector for yellow fever. Reed adopted written agreements (contracts) with all his yellow fever subjects. In addition to obtaining signed permission from all his volunteers, Reed made certain that all published reports of yellow fever cases included the phrase "with his full consent."[42]

Toward the end of the 19th century, women began to play important roles in clinical research. Marie Curie (1867–1934) and her husband, Pierre, won the Nobel prize in physics in 1903 for their work on spontaneous radiation, and in 1911 Marie Curie won a second Nobel prize (in chemistry) for her studies in the separation of radium and description of its therapeutic properties. Marie Curie and her daughter, Irene, promoted the therapeutic use of radium during World War I (Fig. 1-15).[43]

Florence Nightingale (1820–1910), in addition to her famous work in nursing, was an accomplished mathematician who applied her mathematical expertise to dramatize the needless deaths caused by unsanitary conditions in hospitals and the need for reform (Fig. 1-16).[44]

FIGURE 1-16   Florence Nightingale (1820–1910).

7. <u>TWENTIETH CENTURY AND BEYOND</u>

The spectacular advances in medicine during the 20th century would never have happened without the centuries of earlier progress. In the 20th century, medical colleges became well established in Europe and the United States. The great contributions of the United States to medicine in the 20th century are linked to the early commitment to strong medical education. The importance of clinical research as a component of the teaching of medicine was recognized in 1925 by the American medical educator Abraham Flexner, who wrote, "Research can no more be divorced from medical education than can medical education be divorced from research."[45]

Two other dominant drivers of the progress in medicine through clinical research were government investment in biomedical research and private investment in the pharmaceutical industry. These investments, closely linked with academia, resulted in enhanced translation of basic observations to the bedside. Sir Alexander Fleming's discovery of penicillin in 1928 in Scotland spawned expansion of the pharmaceutical industry with the development of antibiotics, antiviral agents, and new vaccines. Banting and Best's discovery of insulin in 1921 in Canada was followed by the discovery of multiple hormones to save lives.

In the 1920s and 1930s, Sir Ronald Aylner Fisher (1890–1962), from the United Kingdom, introduced the application of statistics and experimental design.[46] Fisher worked with farming and plant fertility to introduce the concept of randomization and analysis of variance—procedures used today throughout the world. In 1930, Torald Sollman emphasized the importance of controlled experiments with placebo and blind limbs to a study—a rebirth of the "blinded" or "masked" studies originated by Benjamin Franklin in 1784. Sollman wrote, "Apparent results must be checked by the 'blind test,' i.e., another remedy or a placebo, without the knowledge of the observer, if possible." (Fig. 1-17)[47]

With these approaches many new drugs for treatment of hypertension, cardiovascular disease, manic depression, and epilepsy, to name a few, were developed.

The spectacular advances in the 20th century were associated with troubling events in clinical research that heightened public attention and formalized the field of clinical bioethics. The Nazi's human experimentation led to the "Nuremberg Code" in 1947 that was designed to protect human subjects by ensuring voluntary consent of the human subject and that the anticipated result of the research must justify the performance of the research. The Tuskegee syphilis experiments initiated in the 1930s and continued until 1972 in African American men and the Willowbrook hepatitis studies in the mid-1950s in children with Down syndrome highlighted the need to establish strict rules to protect research patients.

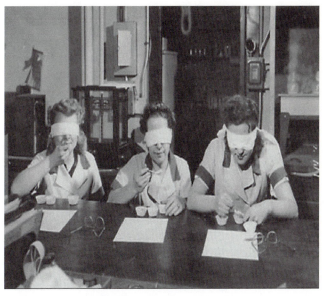

FIGURE 1-17 Testing puddings and gelatins at Consumers Union. Copyright 1945 by Consumers Union of U.S., Inc., Yonkers, NY. Reprinted with permission from the April 1945 issue of *Consumer Reports*.

In 1953, the U.S. National Institutes of Health (NIH) issued "Guiding Principles in Medical Research Involving Humans" that required prior review by medical committee of all human research to be conducted at the newly opened NIH Clinical Center. In 1962, the Kefauver-Harris amendments to the Food and Drug Act stipulated subjects be told if a drug is being used for investigational purposes, and subject consent must be obtained. In 1964, the World Medical Assembly adopted the "Declaration of Helsinki" stressing the importance of assessing risks and determining that the risks are outweighed by the potential benefits of research. In 1966, Henry Beecher pointed out major ethical issues in clinical research.[48] During the same year, the U.S. Surgeon General issued a memo to the heads of institutions conducting research with Public Health Service grants requiring prior review of all clinical research. The purpose was to ensure protection of research subjects, assess the appropriateness of the methods employed, obtain informed consent, and review the risks and benefits of the research; thus institutional review boards were established. In 1967, the Food and Drug Administration added the requirement that all new drug sponsors obtain informed consent for use of investigational drugs in humans.

In the past 50 years, clinical research has become big business. The pharmaceutical industry and the biotechnology industries have engaged university-based clinical investigators in the business of clinical research. Interaction between federal investigators and industry, encouraged by the U.S. Congress when it passed the

Federal Technology Transfer Act in 1986, successfully increased the translation of basic research to the bedside by government scientists. At the same time, however, the relationship between industry and academia grew closer and new ethical, legal, and social issues evolved. Clinical investigators became increasingly associated with real and perceived conflicts. Examples of these issues included promoting an investigator's financial or career goals while protecting the patient, protecting "unborn children" while pursuing the potential use of embryonic stem cells to rebuild damaged organs, and protecting patient confidentiality as a result of gene sequencing. As a result of these issues, the public engaged in debate about the safety of current and future generations of patients who volunteer to partner with the clinical investigator on protocols.

The opportunities for doing clinical research in the 21st century are greater than ever. Today, understanding and meeting public concern are as important for the clinical investigator as performing the clinical study. Principles for conducting clinical research have evolved from centuries of experience. As the science moves forward, ethical, legal, and social issues pose special challenges for the clinical investigator. These challenges are the focus of the following chapters of this book.

Acknowledgment

The author thanks Patricia I. Piringer for outstanding editorial assistance.

References and Notes

1. Association of American Medical Colleges Task Force on Clinical Research 2000. (1999). Vol. 1, p. 3. Washington, DC, Association of American Medical Colleges, 1999.
2. Thorwald J. *Science and Secrets of Early Medicine.* New York, Harcourt, Brace and World, 1962.
3. Garrison FH. *History of Medicine.* Philadelphia, Saunders, 1917, Reprinted 1963, pp. 73–74.
4. Ibid p. 70.
5. Lane B. In Reading the Bible with Understanding (Burgland L., ed) *How We Got the Bible.* St. Louis, MO, Concordia, 1999.
6. Adams F. *The Genuine Works of Hippocrates.* New York, William Wood, 1886.
7. Lyons AS, Petrucellli RJ. *Medicine, An Illustrated History.* New York, Abradale Press, 1987, p. 216.
8. Reference 3, p. 98.
9. Nutton V. Logic, learning, and experimental medicine. *Science* 2002;295:800–801.
10. Reference 3, p. 176.
11. Reference 3, pp. 179–180.
12. Reference 3, pp. 166–157.
13. Witkoski SJ. *The Evil That Has Been Said of Doctors: Extracts from Early Writers,* translation with annotations. Cincinnati, OH, T. C. Minor, 1889. Reprint from the *Lancet-Clinic,* p. 55. Quoted in Lilienfeld AM. Centers Paribus: The evolution of the clinical trial. *Bull History Med* 1982;56:1–18.
14. Da Vinci L. *Copyright in Italy by the Istituto Geografico DeAgostini S.p.A.-Novara.* New York, Reymal & Company, 1956.
15. Knoeff R. Book review of *Soul Made Flesh: The Discovery of the Brain and How It Changed the World* by C. Zimmer [Published in United Kingdom by Heinemann]. *Nature* 2004;427:585.
16. Wintrobe MM. *Blood, Pure and Eloquent.* New York, McGraw-Hill, 1980.
17. Ibid, pp. 661–662.
18. Ibid, p. 663.
19. Nicolson MH. *Pepys' Diary and the New Science.* Charlottesville, University Press of Virginia, 1965. Quoted in reference 13, p. 663.
20. Graunt J. *Natural and Political Observations Mentioned in a Following Index, and Made Upon the Bills of Mortality.* London, 1662. Reprinted by Johns Hopkins Press, Baltimore, 1939. Quoted in Lilienfeld AM. Centeris Paribus: The evolution of the clinical trial. *Bull History Med* 1982;56:1–18.
21. Dobell C. *Antony van Leeuwenhoek and His Little Animals. A Collection of Writings by the Father of Protozoology and Bacteriology.* New York, Dover, 1960. [Original work published in 1932.]
22. Lind J. *A Treatise of the Scurvy.* Edinburgh, UK, Sands, Murray and Cochran, 1753, pp. 191–193. Quoted in Lilienfeld AM. Centeris Paribus: The evolution of the clinical trial. *Bull History Med* 1982;56:1–18.
23. Haagensen CD, Lloyd EB. *A Hundred Years of Medicine.* New York, Sheridan House, 1943.
24. Wood GB. *Practice of Medicine.* Philadelphia, Collins, 1849.
25. Jenner E. *An Inquiry into the Causes and Effects of the Variolae Vaccinae.* London, Sampson Low, 1798.
26. Reference 3, p. 373.
27. Miller G. *The Adoption of Inoculation for Smallpox in England and France.* Philadelphia, University of Pennsylvania Press, 1957, pp. 114–118. Quoted in Lilienfeld AM. Centers Paribus: The evolution of the clinical trial. *Bull History Med* 1982;56:1–18.
28. Plotkin SA. Vaccines: Past, present and future. *Nature Med* 2005;11:S5–S11.
29. Harper DP. Angelical conjunction: Religion, reason, and inoculation in Boston, 1721–1722. *The Pharos* 2000;Winter:1–5.
30. Fenn EA. *Pox Americana. The Great Small Pox Epidemic of 1775–82.* New York, Hill and Wang, 2001.
31. Franklin B. Animal and other commissioners charged by the King of France. Animal Magnetism. 1784. In *An Historical Outline of the "Science" Made by the Committee of the Royal Academy of Medicine in Philadelphia Translated from the French.* Philadelphia, H. Perkins, 1837.
32. *Slater v. Baker & Stapleton.* (1767). 95, Eng. Rep. 860. Quoted in Appelbaum PS, Lidz CW, Meisel A. *Informed Consent. Legal Theory and Clinical Practice.* New York, Oxford University Press, 1987.
33. Reference 3, p. 375.
34. Morabia A. PCA Louis and the birth of clinical epidemiology. *J Clin Epidemiol* 49:1327–1333, 1996.
35. Holmes OW. On the contagiousness of puerperal fever. *N Engl Q J Med* 1842–1843;1:503–530. Quoted in reference 3, p. 435.
36. Reference 3, p. 435.
37. Semmelweiss IP. Die Aetiologie, der Begriff und die Prophylaxis des Kindbettfiebers. Budapest and Vienna, 1861. Quoted in reference 3, p. 436.
38. Reference 3, pp. 505–506.
39. Hirsch JG. Host resistance to infectious diseases: A centennial. In Gallin JI, Fauci AS (eds.) *Advances in Host Defense Mechanisms: Vol. 1. Phagocytic Cells.* New York, Raven Press, 1982.

40. Bendiner E. Gerhard Hansen: Hunter of the leprosy bacillus. *Hospital Practice* 1989, December 15;145–170.

41. Tauber AI, Chernyak L. *Metchnikoff and the Origins of Immunology.* New York, Oxford University Press, 1991.

42. Lederer SE. *Human Experimentation in America before the Second World War.* Baltimore, Johns Hopkins University Press, 1995.

43. Macklis RM. Scientist, Technologist, Proto-Feminist, Superstar. *Science* 295:1647–1648, 2002.

44. Cohen IB. Florence Nightingale. *Sci Am* 1984;250:128–137.

45. Flexner A. *Medical Education. A Comparative Study.* New York, Macmillan, 1925.

46. Efron B. Fisher in the 21st century. *Statistical Sci* 1998;13: 95–122.

47. Sollman T. The evaluation of therapeutic remedies in the hospital. *J Am Med Assoc* 94:1280–1300, 1936.

48. Beecher HK. Ethics and clinical research. *N Engl J Med* 1966;274:1354–1360.

2

Ethical Principles in Clinical Research

CHRISTINE GRADY

Section on Human Subjects Research, Department of Clinical Bioethics, National Institutes of Health Clinical Center, Bethesda, Maryland

Clinical research has resulted in significant benefits for society, yet continues to pose profound ethical questions. This chapter describes ethical principles that guide clinical research and briefly considers the history of clinical research ethics and particular ethical challenges in randomized controlled trials.

1. DISTINGUISHING CLINICAL RESEARCH FROM CLINICAL PRACTICE

Clinical research involves the study of human beings in a systematic investigation of human biology, health, or illness, designed to develop or contribute to generalizable knowledge. Clinical research includes a set of activities meant to test a hypothesis, permit conclusions to be drawn, and thereby contribute to generalizable knowledge useful to others. The goal of clinical research is to generate knowledge useful to improving medical care or the public health and thus serve the common or collective good. The individual subject participating in clinical research may or may not benefit from participation.

Clinical research is distinct from clinical practice in that the purpose and goals of each, although not mutually exclusive, are quite different. The purpose of clinical practice is to diagnose, prevent, treat, or care for an illness or condition in a particular individual or group of individuals with the goal of meeting the needs of and benefiting that individual(s). Clinical practice is a set of activities designed to enhance the patient's well-being and has a reasonable expectation of success. In some cases, participation in clinical research does meet the health needs of, and benefit, individual patient-participants. In fact, through participation in good

clinical research, an individual may receive a very high quality of patient care and treatment, yet that is not the goal of research, and much research does not directly benefit individual participants.

2. WHAT DOES ETHICS HAVE TO DO WITH CLINICAL RESEARCH?

Broadly, ethics is a systematic method of inquiry that helps us answer questions about how we ought to live and behave and why. With respect to clinical research, there are two fundamental ethical questions: (1) Should we do research with human subjects? Why or why not? and (2) If yes, how should it be done? In addressing the first question, two competing considerations are recognized. On the one hand, clinical research is valuable in generating practical knowledge useful for advancing or improving medical care and health. On the other hand, respect for the inviolability, safety, dignity, and freedom of choice of each individual is indispensable. Advancing or improving medical care and/or the public health is desirable as a public good—good for society. Such knowledge is knowledge in "the service of action, [because] health professionals seek knowledge in order to know how to best serve."[1] The pursuit of knowledge through research should be rigorous because false knowledge applied in practice can be harmful. Rigorous clinical research is an important means to the end of progress in medical and health care—progress that would not be possible without research. It has been claimed that conducting clinical research designed to understand human health and illness may be more than a social good; it may be a social imperative.[2] In contrast, it also has been asserted that although progress in medical care and health is good, it is an optional good[3] and that other considerations, such as the primacy of the individual, should take precedence. Even if one accepts that improvement in medical care or health is a social good, and that clinical research is an essential means to that end, limits are necessary as progress is achieved through research with human beings. Human subjects who participate in research are the means to securing practical knowledge. Because human beings should never be used "merely as means to an end, but always as ends in themselves,"[4] the need to respect and protect human participants in research is paramount.

The primary ethical tension in clinical research, therefore, is that a few individuals are asked to accept some burden or risk as research subjects in order to benefit others and society. The beneficiaries of research may sometimes include the subjects themselves but also will include others with similar disorders or risk profiles, as well as future persons and society. Asking human subjects to bear any risk of harm or burden for the good of others creates a potential for exploitation. Ethical requirements for clinical research aim to minimize the possibility of exploitation by ensuring that research subjects are not "merely used" but are treated with respect while they contribute to the social good, and their rights and welfare are protected throughout the process of research. Through history, the perception and acceptance of the methods, goals, and scope of clinical research have shifted significantly along with attention to and appreciation of what respecting and protecting research subjects entails. A brief detour through the history of clinical research illustrates these changing perspectives.

3. HISTORY OF ETHICAL ATTENTION TO CLINICAL RESEARCH

3.1. Benefit to the Individual

For hundreds of years, research was done sporadically. There was little basis for a distinction between experimentation and therapy because most therapy was experimental. Systematic evidence of the effectiveness of medical interventions was rare. Experimental therapy was often used to try to benefit ill patients, but such "therapy" frequently contributed to or caused morbidity or mortality. Most researchers were medical practitioners, motivated to do what they thought best for their patients, and trusted to do the right thing. Fraud and abuse were minimized through peer censorship because there were no specific codes of ethics, laws, or regulations governing the conduct of research.

Early regulations, such as the Pure Food and Drug Act of 1906 in the United States, prohibited unsubstantiated claims on medicine labels. Yet, research began to grow as an enterprise only after the development of penicillin and other early antibiotics and the passage of the Food, Drug, and Cosmetic Act in 1938 that required evidence of safety before a product was marketed.

3.2. Benefit to Society

Around World War II, there was a dramatic shift in clinical research with tremendous growth in research as an enterprise. Pharmaceutical companies were established; large amounts of both public and private money were devoted to research; and research became increasingly centralized, coordinated, standardized in method, and valued. Human subjects research entered what has since been described as an "unashamedly utilitarian phase."[5] During this period, individuals were often included as research subjects because they were available, captive, and possibly considered unimportant, but they were seen as making a contribution to society. Infectious diseases were a significant problem for the armed services. The federal government and the pharmaceutical industry supported intensive research efforts to develop vaccines and antibiotics for infectious diseases to help the soldiers.

A large part of this effort was accomplished through research conducted in prisons, orphanages, homes for the emotionally or developmentally disturbed, and with other institutionalized groups. There was a fairly clear distinction between research and therapy; subjects not necessarily in need of therapy were accepting a personal burden to make a contribution to society. A utilitarian justification was the basis of claims that some individuals could be used for the greater common good. Revelations of the Nazi medical experiments and war crimes raised concerns about research with human subjects.

3.3. Protection of Research Subjects

In the late 1960s and early 1970s in the United States, shock and horror at stories of abuse of human subjects led to intense scientific and public scrutiny and reflection, as well as debate about the scope and limitations of research involving human subjects. A renowned Harvard anesthesiologist, Henry Beecher, published a landmark article in the *New England Journal of Medicine* in 1966[6] questioning the ethics of 22 research studies conducted in reputable U.S. institutions. Accounts of and debate about the hepatitis B studies at Willowbrook, the U.S. Public Health Service Tuskegee syphilis studies, and others all generated intense public attention and concern. Congressional hearings and action led to the passage in 1974 of the National Research Act (EL. 93-348) and the establishment of the National Commission for the Protection of Human Subjects of Biomedical and Behavioral Research. This extremely influential body authored multiple reports and recommendations about clinical research, including reports on research with children and institutional review boards (IRBs). Included in their legacy is the Belmont Report, in which ethical principles underlying the conduct of human subjects research and their application are explicated.[7] The emphasis of the commission's work was the need to protect individuals participating in research from potential exploitation and harm. The commission's work provided the basis for subsequent federal regulations codified in 1981 in Title 45 U.S. Code of Federal Regulations, Part 46, titled "Protection of Human Subjects." These regulations in 1991 became the currently operative Common Rule (45CFR46).[8] The Common Rule governs the conduct of human subjects research funded through any one of 17 U.S. federal agencies. The major thrust of these federal regulations and many of the existing codes of research ethics is protection of subjects from the burdens and harms of research and the possibility of exploitation.

3.4. Research as a Benefit

Events in the late 1980s and 1990s altered some public perspectives on clinical research. Certain very articulate and vocal activists claimed that participation in research can be a benefit that individuals should not be denied rather than a harm to be protected from.[9] According to this perspective, espoused by activists for individuals with the human immunodeficiency virus and breast cancer, among others, participation in research is a benefit, protectionism is discrimination, and exclusion from research can be unjust. Empirical studies have demonstrated that oncology patients, for example, who participate in clinical trials benefit through improved survival.[10,11] Activism and changes in public attitudes about research led to substantive changes in the way research is done and drugs are approved.

In addition to the possible benefits of participation, it was also claimed that certain groups of people traditionally underrepresented in research were being denied the benefits of the application of knowledge gained through research.[12] Since 1994, the U.S. National Institutes of Health requires those who receive research funding to include certain groups of traditionally underrepresented subjects, such as women and ethnic

minorities.[13] Since 1998, NIH guidelines emphasize the importance of including children in research.[14]

3.5. Community Involvement in Research

In recent years, the growth of genetics research and of international collaborative research, in particular, has highlighted an ethical need for more community involvement in research. Clinical research does not occur in a vacuum but is a collaborative social activity that requires the support and investment of involved communities, and it comes with inherent risks and potential benefits for communities. As such, involvement of the community in helping to set research priorities, planning and approving research, evaluating risks and benefits during and after a trial, and influencing particular aspects of recruitment, informed consent, and the form of community benefits demonstrates respect for the community and is likely to promote successful research.

4. CODES OF RESEARCH ETHICS AND REGULATIONS

Throughout this history several influential documents have helped to shape our sense of the contours of ethical research (Table 2-1). Most were written in response to specific crises or historical events, yet all have accepted an underlying assumption that research as a means to progress in medical care or health is good. The Nuremberg Code, a 10-point code on the ethics of human experimentation, was written as the concluding part of the judgment at the Nuremberg Trials (1949).[15] Established in response to Nazi experimentation, the Nuremberg Code recognized the potential value of research knowledge to society but emphasized the absolute necessity of the voluntary consent of the subject. The Nuremberg Code established that to be ethical, the conduct of research must

---

TABLE 2-1 Selected Codes and U.S. Regulations Guiding Research with Human Subjects

- The Nuremberg Code (1949)
- The Declaration of Helsinki (2000)
- The Belmont Report (1979)
- CIOMS *International Ethical Guidelines for Biomedical Research Involving Human Subjects* (2002)
- International Conference on Harmonization Guidelines for Good Clinical Practice (1996)
- Title 45 US CFR, Part 46–The Common Rule
- Title 21 US CFR, Parts 50 and 56

have the rights and welfare of the subject as its utmost priority. Most subsequent codes and guidelines for the ethical conduct of research have maintained this emphasis and incorporated the necessity of informed consent. The Declaration of Helsinki was developed by the World Medical Assembly in 1964 as a guide to the world's physicians involved in human subjects research.[16] The Declaration of Helsinki recognizes that some, but not all, medical research is combined with clinical care and emphasizes that patients' participation in research should not put them at a disadvantage with respect to medical care. The Declaration of Helsinki also recognized as legitimate research with people who cannot give their own informed consent but for whom informed permission would be obtained from a legal guardian. Recognized as "the fundamental document in the field of ethics in biomedical research,"[17] the Declaration of Helsinki has had considerable influence on the formulation of international, regional, and national legislation and regulations. The Declaration of Helsinki has been revised several times (1975, 1983, 1989, 1996), and most recently in 2000. Additions to the 2000 version of the declaration, especially those related to the use of placebo controls and obligations to assure post-trial access to tested interventions, have been the subject of continued debate among international researchers.

The Belmont Report, published by the U.S. National Commission for the Protection of Human Subjects of Biomedical and Behavioral Research, described three broad ethical principles that guide the conduct of research and form the "basis on which specific rules could be formulated, criticized, and interpreted."[7] The three principles are respect for persons, beneficence, and justice. Respect for persons requires respect for the autonomous decision making of capable individuals and protection of those with diminished autonomy. Informed consent is the application of this principle in clinical research. Beneficence requires not deliberately harming others, as well as maximizing benefits and minimizing harms. This principle is applied to clinical research through careful risk–benefit evaluation. Justice requires a fair distribution of the benefits and burdens of research. The application of justice described in the Belmont Report is to the selection of research subjects.

The Council of International Organizations of Medical Sciences (CIOMS) in conjunction with the World Health Organization (WHO) issued *International Ethical Guidelines for Biomedical Research Involving Human Subjects*, first in 1982 and revised in 1993 and 2002,[17] that explored the application of the Helsinki principles to the "special circumstances of many tech-nologically developing countries." The CIOMS guidelines, noting an increase in international research, acknowledge differing circumstances in developing and non-Western countries, where there is generally less of a focus on the individual. CIOMS adopts the three ethical principles spelled out in the U.S. National Commission's Belmont Report and maintains most of the tenets of Nuremberg and Helsinki but provides additional and valuable guidance and commentary on externally sponsored research and research with vulnerable populations.

United States federal regulations found in Title 45 of the U.S. Code of Federal Regulations, Part 46 (45CFR46)[8] were first promulgated in 1981 for research funded by the Department of Health and Human Services (formerly the Department of Health, Education, and Welfare). These regulations were extended in 1991 as the Federal Common Rule, applicable to research funded by any of 17 U.S. federal agencies. Based on the recommendations of the National Commission, the Common Rule stipulates both the membership and the function of IRBs and specifies the criteria an IRB should employ when reviewing a research protocol and determining whether to approve it. The Common Rule also delineates the types of information that should be included in an informed consent document and how consent should be documented. Subparts B, C, and D of 45CFR46 describe additional protections for DHHS-funded research with fetuses and pregnant women, prisoners, and children, respectively.

The U.S. Food and Drug Administration (FDA) regulations[18] found in Title 21, USCFR, Part 50, "Protection of Human Subjects," and Part 56, "Institutional Review Boards," contain regulations that are similar, but not identical, to those found in the Common Rule. Compliance with FDA regulations is required for research that is testing a drug, biologic, or medical device for which FDA approval will ultimately be sought.

5. ETHICAL FRAMEWORK FOR CLINICAL RESEARCH

Based on a synthesis of guidance found in the various ethical codes, guidelines, and literature, a systematic framework of principles that apply sequentially to all clinical research was proposed.[19] According to this framework, clinical research must satisfy the following requirements to be ethical: social or scientific value, validity, fair subject selection, favorable risk–benefit ratio, independent review, informed consent, and respect for the enrolled subject[19] (Table 2-2).

TABLE 2-2 Ethical Framework for Clinical Research

Principles of Ethical Clinical Research	Description
Value	Research poses a clinically, scientifically, or socially valuable question that will contribute to generalizable knowledge about health or be useful to improving health. Research is responsive to health needs and priorities.
Validity	Study has an appropriate and feasible design and end points, rigorous methods, and feasible strategy to ensure valid and interpretable data.
Fair subject selection	The process and outcomes of subject and site selection are fair and based on scientific appropriateness, minimization of vulnerability and risk, and maximization of benefits.
Favorable risk–benefit ratio	Study risks are justified by potential benefits and value of the knowledge. Risks are minimized and benefits are enhanced to the extent possible.
Independent review	Independent evaluation of adherence to ethical guidelines in the design, conduct, and analysis of research.
Informed consent	Clear processes for providing adequate information to and promoting the voluntary enrollment of subjects.
Respect for enrolled participants	Study attends to and shows respect for the rights and welfare of participants both during and at the conclusion of research.

5.1. Value and Validity

The first requirement of ethical research is that the research question be worth asking—that is, have potential social, scientific, or clinical value. Research has value when the answers to the research question might offer practical or useful knowledge to understand or improve health. Critical to value is the usefulness of the knowledge gained, not whether the study results are positive or negative. Value is a requirement because it is unethical to expend resources or to ask individuals to assume risk or inconvenience for no socially valuable purpose.[20] A valuable research question then ethically requires validity and rigor in research design and implementation in order to produce valid, reliable, interpretable, and generalizable results. Poorly designed research—for example, studies with inadequate power, insufficient data, or inappropriate or unfeasible methods—is harmful because human and material resources are wasted and exposed to risk for no benefit.[19]

5.2. Fair Subject Selection

Fair subject selection requires that subjects be chosen for participation in clinical research based first on the scientific question, balanced by considerations of risk, benefit, and vulnerability. As described by the National Commission in the Belmont Report, fairness in both the processes and the outcomes of subject selection prevents exploitation of vulnerable individuals and populations and promotes equitable distribution of research burdens and benefits. Fair procedures means that investigators should select subjects for scientific reasons—that is, related to the problem being studied and justified by the design and the particular questions being asked—and not because of their easy availability or manipulability, or because subjects are favored or disfavored.[7] Extra care should be taken to justify the inclusion in research of vulnerable subjects, as well as to justify excluding those who stand to benefit from participation. Since exclusion without adequate justification can also be unfair, eligibility criteria should be as broad as possible, consistent with the scientific objectives and the anticipated risks of the research. Since distributive justice is concerned with a fair distribution of benefits and burdens, the degree of benefit and burden in a particular study is an important consideration. Scientifically appropriate individuals or groups may be fairly selected consistent with attention to equitably distributing benefits and burdens as well as minimizing risk and maximizing benefit.

Persons are considered vulnerable if their ability to protect or promote their own interests is compromised or they are unable to provide informed consent. Although there remains some disagreement about the meaning of vulnerability in research and who is actually vulnerable,[21] there is support for the idea that among scientifically appropriate subjects, the less vulnerable should be selected first. So, for example, an early drug safety study should be conducted with adults before children, and with consenting adults before including those who cannot consent.

Certain groups, such as pregnant women, fetuses, prisoners, and children, are protected by specific regulations requiring additional safeguards in research. According to U.S. regulations governing research with children, a determination of the permissibility of research with children depends on the level of research risk and the anticipated benefits. Accordingly, research that poses minimal risk to children is acceptable, research with more than minimal risk must either be counterbalanced by a prospect of direct therapeutic benefit for the children in the study, or by the importance of the question in children with the disorder under study, or be approved by a special panel convened by the U.S. Secretary of DHHS.[22] Permission for the research participation of children is sought from their parents or legal guardians, and the child's assent is also sought whenever possible.

Fair subject selection also requires considering the outcomes of subject selection. For example, if women, minorities, or children are not included in studies of a particular intervention, then the results of the study may be difficult to apply to these groups and could actually be harmful. Therefore, study populations recruited for research should be representative of the populations likely to use the interventions tested in the research.[23]

Similarly, it has been argued that justice requires subjects to be among the beneficiaries of research. This means that subjects should be selected as participants in research from which they or others like them can benefit and not be asked to bear the burdens of research for which they can reap no benefits. This understanding of justice has raised important and challenging questions in the conduct of collaborative international research. Some have argued that if a drug or vaccine is tested and found effective in a certain population, there should be prior assurance that that population will have access to the drug or vaccine.[24] Alternatively, subjects or communities should be assured of and involved in negotiation about fair benefits from research that are not necessarily limited to the benefit of available products of research.[25]

5.3. Favorable Risk–Benefit Ratio

The ratio of risks to benefits in research is favorable when risks are justified by benefits to participants or society and research is designed in a way that minimizes risks and maximizes benefits to individual subjects. The ethical principle of beneficence obligates us to (1) do no harm and (2) maximize possible benefits and minimize possible harms. It is a widely accepted principle that one should not deliberately harm another individual regardless of the benefits that might be made available to others. However, as the Belmont Report reminds us, offering benefit to people and avoiding harm requires learning what is of benefit and what is harmful, even if in the process some people may be exposed to some risk of harm. To a great extent, this is what clinical research is about (i.e., learning about the benefits and harms of unproven methods of diagnosing, preventing, treating, and caring for human beings). The challenge for investigators and review groups in clinical research is to decide in advance when it is justifiable to seek certain benefits in research despite the risks, and when it is better to forego the

possible benefits because of the risks. This is called a risk–benefit assessment.

The actual calculation and weighing of risks and benefits in research is complicated. Investigators in designing a study consider whether the inherent risks are justified by the expected value of the information and benefit to the participants. Studies should be designed in a way that risks to participants are minimized and benefits are maximized. When reviewing a study, an IRB must first identify the possible risks and benefits and then weigh them to determine if the relationship of risks to benefits is favorable enough that the proposed study should go forward or should instead be modified or rejected. When reviewing studies with little or no expected benefit for individual subjects, the IRB has the sometimes formidable task of deciding whether the risks or burdens to the subjects in the study are justified only by the potential value of the knowledge to be gained, sometimes a particularly difficult risk–benefit assessment. Prospective subjects do their own risk–benefit assessment to decide whether the risks of participating in a given study are acceptable to them and worth their participation.

Many kinds of risks and benefits may be considered in a risk–benefit assessment, including physical, psychological, social, economic, and legal. For example, in a genetics study, the physical risks may be limited to a blood draw or buccal swab, and assessment of the potential psychological and social risks may be more important. Investigators, reviewers, and potential subjects may not only have dissimilar perspectives about research but also are likely to assign different weights to risks and benefits. For example, IRBs consider only health-related benefits of the research in justifying risks, whereas subjects are likely to consider access to care or financial compensation as important benefits that may tip the balance for them in favor of participation. Acknowledging that risk–benefit assessment is not a straightforward or easy process does not in any way diminish its importance. Careful attention to the potential benefits to individuals or society of a particular study in relation to its risks, as well as consideration of the risks of not conducting the research, is one of the most important steps in evaluating the ethics of clinical research.

5.4. Independent Review

Independent review allows evaluation of the research for adherence to established ethical guidelines by individuals with varied expertise and no personal or business interests in the research. For most clinical research, this independent review is carried out by an IRB or research ethics committee. Using criteria detailed in the U.S. federal regulations,[18] IRBs evaluate the benefits of doing the study, the risks involved, the fairness of the subject selection, and the plans for obtaining informed consent and decide whether to approve a study, with or without modifications, table a proposal for major revisions or more information, or disapprove a study as unacceptable. (See also Chapter 5.)

Independent review of the risks of proposed research by someone other than the investigator has been described as a "central protection for research participants."[26] Nonetheless, many believe the current system of IRBs in the United States is inadequate for protecting subjects, outdated given the current profile of clinical research, beset with conflicts, and in need of reform.[27]

5.5. Informed Consent

Once a proposal is deemed valuable, valid, and acceptable with respect to risks and benefits and subject selection, individuals are recruited and asked to give their informed consent. Through the process of informed consent, prospective subjects are given the opportunity to make autonomous decisions about participating and remaining in research. Respect for persons and their autonomy requires respect for the choices people make and no interference with these choices unless they are detrimental to others. We show lack of respect for persons when we repudiate their considered judgment, deny them the freedom to act on their judgments, or withhold information necessary to make a considered judgment. Inviting people to participate in research voluntarily and with adequate information about the research (i.e., informed consent) demonstrates respect for persons. Informed consent is a process involving three main elements: information, comprehension, and voluntariness.[28] Information provided to subjects about a research study should be adequate, according to a "reasonable volunteer" standard, balanced, and presented in a manner that is understandable to the subject. Information should be provided in the language of the subject, at an appropriate level of complexity given the subject's age and educational level, and culturally appropriate. Attention to the manner and setting in which information is presented is an important aspect of informed consent. The U.S. federal regulations detail the types of information that should be included in informed consent; these essentially include what a reasonable person would need to know to make an informed decision about initial or ongoing research participation. In addition to receiving the necessary information, individuals should be able to process and understand it in the context of their situation and life experiences. Investigators assess the degree to which an individual subject comprehends the particular information provided about a research study and can deliberate and make a choice. After deliberating about information provided, a research subject is asked to make a "voluntary" choice about participation (i.e., a choice about participation free from coercion or undue influence). Informed consent, therefore, is a process that involves presentation of information, discussion and deliberation, assessment of understanding, a choice about participation, and ultimately some form of authorization (Table 2-3).

TABLE 2-3   The Process of Informed Consent

Elements of Informed Consent	Description	Considerations and Challenges
Disclosure of information	Information about the study is disclosed that is based on a "reasonable" person standard. Disclosure takes into account subjects' language, education, familiarity with research, and cultural values. Both written information and discussion are usually provided.	There is a need to balance the goal of being comprehensive with that of attention to the amount and complexity of information in order to give participants the information they need and facilitate understanding.
Understanding	Understanding of the purpose, risks, benefits, alternatives, and requirements of the research.	Empirical data show that participants often do not have a good understanding of the details of the research.
Voluntary decision making	Free from coercion and undue influence. Free to choose not to enroll.	Many possible influences affect participants' decisions about research participation. Avoid controlling influences.
Authorization	Usually given by a signature on a written consent document.	For some individuals or communities, requiring a signature reflects lack of appreciation for their culture or literacy level.

Informed consent is a process that continues throughout someone's participation in research. The process of initial informed consent in research usually culminates with the signing of a document that attests to the fact that the volunteer has given consent to enroll in the study. However, respect for persons requires that subjects continue to be informed throughout a study and are free to modify or withdraw their consent at any time.

Although widely accepted as central to the ethical conduct of research, in reality, achieving true informed consent is challenging. Deciding how much information is adequate is not straightforward. In a complicated clinical trial, written consent documents can be long and complex, and it is not clear the extent to which large amounts of information enhance or hinder subject understanding. The appropriate mix of written and verbal information and discussion varies with the complexity of the study and the individual needs of each subject. Scientific information is often complex; research methods are unfamiliar to many people; and subjects have varying levels of education, understanding of science, knowledge about their diseases and treatments, and are dissimilar in their willingness to enter into dialogue. Besides the amount and detail of information, understanding may be influenced by who presents the information and the setting in which it is given. In some cases, information may be more accessible to potential subjects if presented in group sessions or using print, video, or other media presentations.

Determining whether a subject has the capacity to consent and understands the particular information is also challenging. Capacity to provide consent is study specific. Individuals who are challenged in some areas of decision making may still be capable of consenting to a particular research study. Similarly, individuals may not have the capacity to consent to a particular study, even if generally capable in their lives. Assessing capacity might take into account an individual's educational level and familiarity with science and research, as well as evidence of cognitive or decisional impairment. In some cases, but certainly not all, mental illness, depression, sickness, desperation, or pain may interfere with a person's capacity to understand or process information. Empirical research in informed consent has demonstrated that research participants who give their own consent to participation do not always have a good understanding of the purpose or the potential risks of their research studies.[29]

Informed consent to research should also be voluntary. Life circumstances and experiences provide a context for all decisions, such that decisions are never free from other influences. The expectation in clinical research is that a subject's decision to participate should be free from *controlling* influences.[30] Terminal or chronic illness, having exhausted other treatment options, or having no health insurance may limit a participant's options but do not necessarily render decisions involuntary. Payment and other incentives, trust in health care providers, dependence on the care of clinicians, family pressures, and other factors commonly influence decisions about research participation. Determining the point at which these otherwise acceptable influences become controlling is not straightforward. Given these multiple factors, it is important to ensure that the individual has the option to say no to research participation and to do so with impunity.

Research has demonstrated that active and ongoing dialogue and discussion between the research team and subjects, opportunities to have questions answered, waiting periods between the presentation of information and the actual decision to participate, the opportunity to consult with family members and trusted others, clear understanding of alternatives, and other strategies can serve to enhance the process of informed consent.[31,32]

5.6. Respect for Enrolled Subjects

After enrollment, research participants deserve continued respect throughout the duration of the study and after it is completed. Respect for subjects is demonstrated through appropriate clinical monitoring throughout the study and attention to their well-being. Adverse effects of research interventions and any research-related injuries should be treated. Private information collected about subjects should be kept strictly confidential, and they should be informed about the limits of confidentiality. Research subjects should be reminded of their right to withdraw from the research at any time without penalty. Reevaluation of a decision to participate may be stimulated by a change in clinical status or life circumstances. Information generated by the study or other studies that might become available and could be relevant to a person's decision about continued participation should be expeditiously shared with subjects. Investigators should make plans regarding how to help ensure continued access to successful interventions and to study results after the study is finished.

In summary, ethical clinical research is conducted according to the seven principles in Table 2-2. The exact application of the principles to specific cases will always involve some judgment and specification on the part of investigators, sponsors, review boards, and others involved in clinical research.

6. ETHICAL CONSIDERATIONS IN RANDOMIZED CLINICAL TRIALS

Randomized clinical trials (RCTs) remain the principal method and "gold standard" for demonstrating safety and efficacy in the development of new drugs and biologics, such as vaccines, surgical interventions, behavioral interventions, and systems interventions. An RCT has several characteristic features. It is controlled, randomized, and usually blinded; also, the significance of the results is determined statistically according to a predetermined algorithm. An RCT typically involves the comparison of two or more interventions (e.g., Drug A versus Drug B) to demonstrate the equivalence or the superiority of one intervention over the other in the treatment, diagnosis, or prophylaxis of a specific disorder. Although few existing codes of research ethics, guidelines, or regulations specifically speak to particular issues of moral importance in the conduct of RCTs, the design of the RCT presents a spectrum of unique ethical problems (Table 2-4). "In considering the RCT, the average IRB member must be baffled by its complexity and by the manifold problems it represents."[33]

The ethical justification to begin an RCT is usually described as that of "an honest null hypothesis,"[33] also referred to as equipoise or clinical equipoise.[34] In an RCT comparing intervention A and B, clinical equipoise is satisfied if there is no convincing evidence available to the clinical community about the relative merits of A and B (e.g., evidence that A is more effective than or less toxic than B). The goal of an RCT is by design to disturb this state of equipoise by providing credible evidence about the relative value of each intervention. Equipoise is based on the idea that even in research, patients should receive treatment with a likelihood of success, not one known to be inferior, and they should not be denied effective treatment that is otherwise available. Assigning half or some portion of subjects to each treatment in an RCT is ethically acceptable because patients are not assigned to known inferior treatment. Doubt about which intervention is superior justifies giving subjects an equal chance to get either one. There are many controversies regarding equipoise. Some argue that equipoise is based on a mistaken confluence of research with therapy and therefore should be abandoned.[35]

There are other controversies in RCTs. Universal agreement, for example, about what counts as "convincing" evidence does not exist. The common acceptance of statistical significance at the $p = 0.05$ level, indicating that there is <5% chance that differences noted between interventions in an RCT are due to chance, potentially discounts clinically but not statistically significant observations. There is also disagreement about the extent to which preliminary data, data from previous studies, data from uncontrolled studies and pilot studies, and historical data influence the balance of evidence. In some cases, the existence of these other types of data may make equipoise impossible. However, data from small, uncontrolled studies can also lead to false or inconclusive impressions about safety or efficacy, which likewise can be harmful.

Lack of convincing evidence about which of two or more interventions is superior in terms of long-term outcomes for a group of patients does not necessarily preclude judgments about what is best for a particular patient at a particular time. An individual's unique symptoms, side effects, values, preferences, etc. may suggest that one intervention is better for him or her

TABLE 2-4 Selected Ethical Considerations in Randomized Controlled Trials (RCTs)

Features of RCTs	Description	Questions/Considerations
Equipoise	No convincing evidence that one intervention is better, i.e., more effective or less toxic than the other.	How to factor in early evidence? Is a requirement for equipoise conflating research and therapy?
Choice of control	Appropriate choice of control is necessary for scientific validity and generalizability.	Choice of control is not simply a scientific decision. Placebos as controls require ethical justification.
Randomization	Random assignment decreases bias and controls for many factors.	Random assignment does not allow for autonomous preferences.
Blinding	Either single or double blinding is often used to decrease bias.	Research participants consent to temporarily suspend knowledge of which intervention they are receiving. A blind may need to be broken to treat some clinical problems.
Sharing preliminary information	As evidence accumulates, information about risks and benefits may change and equipoise may be disturbed.	Study monitors, independent data and safety monitoring boards, and others carefully monitor data to help determine when the study should be stopped or information should be shared with participants.

than the other, and if so, the individual may not be a good candidate for participation in an RCT. Clinicians responsible for the care of patients should take these factors into account. When the clinician is also serving as the investigator of a study in which the patient is a subject, tension and role conflict can occur. Being aware of this tension, clearly informing the patient, relying on other members of the team, or, in some cases, separating the roles of clinician and investigator may be necessary so that the patient's needs are not overlooked.[36]

Another important scientific and ethical consideration in RCTs is the selection of outcome variables by which the relative merits of an intervention will be determined. Different conclusions may be reached depending on whether the intervention's efficacy is a measure of survival or of tumor shrinkage, symptoms, surrogate end points, quality of life, or some composite measure. The choice of end points in a clinical trial is never simply a scientific decision.

In an RCT, subjects are assigned to treatment through a process of randomization. This means that each subject has a chance of being assigned to treatment randomly by a computer or the use of a table of random numbers rather than based on individual needs and characteristics. The goal of random assignment is to control for confounding variables by keeping the two or more treatment arms similar in relevant and otherwise uncontrollable aspects. In addition to random assignment, RCTs are often either single blind (subject does not know which intervention he or she is receiving) or double blind (both subject and investigator are blinded to the intervention). Random assignment and blinding are methods used in clinical trials to reduce bias and enhance study validity. Although compatible with the goals of an RCT, random assignment to treatment and blinding to treatment assignment are not necessarily compatible with the best interests or autonomy interests of the patient-subject. It has been shown that in some placebo-controlled blinded studies, both subjects and investigators can guess (more frequently than by chance) whether they are on active drug or placebo.[37] Therefore, the necessity and adequacy of blinding and randomization should be assessed in the design and review of a given research protocol. When randomization and blinding are deemed useful and appropriate for a particular protocol, there are two main ethical concerns: (1) Preferences for an intervention and information about which intervention a subject is receiving may be relevant to autonomous decisions, and (2) information about which intervention the subject is receiving may be important in managing an adverse event or a medical emergency. With respect to the first concern, when consenting to an RCT

subjects are informed about the purpose of the research and asked to consent to random assignment and to a temporary suspension of knowledge about which intervention they are receiving. To balance the need for scientific objectivity with respect for a research subject's need for information to make autonomous decisions, investigators should provide subjects with adequate information about the purpose and methods of randomization and blinding, and investigators should assess their understanding of these methods. Subjects are asked to consent to a suspension of knowledge about their treatment assignment until the completion of the protocol or some other predetermined time point, at which time they are informed about which intervention they received in the clinical trial.

Knowledge of which medications a subject is receiving may in some cases also be important to the treatment of adverse events or other medical emergencies, consistent with a concern about the safety and welfare of subjects. To balance the need for scientific objectivity with concern for subject safety, investigators should consider in advance the conditions under which a blind may be broken to treat an adverse event. Specifically, the protocol should specify where the code will be located, the circumstances (if any) under which the code will be broken, who will break it, how the information will be handled (i.e., will the investigator, the subject, the IRB, and the treating physician be informed), and how breaking of a blind will influence the analysis of data. A research subject should always have information about who to notify in the event of an emergency. The IRB should be satisfied that these plans provide adequate protection of patient safety.

A concern that has received recent attention especially in the international research context is how to ensure that when the trial is over, a subject can continue to access an investigational intervention that is providing benefit.[38] Some argue that those who volunteer for RCTs deserve assurance that they will receive the intervention proven to be superior in the RCT. That is, those subjects randomized to an intervention proven to be superior will continue to receive that intervention, and those randomized to the inferior intervention will be given an opportunity to receive the better one. Considerable disagreement exists regarding the extent of the obligation of the researchers or sponsors to ensure access. Additional dialogue regarding the practicalities and resources needed to ensure continued access to treatment would be very useful.

Consent to randomization may be more difficult for the subject if one of the potential treatment assignments is placebo. Some people perceive randomization to placebo in clinical trials as problematic because it potentially deprives the individual of treatment that he or she may need. On the other hand, if there is clinical equipoise and therefore no proof of the superiority of the experimental treatment, it is just as possible that those randomized to placebo are simply deprived of potentially toxic side effects or of a useless substance.[39] Scientifically, comparing an experimental drug or treatment to placebo allows the investigator to establish efficacy in an efficient and rigorous manner. Alternatively, an RCT involving comparison to another already established therapy, if one exists, may allow the investigator to establish superiority or equivalence (i.e., no difference between the experimental drug and the standard therapy control). Placebo controls in research are justified when there is no standard treatment for a given condition, when new evidence has raised doubts about the net therapeutic advantage of a standard treatment, or when investigating therapies for groups of people who are refractory to or reject standard treatments.[40] In studies that meet these criteria, subjects are not harmed and their rights are not violated by participation in placebo-controlled research. What remains controversial is the use of placebo controls in studies when available alternative therapies do exist. Some authors have argued that the use of placebo controls in these cases is ipso facto wrong and contrary to principles enunciated in the Declaration of Helsinki.[41] Others have argued that the most appropriate choice of a control in an RCT depends on the goals of the study, with considerations of the expected consequences to subjects of randomization to one arm or another, the quality of evidence regarding the effect of existing therapies, the expected variability of spontaneous changes in measured outcomes, and the extent to which a placebo effect may play a role.[42] Some authors have suggested a "middle ground" that considers both scientific design and possible risk to subjects as determinative of the acceptability of placebo.[43] It is widely agreed, however, that if the outcome for the patient of no treatment or placebo treatment is death, disability, or serious morbidity, a placebo control should not be used.[44]

7. CONCLUSION

Ethical principles and guidance for the conduct of human subjects research help to minimize the possibility of exploitation and promote respect and protection of the rights and welfare of individuals who serve as human subjects of research. This chapter reviewed an ethical framework for the conduct of clinical research, some of the historical evolution of research ethics, and ethical considerations of some of the unique features of randomized clinical trials. In addition to adherence to principles, codes of ethics, and regulations, the ethical conduct of human subjects research depends on the integrity and sagacity of all involved.

References

1. Engelhardt HT. Diagnosing well and treating prudently: Randomized clinical trials and the problem of knowing truly. In Spicker SF, Alon I, de Vries A, Engelhardt HT (eds.) *The Use of Human Beings in Research.* Dordrecht, The Netherlands: Kluwer Academic, 1988.
2. Eisenberg L. The social imperatives of medical research. *Science* 1977;198:1105–1110.
3. Jonas H. Philosophical reflections on experimenting with human subjects. In Freund P (ed.) *Experimentation with Human Subjects.* New York, Braziller, 1970.
4. Kant as quoted in Beauchamp T, Childress J (eds.) *Principles of Biomedical Ethics,* 4th ed. New York, Oxford University Press, 1994, p. 351.
5. Rothman D. Ethics and human experimentation—Henry Beecher revisited. *N Engl J Med* 1987;317:1195–1199.
6. Beecher HK. Ethics and clinical research. *N Engl J Med* 1966;274:1354–1360.
7. National Commission for the Protection of Human Subjects of Biomedical and Behavioral Research. *The Belmont Report: Ethical Principles and Guidelines for the Protection of Human Subjects of Research.* Washington, DC, U.S. Government Printing Office, 1979.
8. U.S. Code of Federal Regulations Title 45, Part 46. Available at www.hhs.gov/ohrp/humansubjects/guidance/45cfr46.htm.
9. National Research Council. *The Social Impact of AIDS in the United States.* Washington, DC, National Academy Press, 1993.
10. Herbert-Croteau N, Brisson J, Lemaire J, Latreille J. The benefit of participating to clinical research. *Breast Cancer Treatment Res* 2005;91(3):279–281.
11. Bleyer A, Montello M, Budd T, Saxman S. National survival trends of young adults with sarcoma: Lack of progress is associated with lack of clinical trial participation. *Cancer* 2005;103(9):1891–1897.
12. Dresser R. Wanted: Single, white male for medical research. *Hastings Center Rep* 1992;22(l):21–29.
13. National Institutes of Health. Guidelines for the inclusion of women and minorities as subjects in clinical research. In *NIH Guide for Grants and Contracts.* Bethesda, MD, National Institutes of Health, March 18, 1994.
14. National Institutes of Health. NIH policy and guidelines on the inclusion of children as participants in research involving human subjects. In *NIH Guide for Grants and Contracts.* Bethesda, MD, National Institutes of Health, March 6, 1998.
15. The Nuremberg Code, 1949. Available at www.hhs.gov/ohrp/references/nurcode.htm.
16. World Medical Assembly. Declaration of Helsinki 2000. Available at www.wma.net/e/ethicsunit/helsinki.htm.
17. Council for International Organizations of Medical Sciences. *International Ethical Guidelines for Biomedical Research Involving Human Subjects.* Geneva, CIOMS/WHO, 2002. Available at www.cioms.ch.
18. U.S. Code of Federal Regulations Title 21, Part 50 "Protection of Human Subjects" and Part 56 "Institutional Review Boards." Available at www.fda.gov.
19. Emanuel E, Wendler D, Grady C. What makes clinical research ethical? *J Am Med Assoc* 2000;283(20):2701–2711.

20. Freedman B. Scientific value and validity as ethical requirements for research: A proposed explanation. *IRB Rev Human Subjects Res* 1987;9(5):7–10.

21. Levine C, Faden R, Grady C, Hammerschmidt D, Eckenwiler L, Sugarman J, Consortium to Examine Clinical Research Ethics. The limitations of "vulnerability" as a protection for human research participants. *Am J Bioethics* 2004;4(3):44–49.

22. U.S. Code of Federal Regulations Title 45, Part 46. Subpart D.

23. Weijer C. Evolving ethical issues in selection of subjects for clinical research. *Cambridge Q Healthcare Ethics* 1996;5:334–345.

24. Glantz L, Annas G, Grodin M, Mariner W. Research in developing countries: Taking "benefit" seriously. *Hastings Center Rep* 1998; 28(6):38–42.

25. Participants in the 2001 Conference on Ethical Aspects of Research in Developing Countries. Fair benefits for research in developing countries. *Science* 2002;298:2133–2134.

26. National Bioethics Advisory Commission. *Ethical and Policy Issues in Research Involving Human Participants: Vol. 1. Report and Recommendations.* August 2001. Available at www.bioethics. gov/reports/past_commissions/nbac_human_part.pdf.

27. Emanuel E, Wood A, Fleischman A, *et al.* Oversight of human participants research: Identifying problems to evaluate reform proposals. *Ann Internal Med* 2004;141(4):282–291.

28. Beauchamp T, Childress J. *Principles of Biomedical Ethics*, 4th ed. New York, Oxford University Press, 1994.

29. Pace C, Grady C, Emanuel E. What we don't know about informed consent. *Sci Dev Net* 2003, August 28. Available at www.scidev.net/dossiers/ethics.

30. Faden R, Beauchamp T. *A History and Theory of Informed Consent.* New York, Oxford University Press, 1986.

31. Lavelle-Jones C, *et al.* Factors affecting the quality of informed consent. *Br Med J* 1993;306(6882):885–890.

32. Flory J, Emanuel E. Interventions to improve research participants' understanding in informed consent for research: A systematic review. *J Am Med Assoc* 2004;292:1593–1601.

33. Levine R. *Ethics and Regulation of Clinical Research*, 2nd ed. Baltimore, Urban & Schwarzenberg, 1986.

34. Freedman B. Equipoise and the ethics of clinical research. *N Engl J Med* 1987;317(3):141–145.

35. Miller F, Brody H. A critique of clinical equipoise: Therapeutic misconception in the context of clinical trials. *Hastings Center Rep* 2003;33(3):20–28.

36. Miller F, Rosenstein D, Defense E. Professional integrity in clinical research. *JAMA* 1998;280:1449–1454.

37. Fisher S, Greenberg R. How sound is the double-blind design for evaluating psychotropic drugs? *Nerv Ment Dis* 1993; 181(6):345–350.

38. Grady C. The challenge of assuring continued post-trial access to beneficial treatment. *Yale J Health Policy Law Ethics* 2005; 5(1):425–435.

39. Levine R. The use of placebos in randomized clinical trials. *IRB Rev Human Subjects Res* 1985;7(2):1–4.

40. Freedman B. Placebo controlled trials and the logic of clinical purpose. *IRB Rev Human Subjects Res* 1990;12(6):1–5.

41. Rothman KJ, Michels K. The continuing unethical use of placebo controls. *N Engl J Med* 1994;331(6):394–398.

42. Temple RJ. When are clinical trials of a given agent vs. placebo no longer appropriate or feasible? *Control Clin Trials* 1997; 18(6):613–620.

43. Emanuel EJ, Miller FG. The ethics of placebo-controlled trials— A middle ground. *N Engl J Med* 2001;345(12):915–919.

44. Miller F, Brody H. What makes placebo-controlled trials unethical? *Am J Bioethics* 2002;2(2):3–9.

3

Researching a Bioethical Question

EZEKIEL J. EMANUEL

Department of Clinical Bioethics, National Institutes of Health Clinical Center, Bethesda, Maryland

During the past 35 years, there has been a significant increase in interest in bioethical questions. Common questions include the following: Is it ethical to pay children up to $1400 to participate in clinical research on a new antiasthma drug? Does payment for participation in clinical research lead to having more socioeconomically vulnerable subjects? How large does a payment for participating in clinical research have to be to constitute "undue inducement"? Does payment for participation in clinical research lead to worse informed consent?

Should international clinical research studies offer all participants the best therapy available anywhere in the world? Or is it sufficient to provide subjects in an international clinical research study only the local standard of care? Does a clinical research study have to prospectively include a plan to provide a successfully tested drug to all people in the country in which it is being tested?

Is the current public interest in legalizing euthanasia and physician-assisted suicide the result of advances in life-sustaining technology and improvements in life expectancy? Are patients interested in euthanasia or physician-assisted suicide because they are suffering from excruciating pain? Are vulnerable members of the population likely to be coerced to receive euthanasia or physician-assisted suicide?

Is it appropriate to conduct research on a stored biological sample without the patient's informed consent? Can a stored biological sample be used for a study that is completely unrelated to the original reasons it was collected? Should patients be informed of results that are obtained by studies on their stored biological samples?

These and similar questions are not merely matters of opinion or feelings. They are bioethical questions that require rigorous research. And like other types of clinical research, research into bioethical issues utilizes a variety of methodologies that should adhere to the same standards of rigor. This chapter reviews different types of bioethical issues, different types of research methodologies, examples of how these research methodologies have been utilized to answer important bioethical questions, and special considerations in bioethical research.

1. TYPES OF BIOETHICAL ISSUES

Bioethical issues can be classified into six different types (Table 3-1). Each of these types of issues raises many specific questions that can be subjected to research.

2. TYPES OF BIOETHICAL RESEARCH METHODOLOGIES

There are five main bioethical research methodologies: historical inquiry, conceptual analysis, cross-cultural comparisons, empirical studies, and policy analysis. Conducting an historical inquiry related to a bioethical question requires the same techniques and methods as historical research of any type. It mostly focuses on other historical periods when similar bioethical issues were being considered to discover illuminating insights for current bioethical questions.

**TABLE 3-1 A Typology of Bioethical Issues and Examples
of Each**

The relationship between the physician and the patient
　　Truth telling
　　Confidentiality
　　Informed consent
　　Conflict of interest
The selection of medical interventions
　　Terminating care
　　"Baby Doe" cases
　　Euthanasia
The allocation of medical resources
　　Just health care
　　Patient selection criteria for scarce resources
The application of personally transforming technologies
　　Germline gene transfer
　　Brain tissue transplants
　　Psychosurgery
The use of reproductive technologies
　　Cloning
　　Surrogate motherhood
The conduct of biomedical research
　　Fraud, fabrication, and plagiarism
　　Randomized clinical trials
　　Phase I research

Conceptual analysis of bioethical issues uses the methods of philosophy to make useful distinctions, clarify commonly used concepts, and develop and justify certain positions. Good conceptual analysis is often necessary to clarify the questions subject to empirical research.

Some important bioethical questions can best be answered by cross-cultural studies of practices in different countries or between different cultural groups in the same country. These studies can utilize the methods of anthropology as well as traditional survey methods.

During the past 15 years, one of the most important advances has been rigorous empirical studies of bioethical issues. Empirical studies commonly use the methods of survey research and health services research. Increasingly, qualitative research methods such as grounded theory are being used.

Finally, many policies on bioethical issues are proposed and implemented that can be subjected to policy analysis for their likely impacts.

3. EXAMPLES OF IMPORTANT
BIOETHICAL RESEARCH

Delineating the variety of bioethical research methodologies is relatively dry. However, these abstract points can best be illustrated with specific examples of the use of these different methodologies to illuminate important questions.

3.1. Historical Research Methodology

One of the most interesting uses of historical research on bioethical questions is related to informed consent. Although clinical researchers have embraced informed consent, they remain skeptical about it. Traditionally, in the clinical setting the need for informed consent is dated from the 1957 landmark case of *Salgo v. Leland Stanford, Jr. University Board of Trustees*, in which the term *informed consent* was first used.[1] In the research setting, the Nuremberg Code required subjects to give "voluntary consent."[2] However, historical research has suggested that the notion of providing consent to both clinical care and research participation is much older than 1947 or 1957. Indeed, historical research has suggested that for most of modern history, consent by patients and human research subjects has at least been a well-accepted ideal if not actually standard practice.

In the clinical setting, historical research has revealed that the first reported legal case in the English language involving informed consent was *Slater v. Baker & Stapleton* in 1767.[3–5] A patient sued two surgeons for rebreaking a partially healed leg fracture in an effort to improve its alignment. Relying on the statement of physicians, the court ruled that it is "the usage and law of surgeons" to obtain the patient's consent before performing an operation. The court held that the two practitioners had violated the well-known and accepted rules of consent.

Historical research also demonstrated that consent in the context of clinical research dated to at least the 19th century and the beginnings of clinical research. In the late 19th century, significant efforts were made to identify the etiology of yellow fever. In 1897, the Italian researcher Guiseppe Sanarelli claimed he identified the yellow fever bacillus and, using this bacillus, had produced yellow fever in several patients.[6] William Osier condemned these experiments, saying "to deliberately inject a poison of known high degree of virulency into human beings, unless you obtain that man's sanction, is not ridiculous, it is criminal." When Walter Reed conducted his experiments on the etiology of yellow fever, he developed a written "contract" with the subjects that outlined "the risks of participation in the study as well as the benefits."[6] Examination of correspondence between Reed, members of his research team, and the U.S. Surgeon General Miller Sternberg indicates a keen awareness of the need to ensure that the yellow fever experiments "should not be made upon any individual without his full knowledge and consent."[6]

Meticulous historical research using the traditional methods of historians—examining published articles and government documents, reading correspondence, journals, and notebooks, etc.—reveals that informed consent for clinical care and research is not a recent, post-World War II phenomenon. Indeed, this research reveals that much of the ethical justification and the very mechanisms of implementation—the need to provide human subjects with information and the use of documents signed by the participants—is almost coeval with clinical research. Historical research also reveals that failure to obtain consent was grounds for moral condemnation of clinical research in the strongest possible terms by prominent members of the medical community more than 100 years ago.

Thus, historical research into informed consent has made several important contributions. First, it undercuts the notion that patient consent is foreign to medical practice and is a requirement created and imposed by lawyers or bioethicists. It establishes that patient and human subject consent was a shared and recognized ethical ideal that was used to measure practices. It also shows that there were many particular instances in which obtaining consent and even using written documents were standard practice among doctors and researchers 100 or more years ago.

Another interesting example of the value of historical research for bioethical questions relates to the debate over euthanasia and physician-assisted suicide. The standard view is that advances in technology create interest in and desire for euthanasia and physician-assisted suicide. As the Ninth Circuit Court stated in a recent case, "The emergent right to receive medical assistance in hastening one's death [is the] inevitable consequence of changes in the causes of death, advances in medical science, and the development of new technologies."[7] Rigorous historical research, however, has revealed such a link to be improbable and thoughtless speculation. First, euthanasia and physician-assisted suicide were the subject of significant controversy among medical practitioners in ancient Greece. Indeed, the medical historian Edelstein noted that the Hippocratic Oath contained a prohibition against euthanasia precisely because it opposed the common practice of euthanasia among physicians in ancient Greece.[8] Other researchers have documented that there was a significant debate about legalizing euthanasia and physician-assisted suicide in the United States and Britain in the latter third of the 19th century.[9,10] In 1870, a nonphysician gave a speech urging legalization of euthanasia; this speech was sub-sequently published as a book and debated in many prominent London magazines.[10,11] Examination of the records and publications of state medical societies in the United States shows that in the decades following this speech, many state medical societies, including those of Maine, Pennsylvania, and South Carolina, debated euthanasia at their annual meetings.[10,12] Between 1880 and the early 1900s, many prominent medical journals also published articles about the debate to legalize euthanasia and physician-assisted suicide.[10,13] Indeed, a bill to legalize euthanasia was introduced into the Ohio legislature and was defeated in 1906.[10]

This rich historical research has emphasized that these debates about euthanasia and physician-assisted suicide all took place well before any significant changes in "the causes of death, advances in medical science, and the development of new technologies."[7] This is another case in which what appears recent and contemporary can be shown to have very old roots. It forces reexamination of the reasons for current interest in euthanasia and physician-assisted suicide away from technology to other social factors.

Such historical research can help answer some important bioethical questions, including (1) What bioethical concerns are caused by advances in medical technology and what are inherent in medicine? and (2) How have these bioethical issues been addressed and resolved previously?

3.2. Conceptual Analysis

One of the most important types of bioethical research has been conceptual analysis. Although frequently undervalued and even dismissed, conceptual analysis has been essential to advancing bioethics and, indeed, advancing clinical research. One key example is the development of the justification for randomized controlled trials (RCTs). Many have argued that such trials are justified when a physician can state and believe in a "null hypothesis."[14,15] That is, RCTs are deemed justifiable when physicians have no reason to believe that one therapy is better than a second therapy and that there is no other therapy better than both. This was termed *equipoise*.[16] However, it became quite clear that, as stated, equipoise was very problematic. First, it suggested that the ethical justification of clinical trials depended on the views of individual physicians.[15] More practically, it appeared that in many trials clinicians believed equipoise was not satisfied and failed to enroll patients in the clinical trial.[15] Indeed, some of the most prominent theorists of clinical research design endorsed such a view, arguing that

If a clinician knows, or has good reason to believe, that a new therapy (A) is better than another therapy (B) he cannot participate in comparative trials of therapy A versus therapy B. Ethically, the clinician is obligated to give therapy A.[17]

A significant advance was made in 1987 by Benjamin Freedman when he distinguished theoretical equipoise from what he called clinical equipoise:[15]

Theoretical equipoise exists when, overall the evidence on behalf of two alternative treatment regimens is exactly balanced. . . . Theoretical equipoise is overwhelmingly fragile; that it is disturbed by a slight accretion of evidence favoring one arm of the trial. . . . Theoretical equipoise is also highly sensitive to the vagaries of the investigator's attention and perception. Because of its fragility, theoretical equipoise is disturbed as soon as the investigator perceives a difference between the alternatives—whether or not any genuine difference exists. . . . [T]heoretical equipoise is personal and idiosyncratic. It is disturbed when the clinician has, in Schafer's words, what "might even be labeled a bias or a hunch."

Freedman's advance was to make a careful distinction between theoretical equipoise and clinical equipoise. Clinical equipoise occurs not when the belief of a clinician is in precise balance or when the accumulated evidence is evenly split; rather, clinical equipoise refers to the balance in the views of the community of researchers:

[T]here is a split in the clinical community, with some clinicians favoring [treatment] A and others favoring [treatment] B. Each side recognizes that the opposing side has evidence to support its position, yet each still thinks that overall its own view is correct. There exists . . . an honest professional disagreement among expert clinicians about the preferred treatment.[15]

Clinical equipoise exists when the data are unclear—that is, when there is no consensus among the experts. Clinical equipoise is compatible with an individual investigator or a clinician having a preference or bias for one treatment or another. The insight of Freedman is that equipoise is a communal or social, not an individual, phenomenon.[15,16]

Although Freedman's insight may seem subtle and even trivial, it has been very powerful because it has made clear that the justification of a clinical trial does not depend on any individual's views. Clinical equipoise, for all its own problems, has provided the clearest articulation of the ethical justification for randomized controlled trials and the strongest response to those who argue that physicians cannot ethically enroll patients in randomized clinical trials.

A second example of the importance of conceptual analysis may be found in the issue of coercion, undue inducement, and exploitation. These are critical concepts for research ethics. Coerced consent is involuntary and therefore not valid. Consent rendered in response to undue inducements also is thought to be invalid:

Payment in money or in kind to research subjects should not be so large as to persuade them to take undue risks or volunteer against their better judgment. Payments or rewards that undermine a person's capacity to exercise free choice invalidate consent.[18]

Yet these three concepts are frequently confused and conflated and even mixed up with other concepts, such as misunderstanding or deception:

It is difficult to avoid coercing subjects in most settings where clinical investigation in the developing world is conducted. African subjects with relatively little understanding of medical aspects of research participation, indisposed toward resisting the suggestions of Western doctors, perhaps operating under the mistaken notion that they are being treated, and possibly receiving some ancillary benefits from participation in research, are very susceptible to coercion.[19]

Table 3-2 clarifies the proper definitions of coercion, undue inducement, exploitation, and other concepts with which they are confused.[20] Coercion is a threat that makes people worse off no matter what they choose. The classic example is when the thief says, "Your money or your life." Coercion of this sort is very rare in research, and charges of coercion should be treated with skepticism.[21] Conversely, undue inducement is about offering—not threatening—with too much of a good thing that makes someone expose himor herself to excessive risk. Undue inducement is the irresistible million dollar offer to do something too risky. This should be contrasted with exploitation, which involves giving too little.

These conceptual distinctions are more than merely philosophical casuistry. They are important for designing surveys that ask the right questions. Asking if a person understands a risk is not asking about coercion or exploitation. Similarly, distinguishing these concepts is important for designing the correct remedies to solve these ethical problems.[22] The solution for coercion is to get rid of the threat, that for undue inducement is to lower the offer or reduce the risks, and that for exploitation is to increase the offer and goods to be delivered.

A third example of helpful conceptual analysis is a clarification about the physician–patient relationship. In the 1980s, there evolved a stark polarization in the conception of the physician–patient relationship. Physicians were portrayed as being paternalistic, imposing their own values on patients. Critics and many courts urged an alternative, autonomy-based view in which the physician was supposed to delineate options so that the patient, using his or her values, could choose what to do.[23] One court wrote,

TABLE 3-2   Distinct Ethical Violations and Their Solutions

Ethical Violation	Definition	Classic Example	Solution
Undue inducement	Offer of a desirable good in excess such that it compromises judgment and leads to serious risks that threaten fundamental interests	"I'll pay you $1 million to . . ."	Traditional solution is to reduce the quantity of the desirable good offered. Actual solution is to reduce the risks or improve the risk–benefit ratio.
Coercion	Threats that make a person choose an option that necessarily makes him or her worse off and that he or she does not want to do	"Your money or your life."	Prevent or remove the threat.
Exploitation	Unfair distribution of burdens and benefits from an interaction	"That deal is unfair, you are charging me too much (or you aren't giving me enough)."	Increase benefits to the party receiving the inadequate level of benefits or assuming excessive burdens.
Injustice	Unfair distribution of resources before any interaction, in the background circumstances	Lack access to antiretroviral drugs because of poverty.	Redistribute resources, increasing the resources of the worst off before the interactions.
Deception	Intentional withholding or distortion of essential information to mislead or create a false impression	"This won't hurt at all."	Disclose accurate information.
Inadequate disclosure	Providing insufficient information		Disclose all essential information.
Misunderstanding	Inadequate comprehension of provided essential information	"I did not know I might get a placebo."	Improve comprehension through more discussion between research participant and research team.

It is the prerogative of the patient, not the physician, to determine for himself the direction in which his interests seem to lie. To enable the patient to chart his course, understandably, some familiarity with the therapeutic alternatives and their hazards become essential.[24]

It turned out that it was the role of the physician to delineate these alternatives. Physicians argued that this view made them no more than technicians and did not accurately portray the realities and complexities of their interactions with patients. Unfortunately, the alternatives were characterized as either physician paternalism or patient autonomy. The consequence was that physicians who opposed the autonomy-based view were characterized as advocating paternalism. There seemed to be no middle ground.

Progress was possible only with a more subtle delineation of alternative conceptions of the physician–patient relationship. Four alternative models were characterized (Table 3-3).[25] The alternative models indicated that there is more to the physician–patient relationship than a choice among two options. Patients' values do not come fixed and clear but are in need of elucidation. Furthermore, there is significant consideration of how the options advance these values and how they might require revision of these values. The interpretive and deliberative conceptions of the physician–patient relationship are thought to be descriptively more accurate and also more consistent with the ideal.

3.3. Cross-Cultural Analysis

One of the more important cross-cultural studies in bioethics related to how different cultures in the United States approach explicit discussions of death and dying. For many years, there has been pressure for physicians to be more frank in disclosing a patient's terminal status; data showed that the vast majority of Americans wanted to be told when they were dying. However, there was growing experience that at least some people from other cultures did not desire such frankness about death and dying. Blackhall and colleagues[26] surveyed Mexican Americans and Korean Americans about their preferences regarding end-of-life decision making. They found significant differences between these groups and the dominant white population in the United States (Table 3-4). For instance, only 47% of Korean Americans and 65% of Mexican Americans believed that patients should be told about their terminal diagnosis compared to 87% of Anglo-Americans. Similarly, only 28% of Korean Americans and 41% of Mexican Americans believed that patients should make decisions about using life-sustaining technologies compared to 65% of Anglo-Americans. These data provide important information about different attitudes among different cultural groups regarding end-of-life care. The data imply that the dominant model of end-of-life decision making may not apply to all, and that cultural sensitivity is needed

TABLE 3-3   Four Models of the Physician–Patient Relationship

	Informative	Interpretive	Deliberative	Paternalistic
Patient values	Defined, fixed, and known to the patient	Inchoate and conflicting, requiring elucidation	Open to development and revision through moral discussion	Objective and shared by physician and patient
Physician's obligations	Providing relevant factual information and implementing patient's selected interventions	Elucidating and interpreting relevant patient values as well as informing the patient and implementing the patient's selected interventions	Articulating and persuading the patient of the most admirable values as well as informing the patient and implementing the patient's selected interventions	Promoting the patient's well-being independent of the patient's current preferences
Conception of patient autonomy	Choice of, and control over, medical care	Self-understanding relevant to medical care	Moral self-development relevant to medical care	Assenting to objective values
Conception of physician's role	Competent technical expert	Counselor or adviser	Friend or teacher	Guardian

in approaching different patients about these decisions.

3.4. Empirical Research

Beginning in the mid- to late-1980s, there was increasing understanding that many bioethical questions required rigorous empirical research. This research is important for many reasons. Many ethical norms invoke a "reasonable person" standard; for example, what information would a reasonable person want for informed consent or what safeguards would a reasonable person want? Empirical data help determine what reasonable people want. Similarly, empirical data evaluate claims about what is the case. Are blacks reluctant to participate in research because Tuskegee made them suspicious? Is it people in pain who desire euthanasia? Do research participants really want to know whether their researcher has consulting contracts with the drug company sponsoring the study? Similarly, empirical data help determine whether certain interventions are achieving their objective. Do videos improve the quality of understanding in informed consent?

Various types of empirical studies can be applied to bioethical questions. First, there are descriptive studies. The first area in which such studies occurred and became methodologically rigorous was end-of-life care. A major issue in end-of-life care related to proxy decision making. When patients become incompetent and cannot make decisions about medical interventions, especially about terminating life-sustaining treatments, it was argued that family members should have the authority to decide for them. After all, not only were family members able to make contemporaneous decisions with full knowledge of the medical

TABLE 3-4   Culture and Attitudes Toward End-of-Life Care

Who should decide about whether to put the patient on a life-support machine?

	Patient	Physician	Family
African Americans	60%	16%	22%
European Americans	65%	8%	20%
Korean Americans	28%	15%	57%
Mexican Americans	41%	10%	45%

situation but also, having "unique knowledge of the patient,"[27] the family would "don the mental mantle"[28] of the patient and make "the decision that the incompetent patient would make if he or she were competent."[29] Indeed, this view was used to justify many court decisions that gave family members the right to terminate care for their loved one.

The problem was that such assertions of special family knowledge were an empirical claim that had not been evaluated. In the late 1980s and early 1990s, these assertions were subjected to empirical research.[30–33] Husbands and wives were asked independently what medical treatments they thought the other spouse would want in a variety of clinical circumstances, including in the current state of health and if they became mentally incompetent because of either dementia or a stroke. Agreement between husband and wife was good for interventions that involved the patient's current health, but agreement was no better than chance when it involved any mental incapacity, precisely the circumstance in which the proxy would be called on to make decisions (Fig. 3-1).[30] Indeed, such empirical research has clearly established that proxy decision makers have no special under-

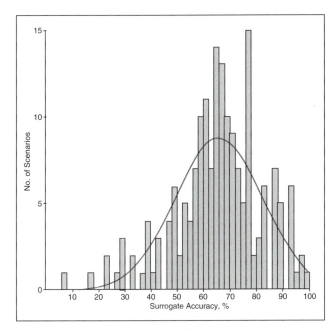

FIGURE 3-1 Distribution of surrogate accuracy in individual scenarios. Each column represents the number of scenarios in which the given percentage of surrogates accurately predicted their patient's treatment preference. The histogram includes 151 scenarios, 2595 surrogate-patient pairs, and 19,526 total paired responses. Adjusted overall accuracy of surrogates, based on meta-analysis, is 68% (95% credible interval, 63–72).

standing of patients' preferences and wishes; the justification for deferring to proxies cannot be their knowledge.[30]

A second type of research involves elucidating predictors of an outcome such as a decision or behavior. In such research, a certain outcome is described and univariate and multivariate analyses indicate whether certain factors are significantly associated with the outcome. This has been done effectively in the area of conflicts of interest. There has been extensive discussion and debate about physician and researcher conflicts of interest and whether these are more matters of appearance or really affect patient care and clinical research.[34] Many defenders of physicians have argued that receiving money for medical services does not affect their medical judgment of appropriateness and that holding stock in or consulting for drug companies does not alter their interpretation of research results. An important series of studies demonstrated that there is a link between physician investment in medical facilities and a higher, even inappropriate, use of the services provided by those facilities. One of the most important studies was conducted by Hillman et al.,[35] who compared the frequency and costs of diagnostic radiologic imagining between physicians who have diagnostic radiologic facilities in their offices and phy-

sicians who refer patients to radiologists for diagnostic imaging. They used the data from health insurance claims of several large U.S. corporations evaluating chest x-ray studies, obstetric ultrasonography, lumbar spine x-ray studies, and excretory urography and cystography. Physicians who self-referred—that is, utilized radiologic services in their offices—performed diagnostic imaging 4 to 4.5 times more often than physicians who referred patients to outside radiologists. For instance, among patients who presented with upper respiratory symptoms, physicians who provided radiologic services obtained a chest x-ray study in 46% of cases, whereas physicians who normally do not perform a chest x-ray study but refer patients out performed x-ray studies in only 11% of cases.[35] Not surprisingly, mean charges per episode were also significantly higher—4.4 to 7.5 times higher—among physicians with their own radiologic services.[35]

This and other empirical research studies of conflicts of interest were so rigorous that they became instrumental in convincing U.S. medical societies to condemn physician ownership of medical facilities and in having Congress establish rules governing the self-referral of patients to facilities in which physicians have ownership interests.[36,37]

There are many studies of conflict of interest among clinical researchers. Probably the most important is a study by Stelfox and colleagues.[38] In the mid-1990s, it was controversial whether the use of calcium channel antagonists as antihypertensive medications was safe. Stelfox and colleagues assessed whether financial ties to drug companies that manufactured calcium channel antagonists influenced researchers' judgments on this controversy. They assessed all the articles in the literature, including 5 original reports, 32 review articles, and 33 letters. As Table 3-5 shows, they found that there was a significant association between a financial interest with a calcium channel manufacturer and views that were supportive of the use of calcium channel drugs. The only exception to this association was for consulting. Why? It is difficult to know, but it may be that drug companies want to know all views and pay even those hostile to their position. Nonetheless, these data suggest a strong link between financial interests and interpretation of data.

Importantly, these studies searching for predictors cannot demonstrate causality, only associations. They are suggestive. With a sufficient number of them, it may be possible to persuade people of the causal connection—as in cigarettes causing cancer based only on associational data—but by themselves they do not prove causation.

Another example in which empirical research on bioethical issues has been important relates to eutha-

TABLE 3-5   Relationship between Financial Interests and Interpretation of Data Related to
Safety of Calcium Channel Antagonists

Variable	Supportive of Calcium Channel Antagonists	Neutral about Calcium Channel Antagonists	Critical of Calcium Channel Antagonists	*p* Value
No. of articles	30	17	23	
No. of authors responding to the survey	24	15	30	
No. of authors with financial ties to any calcium channel manufacturer	24 (100%)	10 (67%)	13 (43%)	<0.001
Honorarium from any calcium channel manufacturer to speak at symposium	75%	40%	17%	<0.001
Research funding from any calcium channel manufacturer	87%	40%	20%	<0.001
Consulting for any calcium channel manufacturer	25%	33%	17%	0.45

nasia and physician-assisted suicide. Cancer patients suffering from extreme pain are the classic example invoked to support euthanasia and physician-assisted suicide.[39–41] The following are the words that one court used in describing cases that would justify euthanasia:

> Americans are living longer, and when they finally succumb to illness, lingering longer, either in great pain or in a stuporous, semicomatose condition that results from the infusion of vast amounts of pain killing medications.... AIDS, which often subjects its victims to a horrifying and drawn-out demise, has also contributed to the growing number of terminally ill patients who die protracted and painful deaths.[7]

This claim can be subjected to empirical research by examining whether the outcomes of interest—desire for euthanasia or physician-assisted suicide or requests and actual attempts—are associated with pain or some other variable. Researchers interviewed patients with HIV/AIDS,[42] amyotrophic lateral sclerosis,[43] and cancer[44] to determine what proportion either could imagine wanting euthanasia or physician-assisted suicide or actually considered these interventions. They also asked about the patients' experience of pain as well as other symptoms. Using multivariate logistic regression analysis, pain was not an independent predictor of patients' interest in or desire for euthanasia or physician-assisted suicide. These studies showed that the factors consistently associated with patients' interest in or desire for euthanasia or physician-assisted suicide are patients' depression or level of hopelessness. Additional empirical data relevant to this question came from different teams of researchers who examined the cases of legalized physician-assisted suicide in The Netherlands, Oregon, and Australia. Physicians in The Netherlands were convinced that patients who requested euthanasia did so after serious

reflection and were not depressed. They studied the matter, comparing cancer patients who requested euthanasia with cancer patients who did not and assessing the proportions who were depressed and the proportions who had other symptoms.[45] To their surprise, cancer patients who requested euthanasia were four times more likely to have depressive symptoms. Similarly, in Australia, seven cancer patients had euthanasia for the brief time it was legal in the Northern Territory.[46] Interestingly, none of these patients suffered any pain (four had controlled pain), but four had depressive symptoms or were frankly suicidal.

A third type of empirical research on bioethical issues is the use of RCTs. Although these are traditionally associated with trials of medical interventions, there have been some trials of bioethical interventions. The largest and most famous is the Study to Understand Prognosis and Preferences for Outcomes and Risks of Treatment (SUPPORT).[47] This study was performed at five teaching hospitals in the United States, enrolling patients admitted with one of nine terminal diagnoses, ranging from metastatic lung and colon cancer to exacerbations of end-stage chronic obstructive lung disease (COPD), congestive heart failure, and cirrhosis, with a prognosis of 6 months or less. The purpose of the study was to evaluate whether an intervention could result in earlier use of do not resuscitate (DNR) orders, fewer intensive care unit (ICU) days, less patient pain, and reduced use of hospital resources. The intervention involved two steps:

1. Providing physicians with detailed prognostic information on each patient, including information on 6-month survival and prognosis for outcome of cardiopulmonary resuscitation (CPR)

2. Having a nurse elicit and document in the chart the understanding of the patient and family regarding the disease and prognosis, their preferences regarding

end-of-life care, including preference for CPR and use of advance directives, and facilitate communication of this information to the physician.

A total of 4804 patients at the five hospitals were randomized to receive either this intervention or "usual" end-of-life care.[47] The result was a wholly negative study. Patients receiving the intervention did not have a shorter time until DNR orders were written, did not have fewer days in the ICU, did not have less pain, and consumed hospital resources at the same level as the control patients.

Although SUPPORT's intervention failed to improve these outcome measures, it was a success in demonstrating that it is possible to conduct an RCT of bioethical interventions.

Finally, as in clinical research, in empirical bioethics research there are systematic reviews or meta-analyses of trials that permit the drawing of stronger conclusions from combining multiple studies. For instance, in the case of surrogate decision making a meta-analysis of 16 studies involving 2595 patient–surrogate pairs has been performed.[48] This meta-analysis showed that surrogates predicted patients' preferences for treatment with only 68% accuracy and that neither patient designation of the surrogate nor prior discussion between the surrogate and patient about the treatment improve this accuracy. Therefore, in approximately one-third of cases, surrogates are not making the decisions that patients would want regarding treatments.

Another example of meta-analyses involves the different approaches that might be used to improve the quality of informed consent in clinical research trials. Do videos or interactive computer modules enhance understanding of the components of research trials? A meta-analysis of the 12 "multimedia" interventions revealed only one showed a statistically significant improvement in understanding using a computerized presentation of information (Table 3-6).[49]

3.5. Policy Analysis

Policy analysis can take a variety of forms, but one that has been useful in relationship to bioethical issues is akin to meta-analysis. It is the collection, summary, and analysis of data on a potential policy choice. An example in which this has been done relates to cost savings from use of hospice care.[50] It has been commonly argued that spending for dying patients is extremely high and that use of a hospice can produce substantial cost savings. Indeed, many health economists and other experts have urged hospice use to lower health care costs and have even suggested that these cost savings could be used to expand health cov-

erage. In an important policy analysis, various studies on the cost and resource utilization of hospice were examined to determine whether any savings occur and the magnitude of the savings. There are seven published reports on cost savings from hospice use, most dating from the early 1980s.[51–56] All deal with cancer patients and the time periods evaluated tend to be one and six months before death. Only one study is a randomized trial; most are retrospective cohort analyses. Overall, the data show substantial savings in the last month of life, with fewer savings when time before death is longer. Indeed, the analysis suggests that during the last year of life, hospice use is associated with a savings of 0–10% in total costs. This analysis provided deeper understanding that hospice certainly does not cost more than conventional care but also that hospice is not likely to generate substantial cost savings even for cancer patients.

4. SPECIAL CONSIDERATIONS IN BIOETHICAL RESEARCH

Many of the requirements for conducting rigorous and reliable bioethical research are no different from the requirements for high-quality clinical research in other areas. However, there are some special considerations that relate to bioethical research. First, some of the methods used are not well-known to clinical researchers and are not ones in which they have been trained. For instance, few clinical researchers have any training or experience in the methods of historical research, such as using original sources and archival material. Similarly, qualitative research methods may not be familiar to many clinical researchers. Collaboration with people who have expertise in these areas can provide a way of obtaining the skills.

Second, clinical researchers may perceive some of the distinctions made in conceptual analysis as "splitting hairs." However, precise distinctions that arise from conceptual analysis are not only indispensable to clarifying ethical judgments but also essential for quality empirical research. The more precise the concepts, the better the empirical research. In this sense, conceptual analysis is frequently an element of good empirical research. For instance, if one wanted to do research on voluntariness in informed consent, then it is important to understand what is essential to voluntariness—not feeling pressure from the researcher, being able to refuse enrollment, and being able to withdraw. However, does feeling pressure from a family member or the unremitting progression of one's disease compromise voluntariness? Does voluntariness require good alternatives?

TABLE 3-6   Results of Trials of Video and Computer Multimedia Interventions

Source	Intervention	Population	Scenario	Methodology	Sample Size	Understanding Scores (%)		p Value
						Control	Intervention	
Dunn et al.,[39] 2002	PowerPoint presentation replaces consent form	Psychiatric outpatients and healthy volunteers	Real	Randomized	99	85	91	0.01
Agre et al.,[43] 2003 (Kass et al. trial)	Supplementary touch-screen presentation on oncology clinical research	Oncology patients	Real	Randomized	87	[a]	[a]	[a]
Agre et al.,[43] 2003 (Mintz et al. trial)	Supplementary video encouraging participant involvement in decision making	Psychiatric patients	Real	Randomized	37	[a]	[a]	[a]
Benson et al.,[21] 1988	Supplementary video prepared by investigator	Psychiatric patients	Real	Nonrandom	44	51	54	NS
Benson et al.,[21] 1988	Revised version of supplementary video	Psychiatric patients	Real	Nonrandom	44	51	58	NS
Llewellyn-Thomas et al.,[24] 1995	Interactive computer program replaces consent form	Oncology patients	Simulated	Randomized	100	81	79	NS
Fureman et al.,[26] 1997	Supplementary video in question-and-answer format	Injecting drug users	Simulated	Randomized	186	81	80	>0.10
Weston et al.,[27] 1997	Supplementary 10-min video	Pregnant women	Simulated	Randomized	90	91	95	NS[b]
Agre and Rapkin,[41] 2003	Video replaces consent form	Patients and healthy volunteers	Real	Randomized	221	68	73	NS
Agre and Rapkin,[41] 2003	Computer presentation replaces consent form	Patients and healthy volunteers	Real	Randomized	209	68	66	NS
Agre et al.,[43] 2003 (Campbell et al. trial)	PowerPoint presentation replaces consent form	Parents of pediatric research participants	Simulated	Randomized	NA	NS	NS	NS
Agre et al.,[43] 2003 (Campbell et al. trial)	Narrated video replaces consent form	Parents of pediatric research participants	Simulated	Randomized	NA	NS	NS	NS

NA, not available; NS, not significant.

[a]Significant improvement reported.

[b]A significant increase in retention of information weeks later was reported; therefore, this intervention was shown to improve memory but not comprehension at the time of disclosure.

A third consideration in research on a bioethical question relates to valid and reliable measures. During the past two decades, much effort has been devoted to creating and validating outcome measures. Thus, a variety of groups have developed reliable measures of pain or quality of life. Unfortunately, for many of the critical bioethical issues there is no "good standard" outcome measure. There is no standard measure of competency, or informed consent, of a good death, interest in euthanasia, good ethics consultation, and voluntariness. Consequently, developing and testing questions and measures are frequently essential elements in conducting empirical research on bioethical questions. Conversely, some of the most sloppy empirical research in bioethics is the result of using bad questions that have not been subjected to rigorous pretesting, may easily be misinterpreted by respondents, have bias, or do not measure what is desired. As in any research, bad methodology generates unreliable results in bioethics. For instance, in the early days of research on euthanasia, there were no standard questions and researchers developed their own questions without any pretesting. Some questions asked, Do you want euthanasia? Others asked, Do you desire death? Still others asked physicians, Has any patient asked you to end their lives? The definition of euthanasia may be unclear to some respondents and can easily be misinterpreted. Desiring death is not the same as euthanasia because some patients may desire death but do not want to intentionally end their lives; conflating these means that the question does not measure euthanasia alone. A life can be ended by fatal injection or turning off a medical intervention; many commentators view these as different. By conflating them, the question leads to biased results. Only after many surveys were conducted did researchers settle on carefully worded questions that asked whether physicians had prescribed or injected medications with the intention of ending a patient's life. This wording fit the definition of euthanasia without using easily misinterpreted words or conflating actions that might result in bias.

For many types of bioethical empirical research, there are methodological skills that can be used to develop and test questions and measures. For instance, there are skills for developing survey questions, methodological standards for how they should be pretested, and protocols for validating such questions. These are arduous and time-consuming procedures. Indeed, validating a quality-of-life instrument or an instrument to evaluate competency to provide informed consent may require years of work. However, such standards are necessary for rigorous empirical research.

Finally, a problem that is common in bioethical empirical research is small numbers and low power. Many studies have been conducted at single institutions with small numbers of participants. Of course, this is not a problem unique to bioethics; rather, it is common in many types of clinical research. What is true at one institution may be the result of unique aspects of that institution or of its patient population. Similarly, small numbers of respondents make it difficult to generalize the results. Fortunately, as empirical bioethical research matures there is greater attention to ensuring larger studies as well as studies at multiple sites. Nevertheless, some research on bioethical questions will inherently involve small numbers of patients. For instance, trying to interview terminally ill patients who want euthanasia involves identifying terminally ill patients and then that subset who desires euthanasia. Since the subset is likely to be very small, less than 10%, to get even 100 respondents means interviewing 1000 terminally ill patients, which is both extremely costly and very difficult. Thus, such studies are likely to be based on small numbers.

References and Notes

1. *Salgo v. Leland Stanford, Jr. University Board of Trustees*, 317 P. 2d 170 (Cal.Ct.App. 1957).
2. The Nuremberg Code. *JAMA* 19.
3. *Slater v. Baker & Stapleton*, 95 Eng. Rep. 860 (K.B. 1767).
4. Appelbaum PS, Lidz CW, Meisel A. *Informed Consent: Legal Theory and Clinical Practice*. New York, Oxford University Press, 1987.
5. Faden RR, Beauchamp TL. *A History and Theory of Informed Consent*. New York, Oxford University Press, 1987.
6. Lederer SE. *Subjected to Science: Human Experimentation in America before the Second World War*. Baltimore, Johns Hopkins University Press, 1995.
7. *Compassion in Dying v. Washington*, 79 F.3d 790 (9th Cir 1996).
8. Edelstein L. The Hippocratic oath: Text, translation, and interpretation. In Temkin O, Temkin CL (eds.) *Ancient Medicine: Selected Papers of Ludwig Edelstein*. Baltimore, Johns Hopkins University Press, 1967.
9. Fye WB. Active euthanasia: An historical survey of its conceptual origins and introduction into medical thought. *Bull Hist Med* 1978;52:492–502.
10. Emanuel EJ. The history of euthanasia debates in the United States and Britain. *Ann Intern Med* 1994;121:793–802.
11. Euthanasia [Editorial]. *The Spectator* 1871;44:314–315.
12. Minutes of the Proceedings of the South Carolina Medical Association, April 7, 1879;14–17.
13. Permissive euthanasia [Editorial]. *Boston Med Surg J* 1884;110:19–20.
14. Levine RJ. *Ethics and Regulation of Clinical Research*, 2nd ed. New Haven, CT, Yale University Press, 1986.
15. Freedman B. Equipoise and the ethics of clinical research. *N Engl J Med* 1987;317(3):141–145.
16. Fried C. *Medical Experimentation: Personal Integrity and Social Policy*. Amsterdam, North Holland, 1974.
17. Shaw LW, Chalmers TC. Ethics in cooperative clinical trials. *Ann NY Acad Sci* 1970;169:487–495.
18. Council for International Organizations of Medical Sciences. *The International Guidelines for Biomedical Research Involving Human Subjects*, guideline 7. Geneva, World Health Organization, 2002.
19. Christakis NA. The ethical design of an AIDS vaccine trial in Africa. *Hastings Center Rep* 1988;18(3):31–37.
20. Emanuel EJ, Currie XE, Herman A. Undue inducement in clinical research in developing countries: Is it a worry? *Lancet* 2005;366:336–340.
21. Hawkins J, Emanuel EJ. Clarifying confusions about coercion. *Hastings Center Rep* 2005;35(4):16–19.
22. Pace CA, Emanuel EJ. The ethics of research in developing countries: Assessing voluntariness. *Lancet* 2005;365:11–12.
23. Dworkin G. *The Theory and Practice of Autonomy*. New York, Cambridge University Press, 1988.

24. *Canterbury v. Spence*, 464 F.2d 772 (DC Cir 1972).
25. Emanuel EJ, Emanuel LL. Four models of the physician-patient relationship. *JAMA* 1992;267:2221–2226.
26. Blackhall LJ, Murphy ST, Frank G, Michel V, Azen S. Ethnicity and attitudes toward patient autonomy. *JAMA* 1995;274: 820–825.
27. Newman SA. Treatment refusals for the critically and terminally ill: Proposed rules for the family, the physician, and the state, 111. *N Y LS Human Rights Annu* 1985;3:35–89.
28. New York State Task Force on Life and the Law. *Life Sustaining Treatment: Making Decisions and Appointing a Health Care Agent.* New York, New York State Task Force on Life and the Law, 1987.
29. In re Jobes 108 NJ 394 (1987).
30. Emanuel EJ, Emanuel LL. Proxy decision making for incompetent patients: An ethical and empirical analysis. *JAMA* 1992;267: 2067–2071.
31. Uhlmann RF, Pearlman RA, Cain KC. Physicians' and spouses predictions of elderly patients' resuscitation preferences. *J Gerontol* 1988;43(Suppl):M115–M121.
32. Seckler AB, Meier DE, Mulvihill M, Cammer Paris BE. Substituted judgment: How accurate are proxy predictions. *Ann Intern Med* 1991;115:92–98.
33. Zweibel NR, Cassel CK. Treatment choices at the end of life: A comparison of decisions by older patients and their physician-selected proxies. *Gerontologist* 1989;29:615–621.
34. Rodwin MA. *Medicine, Money & Morals.* New York, Oxford University Press, 1993.
35. Hillman BJ, *et al.* Frequency and costs of diagnostic imaging in office practice—A comparison of self-referring and radiologist-referring physicians. *N Engl J Med* 1990;323:1604–1608.
36. Mitchell JM, Scott E. Physician ownership of physical therapy services: Effects on charges, utilization, profits, and service characteristics. *JAMA* 1992;268:2055–2059.
37. Council on Ethical and Judicial Affairs, American Medical Association. Conflicts of interest: Physician ownership of medical facilities. *JAMA* 1992;267:2366–2369.
38. Stelfox HT, Chua G, O'Rourke K, Detskyh AS. Conflict of interest in the debate over calcium-channel antagonists. *N Engl J Med* 1998;338:101–106.
39. van der Maas PJ, van Delden JJM, Pijnenborg L, Looman CWN. Euthanasia and other medical decisions concerning the end of life. *Lancet* 1991;338:669–674.
40. Muller MT, van der Wal G, van Eijk JTHM, Ribbe MW. Voluntary active euthanasia and physician-assisted suicide in Dutch nursing homes: Are the requirements for prudent practice properly met? *J Am Geriatr Soc* 1994;42:624–629.
41. van der Maas PJ, *et al.* Euthanasia, physician-assisted suicide, and other medical practices involving the end of life in The Netherlands, 1990–1995. *N Engl J Med* 1996;335:1699–1705.
42. Breitbart W, Rosenbeld BD, Passik SD. Interest in physician-assisted suicide among ambulatory HIV-infected patients. *Am J Psychiatry* 1996;153:238–242.
43. Ganzini L, Johnston WS, McFarland BH, Tolle SW, Lee MA. Attitudes of patients with amyotrophic lateral sclerosis and their caregivers toward assisted suicide. *N Engl J Med* 1998;339: 967–973.
44. Emanuel EJ, Fairclough DL, Daniels ER, Clarridge BR. Euthanasia and physician-assisted suicide: Attitudes and experiences of oncology patients, oncologists, and the public. *Lancet* 1996;347:1805–1810.
45. van der Lee ML, van der Bom JG, Swarte NB, Heintz PM, de Graeff A, van den Bout J. Euthanasia and depression: A prospective cohort study among terminally ill cancer patients. *J Clin Oncol* 2005;23:6607–6612.
46. Kissane DW, Street A, Nitschke P. Seven deaths in Darwin: Case studies under the Rights of the Terminally Ill Act, Northern Territory, Australia. *Lancet* 1998;352:1097–1102.
47. The SUPPORT Principal Investigators. A controlled trial to improve care for seriously ill hospitalized patients: The Study to Understand Prognoses and Preferences for Outcomes and Risks of Treatments (SUPPORT). *JAMA* 1995;274:1591–1598.
48. Shalowitz DI, Garrett-Mayer E, Wendler D. The accuracy of surrogate decision-makers: A systematic review. *Arch Intern Med* 2006;166:493–497.
49. Flory J, Emanuel EJ. Interventions to improve research participants' understanding in informed consent for research: A systematic review. *JAMA* 2004;292:1593–1601.
50. Emanuel EJ. Cost savings at the end of life: What do the data show? *JAMA* 1996;275:1907–1914.
51. Brooks CH, Smyth-Staruch K. Hospice home care cost savings to third party insurers. *Med Care* 1984;22:691–703.
52. Spector WD, Mor V. Utilization and charges for terminal cancer patients in Rhode Island. *Inquiry* 1984;21:328–337.
53. Mor V, Kidder D. Cost savings in hospice: Final results of the National Hospice Study. *Health Serv Res* 1985;20:407–422.
54. Kane RL, Bernstein L, Wales J, Leibowitz A, Kaplan S. A randomized controlled trial of hospice care. *Lancet* 1984;1: 890–894.
55. Kidder D. The effects of hospice coverage on Medicare expenditures. *Health Serv Res* 1992;27:195–217.
56. National Hospice Organization. *An Analysis of the Cost Savings of the Medicare Hospice Benefit.* Miami, Lewin-VHI, 1995.

4

Integrity in Research: Individual and Institutional Responsibility

JOAN P. SCHWARTZ

Office of Intramural Research, Office of the Director, National Institutes of Health, Bethesda, Maryland

1. GUIDELINES FOR THE CONDUCT OF RESEARCH

In the late 1980s, the leadership of the National Institutes of Health (NIH) Intramural Research Program decided to develop a set of guidelines for the conduct of research that could be used as a basis of discussion for, as well as education of, all scientific staff including those in training. The *Guidelines for the Conduct of Research in the Intramural Research Program at NIH*[1] were "developed to promote the highest ethical standards in the conduct of research by intramural scientists at NIH." The intent was to provide a framework for the ethical conduct of research without inhibiting scientific freedom and creativity. The writers of the guidelines tried to take into account the major differences in commonly accepted behaviors among different scientific disciplines. The initial version was issued in 1990 and it has subsequently been revised and reissued twice (the latest in April 1997). The guidelines serve as a framework for the education of NIH scientific staff in research conduct issues, through discussion sessions and more formal courses, as well as a reference book. In 1995, the NIH Committee on Scientific Conduct and Ethics was established for the Intramural Research Program to help set policies on these issues as well as to set in place mechanisms for teaching the principles of scientific conduct and to establish mechanisms to resolve specific cases. This committee has been responsible for the last two versions of the guidelines. In addition, the committee created a computer-based research ethics course[2] that all new scientific staff must complete to ensure that everyone has the same basic understanding of the policies and regulations governing the responsible conduct of research. Finally, the committee selects the topic, and interesting cases, for yearly research ethics case discussions in which all scientific staff participate.

Other institutions have developed comparable sets of guidelines. Books, textbooks, and symposia or colloquia proceedings[3–10] that address scientific conduct and/or misconduct, as well as Internet-based learning programs at many institutions,[11,12] have appeared at an increasing rate during the past two decades. As a result of the mandate from the Office of Science and Technology Policy in the White House for the Office of Research Integrity (ORI), Department of Health and Human Services, to become primarily an educational office, ORI has been funding grants to support institutions in the development of research conduct materials and courses that can be made widely available to any institution interested in using them.[13]

The NIH guidelines cover scientific integrity; mentor–trainee relationships; data acquisition, management, sharing, and ownership; research involving human and animal subjects; collaborative science; conflict of interest and commitment; peer review; and publication practices and responsible authorship. These topics are discussed in the remainder of the chapter.

2. SCIENTIFIC INTEGRITY AND MISCONDUCT

Scientists should be committed to the responsible use of the process known as the scientific method to seek new knowledge. It is the expectation that the research staff in the NIH Intramural Research Program as well as scientists everywhere will maintain exemplary standards of intellectual honesty in designing, conducting, and presenting research. The principles of the scientific method include formulation and testing of hypotheses, controlled observations or experiments, analysis and interpretation of data, and oral and written presentations of all of these components to scientific colleagues for discussion and further conclusions. The scientific community and the general public rightly expect adherence to exemplary standards of intellectual honesty in the formulation, conduct, and reporting of scientific research. Without a high standard of scientific integrity, the scientific community and general public may become victims of scientific misconduct.

The issue of scientific misconduct became one of interest to the public in the 1980s as a result of several cases involving high-profile scientists. In response, the Institute of Medicine convened a committee, under the chairmanship of Dr. Arthur Rubenstein, to examine the issues, and the committee issued its report, *The Responsible Conduct of Research in the Health Sciences*,[14] in 1989. The committee examined the role of each component involved in the handling of allegations of scientific misconduct (i.e., the NIH and other funding agencies, universities and research organizations, professional and scientific societies, and journals) and provided a list of recommendations and best practices to each. The report acknowledged the occurrence of scientific misconduct and the problems that arose when it was not dealt with appropriately. It proposed that each institution develop its own standards for the conduct of research. Responsibility for preventing and handling misconduct was placed on the institutions involved in supporting and overseeing research, as well as on the individual scientists. The report suggested that a distinction be made among three types of behaviors: misconduct in science (Table 4-1), questionable research practices, and other types of misconduct. Questionable research practices include such things as failure to retain data, maintaining inadequate records, honorary authorship, and premature release of results to the public. These clearly do not fall within the rubric of scientific misconduct but have received a lot of attention[15] and will be addressed in the appropriate sections of this chapter. The third category, other types of misconduct, includes financial irregularities, sexual harassment, criminal activities, and other

TABLE 4-1 Federal Definition of Scientific Misconduct and Standards and Process by Which It Is Assessed

Research misconduct defined
   Research misconduct is defined as fabrication, falsification, or plagiarism in proposing, performing, or reviewing research or in reporting research results.
   Fabrication is making up data or results and recording or reporting them.
   Falsification is manipulating research materials, equipment, or processes, or changing or omitting data or results such that the research is not accurately represented in the research record.
   Plagiarism is the appropriation of another person's ideas, processes, results, or words without giving appropriate credit.
   Research misconduct does not include honest error or differences of opinion.
Findings of research misconduct
   A finding of research misconduct requires that
      there be a significant departure from accepted practices of the relevant research community;
      the misconduct be committed intentionally, or knowingly, or recklessly; and
      the allegation be proven by a preponderance of the evidence.
Process for assessing the occurrence of research misconduct
   Allegation assessment—Determination of whether allegations of misconduct, if true, would constitute misconduct and whether the information is sufficiently specific to warrant and enable an inquiry.
   Inquiry—The process of gathering information and initial fact-finding to determine whether an allegation of misconduct warrants an investigation.
   Investigation—The formal examination and evaluation of all relevant facts to determine if scientific misconduct has occurred and, if so, to determine the person(s) who committed it and the seriousness of the misconduct.

behaviors covered by specific rules, regulations, and laws.

The following are the most important recommendations of the report: (1) "Individual *scientists* [italics added] in cooperation with officials of research institutions should accept formal responsibility for ensuring the integrity of the research process"; (2) scientists and research institutions should have educational programs that foster awareness of proper research conduct and what constitutes misconduct; (3) institutions should develop guidelines for the conduct of research; (4) a common definition of misconduct, as well as common policies and procedures for handling allegations of misconduct, should be adopted by institutions and the government; and (5) an independent federal scientific integrity advisory board should be created. In recognition of the range of definitions of scientific misconduct in place in various federal agencies, the White House Office of Science and Technology Policy (OSTP) began a process of consultation with the heads of all the federal science agencies in 1996. These meetings included the National Science Foundation, the

Department of Veterans Affairs, the Department of Agriculture, the Department of Energy, the National Aeronautics and Space Administration, and the National Institutes of Health, each of which had a different definition of scientific misconduct and different policies for how to handle it. In October 1999, the National Science and Technology Council of OSTP issued, on behalf of all federal agencies that supported scientific research, a proposed common statement, "Research Misconduct Defined," with attendant common procedures and policies.[16] The definition and procedures were positively received by the scientific community, and in December 2000 the final federal policy for government-sponsored research was issued.[17] This policy and the definition of scientific misconduct, summarized in Table 4-1, are the first to be universally applicable to federally supported research. The Federal Register issuance with the new definition also included the conversion of ORI into an office primarily responsible for educational activities.

The policy not only defines scientific/research misconduct but also provides certain standards to be adhered to in making a finding of misconduct and it describes a three-step process for assessing and establishing that misconduct has occurred. Beyond that, agencies may handle imposition of sanctions and appeal processes within certain guidelines. For the Department of Health and Human Services, the policy provides for ORI oversight of completed investigations.

3. MENTOR–TRAINEE RELATIONSHIPS

Training depends on the quality of research and mentoring in individual laboratories. The importance of the training role of scientists and the importance of mentoring were formalized by the NIH Committee on Scientific Conduct and Ethics with *A Guide to Training and Mentoring in the Intramural Research Program at NIH*,[18] which describes in detail all the components of a good mentoring and training experience. This is an example of a formal document that provides an explicit set of expectations for the predoctoral and postdoctoral training experience, as well as expectations for the mentors, the institution, and the trainees.

The goals of a mentor–trainee relationship are to ensure that individuals being mentored receive the best possible training in how to conduct research and how to develop and achieve career goals. Mentoring and being mentored are essential lifelong components of professional life. Research supervisors should always be mentors, but trainees should be encouraged to seek out other mentors who may provide additional

expertise: together they form the basis of a professional network. Characteristics of a good mentor include an interest in contributing to the career development of another scientist, research accomplishments, professional networking, accessibility, and past successes in cultivating the professional development of their fellows. The trainees must be committed to the work of the laboratory and the institution and also to the achievement of their research and career goals—they must be active participants in their training.

Among the skills that trainees should acquire during their fellowship period are training in scientific investigation—how to choose a first-rate research project, how to carry out the necessary experiments and analyses in an appropriate and rigorous way, and how to incorporate knowledge of the research field and published literature—with the ultimate goal of developing increasing independence throughout the training period; training in communication skills, both written and oral; training in personal interactions, including negotiations, persuasion, and diplomatic skills, and in networking; and training in scientific responsibility, the legal and ethical aspects of carrying out research. In addition, fellows should be considering career pathways, in consultation with their mentors, being sure to survey the many options available to scientists these days.

4. DATA ACQUISITION, MANAGEMENT, SHARING, AND OWNERSHIP

Scientific data may be divided into three categories: experimental protocols; primary data, which include instrument setup and output, raw and processed data, statistical calculations, photographic images, electronic files, and patient records; and procedures of reduction and analysis. Any individual involved in the design and/or execution of an experiment and subsequent data processing is responsible for the accuracy of the resultant scientific data and must be meticulous in the acquisition and maintenance of them. These individuals may include, in addition to the person responsible for actually carrying out the experiment, the principal investigator, postdoctoral fellows, students, research assistants, and other support staff such as research nurses. Research results should be recorded in a form that allows continuous access for analysis and review, whether via an annotated bound notebook or computerized records. All research data must be made available to the supervisor, as well as collaborators, for immediate review. Data management, including the decision to publish, is ultimately the responsibility of the principal investigator.

Martinson *et al.*[15] carried out a survey that asked respondents to report which, if any, questionable research practices they had engaged in during the previous 3 years. Among those who responded (46% of those surveyed), 27.5% reported that they had kept inadequate research records, suggesting that lack of appropriate record keeping is a serious problem.

At the NIH, data collected, as well as laboratory notebooks, research records, and other supporting materials such as unique reagents, belong to the government and must be retained for a period of time sufficient to allow for further analysis of the results as well as repetition by others of published material. The NIH recommends that all data and laboratory notebooks be retained for 7 years. Once publications have appeared, supporting materials must be made available to all responsible scientists seeking further information or planning additional experiments, when possible. For example, aliquots of any monoclonal antibody that derives from a continuously available cell line must be provided, whereas the final aliquots of a polyclonal antibody, needed by the original lab to finish additional experiments, do not. Many research institutions have required that transgenic or knockout mouse lines be made available through deposition in a commercial mouse facility. Clinical data should be retained as directed by federal regulations. Requests for human samples require institutional review board review and approval prior to sharing to ensure that confidentiality issues are covered.

5. RESEARCH INVOLVING HUMAN AND ANIMAL SUBJECTS

The use of humans and animals in research is essential, but such research entails special ethical and legal considerations. Many chapters in this textbook address the issues related to carrying out human subject research and nothing further will be said in this chapter. The *Guidelines for the Conduct of Research Involving Human Subjects at the National Institutes of Health*[19] are one example of how institutions formalize their approach to human subjects research. Of concern, Martinson *et al.*'s survey[15] reported that 0.3% of those responding said they had ignored major aspects of human subject requirements, whereas 7.6% circumvented certain minor aspects.

The use of laboratory animals is often essential in biomedical research, but in using animals, a number of important points must be kept in mind. Animals must always be cared for and used in a humane and effective way, with procedures conducted as specified in an approved protocol. The use of animals in research

TABLE 4-2   The Three R's in Animal Research

Reduction: Reduction in the numbers of animals used to obtain information of a certain amount and precision

Refinement: Decrease in the incidence or severity of pain and distress in those animals that are used

Replacement: Use of other materials, such as cell lines or eggs, or substitution of a lower species, which might be less sensitive to pain and distress, for a higher species

must be reviewed by an animal care and use committee, in accordance with the Association for Assessment and Accreditation of Laboratory Animal Care International guidelines. Animal care and use committees perform the following functions: review and approve protocols for animal research, review the institute's program for humane care and use of animals, inspect all of the institution's animal facilities every 6 months, and review any concerns raised by individuals regarding the care and use of animals in the institute.

An investigator's responsibilities in using animals for research include humane treatment of animals, following all procedures that were specified in the approved protocol, following the general requirements for animal care and use at the institution, and reporting concerns related to the care and use of laboratory animals. The policies and regulations for the utilization and care of laboratory animals are primarily concerned with minimizing or alleviating the animal's pain and utilizing appropriate alternatives to animal testing when possible. In recent years, great emphasis has been placed on the three R's—reduction, refinement, and replacement (Table 4-2).

6. COLLABORATIVE SCIENCE

Research collaborations facilitate progress and should be encouraged. The ground rules for collaborations, including authorship issues, should be discussed openly among all participants from the beginning. Research data should be made available to all scientific collaborators on a project upon request. Although each research project has unique features, certain core issues are common to most of them and can be addressed by collaborators posing questions in the following areas: overall goals, who will do what, authorship and credit, and contingencies and communicating. Many institutions have formally addressed some of the complex issues related to scientific collaboration. For example, the NIH Ombudsman Office addresses many issues related to collaborations and has created the questions shown in Table 4-3.[20]

TABLE 4-3 Questions for Scientific Collaborators

Although each research project has unique features, certain core issues are common to most of them and can be addressed by collaborators posing the following questions:

Overall goals
1. What are the scientific issues, goals, and anticipated outcomes or products of the collaboration?
2. When is the project over?

Who will do what?
1. What are the expected contributions of each participant?
2. Who will write any progress reports and final reports?
3. How, and by whom, will personnel decisions be made? How, and by whom, will personnel be supervised?
4. How, and by whom, will data be managed? How will access to data be managed? How will long-term storage and access to data be handled after the project is complete?

Authorship and credit
1. What will be the criteria and the process for assigning authorship and credit?
2. How will credit be attributed to each collaborator's institution for public presentations, abstracts, and written articles?
3. How, and by whom, will public presentations be made?
4. How, and by whom, will media inquiries be handled?
5. When and how will intellectual property and patent applications be handled?

Contingencies and communicating
1. What will be the mechanism for routine communications among members of the research team (to ensure that all appropriate members of the team are kept fully informed of relevant issues)?
2. How will decisions about redirecting the research agenda as discoveries are made be reached?
3. How will the development of new collaborations and spin-off projects, if any, be negotiated?
4. Should one of the principals of the research team move to another institution or leave the project, how will data, specimens, lab books, and authorship and credit be handled?

From reference 20.

7. CONFLICT OF INTEREST AND COMMITMENT

Conflict of interest is a legal term that encompasses a wide spectrum of behaviors or actions involving personal gain or financial interest. According to Frank Macrina,[8] "a conflict of interest arises when a person exploits, or appears to exploit, his or her position for personal gain or for the profit of a member of his or her immediate family or household." The existence of a conflict of interest may adversely affect the ability to objectively carry out scientific studies and report their results. Potential conflicts of interest may not be recognized by others unless disclosed; disclosure should include all relevant financial relationships. Disclosure is made to the appropriate organization depending on the activity: to one's research institution while carrying out the research, to the funding agency when involved in peer review of grants, to meeting organizers when giving an invited presentation, and to journal editors when asked to referee articles or when submitting one's own manuscripts for consideration. A total of 0.3% of respondents to the survey[15] on inappropriate research behaviors reported "not properly disclosing involvement in firms whose products were based on their own research," suggesting that this is an issue that needs to be further addressed.

Currently, a major concern is the interaction between industry and clinical researchers in the handling of clinical trials, and the potential for conflict of interest. Given the enormous costs of clinical trials, combined with the desire of clinical investigators to try the latest drugs, which are often only available from drug companies, increasingly companies serve as the sponsors of clinical trials (70% of the funding for such trials) and as such may seek control over the research protocol and publication of the results. To what extent that happens, and is permitted to happen, was the subject of a survey by Mello et al.[21] A survey was sent to the administrator most knowledgeable about and responsible for negotiating clinical trial agreements at each of the 122 U.S. medical schools, asking questions about 17 contractual provisions that might restrict investigators' control over clinical trials that they were involved in or running. Among the findings of interest were that industry sponsors were allowed to: prevent the investigator from changing the study design once an agreement had been executed (68% of surveyed medical schools); insert their own statistical analyses (24%); write the first draft of the manuscript (50%); bar collaborators from sharing data with third parties once the study had been published (41%); own the data after the trial was completed (80%); and delay publication to preview the results prior to media review (62% up to 60 days and an additional 31% up to 90 days). However, 99% were in agreement that an industry sponsor could not prevent results from being published, presumably a reflection of a couple of recent well-publicized cases. Despite signed agreements, 82% of the medical schools reported some type of dispute arising with the industry sponsor, primarily over payment, intellectual property, or control of data.

Although there is currently great interest in conflict of interest issues, conflict of commitment can be equally important. This refers to the idea that someone has agreed to do more things than possible, especially nonofficial duty activities that have no direct bearing on their employment responsibilities. In contrast to conflict of commitment, there is also the issue of overcommitment, when someone takes on too many trainees, thereby not giving the best effort to all of them. When an advisor cannot find the time to meet with a fellow, or to review and critique the first draft of a manuscript, within a few days or a week, that is a strong sign of overcommitment.

8. PEER REVIEW

Peer review is defined as a critical evaluation, conducted by one or more experts in the relevant field, of either a scientific document (e.g., a research article submitted for publication, a grant proposal, or a study protocol) or a research program. One requisite element for peer review is the need for reviewers to be experts in the relevant subject areas. At the same time, real or perceived conflict of interest arising as a result of a direct competitive, collaborative, or other close relationship with one of the authors of the material under review should be avoided. All evaluations should be thorough and objective, fair and timely, and based solely on the material under review: information not yet publicly available cannot be taken into consideration. The use of multiple reviewers mitigates to some extent one inappropriate review, but nevertheless reviewers should strive to provide constructive advice and avoid pejorative comments. Since reviews are usually conducted anonymously, it is incumbent on the reviewer to protect the privileged information to which he or she becomes privy. No reviewer should share any material with others unless permission has been requested and obtained from those managing the review process. One of the marks of a good mentor is someone who teaches trainees how to handle peer review by asking them to review a submitted manuscript, but it is incumbent on the mentor to notify the journal that he or she plans to do so and get explicit permission before doing so. A reviewer should never copy and retain any of the materials unless specifically permitted to do so, yet 1.7% of those surveyed by Martinson et al.[15] reported that they had done so.

9. PUBLICATION PRACTICES AND RESPONSIBLE AUTHORSHIP

Publication of results fulfills a scientist's responsibility to communicate research findings to the scientific community, a responsibility that derives from the fact that much research is funded by the federal government using taxpayers' money. Publication of clinical studies also fulfills the responsibility to have a scientific benefit in return for putting human subjects at risk. Other than presentations at scientific meetings,

publication in a scientific journal should normally be the mechanism for the first public disclosure of new findings. An exception may be appropriate when serious public health or safety issues are involved. However, publication can generate some of the most difficult disputes among scientists because it is so important for their careers. Publications share findings that benefit society and promote human health, but they also establish scientific principles. Credit for a discovery belongs to the first to publish, and reputations and research funding are based on the number and impact of publications. Furthermore, prestigious positions are gained through reputation and publications. A study by Benos et al.[22] addresses many issues related to the ethics of scientific publication.

Although each paper should contain sufficient information for the informed reader to assess its validity, the principal method of scientific verification is not review of submitted or published papers but, rather, the ability of others to replicate the results. Therefore, each paper should contain all the information necessary for other scientists to repeat the work. Failing to do so was reported by 10.8% of respondents in Martinson et al.'s survey,[15] suggesting either carelessness or a significant attempt by some scientists to delay or prevent others from repeating and advancing their findings. Timely publication of new and significant results is important for the progress of science, but each publication should make a substantial contribution to its field. Fragmentary publication of the results of a scientific investigation or multiple publications of the same or similar data are not appropriate, yet 4.7% of those surveyed stated that they had done so.[15]

Authorship is the primary mechanism for determining the allocation of credit for scientific advances and is thus the primary basis for assessing a scientist's contributions to developing new knowledge. As such, it potentially conveys great benefit, as well as responsibility. Authorship involves the listing of names of participants in all communications (oral or written) concerning experimental results and their interpretation, as well as making decisions about who will be the first author, the senior author, and the corresponding author. Authorship is justified by a significant contribution to the conceptualization, design, execution, and/or interpretation of the research study and a willingness to assume responsibility for the study. Other ways to establish credit for contributions besides authorship include acknowledgments and references. Acknowledgments provide recognition of individuals who have assisted the research by their encouragement and advice about the study; editorial assistance; technical support; or provision of space, financial support, reagents, or specimens. References acknowledge others' discoveries, words, ideas, data, or analyses and must be cited in a way that others can find the reference and see the contribution. According to results from the survey on questionable practices,[15] 1.4% reported that they had used others' ideas without obtaining permission or giving credit.

When should authorship issues be discussed? Although there is no universal set of standards for authorship, each research group should freely discuss and resolve questions of authorship before and during the course of a study. Each author should fully review material that is to be presented in a public forum or submitted (originally or in revision) for publication. Each author should indicate a willingness to support the general conclusions of the study before its presentation or submission. Since a significant fraction of allegations of misconduct turn out to be authorship disputes, including use of data, plagiarism, and conflicts over credit for scientific work, settling authorship issues as early as possible is important. With the recent increase in numbers of authors on publications, the problem has increased in magnitude. The NIH ombudsman has reported that authorship disputes constitute the single largest group of scientific complaints with which the office deals. The ORI has determined that any authorship dispute involving present or past collaborators cannot qualify as research misconduct, thereby leaving resolution of such disputes to the authors or their institution or office of the ombudsman.

Drummond Rennie, deputy editor of the *Journal of the American Medical Association*, has been interested in the misuse of authorship for a long time. He has defined a number of categories of irresponsible authorship.[23] These include honorary authorship—an author who does not meet the criteria; ghost authorship—failure to include as an author someone who made substantial contributions to the article; refusal to accept responsibility for an article despite ready acceptance of credit; and duplicate and redundant publications. Rennie and colleagues[23] carried out a study based on the following hypotheses: Research articles in large-circulation prestigious medical journals would be more likely to have honorary authors, whereas review articles in smaller circulation journals that publish symposia proceedings would be more likely to have ghost authors. The results of the study, shown in Table 4-4, were just the opposite.

Despite disproving the hypotheses, however, the study showed significant misuse of authorship in biomedical journals. This is supported by the findings in Martinson *et al.*'s survey,[15] in which 10% reported that they had inappropriately assigned authorship. This has led to a number of changes, many spearheaded by the International Committee of Medical Journal Editors

TABLE 4-4   Authorship Analysis[a]

	Research Articles	Reviews
Honorary authorship	79 (16%)	61 (26%)
Ghost authorship	65 (13%)	23 (10%)

[a]The corresponding authors of 492 research articles and 240 reviews from *American Journal of Cardiology, American Journal of Medicine, American Journal of Obstetrics and Gynecology, Annals of Internal Medicine, Journal of the American Medical Association*, and *New England Journal of Medicine* were surveyed.

Data excerpted from reference 23.

TABLE 4-5   International Committee of Medical Journal Editors—Criteria for Authorship

Authorship should be based on
    Substantial contributions to conception and design, or acquisition of data, or analysis and interpretation of data
    Drafting the article or revising it critically for important intellectual content
    Final approval of the version to be published
Authors should meet all three conditions. Furthermore, all persons designated as authors should qualify for authorship, and all those who qualify should be listed.

Based on reference 24.

(ICMJE). ICMJE issued a set of uniform requirements for manuscripts submitted to biomedical journals, revised in October 2004,[24] to address requirements for authorship and, more recently, "ethical principles related to publication in biomedical journals." They define an author as someone "who has made substantive intellectual contributions to a published study" and provide a set of criteria for authorship as shown in Table 4-5. In addition, ICMJE has stated that if someone is involved only in acquisition of funding, collection of data, or general supervision of a research group, that does not justify authorship. Furthermore, each author should have participated sufficiently in the work to take public responsibility for appropriate portions of the content.

The *Journal of the American Medical Association* authorship policy more specifically states that all authors must describe their specific contributions as well as the contributions of those listed in the acknowledgments. Authors must decide who should be an author, and who should be acknowledged, by discussions among themselves. Authors should be listed in order of actual degree of contribution, to be decided by the authors. The *Annals of Internal Medicine* further notes that the following, by themselves, are not criteria for authorship: holding a position of administrative leadership, contributing patients or reagents, or collecting and assembling data. Adhering to these criteria will result in a significant change to the way authorship is determined for clinical studies and may require a culture change.

Acknowledgment

The previous version of this chapter, by Dr. Alan Schechter, informed my writing of the current version.

References

1. *Guidelines for the Conduct of Research in the Intramural Program at NIH.* Available at www.nih.gov/campus/irnews/guidelines. htm.
2. http://researchethics.od.nih.gov.
3. National Academy of Sciences. *On Being a Scientist.* Washington, DC, National Academy Press, 1989.
4. Sigma Xi. *Honor in Science,* 2nd ed. New Haven, CT, Sigma Xi, 1989.
5. Frankel MS, Teich AH. *Good Science and Responsible Scientists: Meeting the Challenge of Fraud and Misconduct in Science.* Washington, DC, American Association for the Advancement of Science, 1992.
6. Korenman SG, Shipp AC. *Teaching the Responsible Conduct of Research through a Case Study Approach.* Washington, DC, Association of American Medical Colleges, 1994.
7. Bird SJ, Dustira AK. Scientific misconduct. *Sci Eng Ethics* 1999;5:129–304.
8. Macrina FL. *Scientific Integrity: An Introductory Text with Cases.* Washington, DC, ASM Press, 2000.
9. Guston DH. *Between Politics and Science: Assuring the Integrity and Productivity of Research.* New York, Cambridge University Press, 2000.
10. Jones AH, McLellan F. *Ethical Issues in Biomedical Publication.* Baltimore, Johns Hopkins University Press, 2000.
11. Responsible Conduct of Research Education Consortium, rcrec. org.
12. Poynter Center for the Study of Ethics and American Institutions, poynter.indiana.edu/index.shtml.
13. Office of Research Integrity, www.ori.dhhs.gov.
14. Institute of Medicine. *The Responsible Conduct of Research in the Health Sciences.* Washington, DC, National Academy Press, 1989.
15. Martinson BC, Anderson MS, de Vries R. Scientists behaving badly. *Nature* 2005;435:737–738.
16. Office of Science and Technology Policy. Proposed federal policy on research misconduct to protect the integrity of the research record. *Fed Reg* 1999;64:55722–55725.
17. Office of Science and Technology. *Federal Policy on Research Misconduct.* 2000. Available at www.ostp.gov/html/001207. html.
18. National Institutes of Health. *A Guide to Training and Mentoring in the Intramural Research Program at NIH.* Available at www1. od.nih.gov/oir/sourcebook/ethic-conduct/TrainingMentoring Guide_7.3.02.pdf.
19. National Institutes of Health. *Guidelines for the Conduct of Research Involving Human Subjects at the National Institutes of Health.* Available at ohsr.od.nih.gov/guidelines/GrayBooklet82404. pdf.
20. National Institutes of Health. *Questions for Scientific Collaborators.* Available at www1.od.nih.gov/oir/sourcebook/ethic-conduct/ scicollaboratorsquestions.htm.
21. Mello MM, Clarridge BR, Studdert DM. Academic medical centers' standards for clinical-trial agreements with industry. *N Engl J Med* 2005;352:2202–2210.
22. Benos DJ, Fabres J, Farmer J, *et al.* Ethics and scientific publication. *Adv Physiol Educ* 2005;29:59–74.
23. Flanagin A, Carey LA, Fontanarosa PB, *et al.* Prevalence of articles with honorary authors and ghost authors in peer-reviewed medical journals. *JAMA* 1998;280:222–224.
24. International Committee of Medical Journal Editors. *Uniform Requirements.* Available at www.ICMJE.org.

5

Institutional Review Boards

ALISON WICHMAN

Office of Human Subjects Research, Intramural Research Program, National Institutes of Health, Bethesda, Maryland

In the United States, the rights and welfare of human research subjects take precedent over the advance of scientific knowledge. Ethical guidelines, federal regulations, local institutional policies and procedures, and the knowledge and integrity of researchers and research staff all contribute to promoting the protection of human subjects. Research investigators have the primary responsibility to protect the rights and safeguard the welfare of the people participating in their research activities. In addition, our society has decided by law that objective, ongoing review of research activities by a group of diverse individuals is most likely to protect human subjects and promote ethically sound research. Prospective review of research by institutional review boards (IRBs) is an important assurance that the rights and welfare of human subjects are given serious consideration. This chapter focuses on the development of U.S. federal regulations concerning research involving human subjects and the roles and responsibilities of IRBs.

1. HISTORICAL, ETHICAL, AND REGULATORY FOUNDATIONS OF CURRENT REQUIREMENTS FOR RESEARCH INVOLVING HUMAN SUBJECTS

1.1. Historical Foundations

Concerns about the ethics of the practice of medicine have a long history, but until the mid-20th century, they were mostly centered around the practice of therapeutic medicine, not research medicine. In 1946, 23 Nazi physicians went on trial at Nuremberg for crimes committed against prisoners of war and in concentration camps. These crimes included exposure of humans to extremes of temperature, performance of mutilating surgery, and deliberate infection with lethal pathogens. During the trial, fundamental ethical standards for the conduct of research involving humans were codified into the Nuremberg Code,[1] which sets forth 10 conditions that must be met to justify research involving human subjects. Two important conditions are the need for voluntary informed consent of subjects and a scientifically valid research design that can produce fruitful results for the good of society.

The Nuremberg Code was reflected in the Declaration of Human Rights and accepted in principle by each of the 51 original signatory nations of the Charter of the United Nations. However, in the United States, the existence of the Nuremberg Code was not widely appreciated. Researchers and physicians who were familiar with it generally believed that its requirements narrowly applied to research conducted by German researchers and that it had little applicability or relevance to research conducted in the United States.[2] In fact, implementation of the first condition of the code in the United States—the voluntary consent of subjects who are able to exercise free power of choice—would have severely curtailed, if not eliminated, research involving prisoners. In the United States during the 1950s through the mid-1970s, many chemotherapeutic agents for cancer and other diseases/disorders were tested initially in healthy prisoners; in fact, some pharmaceutical companies had research buildings located on or near prisons to facilitate their research activities. Therefore, implementation of the code would have had major, dramatic effects on the conduct of research in the United States.

Also, most countries accepting the principles of the code, including the United States, had no mechanism for implementing its provisions. In 1953, the National Institutes of Health (NIH) opened the Clinical Center (CC), its major research hospital in Bethesda, Maryland, which produced the first U.S. federal policy for the protection of human subjects. This policy was consistent with the Nuremberg Code in that it gave special emphasis to the protection of healthy, adult research volunteers who had little to gain directly from participation in research. The CC's policy was innovative not only for its existence but also for providing a mechanism for prospective review of research by individuals who had no direct involvement or intellectual investment in the research. This was the beginning of the research review mechanism—the IRB—that is now fundamental to the current system of human subject protections throughout the United States. In fact, the first two research protocols submitted to the CC's research review committee were disapproved because it judged that research-related risks to the healthy volunteers were too high.[3] However, the CC requirements for prospective review of research and obtaining subjects' informed consent were applicable only to research involving healthy volunteers, not patients. In excluding research involving patients from these requirements, the policy was consistent with contemporaneous thinking of U.S. physician/researchers; most were reluctant to set forth explicit rules for the conduct of research involving patients, arguing that they would impede research and undermine trust in the physician.[4]

In the 1960s, federal funding of clinical research expanded, with a concomitant increase in the number of individuals participating as subjects. Interest in the rights of research subjects grew not only because of a general increase in U.S. attention to human rights but also because of a number of highly publicized clinical research abuses. For example, there were newspaper reports of investigators in New York injecting elderly, indigent people with live cancer cells, without their consent, to learn more about the human immune system. Although no apparent harm to subjects occurred, the investigators were cited for fraud, deceit, and unprofessional conduct. In 1966, Henry Beecher, a highly respected physician–investigator from Harvard University, shocked the medical community when he reported that unethical and questionably ethical practices were common in the conduct of human subjects research in many of the premier research institutions of the United States.[5]

The World Health Organization recognized a need for guidelines that were broader in scope than the Nuremberg Code. The Declaration of Helsinki: Recommendations Guiding Medical Doctors in Biomedical Research Involving Human Subjects[6] was adopted by the World Medical Society in 1964. These guidelines have been revised a number of times and currently are in use throughout the world.

The NIH, under the directorship of Dr. James Shannon, promoted the development of the first Public Health Service Policy on the Protection of Human Subjects, issued in 1966. The policy, which applied to research conducted or supported by the Department of Health, Education, and Welfare (HEW), including the NIH, required prospective review of human subjects research, taking into account the rights and welfare of the subjects involved, the appropriateness of the methods used to secure informed consent, and the risks and potential benefits of the research. The elements of informed consent included the requirement that consent be documented by the signature of subjects or their representatives.

Several events in the early 1970s led to renewed and intense efforts in the United States to protect human subjects. Most notable was the revelation that, since the 1930s, more than 400 black men in Tuskegee, Alabama, had been involved, without their knowledge, in a lengthy study (the Tuskegee Syphilis Study) on the natural history of syphilis.[7] These men were systematically denied penicillin even after its introduction as the standard treatment for the disease. The U.S. Senate Committee on Labor and Human Resources held hearings on this study and on other alleged health care abuses of prisoners and children. The outcomes of these hearings were (1) enactment of the National Research Act of 1974 requiring HEW to codify its policy for the protection of human subjects into federal regulations, which it did in 1974; (2) formation of the National Commission for the Protection of Human Subjects of Biomedical and Behavioral Research; and (3) imposition of a moratorium on research conducted or supported by HEW involving live human fetuses until the National Commission could study and make recommendations on it.

The National Commission, which functioned from 1974 to 1978, evaluated the existing HEW regulations; recommended improvements to the Secretary of HEW; and issued reports on research involving pregnant women, live human fetuses, prisoners, children, the mentally disabled, and the use of psychosurgery. The National Commission also issued the *Belmont Report: Ethical Principles and Guidelines for the Protection of Human Subjects of Research.*[8] A major advancement in the development of public policy, the Belmont Report provided guidance for distinguishing therapeutic medicine from research, identified three fundamental ethical principles for the protection of human subjects, and illustrated how the ethical principles should be applied to the conduct of human subjects research.

In 1979, HEW began the process of revising the 1974 regulations, but it was not until January 1981 that final department (renamed the Department of Health and Human Services [DHHS]) approval was given to Title 45 section 46 Code of Federal Regulations (CFR) governing protection of human subjects (45 CFR 46).[9] Initially, these regulations were applicable only when research was conducted or supported by DHHS, but in June 1991, the core of the regulations (Subpart A)—referred to as the Common Rule—was adopted by 16 other federal department agencies.[10]

1.2. Ethical Foundations

The ethical foundation for the current laws governing human subject research protections is enunciated in the Belmont Report, which was issued in 1979. It establishes three fundamental ethical principles that are relevant to all research involving human subjects—respect for persons, beneficence, and justice—and demonstrates how they are applied to the conduct of research involving human subjects.[11]

Respect for persons acknowledges the dignity and autonomy of individuals and requires that subjects give informed consent to participation in research. However, not all individuals are capable of self-determination, and the Belmont Report acknowledges that people with diminished autonomy are entitled to additional protection. For example, some individuals may need extensive protection, even to the point of excluding them from research activities that may harm them, whereas others require little protection beyond making sure they undertake research freely, with awareness of the possible adverse consequences.[12]

Beneficence requires that the benefits of research be maximized and possible harms minimized. This principle finds expression in a careful analysis by researchers and IRBs of the risks and benefits of particular research protocols.[13]

Justice requires fair selection and treatment of research subjects. For example, subjects should be equitably chosen to ensure that certain individuals or classes of individuals are not systematically selected for or excluded from research, unless there are scientifically or ethically valid reasons for doing so. Also, unless there is careful justification for an exception, research should not involve people from groups that are unlikely to benefit directly or from subsequent applications of the research.[14]

These three principles are not mutually exclusive. Each principle carries strong moral force, and difficult ethical questions arise when they conflict. However,

understanding and applying the principles of the Belmont Report helps promote the respectful and ethical treatment of research subjects.

1.3. Regulatory Foundations

Biomedical and behavioral research funded or supported by DHHS, including NIH, is under the purview of regulations for the protection of human subjects at 45 CFR 46.[15] These regulations embody the principles of the Belmont Report. Taken together, the Belmont Report and 45 CFR 46 articulate the minimal ethical standards and legal obligations of those who conduct, review, and oversee research. Also, regardless of the funding source, all clinical trials in the United States involving investigational drugs or devices are under the regulatory purview of the Food and Drug Administration (FDA), which endorses 45 CFR 46. Additional FDA regulations contained in Title 21 sections 50 and 56 CFR govern the development and approval of drugs, biologies, and devices regardless of the funding source.[16] FDA and DHHS regulations on the protection of human subjects and IRBs are consistent in many elements, although there are some differences.[17]

DHHS is the primary federal funding agency of biomedical and behavioral research. In 1998, it provided $5 billion for clinical research activities,[18] and in 2004 it provided approximately $8.5 billion for these activities (personal communication with the NIH Office of Extramural Research). All research involving human subjects conducted or supported by DHHS must be performed in keeping with the requirements of 45 CFR 46. DHHS's regulatory apparatus for overseeing the protection of human subjects involved in the research that it funds consists of two major tiers of review—one at the federal level and the other at the institutional level. For example, as a condition for receipt of NIH research funds, institutions must assure in writing that personnel will abide by ethical principles of the Belmont Report and the requirements of 45 CFR 46. These written documents are referred to as assurances of compliance. They are contract-like agreements that are negotiated and approved by the Office for Human Research Protections (OHRP)[19] on behalf of the Secretary of DHHS. In January 2006, OHRP estimated that it held 9350 assurances with entities in the United States and abroad (personal communication with OHRP).

All assurances set forth the institution's policies and procedures for the review and monitoring of human subject research activities, including IRB membership requirements and review and record-keeping procedures. A variety of administrative actions can be taken by OHRP for violation of the requirements of

45 CFR 46 or the terms and conditions of an institution's assurance of compliance. Compliance oversight investigations conducted from 1990 through mid-2000 resulted in restrictions of clinical research activities or corrective measures in 38 U.S. research institutions. Actions included temporary suspension of all DHHS-funded clinical research in some institutions, the requirement that some or all investigators conducting research in these institutions receive appropriate additional education concerning the protection of human subjects, and quarterly reports to DHHS of the institution's progress in correcting identified deficiencies. In particularly serious cases, OHRP may recommend to DHHS officials that institutions or investigators be declared ineligible to participate in DHHS-supported research (i.e., debarment or suspension).

Most research conducted in the United States falls under federal regulatory purview either because it is funded by the NIH or other government agencies or because it involves investigational drugs or devices and, therefore, is regulated by the FDA. Some clinical research conducted in the United States does not fall under federal human subject protection regulations either because it is not funded by the federal government or because it does not involve compounds under the FDA's jurisdiction. The amount of such research and the settings in which it is being conducted are not known. Efforts have been made to bring all U.S. clinical research under the purview of federal regulations, but none has succeeded so far.

2. INSTITUTIONAL REVIEW BOARDS

DHHS and FDA regulations require most proposed clinical research to undergo prospective review by an IRB. IRBs are important because clinical investigators have an inherent conflict of interest. As health care professionals, they are dedicated to promoting the welfare of individual patients; as researchers, they seek generalizable knowledge applicable to persons other than their individual patients. Because the second goal may conflict with the first, our society has decided by law that an objective review of human subjects research by a group of diverse individuals is most likely to protect human subjects and promote ethically sound research. Although the IRB system is not perfect, conscientious IRBs reassure the U.S. public that the rights and welfare of human subjects are seriously considered by people not directly involved in the research. It is through this process of research review and approval that investigators, research institutions, IRB members, and others are held publicly accountable for their decisions and actions.

Definitions of terms, record-keeping requirements, and requirements for IRB review and approval of research involving human subjects are provided in 45 CFR 46. In institutions with OPRR-approved assurances, they are also provided in the institutions' written assurance documents. Some of them are reviewed here.

2.1. Definitions

Research is any systematic investigation designed to develop or contribute to generalizable knowledge (45 CFR 46.102{d}). A *human subject* is a living individual about whom an investigator obtains either (1) data through interaction or intervention with the individual or (2) identifiable private information (45 CFR 46.102{f}). For example, consider the situation in which a physician asks the hospital medical records department to make available for review the medical records of all patients with a diagnosis of HIV infection. The physician wants to learn about the medical management of these patients treated in the hospital and its clinics during the past 5 years. According to the preceding definitions, if the physician reviews medical records of patients who are no longer living, he or she is conducting research, but it does not involve human subjects (defined as living individuals). However, if the physician reviews medical records of patients who are still living, he or she is conducting research involving human subjects. Therefore, before reviewing the medical records of living patients, a decision needs to be made if the research requires prospective IRB review and approval or if it is exempt from this requirement.

2.2. Exempt Research Activities

Not all research involving human subjects requires prospective IRB review and approval. There are six categories of research that, although they involve human subjects, are exempt from the requirements of 45 CFR 46 for IRB review. The general rationale behind the six categories of exemptions is that although the research involves human subjects, it does not expose them to physical, social, psychological, or other risks beyond those of daily life. One example of exempt research is the study of existing records (e.g., pathologic samples and medical records) if these sources are publicly available, or if the information is recorded by the investigator so that subjects cannot be identified directly or through identifiers linked to the subjects. Therefore, in the preceding example in which the researcher wants to study existing medical records, the research may be exempt from the requirement for IRB review and approval if the researcher records information from the medical charts in an anonymous fashion (no links or codes identifying patients). However, many hospitals have more restrictive policies concerning the research use of medical records and pathologic samples, and researchers should be familiar with relevant institutional policies.

Also, survey and questionnaire research is frequently conducted in the United States. Such research may be exempted unless the information elicited, if disclosed outside the research, could reasonably place the subjects at risk of criminal or civil liability or be damaging to the subjects' financial standing, employability, or reputation. Therefore, a questionnaire or survey ought not be exempted if, for example, it elicits information about illegal behaviors, such as drug use, child or spousal abuse, or other sensitive issues such as sexual and other private behaviors.

Institutional procedures vary for making determinations about whether proposed research activities are exempt. For example, in some institutions the IRB makes these determinations; in others, an office for research regulation or its equivalent makes these determinations.[20] However, research investigators are not authorized to make final determinations about whether their proposed research activities are exempt from the requirement for prospective IRB review and approval.[21] Researchers should be familiar with institutional procedures for requesting and receiving exemptions before their research begins.

2.3. Minimal Risk and Expedited Review Procedures

Minimal risk means that "the probability and magnitude of harm or discomfort anticipated in the research are not greater in and of themselves than those ordinarily encountered in daily life or the performance of routine physical or psychological examinations or tests" (45 CFR 46.102{i}). A regulatory definition of "minimal risk" is provided because some minimal risk research activities are eligible for IRB review through expedited review procedures. This means that the IRB chair, and/or other experienced IRB members designated by the chair, may approve (but not disapprove) the research on behalf of the IRB. The expedited review process was put into place to streamline and hasten IRB review of certain minimal risk research activities.

2.4. IRB Review of Research

When a researcher proposes to do research that is neither exempt nor meets criteria for expedited review, he or she submits a research protocol for review by the full IRB. A protocol is the researcher's written description of the research including issues related to the protection of the subjects. The following sections provide some of the regulatory requirements for IRB composition and criteria for IRB review and approval of research involving human subjects.

2.4.1. IRB Membership

Federal regulations set minimal IRB membership standards. All IRBs must have at least five members who have expertise in scientific and nonscientific areas (45 CR 46.107). Diversity in the professional and cultural backgrounds and gender of IRB members is expected to foster a comprehensive approach to, and promote respect for, the IRB's advice and counsel in safeguarding the rights and welfare of subjects. Because IRBs normally are situated at the site of the research, members are expected to have expertise in and sensitivity to specific conditions affecting the conduct of the research and the protection of the participants. For example, research institutions vary in their geographic locations and often draw from culturally dissimilar groups. Each IRB shall include at least one member whose primary concerns are in scientific areas, one whose primary concerns are in nonscientific areas, and one who is not otherwise affiliated with the institution. Also, when in its judgment the IRB requires expertise beyond or in addition to that available through its members, it may invite individuals with competence in special areas to assist in its reviews. These requirements for membership are grounded in the belief that the protection of human subjects is promoted by an objective review of research activities by a group of diverse individuals who have no direct involvement in the research.

2.4.2. Criteria for IRB Review of Research

To approve research, an IRB must determine that it meets minimal requirements. Table 5-1 lists the minimal regulatory criteria for IRB review and approval (45 CFR 46.111) and questions that IRBs in NIH's Intramural Research Program often consider when reviewing research protocols. All clinical researchers, particularly principal investigators, must be familiar with these requirements and understand how they apply to their research protocols.

The Proposed Research Design Is Scientifically Sound and Will Not Unnecessarily Expose Subjects to Risk (Table 5-1, #1)

Certainly, the nature, content, and scientific design of the research protocol are important concerns. At a minimum, the IRB should determine that the hypoth-

TABLE 5-1   IRB Protocol Review Standards: Regulatory Requirements for IRB Review and Documentation in the Minutes

Regulatory Review Requirement (46.111)	Possible Questions for IRB Discussion
1. The proposed research design is scientifically sound and will not unnecessarily expose subjects to risk.	(a) Is the hypothesis clear? It is clearly stated? (b) Is the study design appropriate? (c) Will the research contribute to generalizable knowledge and is it ethically permissible to expose subjects to risk?
2. Risks to subjects are reasonable in relation to anticipated benefits, if any, to subjects and the importance of knowledge that may reasonably be expected to result.	(a) What does the IRB consider the level or risk to be? (See risk assessment in Table 5-2.) (b) What does the principal investigator consider the level of risk/discomfort/inconvenience to be? (c) Is there prospect of direct benefit to subjects? (See benefit assessment in Table 5-2.)
3. Subject selection is equitable.	(a) Who is to be enrolled? Men? Women? Ethnic minorities? Children (rationale for inclusion/exclusion addressed)? Seriously ill persons? Healthy volunteers? (b) Are these subjects appropriate for the protocol?
4. Additional safeguards required for subjects likely to be vulnerable to coercion or undue influence.	(a) Are appropriate protections in place for vulnerable subjects (e.g., pregnant women, fetuses, socially or economically disadvantaged, and decisionally impaired)?
5. Informed consent is obtained from research subjects or their legally authorized representative(s).	(a) Does the informed consent document include the eight required elements? (b) Is the consent document understandable to subjects? (c) Who will obtain informed consent (principal investigator, nurse, or other?) and in what setting? (d) If appropriate, is there a children's assent? (e) Is the IRB requested to waive or alter any informed consent requirement?
6. Risks to subjects are minimized.	(a) Does the research design minimize risks to subjects? (b) Would use of a data and safety monitoring board or other research oversight process enhance subject safety?
7. Subject privacy and confidentiality are maximized.	(a) Will personally identifiable research data be protected to the extent possible from access or use? (b) Are any special privacy and confidentiality issues properly addressed (e.g., use of genetic information)?

esis is clear and that the study design is appropriate. If a research protocol is poorly designed and not likely to obtain meaningful information, it is not ethically justifiable to expose subjects to any risk, discomfort, or inconvenience. However, although IRBs have some members with scientific expertise, they are not constituted to act as primary scientific review committees. In many institutions, protocols receive pre-IRB scientific review to ensure that protocols sent to the IRB for review are well designed. This is a desirable approach because it allows the IRB to focus its major attention on the protection of the subjects. In any event, an IRB should not approve a research protocol that it does not believe to be scientifically sound.

Risks to Subjects Are Reasonable in Relation to Anticipated Benefits, if Any, to Subjects and the Importance of Knowledge That May Reasonably Be Expected to Result (Table 5-1, #2)

The IRB is required to determine the risks, discomforts, and burdens of participation in the protocol under consideration. "Risk" is the probability of harm or injury (physical, psychological, social, and economic) occurring as a result of participation in a research study. Risk varies in magnitude but only "minimal risk" is defined by federal regulations. Also, the IRB is expected to identify research-related benefits. Benefit is not defined in the regulations but may be considered a valued or desired outcome. Generally, the benefits of research fall into two major categories: (1) direct benefits to individual subjects, for example, in the form of cure or diminution of symptoms of a disease/disorder, and (2) benefits to others (e.g., society at large and future patients) because of advancements of knowledge through research. If research subjects stand to benefit directly from participation in the research, because they are receiving treatment or diagnostic procedures, higher risks and discomforts may be justifiable. However, in any trial of a new or not-yet-validated treatment, the ratio of benefits to risks should be similar or comparable to those presented by any alternative therapy. On the other hand, in research for which there is no prospect of direct benefits to individual subjects, such as research involving healthy volunteers, the IRB must evaluate whether the risks to subjects presented by research-related procedures/interventions solely to obtain generalizable knowledge are ethically acceptable. For example, in the NIH's Intramural Research Program, IRBs are expected to categorize research-related benefits and risks according to the criteria in Table 5-2.

Subject Selection Is Equitable (Table 5-1, #3)

The ethical principle of justice, which requires fair distribution of both the burdens and the benefits of research, underlies the requirement for the equitable selection of research subjects. On the one hand, when the NIH funds research, it expects the findings to be of benefit to all persons at risk of the disease, disorder, or condition under study. Therefore, adequate representation of women and minorities is particularly important in studies of diseases, disorders, and conditions that affect them.[22] On the other hand, IRBs are required to ensure that subjects (e.g., indigent persons, racial and ethnic minorities, or persons confined to institu-

tions) are not being systematically selected merely because of their easy availability, their compromised position, or their manipulability rather than for reasons directly related to the problem being studied. When defining the appropriate group of subjects to be studied in a research protocol, researchers take into account scientific design, potential subjects' susceptibility to risk, the likelihood of direct benefits to them, and considerations of practicability and fairness. Generally, the rationale for the subject selection takes into account the gender/ethnic/race categories at risk for the disease or condition being studied in the protocol. IRBs are expected to determine that the subject selection as proposed by the researcher in his or her research protocol is scientifically and ethically appropriate.

Informed Consent Is Obtained from Research Subjects or Their Legally Authorized Representative(s) (Table 5-1, #5)

Although the requirement to obtain informed consent has substantial foundations in law, it is essentially an ethical imperative. It is through informed consent that researchers make operational their duty to respect the rights of prospective subjects to be self-determining; for example, to be left alone, to make free choices, and to have private information about them shared only in ways for which they give permission.[23] When IRBs review protocols, they spend considerable time reviewing the written informed consent document(s). The IRB's role is to ensure that the consent document contains required elements of consent (Table 5-3) and that it is written at a reading level, and in a format, understandable to prospective subjects. However, in practical terms, signing the consent document is only one element in a subject's decision-making process about participating in a research protocol. A prospective subject's decision-making process is influenced by a number of factors: (1) the written consent document, (2) the knowledge and skills of the professionals involved in the process (e.g., researchers and nurses), (3) the prospective research subject (e.g., medical and emotional state, his or her primary language, ethnic/cultural background, financial considerations, and other personal factors), and (4) the circumstances in which the process takes place (e.g., an emergency room, private practice setting, and academic institution). In addition to reviewing the consent document, IRBs can influence the informed consent process by ensuring that the individuals obtaining consent are qualified to take on this important responsibility. For example, an IRB should know who will obtain informed consent to participation in the protocol and in what circumstances. Depending on the complexity and risks associated with a research study, it may require an experienced senior researcher, rather

TABLE 5-2 IRB Assessment of Research-Related Risks and Benefits

Risk

Definition of minimal risk: Minimal risk means that the probability and magnitude of harm or discomfort anticipated in the research are not greater in and of themselves than those ordinarily encountered in daily life or during the performance of routine physical or psychological examinations or tests [45 CFR 46.102(i)].

What is the appropriate risk category for the protocol under consideration?

_____The research involves no more than minimal risk to subjects.

_____The research involves more than minimal risk to subjects.

_____The risk(s) represents a minor increase over minimal risk, or

_____The risk(s) represents more than a minor increase over minimal risk.

Benefit

Definition: A research benefit is considered to be something of health-related, psychosocial, or other value to an individual research subject, or something that will contribute to the acquisition of generalizable knowledge. Money or other compensation for participation in research is not considered to be a benefit but, rather, compensation for research-related inconveniences.

The appropriate benefit category for the protocol under consideration is:

_____The research involves no prospect of direct benefit to individual subjects but is likely to yield generalizable knowledge about the subject's disorder or condition.

_____The research involves the prospect of direct benefit to individual subjects.

TABLE 5-3   General Requirements for Informed Consent (45 CFR 46.116)

In seeking informed consent, the following information shall be provided to each subject:
1. A statement that the study involves research, and
    An explanation of the purposes of the research;
    The expected duration of the subject's participation;
    A description of procedures to be followed; and
    Identification of any procedures that are experimental;
2. A description of any foreseeable risks or discomforts to the subject;
3. A description of any benefits to subjects or to others that may reasonably be expected from the research;
4. A disclosure of appropriate alternative procedures of courses of treatment, if any, that might be advantageous to the subject;
5. A statement describing the extent, if any, to which confidentiality of records identifying the subject will be maintained;
6. For research involving greater than minimal risk, an explanation as to whether any compensation and an explanation as to whether any medical treatments are available if injury occurs and, if so, what they consist of and where further information may be obtained;
7. An explanation of whom to contact for answers to pertinent questions about the research and research subjects' rights, and whom to contact in the event of a research-related injury to the subjects; and
8. A statement that participation is voluntary, refusal to participate will involve no penalty or loss of benefits to which the subject is otherwise entitled, and the subject may discontinue participation at any time without penalty or loss of benefits to which the subject is otherwise entitled.

TABLE 5-4   Vulnerable (or Potentially Vulnerable) Research Subjects

This is a noninclusive list of research subjects who have limitations to their abilities to provide informed consent and/or who may be susceptible to coercion or undue influence in decisions about research participation:
    Comatose people
    Critically ill people
    Mentally retarded people
    People with dementias/some psychiatric diseases
    Children
    Non-English speaking people
    Educationally/economically deprived people
    Prisoners
    Seriously/terminally ill people
    Paid research volunteers

pregnant women, mentally disabled persons, or economically or educationally disadvantaged persons, additional safeguards have been included in the study to protect the rights and welfare of these subjects. (45 CFR 46.111{b})

However, little additional practical guidance is provided except when the subjects of research are pregnant women (45 CFR 46, Subpart B), prisoners (Subpart C), and children (Subpart D). Otherwise, IRBs, in consultation with investigators, are expected to determine when subjects are likely to be vulnerable to coercion or undue influence and to provide additional safeguards appropriate to the particular research protocol under consideration. Vulnerable research subjects are people who are relatively or absolutely incapable of protecting their interests. In other words, "they have insufficient power, prowess, intelligence, resources, strength, or needed attributes to protect their own interests through negotiations for informed consent."[25] Table 5-4 is a noninclusive list of vulnerable or potentially vulnerable research subjects. It lists individuals who have no, or limited, ability to provide informed consent, as well as people who may be particularly susceptible to undue influence or coercion. For example, people suffering from prolonged or serious illnesses that are refractory to standard therapies, or for which there are no standard therapies, should be considered vulnerable. Although these sick people may have the intellectual capacity to provide informed consent, attention must be paid to the validity of the consent. Because of the severe restriction of their choices, out of desperation they may be willing to take serious risks even for a highly remote prospect of direct benefit. Although this is not necessarily inappropriate, researchers and IRBs need to give careful attention to the informed consent process in protocols studying terminally ill or very sick people. For example,

than a junior person, to obtain consent. Also, IRBs may exercise their authority to observe or have a third party observe the consent process and the research, although they rarely do so.

Informed consent to research participation is a complex process; therefore, it has been a topic of interest, discussion, and publication for many years. In 1966, Dr. Henry Beecher wrote that the two most important elements in ethical research involving human subjects are informed consent (which he acknowledged that in some cases was difficult, if not impossible, to obtain) and the "presence of an intelligent, informed, conscientious, compassionate, responsible investigator."[24] His ideas still ring true today. Even though the IRB's role in promoting subjects' informed consent is important, it is primarily the responsibility of the investigator obtaining the consent to ensure that it is, in fact, informed and valid.

Additional Safeguards Required for Subjects Likely to Be Vulnerable to Coercion or Undue Influence (Table 5-1, #4)

Federal regulations direct IRBs to ensure that

when some or all of the subjects are likely to be vulnerable to coercion or undue influence, such as children, prisoners, the IRB may ask that an "uninterested" individual, such as a social worker, a physician not involved in the research, or a research subject advocate, discuss with prospective subjects the research study and other clinical or research alternatives.[26] Attention has been given to additional protections for research involving people with mental disorders[27] and research conducted in emergency circumstances.[28]

Subject Privacy and Confidentiality Are Maximized (Table 5-1, #7)

Confidentiality refers to the management of information that an individual has disclosed in a relationship of trust; an expectation is that it will not be divulged to others in ways that are inconsistent with the understanding of the original disclosure without the person's permission. Privacy is defined in terms of having control over the extent, timing, and circumstances of sharing information about oneself (physical, intellectual, or behavioral) with others. Biomedical and behavioral research may invade the privacy of individuals or result in a breach of confidentiality. In certain circumstances, a breach of confidentiality may present a risk of serious harm to subjects, for example, when a researcher obtains information about subjects that, if disclosed by the researcher, would jeopardize their jobs or lead to their prosecution for criminal behavior. In other circumstances, such as observation and recording of public behavior, the invasion of privacy may present little or no harm. However, the need for confidentiality exists in virtually all studies in which data are collected about identified subjects. In most research, ensuring confidentiality is a matter of following some routine practices, such as substituting codes for personal identifiers, properly disposing of computer sheets and other papers, limiting access to identified data, and/or storing research records in locked cabinets. Most researchers are familiar with these routine precautions taken to maintain the confidentiality of data. At a minimum, IRBs should assure themselves that adequate protections will be taken in the protocol under review to safeguard the confidentiality of research information to the extent possible. The types and stringency of measures depend on the type of information to be gathered in the study. In any case, guarantees of "absolute" confidentiality should be avoided; in fact, the limits of confidentiality should be clarified. For example, federal officials have the right to inspect research records, including informed consent documents and individual medical records, to ensure compliance with the rules and standards for their programs (e.g., FDA inspections of clinical trial records). More elaborate procedures may be needed in studies in which data are collected on sensitive matters such as sexual behavior, criminal activities, and genetic predilection to disease.

Other federal, state, or local laws may deal with the confidentiality and maintenance of health-related information. For example, the Health Insurance Portability and Accountability Act of 1996 (HIPAA), which went into effect in 2003, gives patients rights over their health information and sets rules and limits on who can look at and receive this information. HIPAA, also referred to as the "Privacy Rule," was a federal legislative response to public concern over potential abuses of the privacy of health information. The Privacy Rule establishes a category of health information, referred to as PHI (private health information), which may be used or disclosed to others only in certain circumstances or under certain conditions. PHI includes what physicians and other health care professionals typically regard as a patient's personal health information, such as information in a patient's medical chart or a patient's test results. The rule applies to identifiable health information about subjects of clinical research gathered by researchers who qualify as "covered health care providers." Therefore, for researchers covered under HIPAA, familiarity with its requirements is important to protecting the confidentiality of research subjects.[29]

2.4.3. Continuing Review of Research

IRBs are required to conduct continuing review of approved research at least annually or sooner if they determine that the research presents significant physical, social, or psychological risks to subjects. Continuing review is required to ensure IRBs, investigators, research subjects, and the public that appropriate and ongoing measures are being taken to protect the rights and welfare of subjects. Requirements for what information investigators must submit to an IRB at the time of its continuing review vary. For example, in the Intramural Research Program (IRP) of the NIH, investigators are required to submit for review by the IRB a copy of the currently approved protocol consent document; a concise summary of the protocol's progress to date; the reason(s) for continuing the study; the gender/ethnic breakdown of subjects recruited to date; and any scientific developments that bear on the protocol, especially those that deal with risk(s), burdens, or benefits to individual subjects. Also, at the time of continuing review, protocol investigators must report any new equity, consultative, or other relationships with non-NIH entities that might present a real or apparent conflict of interest in the conduct of the protocol (see Chapter 11). At its continuing review, or at any other time, an IRB may suspend, modify, or terminate approval of research that has been associated with serious harm to subjects or is not being conducted in accord with federal regulatory requirements, ethical guidelines, and/or institutional policies.

3. CLINICAL RESEARCHERS AND IRBs

Good clinical researchers know that strong ethical practices go hand in hand with scientifically valid research involving human subjects. Therefore, researchers who conduct high-quality clinical research understand the IRB's mandate to protect human subjects and strive to work effectively with them. Researchers' knowledge of and expertise in the ethical dimensions of their research activities are important to IRBs for several reasons. First, clinical researchers can help educate IRBs about the human subject protections issues related to their research protocols. It helps IRBs to understand and resolve human subject protection issues if principal investigators (PIs) are knowledgeable about the IRB review standards and expert in applying them to their protocols. For example, when writing a protocol to test an investigational drug in people with Alzheimer's disease, the researcher should provide clear scientific justification in the protocol for using demented people in the research. He or she should give procedures for determining if subjects have the intellectual capacity to provide consent, who will act as the legally authorized representatives for subjects who cannot provide consent, and what, if any, additional protections will be afforded subjects. The PI may propose that a person otherwise not involved in the research monitor the informed consent process to ensure that subjects and/or their representatives understand the investigational nature of the study and its risks. This approach assists the IRB greatly by providing it with a thorough overview of, and the PI's proposed resolution to, the human subject protection issues specific to the protocol under review.

Second, in the early phases of scientifically innovative research, the ethical and human subject protection issues may be unique and/or unclear; researchers who are experts in the scientific and ethical aspects of their research can provide IRBs with invaluable guidance in areas of uncertainty. IRB decisions are matters of judgment, and when reviewing highly innovative research, it is particularly important that such judgments take into account the breadth of contemporaneous ethical thinking and scientific knowledge.

4. THE CURRENT IRB SYSTEM UNDER EVALUATION

In the past 20 years, significant advances have been made in implementing protections for research subjects in the United States and abroad. Although the 1998 Government Accountability Office report criticized a number of aspects of the U.S. IRB system, it acknowledged that the review activities of the estimated 3000–5000 IRBs in U.S. universities, hospitals, and private and public research facilities had played an important role in educating researchers about, and overseeing compliance with, regulatory requirements.[30] However, despite its successes, the IRB system is currently under considerable criticism, some deserved and some not deserved. IRBs tend to be a convenient lightening rod for identifying what is wrong with an increasingly complex and regulated system of clinical research. Throughout the years, IRBs have been given more responsibilities and increasingly they are faced with the review of challenging research activities that have broad societal impact, such as genetics research.[31] However, since the current IRB system was put into place, the research enterprise and its funding mechanisms have changed considerably. For example, the pharmaceutical industry's share of total funding of biomedical research has increased from 32% in 1980 to 62% in the early 2000s, whereas the federal government's share fell.[32] Also, previously most clinical research protocols were conducted in single, academic institutions where one IRB was responsible for overseeing the protection of the subjects. Today, it is routine to have a single protocol being conducted at many different sites throughout the United States and abroad involving hundreds or thousands of subjects. Such multisite research offers significant challenges to the local IRB system and has been an impetus for establishment of data safety and monitoring boards. Also, pharmaceutical companies frequently support research conducted in physicians' private practices in addition to, or in place of, academic medical centers. Such research may offer the advantage of broader subject recruitment; however, protocols conducted in private practices are usually reviewed and approved by "central" IRBs, which may be located far away from the site of the research.

The current IRB system deserves serious reevaluation; its strengths should be acknowledged and supported, and its weaknesses should be addressed. However, some of the strengths of the IRB system also contribute to its potential weaknesses. For example, having IRBs situated at the site of the research has the advantage that research is reviewed by people most likely to be familiar with the researchers and with institutional and other local factors relevant to the protection of research participants. It also can have an important educational role within the organization. For example, the NIH's IRP has 14 IRBs consisting of approximately 200 members who provide a significant educational resource to the NIH research community. However, on-site IRB review also introduces some potential problems. For example, a busy IRB may not engage in ongoing educational efforts to ensure that members are kept abreast of complex ethical and regulatory issues concerning the protocols it reviews. If not properly staffed and supported by its institution, the IRB may be given additional responsibilities or get bogged down with paperwork and other requirements that divert its focus away from meaningful human subject protections. In addition, its members may be predominantly employees of the research institution, giving them a real or apparent conflict of interest, particularly when reviewing research protocols involving large amounts of grant or other support money. Many organizations take steps to address and minimize these real or potential conflicts of interest of IRB members.

An IRB's ability to fulfill its mandate is influenced by a number of factors, including the knowledge and experience of the members and institutional resources and commitment. IRB decisions are matters of judgment and therefore depend on an understanding and wise application of ethical guidelines and regulatory requirements, as well as an appreciation of local influences such as cultural considerations. Improving IRBs' abilities and procedures should be aimed at promoting consistency and thoroughness of the review process within, and between, IRBs.

Although IRBs have been a primary element in the protection of human subjects for many years, there is relatively little published research on them compared, for example, with published literature on the oversight and self-evaluation procedures of hospital-based clinical ethics committees. Most studies of IRB performance examine only IRB records and procedures and IRB members' knowledge and attitudes; little published work evaluates IRBs' protocol review activities in their convened meetings. As a consequence, in the NIH IRP efforts are under way to develop a reliable instrument to evaluate the protocol review activities of the convened meetings of its IRBs. When more fully developed, the instrument will be used to evaluate IRBs' convened activities in fulfilling their regulatory mandate and as an educational tool for IRB members and researchers.[33]

The NIH also is attempting to address the lack of empiric information on IRBs through various funding mechanisms including grants.[34] Such efforts to learn how to improve the protection of human subjects are

appropriate and timely. The U.S. system for protecting human subjects was reviewed in 1975 by the congressionally mandated National Commission for the Protection of Human Subjects of Biomedical and Behavioral Research. Based on recommendations of the commission, the system was substantially revised in the late 1970s and early 1980s. It was only in the mid-1990s that other systematic evaluations were begun to examine the extent to which the current system provides adequate protection for the rights and welfare of human subjects. In 1998, results were released of two well-publicized evaluations, one conducted by DHHS's OPRR[35] and the other by DHHS's Office of the Inspector General.[36] These evaluations, as well as the President's Advisory Committee on Human Radiation Experiments[37] and a report by the Institute of Medicine (IOM),[38] have provided suggestions for improving the IRB system. Recommendations made to date generally are aimed at improving the education of researchers, IRB members, and institutional officials overseeing research involving human subjects; ensuring that IRBs have sufficient time and resources; and strengthening federal oversight of research.

In the United States, there is growing support for independent evaluation and accreditation of organizations that conduct clinical research. In one of its reports, IOM encouraged the development of accreditation standards that build upon federal regulations and urged that accrediting organizations be nongovernmental entitites.[39] Standards have been published and they acknowledge that IRBs are one of several important elements (or domains) in an institution's overall Human Research Protection Program (HRPP). Other critical domains addressed by the standards are the roles and responsibilities of, and educational requirements for, institutional/organizational officials, researchers, research staff and sponsors, and research subjects.[40] The process of accreditation includes organizational self-evaluation and site visits by independent human subject protection experts. As of February 2007, 97 organizations had received full accreditation by the Association for the Accreditation of Human Research Protection Programs.[41]

5. CONCLUSION

Research involving human subjects, even if they may benefit directly from participation, is a different kind of enterprise from the routine practice of medicine. In research, physician/researchers' goals include not only the welfare of individual subjects but also the gathering of scientific data for application in the future. Therefore, our society has granted a conditional privilege to perform research with human subjects; the condition is that the research must be scientifically sound and conducted in a manner that protects the rights and safeguards the welfare of the participants.

The current U.S. system for protecting human research subjects, including the role of IRBs, is undergoing serious evaluation. The IRB system is well developed but ever-evolving. Successful evolution of the system depends on learning from the past, understanding current and future needs, and applying the knowledge to implement meaningful improvements. Researchers, research participants and institutions, and others, particularly the American people, who bear the burdens of research and to whom the benefits accrue, all have a stake in the process.

References and Notes

1. Levine J. The Nuremberg Code. In *Ethics and Regulation of Clinical Research*. New Haven, CT, Yale University Press, 1988.
2. Katz J. The Nuremberg Code and the Nuremberg trial: A reappraisal. *JAMA* 1998;276:1662–1666.
3. Faden RR, Beauchamp TL. *A History and Theory of Informed Consent*. New York, University Press, 1986.
4. Katz, p. 1663.
5. Beecher HK. Ethics and clinical research. *N Engl J Med* 1966;274:1354–1366.
6. Levine, pp. 427–429. See also World Medical Association at www.wma.net/e/policy/b3.htm.
7. Levine, pp. 69–70.
8. National Commission for the Protection of Human Subjects of Biomedical & Behavioral Research. *The Belmont Report: Ethical Principles and Guidelines for the Protection of Human Subjects of Research*, publication No. 887-809. Washington, DC, Government Printing Office, 1979.
9. Regulations for Protection of Human Subjects, 45 CFR. Part 46, 1981. Most recent revision, June 23, 2005.
10. In addition to HHS and the FDA, the following federal departments and agencies adopted the Common Rule: the Departments of Agriculture, Energy, Commerce, Housing and Urban Development, Justice, Defense, Education, Veterans Affairs, and Transportation; the National Aeronautics and Space Administration; the Consumer Product Safety Commission; the Agency for International Development; the Environmental Protection Agency; the National Science Foundation; the Central Intelligence Agency; and the Social Security Administration.
11. Belmont Report.
12. Belmont Report, pp. 4–8.
13. Belmont Report, pp. 4–8.
14. Belmont Report, pp. 4–8.
15. 45 CFR 46, see note 9.
16. FDA Regulations for the Protection of Human Subjects, 21 CFR. Parts 50, 56, 1996.
17. Food and Drug Administration. *Guidance for Institutional Review Boards and Clinical Researchers*. Information sheet on "Significant Differences in FDA and HHS Regulations," 1998 update. Available at www.fda.gov/oc/oha/IRB/toc.html or by calling the FDA Office of Health Affairs at 301-827-1685.
18. U.S. Accounting Office. Report to the Ranking Minority Member, Senate Commission on Governmental Affairs. Continued Vigilance, Critical to Protecting Human Subjects, GAO/HEHS publication No. 96-72, p. 6, 1998.
19. In June 2000, OHRP was established when the previous office, the Office for Protection from Research Risks (OPRR), was reorganized, renamed, and administratively moved from the NIH to the Department of Health and Human Services.
20. Wichman A, Mills D, Sandier AL. Exempt research: Procedures in the intramural research program of the National Institutes of Health. *1KB Rev Human Subjects Res* 1996;March–April:3–5.
21. OPRR Dear Colleague Letter (1995, May 5). Exempt research and research that may undergo expedited review. Available at http://www.hhs.gov/ohrp/humansubjects/guidance/hsde95-02.htm or by calling OPRR at 301-496-7005.
22. NIH guidelines on the inclusion of women and minorities in clinical research. *Fed Reg* 1994;59:11146–11151.
23. Levine, p. 96.
24. Beecher, p. 1360.
25. Levine, p. 72.
26. American Academy of Neurology. Position statement, American Academy of Neurology. Ethical issues in clinical research in neurology. *Neurology* 1998;50:592–595.
27. National Bioethics Advisory Commission. *Research Involving Persons with Mental Disorders That May Affect Decision Making Capacity*. Washington, DC, Government Printing Office, 1998.
28. Food and Drug Administration. Protection of human subjects: Informed consent and waiver of informed consent in emergency research—Final rule. *Fed Reg* 1996;61:51498–51531.
29. Standards for Privacy of Individually Identifiable Health Information, 45 CFR 46 Parts 160, 165. *Fed Reg* 2000;65:82462–82829. See also HHS Office for Civil Rights, http://www.hhs.gov/ocr/hipaa. For effect of the Privacy Rule on clinical research, see http://privacyruleandresearch.nih.gov/clin_research.asp.
30. GAO report, see note 18.
31. Edgar H, Rothman DJ. The institutional review board and beyond: Future challenges to the ethics of human experimentation. *Milbank Q* 1995;489:498–501.
32. Bekelman JE, Li Y, Gross CP. Scope and impact of financial conflicts of interests in biomedical research. *JAMA* 2003;289:454–465.
33. Wichman A, Kalyan DN, Abott LJ, Wesley R, Sandler AL. Protecting human subjects in the NIH's intramural research program: A draft instrument to evaluate convened meetings of its IRBs. *Ethics Human Res* 2006;28:7–10.
34. See PA-99-079 at grants.nih.gov/grants/guide/pa-files/PA-02-103.html.
35. Final report (1998, June 15). *Evaluation of NIH Implementation of Section 491 of the Public Health Service Act, Mandating a Program of Protection for Research Subjects*. Available at grants.nih.gov/grants/oprr/library_human.htm or by calling the Office for Human Research Protections at 301-496-7005.
36. GAO report, see note 18, p. 7.
37. *Advisory Committee on Human Radiation Experiments: Final Report*, Publication No. 061-000-00-848-9. Washington, DC, U.S. Government Printing Office, 1995.
38. Institute of Medicine Board on Health Sciences Policy. *Responsible Research: A Systems Approach to Protecting Research Participants*. Washington, DC, National Academy Press, 2002. Also available at www.iom.edu.
39. Institute of Medicine Board on Health Sciences Policy. *Preserving Public Trust: Accreditation of Human Research Protection Programs*. Washington, DC, National Academy Press, 2002. Also available at www.iom.edu.
40. Association for Accreditation of Human Research Protection Programs, www.aahrpp.org. Accessed February 2007.
41. See note 40.

6

Data and Safety Monitoring Boards

LAWRENCE M. FRIEDMAN

Formerly, National Institutes of Health, Bethesda, Maryland

All trials, regardless of the type or phase, require monitoring. The nature and degree of monitoring, and who performs the monitoring, vary depending on many factors, but monitoring is essential for several reasons. First and foremost, it is necessary to ensure, to the extent possible, the safety of the study participants. Second, it enhances the scientific integrity of the study. Third, if important results are evident before the scheduled end of the study, it allows those results to be reported more quickly than otherwise would occur.[1]

For many phase I and phase II clinical trials, and other single center studies, the investigator monitors participant safety and study progress. These studies are small and typically short term. However, if the phase II trial is randomized, with assignment to study intervention or a control being blinded (sometimes called "masked"), then the investigator would not be able to perform the monitoring function. Another person or group of people would need to do that in order for the investigator to remain blinded to the assignment. Even other early phase studies may benefit from external monitoring. In all studies, study progress and major adverse events would be reported to the institutional review board (IRB).

Late phase clinical trials generally require monitoring by a person or group other than the investigator for several reasons. First, these studies are more likely to be double-blind, making it impossible for the investigator to perform monitoring. Second, even if the study is not blinded, many late phase trials are conducted in multiple centers, making monitoring by a single investigator difficult. This factor also means that an IRB, which usually oversees a single center, would have great difficulty in performing this monitoring

function. Third, the intent of a late phase trial is to provide a clear answer to the balance of possible benefits and harms from an intervention, and thus affect clinical practice. Therefore, avoiding the conscious or subconscious desire on the part of the investigator to see a certain result is essential. A tendency to stop the study sooner than appropriate, or continue longer than reasonable, can be minimized by having someone uninvolved in the research make that decision. Fourth, if changes to the protocol are made, they are best made by an investigator who is unaware of the direction in which the data are trending.

1. DESCRIPTION OF THE DATA AND SAFETY MONITORING BOARD

The first data and safety monitoring board (DSMB) for a study sponsored by the National Institutes of Health (NIH) was established almost 40 years ago.[2] Over time, the use of such boards at NIH has expanded and is now common practice for many clinical trials.[3,4]

As a consequence of the need for external monitoring, the concept of the DSMB was developed.[2,3] The monitoring committee may go by various other names, such as data monitoring committee or safety and monitoring efficacy committee. Although there may be differences in how various boards operate,[4,5] and the guidelines from different groups may differ, a key factor in all is that they are entirely, or almost entirely, made up of people external to the study investigative group. Furthermore, they are generally independent of the sponsor of the study and of the manufacturer of any product that is being evaluated or of direct com-

59

petitor manufacturers. Thus, the members of these boards have no financial interest in whether the studies they are monitoring continue. The members receive no scientific recognition or glory in the form of publications or promotions as a result of the study results. Another key feature is that the members, in the aggregate, have relevant expertise. That is, they would be knowledgeable about the scientific question the study seeks to answer, study design, and biostatistics. Often, an ethicist and a patient advocate are appointed to the board.

Who appoints these boards and to whom do they report? This varies, depending on the sponsor of the clinical trial, the phase of the trial, whether it is single or multicenter, and other factors. In NIH-sponsored early phase trials and others that are small and single center, the principal investigator typically appoints the members of the board. For multicenter late phase trials, the funding institute often does so. For industry-supported trials, the patterns vary greatly. If the company sponsoring the study adopts a "hands-off" attitude, the responsibility for appointing the members falls to a data center or coordinating center or other group external to the company. Frequently, however, the company appoints the board. Typically, the board reports to and advises the person or organization that appointed it. NIH guidelines also stipulate that, for multicenter trials, summaries of the DSMB's deliberations be provided to each participating institution's IRB.[6]

In some settings, a single DSMB may monitor more than one clinical trial. This may happen if there are several studies with common themes being conducted simultaneously by the funding organization. Examples are the clinical trial networks established by the Division of AIDS of the National Institute of Allergy and Infectious Diseases and the Clinical Trials Cooperative Groups of the National Cancer Institute. If a DSMB monitors a single study, the members would usually serve for the duration of the trial. If it monitors several studies that may start at different times, and would not have the same ending time for all studies, members would be appointed for fixed durations and would rotate off, with new members appointed to replace them.

In addition to the members, who attends meetings of the DSMB? This depends on the phase of the study and on what data are being discussed. DSMB meetings can often be divided into open, closed, and executive sessions. During the open session, study progress is reviewed. One or more investigators may be present for this discussion, as well as others who have a role in the trial and can help answer any questions posed by the DSMB members. During the closed session, when study outcome data are reviewed, the investiga-

tors and those not privy to such data are excused. The intent is to avoid unblinding the investigators and to prevent them from developing a treatment preference based on preliminary and highly changeable data. Similarly, industry representatives might be present during an open session but, particularly in late phase trials, would not be present during the closed session. This is generally the case even if a company is sponsoring the trial. In NIH-sponsored trials, representatives of the funding institute are often present during the closed session. This is controversial, however, with many questioning their attendance.[5,7] Obviously, the person or center that analyzes the data, whether in a single or multicenter trial, must attend to present the data and answer questions about the data. In some models, this person is separate from those responsible for data entry and other day-to-day trial activities, in order to avoid having accumulating data influence the conduct of the trial investigators. During the executive session, only the voting members, and perhaps an executive secretary, would be present. It is probably prudent for staff of regulatory agencies, such as the Food and Drug Administration, not to be present at DSMB meetings. Because their role is to decide on approval of drugs, devices, or biologics, having a role in the study that can possibly lead to such approval may be seen as a conflict.

2. DATA AND SAFETY MONITORING BOARD FUNCTIONS

The DSMB typically gets two types of reports to review. One involves process, and the other involves outcome. The former might consist of participant accrual status, comparability of study groups at baseline, compliance to protocol (by investigators and participants), and quality of data. These address how well the study is being conducted and whether it will be able to answer the questions posed. If participant accrual or adherence to the intervention regimen is poor, the study's power may be inadequate, and study continuation may not be scientifically or ethically appropriate. The latter type of report would consist of primary outcome variables, adverse events, other outcome measures such as laboratory tests, and interim variables that assess if the intervention is having the postulated physiologic or biochemical effects. The frequency of data reports depends on the duration of the trial and the rapidity with which data accumulate. In addition to reports that are prepared for meetings of the DSMB, there might be interim reports to monitor adverse events or other concerns. The frequency and nature of the data reports are ideally established in advance, but safety of the participants often requires modifications as the trial progresses.

There are many issues with which DSMBs concern themselves with in the monitoring of trials, a few of which are reviewed in the examples of monitoring discussed in this chapter. One issue involves repeated testing for statistical significance. To fulfill its function of safeguarding participants, the DSMB must review the data regularly; however, this carries a penalty.

> If the null hypothesis, H_0, of no difference between two groups is, in fact, true, and repeated tests of that hypothesis are made at the same level of significance using accumulating data, the probability that, at some time, the test will be called significant by chance alone will be larger than the significance level selected.[1]

If there are many looks at the data, the increase in the so-called "type 1 error" may be several times the preselected significance level. To correct for this, DSMBs use stopping guidelines that require more evidence than the usual test for significance. Several statistical approaches have been developed to control for this inflated type 1 error. The examples illustrate the use of group sequential monitoring techniques and curtailed sampling, or conditional probability. In group sequential techniques, the data are analyzed after a more or less specified number of events or periods of time. Boundaries are created such that if the difference between the treatment groups exceeds the boundaries before the study is scheduled to end, the result is statistically significant, taking into account the multiple looks at the data. There are various ways to create these boundaries, but all maintain the overall prespecified alpha level. It needs to be emphasized that these boundaries are advisory. They are guidelines to help the DSMB evaluate the strength of evidence. They do not replace thoughtful consideration of all aspects of the study, not least because they do not encompass all of the possibly relevant adverse and beneficial effects of the intervention. Another data monitoring technique uses the concept of curtailed sampling. This approach uses data existing at the time of the analysis and assumptions about the data yet to be collected. That is, one calculates the conditional probability of rejecting the null hypothesis at the end of the trial. If, under various reasonable assumptions regarding data yet to be obtained, the study conclusions would not change, then consideration might be given to stopping the trial early.

A second issue concerns asymmetry in monitoring. Although it is reasonable to proceed with a trial to show that a new intervention is superior to placebo or standard therapy, it may be inappropriate to do so to prove that the new intervention is harmful. Therefore, less evidence may be required in one monitoring direction than in another. This has implications for the traditional two-tailed test of significance.

A third issue concerns the use of external information—that is, information from outside the trial, perhaps from other trials—in monitoring.

3. DATA AND SAFETY MONITORING BOARD DECISION MAKING

When the DSMB reviews data, it can, in essence, make four decisions. First, it can recommend that the trial continue as is. The study is going well, as planned, and no changes are needed. Second, it can recommend that the trial protocol be modified in any number of ways. One modification might involve dropping a subset of participants who are having an undue number of adverse events. The DSMB might recommend that entry criteria be altered to enhance participant safety. It might recommend changes in study forms or procedures. It might recommend that the consent form be modified. For studies with more than two arms, the DSMB might recommend that one arm be stopped while the others continue.

Third, the DSMB can recommend that the study stop early. This may be done because the intervention has been shown clearly to be beneficial, because the intervention has been shown to be harmful, or because there is no realistic chance of detecting a significant or meaningful difference between groups were the study to continue. Because early stopping is irrevocable, it must be done carefully and with considerable discussion. In addition to observing that the results have crossed some predetermined monitoring boundary, or are unlikely to do so, the DSMB needs to consider other factors. Are the results possibly due to imbalance between the groups in baseline characteristics? Is there ascertainment bias for the primary outcome? Are the results consistent across subgroups of participants? Are the results for the primary outcome consistent with those for other outcomes that would be expected to respond similarly to the intervention? What are the overall risks and benefits of the intervention? Are the results due to poor adherence to intervention or to differential concomitant therapy? How likely is it that the current conclusions would change? How much additional information or precision would be obtained by continuing? Is there other ongoing research that might affect the conclusions? Will the impact of the results be persuasive to the medical and scientific communities?[1,8–10]

A fourth option is for the DSMB to recommend that the study be extended.

4. EXAMPLES

The following examples provide a sample of the kinds of issues that DSMBs discuss. For a fuller presentation of these and many others, the reader is referred to the book edited by DeMets *et al.*[11]

An example of dropping a subgroup of participants because of observed harm in that subgroup comes from the National Emphysema Treatment Trial (NETT). NETT compared lung volume-reduction surgery, on top of optimal medical treatment, against optimal medical treatment alone in 1218 patients with advanced emphysema.[12] After 1033 patients had been enrolled, the study's DSMB noted that in a subset of the patients, the surgical group was doing worse than the medical treatment-alone group.[13] Specifically, among the 140 patients (70 in each group) who were at particularly high risk, 30-day mortality was significantly worse in those receiving lung volume-reduction surgery (16% vs. 0%). As a result, further enrollment of patients meeting the criteria of low forced expiratory volume at one second and either a low carbon monoxide diffusing capacity or a homogeneous distribution of emphysema was discontinued. The investigators and the medical community were rapidly notified of this finding. This subgroup was not prespecified in the protocol, complicating the DSMB's discussion. However, even though the finding may have been due to chance, given the many possible subgroups, the DSMB chose to act in the interests of participant safety.[14]

The Heart and Estrogen-Progestin Replacement Study[15] provides an example of the need to inform participants of an unexpected adverse event. This trial enrolled 2763 women with known coronary heart disease. They were randomly assigned to either a combination of conjugated equine estrogen plus progestin or placebo. Midway in the trial, the DSMB observed an increase in venous thromboembolic events in those taking the estrogen-progestin replacement compared with those on placebo. This was not clearly stated in the consent form as a possible adverse event. Even though the frequency of the event was not such as to require stopping the study, the DSMB did think that all women in the trial needed to be made aware of it. In addition to informing the women, a letter to the editor was published prior to the end of the trial.[16]

An example of a trial stopping early for benefit is the Beta-Blocker Heart Attack Trial (BHAT).[17] This trial compared the beta-blocker, propranolol, against placebo in 3837 people who had recently suffered a myocardial infarction. All-cause mortality was the primary outcome. Nine months before the scheduled end of the trial, there were 183 deaths in the placebo group (9.5%) and 135 deaths in the propranolol group (7%). This was highly statistically significant, with a *z* value of 2.82. This had crossed the prespecified group sequential monitoring boundary.[18] There was a small amount of uncertainty because the vital status of 20 subjects was unknown at the time, and there was also a small lag in reporting deaths. Neither of these was great enough to reverse the boundary crossing. In addition, there was a low likelihood that the conclusions would be changed if the trial were to continue, given the expected number of additional deaths that would occur in the remaining nine months of the trial. The DSMB for the trial considered these and the other questions mentioned previously. Furthermore, the results of BHAT were generally consistent with those of other studies of beta-blockers in myocardial infarction patients. Arguments against stopping the trial early were that the results might be less accepted by the medical community and there would be some loss of long-term data. After a long discussion, the DSMB decided that the benefit to those suffering a myocardial infarction in the community outweighed the remaining uncertainties. Therefore, the trial was stopped early.[18]

An example of a trial that was stopped early for harm is the Cardiac Arrhythmia Suppression Trial (CAST).[19,20] The objective of this trial was to determine if antiarrhythmic therapy given to people who had had a myocardial infarction and who had frequent ventricular premature beats would reduce the incidence of death resulting from arrhythmia. The projected sample size was 4400. Eligible participants were initially randomized to open-label assessment on one or more of three antiarrhythmic drugs (encainide, flecainide, or moricizine). If the ventricular arrhythmias were suppressed, as judged by Holter monitoring, the participants were randomized to the drug and dose that worked best or to matching placebo. Early in the study, with approximately 1100 participants randomized, the DSMB noted a strong trend in mortality between the two groups (active vs. placebo). There were 19 arrhythmic deaths in one group and 3 in the other. Total mortality also showed an impressive difference. The DSMB, which knew the groups only as X and Y, but not by the actual treatment, decided to remain blinded to the identity of the groups. Even though the trends were strong, the numbers were still quite small and only represented a tiny fraction of the expected number of events. Several months later, because the difference in mortality (both overall and arrhythmic) persisted, the DSMB members asked to be unblinded and to discuss the data via a conference call. The study group doing more poorly was the one on active medication. Six months after the early trends were noticed, by the time of the subsequent meeting of the DSMB, the results had crossed the prespecified advisory boundary for harm. All of the harm was concentrated in two of the three antiarrhythmic drugs (encainide and flecainide). Therefore, participants were removed from these drugs. The trial was continued comparing the third drug (moricizine) against placebo. However, two years later, that drug was also stopped. During the short-term phase of the study when the drug was being assessed to determine if it would suppress arrhythmias, there were 15 arrhythmic deaths in the moricizine group and only 3 in the placebo group. Among the participants who had had their arrhythmia suppressed and were then randomized to long-term study drug or matching placebo, there was a nonsignificant trend against moricizine.[21] The findings from the portion of the phase of the trial when arrhythmia suppression was assessed and the very small likelihood (low conditional probability) that moricizine would turn out to reduce mortality in the longer term were the study to continue to its scheduled end, in addition to the earlier experience with encainide and flecainide, led the DSMB to recommend stopping CAST entirely.[20]

An example of stopping at least partly for futility is the aspirin component of the Physicians' Health Study.[22,23] This was a two-by-two factorial design study of aspirin and beta carotene in more than 22,000 healthy U.S. male physicians. The primary objective of the beta carotene intervention was to determine whether it reduced the incidence of cancer compared with placebo. The primary objective of the aspirin intervention was to determine whether it reduced death from cardiovascular causes, again compared with placebo. After a few years, it was noted that there was a trend in favor of aspirin, compared with placebo, with respect to incidence of nonfatal myocardial infarction, but the overall and cardiovascular death rates were much lower than predicted. During the next couple of years of the study, the difference with respect to myocardial infarction increased and became highly statistically significant. The cardiovascular death rate, however, remained quite low, with little or no difference between groups. At the same time, there was an adverse trend for hemorrhagic stroke, although the number of these events was small. Eventually, the monitoring committee recommended stopping the aspirin component of the trial. A major reason was the extremely low conditional probability, even under various assumptions of future events, that a significant difference would be seen for the outcome of cardiovascular death. To obtain a sufficient number of events, the study would have needed to be extended for many more years. An additional reason was that a clear answer had been obtained for the outcome of myocar-

dial infarction, which was primarily nonfatal. The adverse trend for stroke also was a factor in the recommendation to stop.[23]

The beta carotene component of the Physicians' Health Study continued and, in fact, was extended beyond its originally scheduled duration. At the end, no difference in cancer outcome was seen.[24] The primary reason why a DSMB might recommend extending a trial is lower than expected power. This could occur as a result of slower than anticipated enrollment of participants or, as in the case of the Physicians' Health Study, lower than expected event rate. An inappropriate reason for extension would be the observation of an encouraging but not quite significant trend, especially late in the trial. Extension on that basis affects the test of significance and is therefore strongly discouraged.

As noted previously, if a new intervention is being compared against a standard therapy, no therapy, or placebo, it may be thought inappropriate to prove, at the usual level of significance, that the new intervention is harmful. This is particularly the case when the outcome of interest is a serious or irreversible event. Therefore, although the monitoring boundary for showing benefit may be set so that the overall alpha level is the traditional 0.05 or 0.025, the boundary for harm may be set at a less extreme level.[25] The monitoring boundaries may be asymmetric, even if the study is designed as a two-sided test of the hypothesis. Sometimes, instead of a formal two-sided test, the study will be designed as one-sided. This does not necessarily mean that there is no expectation that the new intervention might be harmful, but that the study would be stopped long before harm is conclusively shown. In such instances, there would still be an advisory boundary in the harmful direction.[26] In addition, because in the classic two-sided test, with an overall alpha of 0.05, each direction would have an alpha of 0.025, the one-sided test might employ an alpha of 0.025 to declare significant benefit. This is what occurred in CAST. In the first part of CAST (before encainide and flecainide were discontinued), the test was one-sided but with a symmetric advisory boundary for harm. In the second part of CAST, once two drugs had been seen to be harmful, the advisory boundary for harm was less extreme than the boundary for benefit (i.e., it was asymmetric).[20]

External information may sometimes be used by the DSMB in its deliberations. At the same time the Physicians' Health Study was being conducted in the United States, a similar trial of aspirin was being performed in Britain.[27] The results of that trial, which were neutral with respect to cardiovascular death and myocardial infarction but which showed an adverse trend for stroke, as did the U.S. trial, became known to the Physicians' Health Study monitoring committee. Although not a deciding factor in the recommendation to stop the aspirin component of the Physicians' Health Study, it contributed to the deliberations of the monitoring committee.[23]

Several clinical trials were conducted at approximately the same time, all examining the effects of warfarin on stroke in patients with chronic atrial fibrillation.[28] One trial that was done somewhat later than three others was the Canadian Atrial Fibrillation Anticoagulation (CAFA) study.[29] The projected sample size was 660 participants. By the time 383 had been enrolled, the reports of the other three trials had clearly shown benefit of warfarin. The CAFA study was stopped by the trial's steering committee, without even bringing it to the monitoring committee, because regardless of the data, the investigators saw no need to continue the trial. Another similar trial being conducted by the Department of Veterans Affairs also ended early, at least in part because of the previously reported studies.[30]

In 1994, the results of the Finnish Alpha-Tocopherol, Beta Carotene Cancer Prevention study were released, providing evidence of increased lung cancer in the group receiving beta carotene.[31] Two ongoing trials of beta carotene, the Physicians Health Study[24] and the Beta Carotene and Retinol Efficacy Trial (CARET),[32] continued to follow the participants but alerted them to the results of the Finnish trial. In 1996, however, CARET stopped ahead of schedule.[32,33] By themselves, the data from CARET might not have led to early stopping. The monitoring committee and a second independent review group, however, decided that the CARET data on lung cancer were sufficiently similar to those from the Finnish study that stopping the trial early was appropriate.

These examples show that investigators and external monitoring groups need to be aware of information from other ongoing research. If the question being addressed by the clinical trial has already been answered, there need to be very strong reasons not to stop.

5. CONCLUSIONS

External monitoring groups such as DSMBs play important roles in reviewing accumulating data. Their primary function is ensuring, to the extent possible, the safety of the trial participants. The DSMBs also help ensure study integrity. Numerous statistical and nonstatistical approaches are used by DSMBs. Because they need to consider unexpected, as well as expected,

adverse events and other outcomes, a simple algorithm for deciding whether to continue a study, or whether to make a protocol change, is probably not possible. There is no substitute for experienced, thoughtful members deliberating the many complex issues and factors that enter into decisions to stop or modify a clinical trial.

References

1. Friedman LM, Furberg CD, DeMets DL. *Fundamentals of Clinical Trials*, 3rd ed. New York, Springer, 1998.
2. Organization, review and administration of cooperative studies (Greenberg Report): A report from the Heart Special Project Committee to the National Advisory Heart Council, May 1967. *Control Clin Trials* 1988;9:137–148.
3. Proceedings of Practical Issues in Data Monitoring of Clinical Trials, Bethesda, Maryland; January 27–28 1992 (Ellenberg S, Geller N, Simon R, Yusuf S, eds.). *Stat Med* 1993;12:415–616.
4. National Institutes of Health. *NIH Guide. NIH Policy for Data and Safety Monitoring*, June 10, 1998. Available at grants.nih.gov/grants/guide/notice-files/not98-084.html.
5. Ellenberg S, Fleming T, DeMets D. *Data Monitoring Committees in Clinical Trials: A Practical Perspective*. West Sussex, UK, Wiley, 2002.
6. National Institutes of Health. *NIH Guide. Guidance on Reporting Adverse Events to Institutional Review Boards for NIH-Supported Multicenter Clinical Trials*, June 11, 1999. Available at grants.nih.gov/grants/guide/notice-files/not99-107.html.
7. U.S. Food and Drug Administration. *Guidance for Clinical Trial Sponsors: Establishment and Operation of Clinical Trial Data Monitoring Committees*, 2006. Available at www.fda.gov/cber/gdlns/clintrialdmc.pdf.
8. Canner PL. Monitoring of the data for evidence of adverse or beneficial treatment effects. *Control Clin Trials* 1983;4:467–483.
9. DeMets D. Stopping guidelines vs. stopping rules: A practitioner's point of view. *Commun Statis Theor Methods A* 1984;13:2395–2417.
10. Fleming T, DeMets DL. Monitoring of clinical trials: Issues and recommendations. *Control Clin Trials* 1993;14:183–197.
11. DeMets DL, Furberg CD, Friedman LM (eds.). *Data Monitoring in Clinical Trials: A Case Studies Approach*. New York, Springer, 2005.
12. National Emphysema Treatment Trial Research Group. A randomized trial comparing lung-volume-reduction surgery with medical therapy for severe emphysema. *N Engl J Med* 2003;348:2059–2073.
13. National Emphysema Treatment Trial Research Group. Patients at high risk of death after lung-volume-reduction surgery. *N Engl J Med* 2001;345:1075–1083.
14. Lee SM, Wise R, Sternberg AL, Tonascia J, Piantadosi S, for the National Emphysema Treatment Trial Research Group. Methodologic issues in terminating enrollment of a subgroup of patients in a multicenter randomized trial. *Clin Trials* 2004;1:326–338.
15. Hulley S, Grady D, Bush T, *et al.* Randomized trial of estrogen plus progestin for secondary prevention of coronary heart disease in post-menopausal women. Heart and Estrogen/Progestin Replacement Study (HERS). *JAMA* 1998;280:605–613.
16. Grady D, Hulley SB, Furberg C. Venous thromboembolic events associated with hormone replacement therapy. *JAMA* 1997;278:477.

17. Beta-Blocker Heart Attack Trial Research Group. A randomized trial of propranolol in patients with acute myocardial infarction: I. Mortality results. *JAMA* 1982;247:1707–1714.

18. DeMets DL, Hardy R, Friedman LM, Lan KKG. Statistical aspects of early termination in the Beta-Blocker Heart Attack Trial. *Control Clin Trials* 1983;5:362–372.

19. The Cardiac Arrhythmia Suppression Trial (CAST) Investigators. Preliminary report: Effect of encainide and flecainide on mortality in a randomized trial of arrhythmia suppression after myocardial infarction. *N Engl J Med* 1989;321:406–412.

20. Friedman LM, Bristow JD, Hallstrom A, *et al.* Data monitoring in the Cardiac Arrhythmia Suppression Trial. *Online J Curr Clin Trials*, 1993, document 79.

21. The Cardiac Arrhythmia Suppression Trial II Investigators. Effect of the antiarrhymic agent moricizine on survival after myocardial infarction. *N Engl J Med* 1992;327:227–233.

22. Steering Committee of the Physicians' Health Study Research Group. Final report on the aspirin component of the ongoing Physicians' Health Study. *N Engl J Med* 1989;321:129–135.

23. Cairns J, Cohen L, Colton T, *et al.* Issues in the early termination of the aspirin component of the Physicians' Health Study. Data Monitoring Board of the Physicians' Health Study. *Ann Epidemiol* 1991;1:395–405.

24. Hennekens CH, Buring JE, Manson JE, *et al.* Lack of effect of long-term supplementation with beta carotene on the incidence of malignant neoplasms and cardiovascular disease. *N Engl J Med* 1996;334:1145–1149.

25. DeMets DL, Ware JH. Asymmetric group sequential boundaries for monitoring clinical trials. *Biometrika* 1982;69:661–663.

26. Lan KKG, Friedman L. Monitoring boundaries for adverse effects in long-term clinical trials. *Control Clin Trials* 1985;7:1–7.

27. Peto R, Gray R, Collins R, *et al.* Randomized trial of prophylactic daily aspirin in British male doctors. *Br Med J* 1988;296:313–316.

28. Hart RG, Benavente O, McBride R, Pearce LA. Antithrombotic therapy to prevent stroke in patients with atrial fibrillation: A meta-analysis. *Ann Intern Med* 1999;131:492–501.

29. Connolly SJ, Laupacis A, Gent M, Roberts RS, Cairns JA, Joyner C. Canadian Atrial Fibrillation Anticoagulation (CAFA) Study. *J Am Coll Cardiol* 1991;18:349–355.

30. Ezekowitz MD, Bridgers SL, James KE, *et al.* Warfarin in the prevention of stroke associated with nonrheumatic atrial fibrillation. Veterans Affairs Stroke Prevention in Nonrheumatic Atrial Fibrillation Investigators. *N Engl J Med* 1992;327:1406–1412.

31. The Alpha-Tocopherol, Beta Carotene Cancer Prevention Study Group. The effect of vitamin E and beta carotene on the incidence of lung cancer and other cancers in male smokers. *N Engl J Med* 1994;330:1029–1035.

32. Omen GS, Goodman GE, Thornquist MD, *et al.* Effects of a combination of beta carotene and vitamin A on lung cancer and cardiovascular disease. *N Engl J Med* 1996;334:1150–1155.

33. Bowen DJ, Thornquist M, Anderson K, *et al.* Stopping the active intervention: CARET. *Control Clin Trials* 2003;24:39–50.

7

Data Management in Clinical Trials

ANNE TOMPKINS

Division of Cancer Prevention, National Cancer Institute, National Institutes of Health, Bethesda, Maryland

What are data and why are data so important? Data are facts, such as baseline observations, imaging study results, drug doses given, lesion measurements, vital signs, and adverse events. The data collected in a clinical trial constitute an accounting of the trial. Rules and guidelines that govern research include the Code of Federal Regulations,[1] the Good Clinical Practices (GCPs)[2] guidelines from the International Conference on Harmonisation, state laws, sponsor standard operating procedures (SOPs), and institutional SOPs. The GCPs are an international ethical and scientific quality standard for clinical trial conduct.[2] A trial conducted under good clinical practices is the basis for demonstrating that the trial was conducted according to protocol.

Plans for data management should be set up early during the development phase of a clinical trial. Included in the plan are the appropriate mix of research personnel and resources such as staff time, workspace, computer equipment, and secure storage facilities for both paper and electronic equipment.

1. THE RESEARCH TEAM

The research team consists of individuals who possess the expertise specific for the study. The number of members on the research team usually reflects the sample size of the clinical trial and whether the institution conducting the trial has a dedicated research department. Regardless of the size of the institution or the size of the trial, each member of the team must be educated in the conduct of clinical trials, the regulations that govern trials, and the protocol document that describes the trial to be conducted.

1.1. Sponsor

A sponsor can be an individual, such as a physician, or an organization, such as a pharmaceutical company, an academic center, or a government agency such as an institute or a center within the National Institutes of Health (NIH). For some studies, the sponsor may provide financial support; however, in general the sponsor is responsible for the following activities:

Selection of qualified investigators
Verification that regulatory issues are met
Submission of an investigational new drug (IND) application to the Food and Drug Administration (FDA)
Monitoring the study to verify that it is being conducted according to the approved protocol
Informing investigators at all sites of significant new adverse events
Reporting of serious adverse events to the FDA

1.2. Principal Investigator

The principal investigator (PI) is usually the author of the protocol document and is the person who is primarily responsible for ensuring that the trial is conducted according to good clinical practices. The investigator signs the Statement of the Investigator, FDA Form 1572, which is an agreement to comply with FDA regulations in the use of the investigational agent. In addition, the PI also agrees to conduct the trial according to the written protocol, obtain approval of the institutional review board (IRB) prior to initiating the trial and at any time the protocol is amended, maintain adequate records of the trial, protect subjects through the informed consent process, and notify the sponsor and the IRB of adverse events. Although the investigator can delegate authority to other members of the team to perform various functions, the ultimate responsibility for the study cannot be delegated. The PI has the final responsibility for the conduct of the trial and must instruct all members of the research team of their responsibilities in the conduct of the trial.

1.3. Coinvestigators/Associate Investigators/Subinvestigators

There are several additional members of the research team, including other physician–investigators, clinical trial nurses, data managers, statisticians, pharmacists, bioethicists, and social workers. In addition, for multicenter trials, a PI at a clinical site may also be considered an integral member of the team. It is important to document which individuals listed as members of the research team have responsibility for patient care. Each physician–investigator having responsibility for patient care must file a FDA Form 1572 with the sponsor. In addition, they might also obtain informed consent, order investigational agent, and monitor study participants for adverse events.

1.4. Study Coordinator or Clinical Trial Nurse

The study coordinator is often a nurse with a baccalaureate or master's degree with experience in clinical trial management. With the advent of computerized systems for clinical trial management, coordinators should also have experience with automated systems. The study coordinator usually is responsible for the following activities:

Provide education for the research team and other staff about the general conduct of clinical trials and training for specific trials at the site
Provide education for the participant and family to help with the decision to participate in the clinical trial and to assist with care during the continuum of the trial
Check eligibility criteria
Arrange for study tests
Collect results of these tests.

The clinical trial nurse (CTN) monitors the participant's use of the investigational agent and interviews the participant about possible adverse event experiences. The CTN might also be responsible for drawing pharmacokinetic samples.

1.5. Data Manager

This role has changed as electronic systems for clinical trials have evolved. The data manager is often expected to have extensive knowledge of computer systems, remote data capture, and quality assurance. Activities of the data manager include abstracting data from the source documents into the research record, performing quality checks on data, preparing routine reports for patient care, interim monitoring of the trial, and regulatory reporting. As part of the quality assurance activity, the data manager reports missing data and reports discrepancies to the study coordinator and PI. Often, the data manager may be initially aware of study trends and plays a pivotal role in informing the study team of these trends.

1.6. Statistician

The statistician works closely with the PI early in the writing phase of the protocol to ensure that the trial design is appropriate for the study and that the study is powered to address the study questions. In addition, the statistician may serve as a reviewer when needed if the protocol document is amended during the implementation of the trial. Statistical expertise is essential during the analysis phase of the study, and the statistician is often asked to assist in the written final report of the study.

1.7. Other Team Members

Depending on the nature of the research, other members of the team could include bioethicists, pharmacists, social workers, dieticians, radiation specialists, pathologists, and other experts as needed. At the invitation of the PI, these other team members may be considered associate investigators. In addition, they should be informed of amendments to the protocol or a change in SOPs that are required for the specific care of a participant enrolled in a clinical trial.

1.8. The Study Participant

The study participant may be referred to as a study subject, participant, normal volunteer, or a patient. The safety and privacy of the study participant should be protected throughout the trial. It is well recognized that the person enrolled in a clinical trial is the focus of the research and is offering his or her time and effort in the search for increased knowledge in preventing, treating, or palliating disease. The entire team depends on an educated and dedicated participant to complete the research study since compliance with the study regimen and early notification of potential adverse events is essential for the participant's safety and the integrity of the trial.

2. PLANNING THE TRIAL

Data management is proactive; it begins while the protocol is being written. It includes a plan for recruitment of study participants and the management of staff and monetary resources. In addition, the identification of data fields for case report forms (CRFs) should be initiated. The CRFs should be carefully reviewed referencing the protocol document to be sure that the questions and required fields on the CRFs are clear and unambiguous. The events assigned on the study calendar should include required tests and responses, study drug administration, adverse event monitoring, and time points for evaluating response to the study intervention. These data points should be captured on the CRFs.

Designing CRFs for each trial is time- and resource-intensive, which contributes to the cost of conducting a clinical trial. To minimize these costs and to provide consistency for sharing data, efforts are being made to standardize the structure and reporting of clinical trial data. The FDA is working with several members of the research and standards communities such as the National Cancer Institute (NCI), Health Level Seven, pharmaceutical agencies, and the Clinical Data Interchange Standards Committee to establish structure for representing data and reporting research results. Standardization would also facilitate data mining of study results.

3. WHERE ARE DATA?

Data are found in the source documents. What is a source document? The designation of a source document for clinical research has several interpretations; however, it is defined as the first recording of any information about the participant or as a certified copy of an original document.[3] In addition, these initial recorded data should be signed and dated for designation as a source document. For example, a blood pressure reading recorded directly onto the participant's medical record is the first recording of that blood pressure reading. The page in the medical record containing that recording would be considered as the source document.

However, the medical record may not always be considered as a source document. For example, the clinic nurse could record a blood pressure result on a

TABLE 7-1   Source Documents

Original lab reports
Pathology reports
Surgical reports
Physician progress notes
Nurses notes
Medical record
Letters from referring physicians
Original radiological films
Tumor measurements
Patient diary
Patient notes
Patient interview
Hospital records/discharge summary/emergency room visit

clipboard vital signs sheet and then later note that result on the medical record. In this example, the clipboard vital signs sheet, rather than the medical record, would be considered the source document because it was first recorded data for the blood pressure. If the data manager later records the blood pressure reading onto a CRF, the blood pressure result on the CRF would not be considered a source document. The blood pressure result recorded on the vital signs sheet would be considered a source document. The blood pressure reading in the medical record could be considered a source document if the nurse signed that she was certifying it as an original copy. When a trial is audited, the auditor will refer to the source documents to verify the data recorded in the CRFs. Table 7-1 provides examples of source documents.

Trial sites may create a "research record" for each participant in addition to the medical record. The research record may contain copies or original case report forms (CRFs) and copies of source documents. It may also contain source documents that are not kept in the participant's medical record. An example of this latter type of document would be the results of a patient interview by the CTN asking about possible adverse events that the participant may have experienced between clinic visits. The source document must be signed and dated by the recorder to be valid. Another example of a source document would be a patient diary, in which the participant documents that the investigational agent was taken on a daily basis. The patient diary must also be signed and dated by the recorder, who could be the participant, the parent, spouse, or significant other.

4. WHO CAN COLLECT DATA?

Members of the research team, the treating physician, the referring physician, the participant, or the participant's family member all may collect data. It is important that each person who makes a recording also dates and signs the data entry. If the SOPs for the site allow initials for certain documents, a signature log with each person's initials and signature should be kept for the study file.

Patient diaries can be used to document administration of investigational agents, concomitant medications, and adverse events. Prior to using a diary, the patient and, if possible, a family member should be instructed in the importance of the diary and how to use it. The use of the diary should be simplified as appropriate or the participant might not use it, especially if the trial extends over a long period of time. Participants may begin the study by carefully recording drug doses and adverse events; however, they may lose interest since routine entry of data into a diary is time-consuming. It is important to review the diary while interviewing the trial participant at each clinic visit to clarify notations and to monitor adherence to protocol. This interview also validates the importance of the diary for the participant. An interview substantiated with a diary can help the participant recall exact symptoms experienced or drug doses missed, which is essential for a full accounting of the trial.

5. SITE INITIATION VISIT

The sponsor holds a site initiation visit at the designated clinical site just prior to the start of the study. All site personnel involved in the study should be present for this important meeting. The sponsor representative, usually a clinical research associate (CRA), reviews the plan for the study with the site personnel to be sure that all study team members understand study procedures. Other sponsor personnel, such as a medical monitor, may also attend this meeting. A visit to the pharmacy is another component of the site initiation visit and is done to assure agent security and drug accountability procedures. The sponsor also reviews investigator responsibilities with the PI. The presentation by the sponsor includes a detailed review of the protocol, including:

Eligibility criteria
Randomization and blinding
Study procedures
Study agent administration
Adverse event recording
Review of CRFs and data entry.[4]

The site initiation visit includes an educational component promoting discussions with site personnel so that questions can be answered, resulting in a staff that has a good understanding of study procedures. Educated study personnel help to ensure compliance with protocol and accurate data collection and management.

6. INFORMED CONSENT

Informed consent is a process that should be followed prior to requesting the signature from the participant and may begin before the participant is designated as eligible for the study. The process of informed consent includes an explanation of the research study to the participant, a discussion of the participant's review of the consent document, and a time for questions and answers.

Once the initial component of the process has been completed, the last phase in the process is the request for the participant's signature and date of signing. The consent document is then witnessed by the investigator and additional witnesses as required by the SOPs of the sponsor or the site. The informed consent process is then documented in the patient's medical record. The study participant should also receive a copy of the consent document. Table 7-2 lists the essential requirements of the consent document.

If the informed consent document is amended later in the study, the patient should be re-consented. The informed consent process is again initiated to explain the reasons for the new consent. Both the original and any additional consent documents that the patient has signed should be kept in the research record and a copy placed in the medical record.

7. ELIGIBILITY

Eligibility criteria describe the specific parameters of the population to be studied, including the age range, the diagnosis, prior therapy allowed, and organ function requirements. Protocol-specific checklists are often used as a reference to verify the patient's eligibility to participate in a specific clinical trial. Strict adherence to the eligibility criteria is necessary in order to report the findings as they relate to the study population. The inability to replicate the study for other populations can limit the generalizability of the results. Disregard for eligibility criteria may suggest a protocol violation that should be reported to the sponsor and the IRB.

TABLE 7-2   Elements of Informed Consent

Basic Elements of Informed Consent
45 CFR 46.116 (a)

The following information shall be provided to each subject:

1. A statement that the study involves research, an explanation of the purposes of the research and the expected duration of the subject's participation, a description of the procedures to be followed, and identification of any procedures that are experimental.
2. A description of any reasonably foreseeable risks or discomforts to the subject.
3. A description of any benefits to the subject or to others that may reasonably be expected from the research.
4. A disclosure of appropriate alternative procedures or courses of treatment, if any, that might be advantageous to the subject.
5. A statement describing the extent, if any, to which confidentiality of records identifying the subject will be maintained.
6. For research involving more than minimal risk, an explanation as to whether any compensation and an explanation as to whether any medical treatments are available if injury occurs and, if so, what they consist of, or where further information may be obtained.
7. An explanation of whom to contact for answers to pertinent questions about the research and research subjects' rights, and whom to contact in the event of a research-related injury to the subject.
8. A statement that participation is voluntary, refusal to participate will involve no penalty or loss of benefits to which the subject is otherwise entitled, and the subject may discontinue participation at any time without penalty or loss of benefits to which the subject is otherwise entitled.

Additional Elements of Informed Consent
45 CFR 46.116 (b)

When appropriate, one or more of the following elements of information shall also be provided to each subject:

1. A statement that the particular treatment or procedure may involve risks to the subject (or to the embryo or fetus, if the subject is or may become pregnant) that are currently unforeseeable.
2. Anticipated circumstances under which the subject's participation may be terminated by the investigator without regard to the subject's consent.
3. Any additional costs to the subject that may result from participation in the research.
4. The consequences of a subject's decision to withdraw from the research and procedures for orderly termination of participation by the subject.
5. A statement that significant new findings developed during the course of the research that may relate to the subject's willingness to continue participation will be provided to the subject.
6. The approximate number of subjects involved in the study.

7.1. Eligibility Checklist

The eligibility checklist is reviewed by the CTN or study coordinator and signed by the PI, whose signature attests to the patient's eligibility to participate. Once the checklist is signed, it is the formal verification that all eligibility requirements have been met, including pathologic verification of tissue samples and completion of baseline laboratory and imaging studies performed within the protocol-specified time frame. For example, if a baseline computed tomography scan of the chest must be performed within four weeks of entry into study, then the date of the scan cannot be five weeks prior to study entry. In that case, the scan would need to be repeated and interpreted, and if target lesions will be followed, new tumor measurements must be made and recorded. If entry into study is delayed because a new scan is needed, other baseline eligibility tests, such as blood chemistries that might need to be done within seven days of starting in study, may also need to be repeated and reviewed. The CTN is often responsible for coordinating these tests to ensure that they are performed within the appropriate time frame described in the protocol. Form 7-1 is an example of an eligibility checklist.

8. REGISTRATION

The patient is registered using a procedure that is described in the protocol document. Multicenter trials often have a central registration office. Eligibility will be checked and documented prior to the start of the trial. Each participant will be assigned a unique identification number that should be used on all CRFs, adverse event reports, and other reports. Randomization is often performed through the central registration office according to the specifications described in the protocol. See Forms 7-2 and 7-3 for samples of registration and randomization CRFs.

9. WHAT DATA DO YOU COLLECT?

Biographical data, such as date of birth, sex, ethnicity, and race

INSTITUTION CODE	PARTICIPANT ID	VISIT TYPE	VISIT DATE (MM/DD/YYYY)
_____	_____	_____	__ __ / __ __ / __ __ __ __

Answers to questions 1-10 must be **YES** for the subject to be eligible.
Criteria 4-8 may be evaluated using laboratory test results obtained during a time not to exceed four weeks prior to going on study.

	Criteria	Yes	No
1	The participant is male, and has localized, biopsy-proven adenocarcinoma of the prostate and planned radical prostatectomy	☐	☐
2	The participant is ≥ 18 years of age	☐	☐
3	ECOG performance status ≤ 2 (Karnofsky ≥ 60%)	☐	☐
4	Leukocytes are ≥ 3,000/μL	☐	☐
5	Platelets are ≥ 100,000/μL	☐	☐
6	Total bilirubin is within normal institutional limits	☐	☐
7	The AST (SGOT)/ALT (SGPT) ≤ 2.5 X institutional ULN	☐	☐
8	Creatinine is within normal institutional limits	☐	☐
9	The participant has agreed to use adequate contraception (barrier method of birth control or abstinence) prior to study entry and for the duration of study participation	☐	☐
10	Participant has the ability to understand and willingness to sign the informed consent	☐	☐

FORM 7-1   Inclusion criteria.

Eligibility
History and physical exam
Prior conditions, surgeries, and therapies
Concurrent therapies
Agent or therapies administered
Adverse events
Assessments: exams, medical tests, laboratory tests, and tumor measurements
Response to intervention
Off-study information

10. TREATMENT PLAN

Treatments in the plan may include administration of investigational agents, commercial agents, surgery, radiation, or combinations of these. Treatment must be given "according to protocol."[5] Documentation of administration of the study agent should include information regarding the dose, route, date and time, and duration. Any dose reductions for adverse events or weight change would require documentation in the medical and research records. Deviations from the treatment plan described in the protocol should be documented.

11. CONCURRENT THERAPY

Concurrent therapy for other medical conditions, such as diabetes or hypertension, should be documented and captured in the medical record and abstracted onto the CRFs. Concurrent therapies should include not only prescription drugs but also

INSTITUTION CODE	PARTICIPANT ID	VISIT TYPE	VISIT DATE *(MM/DD/YYYY)*
_____	_____	_____	__ __ / __ __ / __ __ __ __

Gender: ☐ Male ☐ Female ☐ Unknown | Year of Birth *(YYYY)*: __ __ __ __

Race: *check one or more*
☐ White
☐ Black or African American
☐ Native Hawaiian or Other Pacific Islander
☐ Asian
☐ American Indian or Alaska Native
☐ Unknown

Ethnicity: ☐ Hispanic or Latino
☐ Not Hispanic or Latino
☐ Unknown

Date Informed

Consent Signed: __ __ / __ __ / __ __ __ __

(MM/DD/YYYY)

Date of Registration: __ __ / __ __ / __ __ __ __
(MM/DD/YYYY)

☐ Not Applicable

Does the participant satisfy all of the eligibility criteria? ☐ Yes ☐ No

FORM 7-2 Registration.

INSTITUTION CODE	PARTICIPANT ID	VISIT TYPE	VISIT DATE *(MM/DD/YYYY)*
_____	_____	_____	__ __ / __ __ / __ __ __ __

Date Run-In Started: __ __ / __ __ / __ __ __ __
(MM/DD/YYYY)

Date Run-In Ended: __ __ / __ __ / __ __ __ __
(MM/DD/YYYY)

Does the participant satisfy all of the randomization criteria? ☐ Yes ☐ No

Date Participant Randomized: __ __ / __ __ / __ __ __ __
(MM/DD/YYYY)

Randomization Number: __ __ __ __ __

Agent Name: _____

Agent Dose: _____ Units: _____ Frequency: _____

Date Agent Provided (to participant): __ __ / __ __ / __ __ __ __
(MM/DD/YYYY)

Date Agent Started: __ __ / __ __ / __ __ __ __
(MM/DD/YYYY)

FORM 7-3 Randomization.

those purchased without a prescription (over-the-counter drugs), complementary or alternative therapies, and food or vitamin supplements. Any other medications, such as those obtained from a family medicine cabinet (often forgotten), should be recorded. This inventory of other therapies is important for analysis of response to treatment or to analyze adverse events since other medicinals may enhance or interfere with study drug availability and possibly lead to a negative response or increase in the occurrence of adverse events.

12. ADVERSE EVENT MONITORING

An adverse event (AE) is any unexpected decline from baseline that is temporally associated with the use of the investigational agent. *Adverse event* has replaced the older term *toxicity*. All AEs experienced by the participant in a clinical trial must be documented in the research record and in the participant's medical record, even if the AE is thought not to be related to the study agent.

Standard terminology in reporting AEs leads to better communication between sponsors, other investigators, research personnel at other sites using the same investigational agent, and regulatory agencies. Standardization facilitates safety monitoring, analysis, and drug development. The Common Terminology Criteria for Adverse Events,[6] developed by the Cancer Therapy Evaluation Program of the NCI, consists of more than 1000 terms describing AEs categorized by organ and grade of severity.

The Code of Federal Regulations requires expedited reporting of serious and unexpected AEs associated with the use of the drug. Serious adverse events (SAEs) are those that are considered life threatening or cause death, inpatient hospitalization or the prolongation of hospitalization, persistent or significant disability or incapacity, or congenital anomaly or birth defect.[7] Unexpected events are those that are not listed in the investigator brochure, the protocol, or the informed consent document. Prompt notification of the SAE to the FDA is mandatory, and the reporting times must be stated in the protocol. The investigator must also report these events to the sponsor, who in turn provides a written IND safety report to the FDA and to all other investigators conducting trials using that agent. The investigator must notify the local IRB or the central IRB of record. The sponsor or the IRB may require an amendment to the protocol and informed consent document.

Adverse events must be captured on CRFs or entered directly into an electronic database. Since there are several terminological systems used to identify AEs, the protocol should state the specific terminology and version that will be used for the reporting of AEs.

13. ROUTINE MONITORING VISITS

The sponsor sends a representative to the investigative site for routine monitoring visits at regular intervals throughout the study to monitor progress and data management procedures. Following the site initiation visit, the monitor may return after the first two or three participants are enrolled to validate that the site personnel are conducting the study according to protocol. This early monitoring visit is a good time to review the CRFs to determine if site personnel have understood the instructions for data entry. Collecting data "as it happens" is easier than collecting it retrospectively since it may be impossible to collect a data point that was missed. If problems have not occurred, the monitoring visits can be scheduled on a routine basis.

The routine monitoring visits are a quality assurance tool. The CRA checks to determine that the subjects met eligibility criteria, signed the informed consent document, received the study agent, and performed assessments on time. The CRFs are reviewed for completeness, legibility, and accuracy, as verified against the medical record or other source document. Discrepancies between the source documents and the CRFs will be listed, and they must be corrected by site personnel. Corrections to CRFs are made with one line drawn through the incorrect information, with a correction that is signed and dated on paper CRFs. Electronic CRFs should have an audit trail for each entry. Figure 7-1 demonstrates fewer steps in data collection with electronic data capture. If both paper and electronic CRFs are used, data in each format should be a mirror image of the other.

The drug accountability form is reviewed to verify that drug has been signed out in correct quantities only for those subjects who are eligible and who have been consented. The monitoring report should note any inconsistencies and the site should develop a plan for correcting the problems.

An audit can be routine or "for cause." It can be conducted by a sponsor or the FDA. The sponsor may decide to audit a site if the sponsor anticipates an audit by the FDA. This could be a routine audit in preparation for a new drug approval. If the sponsor has reason to think that there are problems at a site, a for cause audit may be indicated. The sponsor could be alerted by a routine monitoring report from the CRA that there

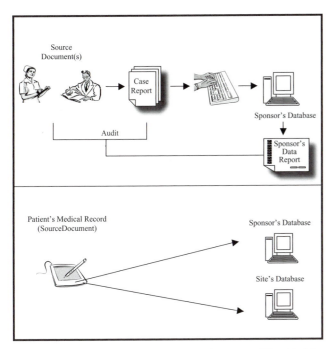

FIGURE 7-1 Electronic data capture. Courtesy of Troy Budd.

are problems. The FDA could also decide to audit, either for cause or routinely.

The following scenarios might result in an audit:

Areas with an unusually high volume of research
Unusually large study population
Data inconsistent with those of other sites that are studying the same intervention
Unusual publicity[8]

14. AUDIT TRAIL

Data collected for the study should show an accounting or reference source for the data field so an audit trail can be verified. The use of an audit trail indicates that appropriate data collection and management practices are in place so that the inspector can reconstruct the study and show that the study was conducted according to protocol.

15. ELECTRONIC DATABASE

An electronic database facilitates rapid data analysis, both at the investigative site and for the sponsor. Information provided on automatically generated reports can alert the investigator to issues and trends such as unanticipated numbers of adverse events, data

discrepancies, or slow accrual. The investigator could then initiate an early intervention plan. Necessities for an electronic database for clinical trial management include the following:

A coding system: A well-defined coding system with prescribed terminology and business rules facilitates consistent data entry resulting in easier data analysis, both for interim analysis and at the conclusion of the study.
Security: Computer security includes such basics as a unique password assigned to each individual, scheduled password changes, secured computers, backup tapes stored in a separate location, defined user roles, firewalls, and virus protection. In addition, encryption and secure transfer mechanisms are necessary to protect data.
Edit checks: The database should have edit checks for data entry where an alert or constraint is triggered when data are entered that do not conform to programmed parameters or mandatory fields are not completed.
On-site computer support and help desk availability: Both of these are essential for increasing efficiency in the conduct of a clinical trial for the investigative site, the sponsor, and the regulatory agency by providing technical expertise to solve problems and maintain current systems.

16. SUMMARY

Data management includes the entire spectrum from data collection and entry to data analysis and reporting. Even as automated systems are employed to facilitate clinical trial data management, the central themes remain: Was the trial conducted according to good clinical practices? Was the study carried out according to protocol? Was the participant treated according to protocol? Was the participant assessed according to protocol? These questions should always be asked and affirmed to assure the integrity of the research and the protection of the safety of the participant.

References

1. 45 Code of Federal Regulations § 46.116 and 21 Code of Federal Regulations § 312.50.
2. International Conference on Harmonisation, E6.
3. International Conference on Harmonisation, E6, Section 1.52.
4. Woodin K, Schneider J. *The CRA's Guide to Monitoring Clinical Research.* Boston, Thomsom Centerwatch, 2003, p. 126.

5. Cassidy J. Data and data management in clinical trials. In *Principles and Practice of Clinical Research*. San Diego, Academic Press, 2002.

6. Cancer Therapy Evaluation Program. *Common Terminology Criteria for Adverse Events*, Version 3.0. DCTD, NCI, NIH, DHHS, March 31, 2003.

7. 21 Code of Federal Regulations § 312.32.

8. Engelbach I. *et al. Sponsor Visits and Regulatory Audits: What You Need to Know*. Media, PA, Barnett International, 1993, p. 53.

8

Unanticipated Risk in Clinical Research

STEPHEN E. STRAUS

Laboratory of Clinical Infectious Diseases, National Institute of Allergy and Infectious Diseases and Office of the Director, National Center for Complementary and Alternative Medicine, National Institutes of Health, Bethesda, Maryland

Each of us engages in clinical research, as a participant or as an investigator, for personal reasons. The decision to do so may involve an investigator's aspirations to extend current knowledge and therapeutic options or his or her personal ambitions, a patient's gesture to future generations, or an act of desperation. Whatever the blend of considerations that lead to one's decision, they all distill down to one thing: a desired outcome. Sometimes these are fulfilled; other times, they are not. When the research is successful, everyone benefits. When it fails, the motivation for undertaking the study in the first place is called into question.

This cycle of clinical research is by now a very familiar one, but the process is serious and charged with risk, both for the subject and for the investigator. Some adverse outcomes in clinical studies are predictable, based on what is known of the underlying medical condition of the research subjects and the nature of the experimental intervention, whereas others may not be. These latter, unanticipated risks are particularly challenging in that they erode public trust in research and lead to progressive changes in research regulation, oversight, and conduct.

Two examples of unanticipated risks and their broader implications for clinical research are worth considering here. In 1999, investigators at the University of Pennsylvania and their partners in a biotechnology company initiated, with full approval of all institutional and national regulatory entities, a first phase study of gene therapy for the rare genetic disorder known as ornithine transcarbamylase (OTC) defi-ciency. The human gene encoding OTC was engineered into a human adenovirus vector with the intent, per protocol, to administer it to humans with OTC deficiency, with the hope that expression of the gene from its viral vector would reconstitute normal enzyme function and ameliorate the severe metabolic consequences of the disease. In what is now a well-reported episode in biomedical research, research subject Jessie Gelsinger died from the experimental intervention, triggering congressional investigations, lawsuits, and serious repercussions for the study's principal investigator, Dr. James Wilson.[1-4]

Unanticipated risks emerging in early human gene therapy trials for severe combined immunodeficiency have led to patient deaths and reassessment of the hazards of viral vectors for human gene replacement, but far fewer investigations and regulatory adjustments. Investigators in France used a retroviral vector to replace a gene critical for the proper maturation of the cellular immune response.[5] It turned out that the gene inserted itself into a region of human chromosomes responsible for regulation of cell division, thereby precipitating malignant transformation of lymphocytes and the development of leukemia. Certain risks cannot be readily foreseen but become evident only during the course of the research. Only by prohibiting all studies of novel clinical interventions can one reliably prevent all unforeseeable risks to research subjects.

The rationale for the present textbook is to illuminate the mechanics of clinical research: how one designs a protocol, how the sample size should be calculated, optimal ways of managing data, which interventions are ethical, and which interventions are proscribed. The heft of this book attests to the fact that clinical research is a complex undertaking—so complex that, as in all things in medicine today, one acquires expertise only through prolonged practice.

The complexity of clinical research arises because it involves more than just a contract between subject and investigator brokered over an informed consent document. There is an almost sacred trust between partners who each commit to fulfill their end of an agreement. One subjects him- or herself to the demands of the protocol. The other commits to engage in important rather than trivial work and to incite as little harm as possible in the process. Overseeing this relationship is an elaborate hierarchy of committees and agencies whose responsibilities include ensuring that the risk to the subject is minimized and at all times justified.

The language of clinical research revolves around risk. For example, have the preclinical studies shown that a healthy volunteer is likely to tolerate a new drug well, a drug that will afford him or her no benefit? Is the risk to him or her in helping to reveal the distribution and metabolism of the drug an acceptable one? For the patient who has failed all other chemotherapeutic agents, is it justified to administer a new retroviral vector carrying a tumor suppressor gene? The issue of risk is so fundamental to clinical research that major parts of the Code of Federal Regulations are dedicated to it.[6] Several different federal government offices and agencies and countless institutional review boards (IRBs) routinely opine on risks to experimental subjects. Simply thumb through the present text and estimate the proportion of chapters that deal with such risk.

Considering that the research endeavor involves a partnership, it is surprising that all of the discussions of risk concern only one of the partners, the research subject. Virtually nothing is said of the risks investigators face, not to say that they are the same as those faced by the subjects or potentially as grave. Investigators face many risks in undertaking clinical research. Inherent to the formidable review process for research protocols is the possibility that the investigator's ambition to test some new therapy will be disapproved. Just because something works in the mouse does not mean that one can justify doing it in humans. However, ego deflation is not the most serious risk that an investigator might face. That the study may fail is also a risk. Negative studies are never as exciting as positive ones. That, too, is of little consequence. Such is the nature of science.

The important risk to the investigator is that something will go terribly wrong; patients could be hurt or even die, and one's judgment in designing the study will be questioned. More formally, one might be accused of misconduct or fraud. Although intentional deception and data falsifications in clinical research have occurred—and likely will continue to occur—not all instances in which investigators have been charged with such misconduct are valid. The elaborate mechanisms that are structured to protect the subject do not protect the investigator. Due process is a more remote and theoretical concept in academia than in the commercial world. Consider eminent scientists whose work was pilloried in the press and in the hearing rooms of Congress during the past decade. Investigator James Wilson of the University of Pennsylvania was accused of bad judgment in the design and conduct of his OTC gene therapy trial because he held proprietary interests in the gene vector technology.

In formulating the first year's schedule for a new National Institutes of Health (NIH) Clinical Center clinical research curriculum in 1995, the course director, John Gallin, asked me to speak of my then recent experiences with a new drug for hepatitis, a drug that appeared to be promising until five people who took it died, setting off a national dialogue about the studies and nearly ending the careers of several highly respected investigators, as well as my own. My lecture that first year, and each subsequent year, has formed the basis for this chapter. The process of preparing the lectures proved cathartic at first, but with increased distance has come a greater clarity and balance that both the students and I appreciate.

The issues surrounding our hepatitis studies were mired for two years in government inquiry, litigation, media speculation, and calumny. The history of the affair is very complex,[7–9] but this is not the place for all of the minutia of that moment. In this chapter, I comment from as broad and as neutral a perspective as I can achieve because the purpose of the chapter is to illustrate how clinical research can be a risky undertaking not only for the subject but also for the investigator.

1. THE REASONS

Why do we do it? There are many reasons why people participate in clinical research. Also, the reasons we investigators undertake such studies are equally diverse.

The most venal of reasons to volunteer for a research study is that one can earn a lot of money. Maybe it is not the amount accorded by a good steady job, but it could make a difference to a student or a homeless person. Money explains the willingness of some volunteers to be inoculated with influenza virus or to undergo a bone marrow biopsy. Transient pain and inconvenience seem justified when the check arrives.

Money aside, the primary reason for volunteering for clinical studies is the hope of contributing in some incremental manner to the growth of knowledge and therapeutics that could help a family member or society in general. Patients who have failed all other therapies, be they mainstream or alternative, may offer themselves to science. Maybe they would be the lucky recipient of a wonderful new drug, or the next research subject might benefit from what is learned from studies they undergo.

All subjects of clinical research develop their own personal calculus in forming the decision to participate. What benefit may there be, and at what risk? All risks cannot be known in advance, and that which is known is better understood by the investigator than by the subject. However, the subject trusts that the investigator and the research enterprise surrounding him or her, including the IRBs and the Food and Drug Administration (FDA), insist on honesty and care in revealing the known risks. As better articulated in other chapters in this book, though, the process of informed consent is an imperfect one. The subject may be well informed, but when something bad occurs, it is clear that the risks were not foremost in his or her mind.

Investigators engage in clinical research, in part, because it is a noble adventure. Youthful fantasies of being another "Microbe Hunter," another "Arrowsmith," a Nobel Laureate, or another "Osler" drive us. Yet, we quickly recognize these fantasies for what they are, and our motives can be self-serving. It takes so many to achieve so little. Science creeps forward, punctuated by new insights and technologies. For many of us, it is sufficient just to get one paper in the *New England Journal of Medicine.* If we are so fortunate, we aim for more. Our academic appointments, promotions, salaries, and fame all hang on our research accomplishments.

This is admittedly cynical. Yet, we physicians have been taught to believe in miracles. We have seen our patients spiral helplessly downward and then have seen others released from their misery by new drugs. Any of us who has treated a severe anaphylactic reaction with epinephrine knows what a real miracle drug is. Likewise this is true of quinine for malaria, aspirin for fever, insulin, and penicillin.

Our ambition, though, needs to be channeled properly. By doing clinical research, and now through formal coursework and textbooks such as this one, we learn its ground rules. They require us to be both the physician and the scientist, a potential conflict. Yet we work in a team of individuals and organizations that help maintain our perspective. And the patients sign informed consent documents, some of which should make any rational person reject participation in a study, given their catalog of potential reactions: disfigurement, bone marrow toxicity, and even death.

We conduct clinical research for the same reasons that subjects participate in them: we are inveterate optimists. And if anything goes wrong, the institutions that oversee our research to protect our patients will pardon our errors because we followed the rules and attempted to do things correctly. The following is a cautionary tale that instructs us to the contrary.

2. THE DRUG

During the late 1970s, Jack Fox of Memorial–Sloan Kettering Hospital in New York synthesized a series of fluoropyrimidine analogs of natural nucleosides and demonstrated them to be potent and specific antiviral compounds. Recall that this was an era in which antiviral therapy first emerged from academic obscurity into mainstream practice. Amantadine had been shown to be effective for influenza, and large-scale collaborative trials were finding intravenous vidarabine to reduce the mortality of herpes simplex encephalitis and severe herpes zoster infections in cancer patients. Although vidarabine caused neuromuscular and hematologic toxicity, that it could favorably alter the outcome of life-threatening viral infections infused optimism that even better antiviral drugs were feasible.

Fox and colleagues in New York recognized one of their compounds, fluoroiodoarabinosylcytosine, or fiacitabine (FIAC) (Fig. 8-1), to be a particularly promising candidate as an inhibitor of herpes simplex and varicella zoster virus replication, meaning that it might prove beneficial for severe herpes, chicken pox, and shingles. *In vitro* and animal studies suggested it to be far more potent than vidarabine.[7] During the early 1980s, they conducted a series of exploratory phase I and II clinical studies that confirmed their suspicions.[8] In one controlled trial, they demonstrated that FIAC was superior to vidarabine for herpes zoster in patients with advanced cancer.[9] Like vidarabine, it too showed bone marrow toxicity but hinted that cardiac, neurologic, and hepatic toxicities might occur as well.

The competitive world of drug development, however, tarnished the early luster of FIAC. Just as it was appearing useful, a far better drug emerged from the laboratory of Gertrude Elion at the Burroughs–Wellcome Company. Elion and her long-term collaborator George Hitchings had earned reputations (and a later Nobel prize) for the synthesis of novel drugs based on nucleoside chemistry: allopurinol, 6-mercaptopurine, and others.

Their most stunning discovery was acyclovir, a novel guanosine analog that revolutionized antiviral drug therapy and established the strategy that led to zidovudine (AZT) and other contemporary mainstays of HIV management. Acyclovir proved to be dramatically more effective than vidarabine; it could be administered orally, and toxicity was negligible.[10]

It was clear that there could be no role for FIAC as a means of treating severe herpes simplex or varicella zoster virus infections, but in 1981, New York City became an epicenter of a bewildering new syndrome among promiscuous homosexual men who developed sight- and life-threatening cytomegalovirus (CMV) opportunistic infections.[11]

In vitro, FIAC proved to be very active against CMV, whereas vidarabine and acyclovir were essentially inactive. Could it work in people? Early tests involving intravenous FIAC doses of up to 1 g or more per day for 10 days in desperately ill patients

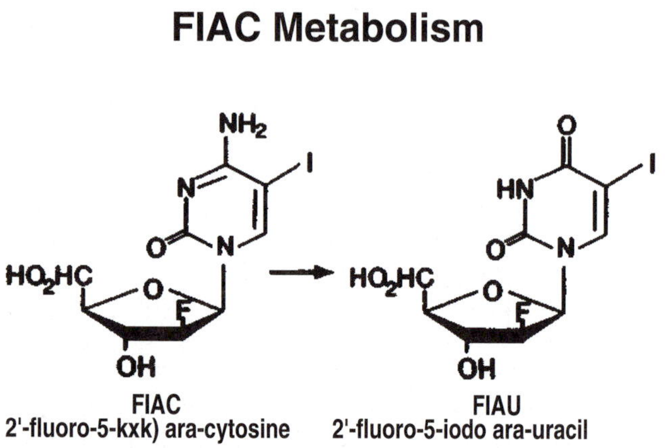

with the then recently recognized acquired immunodeficiency syndrome (AIDS) suggested that it could.[8]

FIAC was licensed to the Bristol-Myers Company for further development. Apparently, their internal assessments of the compound yielded mixed results because they opted, in time, not to develop it for CMV since another company's compound, ganciclovir, was already proving effective.[12] The market for CMV drugs was seen as too small to justify the nearly $200 million needed to bring a novel drug to market.

In the late 1980s, a small company in the San Francisco Bay area, Oclassen Pharmaceuticals, acquired the rights to develop FIAC and its congeners. Their consultants reviewed all of the preclinical and clinical data on the drugs and proposed that, as an orally bioavailable agent, FIAC may be an effective product for serious CMV infections. A team of investigators—including Douglas Richman of the University of California at San Diego, Lawrence Corey of the University of Washington, and I—proposed a phase I trial protocol that would be conducted independently under the aegis of the nationwide NIH-sponsored AIDS Clinical Trials Group (ACTG) (Table 8-1). The goal was to administer FIAC orally to HIV patients with positive urine CMV cultures. It was reasoned that if oral FIAC proved as active as ganciclovir, and no more toxic than it, FIAC would represent a therapeutic advance. Once begun, treatment of CMV infection in AIDS patients is an essentially lifelong undertaking. Because ganciclovir could only be administered intravenously, the patient required a permanent indwelling line and an endless cycle of infusions.[12] For the proposed study, two weeks of FIAC liquid would be administered in doses ranging from 0.6 to 5 mg/kg/day in six patients each, with escalation depending on how well it was tolerated.

Even at the lower FIAC does range, however, nausea and fatigue proved unacceptable with no obvious effect on CMV shedding in the urine. It was apparent

FIAC Metabolism

FIAC
2'-fluoro-5-kxk) ara-cytosine

FIAU
2'-fluoro-5-iodo ara-uracil

FIGURE 8-1 The chemical structures of fiacitabine (FIAC) and fialuridine (FIAU).

TABLE 8-1   FIAU and FIAC Clinical Trials

Principal Investigator	Location	Patients	Planned Duration	Study Dates
Richman	UCSD	10 HIV+/CMV+	35 days	11/89–03/90
Corey	UW	2 HIV+/CMV+	35 days	03/90–05/90
Corey	UW	25 HIV+/HBV+	14 days	10/90–06/92
Straus	NIH	14 HIV+/HBV+	14 days	04/91–06/92
Richman	UCSD	4 HIV+/HBV+	14 days	05/91–05/92
Hoofnagle	NIH	24 HBV+	28 days	04/92–09/92
Hoofnagle	NIH	15 HBV+	6 months	03/93–06/93

UCSD, University of California (San Diego); UW, University of Washington (Seattle); NIH, National Institutes of Health (Bethesda, MD).
Modified from reference 20.

by 1990 that FIAC had no place in the treatment of herpes viruses. However, it was reasoned that the toxicity of FIAC might not extend to some of its analogs. It was known that in humans most of a dose of FIAC was converted to a similar molecule called fluoroiodoarauracil, fialuridine, or FIAU (Fig. 8-1). FIAU possessed all of FIAC's antiviral activity. The exploratory ACTG trial was revised to test escalating doses of FIAU in patients with HIV and CMV coinfection (Table 8-1). Tests in the first 13 such patients again revealed nausea at doses above 1 mg/kg/day and still no anti-CMV activity. It was now clear that neither FIAC nor FIAU would be an effective anti-CMV drug.

Before abandoning this family of drugs, the collaborative research team decided to pursue its possible use for hepatitis B virus (HBV) infection. Both FIAC and FIAU were very potent inhibitors of the enzyme on which the HBV depends for its replication, the viral DNA polymerase,[13] and FIAC had shown activity in woodchucks chronically infected with a virus closely related to HBV.[14] Chronic hepatitis is an important human disease.

3. THE TARGET

HBV produces a common acute infection of the liver.[15] Most people resolve the infection, but it remains active for years in approximately 5% of all humans, including approximately 1% of all Americans. Chronic HBV infection can result in gradual scarring of the liver, a process known as cirrhosis; it can lead to liver failure and the need for a transplanted replacement liver. After decades of uncontrolled chronic infection, liver cancer develops. HBV is the major cause of cancer deaths in areas of Asia.

A very effective vaccine can be given to prevent HBV infection, but it is of no value for the estimated 300 million people who are already chronically infected. During the late 1970s, daily or thrice weekly injections of interferon-α for 4–6 months were shown to suppress HBV infection in most people but to provide sustained benefit for only 25–40% of recipients.[16,17] The treatment is inconvenient, expensive, and toxic, leading to low blood counts, depression, and many more problems that have greatly limited interferon's acceptance. Hepatitis remains an unmet therapeutic target. The decision was made to test FIAU.

4. THE TRIALS

In the spring of 1991, I treated the first patient with HIV and HBV infection with FIAU at a dose of

1 mg/kg/day under a new research protocol (Table 8-1). The patient tolerated the two weeks of treatment well and, remarkably, his HBV blood levels fell approximately 10-fold. That degree and speed of HBV inhibition had never before been achieved with an antiviral drug, and the research team became energized. In quick succession, additional patients were treated, all of whom had responses (Fig. 8-2). However, there was still occasional nausea and the potential for other troubling side effects, so the protocol was revised to allow us to test successively lower doses of FIAU. Over the next year, a total of 43 patients received FIAU in San Diego, Seattle, and Bethesda.[18] The drug proved active at doses as low as 0.1 mg/kg/day and was well tolerated for two weeks.

The prospects for FIAU as a treatment for HBV were encouraging, yet its real value would not be in the modest number of people who are dually infected with HIV and HBV but, rather, in the larger population of people infected with HBV alone. The decision was made to design a new series of studies in otherwise healthy people with chronic HBV infection.

These further studies required, more than ever, the advice and assistance of expert hepatologists experienced in the diagnosis and management of patients with chronic HBV infection. I was fortunate to enlist a long-standing colleague and collaborator, Jay Hoofnagle, of NIH's National Institute of Diabetes,

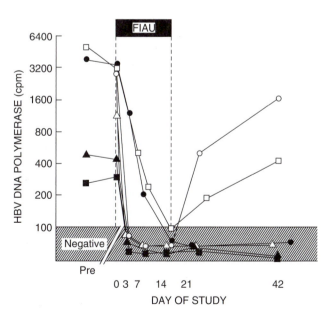

FIGURE 8-2 Inhibition of serum hepatitis B virus (HBV) DNA polymerase levels in the first six HIV-positive, HBV-positive recipients of FIAU, 1 mg/kg/day by mouth, for 14 days. The hatched area of the graph shows the level below which the assay cannot reliably detect the viral enzyme in serum. Data from reference 18.

Digestive and Kidney Diseases. He advised that any treatment likely to be effective for HBV would need to be prolonged (a fact subsequently proven with newer drugs for HBV). We needed to conduct a series of progressively longer trials until we knew whether FIAU would remain well tolerated and lead to sustained clearance of the virus.

Hoofnagle's group assumed the leadership of a second study in which 24 chronically HBV-infected (but HIV-negative) patients would receive FIAU for 28 days each. There would be four groups of 6 patients each, randomly assigned to doses of 0.05, 0.1, 0.25, and 0.5 mg/kg/day. No one would actually receive more total FIAU than in our prior study in HIV patients. All of the patients were enrolled and treated by mid-1992. FIAU was well tolerated, and there appeared to be a dose response, with slightly less suppression of blood virus levels at the 0.05 mg/kg dose level than at the higher levels. At doses of 0.1 mg/kg/day, HBV DNA levels in the blood dropped by an average of approximately 90%: 9 of 24 patients lost all detectable viral DNA.[19]

As these exciting early results were emerging from the FIAU trial, a few storm clouds appeared on the research horizon—ones whose portent would not be appreciated for another year or more. The studies of FIAU in people dually infected with HBV and HIV were wrapping up as the studies in normal hosts were beginning. The final stages in the HIV cohort trial involved exploratory retreatments of four patients who had responded to their initial two-week courses of FIAU but who then relapsed weeks to months thereafter. It was argued that prolonged treatment or retreatments would become necessary in later studies of otherwise healthy patients to affect viral clearance in as large a percentage of them as possible.

The four patients who were retreated were, of course, further along in their HIV disease and were requiring antiretroviral therapies and drugs to prevent and treat opportunistic infections. Their FIAU retreatments were for another 2 weeks at 1 mg/kg/day, as before, or at 0.5 mg/kg/day for 4 weeks, beginning 2–10 months after completion of their first courses of FIAU. In the 3–5 months after these second courses of FIAU, however, all four of these patients developed serious problems. Two patients developed pancreatitis, which proved fatal in one. They were on other drugs such as didanosine, an antiretroviral known to cause pancreatitis. The other two developed progressive liver failure from which they eventually died. Extensive consultations and liver biopsies led us to conclude at the time that the liver failure was a manifestation of progressive hepatitis and cirrhosis in one of the patients and a result of the known toxicity of a different drug the second patient was taking. We could not attribute these deaths to FIAU because the problems emerged only long after the treatments were completed and each of the patients had tolerated prior courses with the same total amounts of FIAU. Subsequent, independent reviews of these cases[20] and their autopsies supported our impressions, but we were never sure what had really happened.

Of the 24 otherwise healthy patients with chronic hepatitis, two developed some delayed medical problems. One reported pain and tingling of his feet four months after completing FIAU. These neuropathic symptoms were similar to ones attributed to his alcoholism five years earlier. Another patient described fatigue and nausea starting one month after completing FIAU. Over the next month, his liver enzymes rose, he noted tingling in his feet, and physicians elsewhere opted to remove his gallbladder against our advice. No gallstones were found; however, one week afterward, ascitic fluid accumulated in his abdomen, and liver failure progressed to death over the next six weeks. His autopsy was reviewed with multiple consultants, including ones from the Armed Forces Institute of Pathology. In addition to the severe viral hepatitis, we found microsteatosis of the liver—that is, the accumulation of microscopic fat droplets. A rare process, microsteatosis, was known to occur in diverse settings including drug toxicity. Although we could not understand how it would arise many weeks after stopping FIAU therapy, we alerted all of our future patients to the problem. This turned out to be the crucial clue to the disaster that befell our subsequent studies.

5. CASSANDRA REVEALED

During the early decades of antiviral drug development, there was a chorus of critics who declaimed that a drug for viruses that is both safe and effective could never be identified. The replication of viruses is so inextricably linked to that of the host cell, they argued, that any compound that interferes successfully with virus growth would necessarily impair that of the cell, a formula for toxicity. Like Cassandra of classic legend, the daughter of King Priam and Queen Hecuba of Troy, they predicted doom in vain.

Any lingering doubts about the feasibility of antiviral therapy were summarily dispatched with the synthesis of acyclovir. Yet there remained (and remain still) aspects of the mechanisms by which nucleoside and other analogs of essential cellular processes act that are not fully understood. That such ignorance could prove fatal was revealed in the course of the last FIAU trial.

6. EXTENDED STUDIES

In early 1993, the overall prospects for FIAU were excellent. The ability to suppress blood levels of a major human viral pathogen with a simple oral medication had enormous market implications. Oclassen Pharmaceuticals realized that the further development of FIAU required extended and very costly studies, ones that dictated the assistance of a corporate partner. From among the several potential suitors, the Eli Lilly Company, one of the world's largest drug companies, was chosen to assume leadership of the further testing of FIAU. Lilly's plan was to formulate FIAU into a pill rather than the liquid suspension we had used up to that point, to extend treatment to one year, and to expand the studies into many medical centers in the United States and in Asia, where a huge need for HBV treatment was appreciated.

While the Lilly studies were beginning elsewhere, we at NIH decided that another careful study of six-months' duration was needed before expanding to larger, year-long studies (Table 8-1). To this end, we began our third trial in March 1993. It was designed to treat 24 otherwise healthy HBV-infected patients with 0.1 or 0.25 mg/kg/day. After the initial 8–10 weeks of treatment, a few patients began to report nausea and fatigue, and the doses were reduced or stopped in them, according to protocol guidelines.

The very first patient in this study noted tingling in his toes after four weeks of treatment. Nerve conduction studies proved normal, but his FIAU dose was reduced in early June, nonetheless, and then stopped entirely one week later, when the symptoms persisted. Two weeks afterward, the nausea and fatigue became progressively severe. Late in the evening of June 25, 1993, he was taken to an emergency room in Virginia and found to be hypotensive and acidotic. Although we did not understand what had happened to this patient, our nagging concerns about prior adverse events and the gravity of the present one left us only one decision: to contact all of the other study patients and ask them to immediately stop taking FIAU. Meanwhile, this first seriously ill man was transferred to the NIH Clinical Center intensive care unit, where severe liver failure was documented. Failing any sign of improvement, he was transferred to the University of Virginia at Charlottesville four days later for emergency liver transplantation. He died on July 6 of progressive acidosis and shock.

Of the 24 patients projected for this six-month trial of FIAU, 15 had already been enrolled by late June. Eleven of the 15 had been participants in the prior year's month-long study. They had experienced significant but only transient reductions in HBV blood level in that previous study, and a longer course of treatment was seen as a way of achieving even more substantive and sustained results. These 11 patients were enrolled first and had completed eight or more weeks of FIAU treatment when we terminated the study. The other four were new patients who had received three weeks or less of FIAU to that point.

As the other study participants were evaluated, we found that most of them had some sign of toxicity. A few felt nauseated or fatigued, but we and they had thought little of it before because we were monitoring their blood tests every two weeks, they remained fairly stable, and the hepatitis treatment with interferon that they all had failed was associated with adverse effects at least as severe as these.[16,17] But now, a few of the patients were showing more serious toxic reactions, and blood test results deteriorated progressively despite their having stopped treatment. At one point, 9 of the 15 were inpatients on our research unit at the same time. Reviewing their status on rounds every few hours brought increasingly frightening and perplexing findings. Over the ensuing months, a total of 5 patients died, 2 survived only with emergency liver transplantation, 3 recovered fully, and 3 who had received the least FIAU showed no definite adverse reactions.

7. FIAU TOXICITY

The cause of the decline of our patients was realized only gradually over the next two years through molecular, biochemical, toxicologic, and animal model studies by several teams of collaborating investigators at multiple institutions.[21] The nature of the acute reactions we had seen provided the necessary clues that informed this work.

The FIAU recipients who were fated to die exhibited greatly elevated blood levels of lactic acid, with blood pHs below 7.0 in several instances. Liver failure was marked by the complete loss of hepatocellular synthetic function with hypoalbuminemia, hypoprothrombinemia, and preterminal rises in bilirubin to 20 mg/dl with surprisingly little increases in aminotransferase levels. Serum amylase and lipase levels rose. We consulted authorities worldwide, convened several scholarly task forces to advise us, and attempted every possibly useful treatment to reverse the process, with minimal success.

Realizing that the toxicity emanated from a nucleoside analog, we infused thymidine and uridine in the hope of displacing the FIAU molecules from synthetic pathways in the cells.[22,23] We infused high-dose dextrose and enormous volumes of bicarbonate to correct the acidosis. Permission was obtained from the FDA late one evening to use an experimental device for treatment of liver failure in one subject. A column in which 200 g of hepatoblastoma cells were growing within capillary tubes was in its earliest stages of testing as an artificial liver.[24] The patient's circulation was diverted through the device while awaiting a donor liver. He died nonetheless.

Even now, we have no clear sense that any of these desperate treatments were beneficial. The one treatment that we are convinced was life saving was liver transplantation.[25] It was attempted in five patients, but three were far too sick when it was performed. Only two patients made it through the first weeks after transplantation, and they lived for at least several more years with fairly normal hepatic function.

The cause of the complications we fought became clear as we examined patient tissues, of which we had many. We biopsied every affected tissue that we could, and, unfortunately, we had available for study a mounting number of autopsies and livers removed at the time of emergency transplantation. The first specimen of liver we examined gave us an eerie, sinking feeling because we had seen something like it once before. When tissue sections are stained in their usual fashion, the transparent pink liver cells appear as a smooth mosaic separated by bile ducts and blood vessels. Of course, chronic HBV infection disrupts this mosaic with infiltrating inflammatory cells and dense bridges of fibrotic reaction, but the liver cells still appear fairly pink.

The liver cells of the patient who died after nearly 3 months of FIAU treatment looked pale and foamy.[21] They were filled with tiny droplets of fat—the very same microsteatosis we had seen in the autopsy of the patient in our earlier, month-long FIAU treatment study. Whatever it was that killed the earlier patient was now killing others.

The spectrum of adverse reactions we had documented in the FIAU recipients was broad but suggested a single underlying theme. The initial symptoms of nausea and fatigue were followed by a relentless cascade of lactic acidosis, hepatic failure, pancreatitis, peripheral neuropathy, and skeletal myopathy. This constellation of reactions suggested an underlying injury to the mitochondria that are responsible for converting sugar and lipid molecules into energy in every living cell. When mitochondrial enzymes are inhibited, cells accumulate lactate and long-chain fat molecules. The normal functions of the cell cease.

It seemed as if FIAU had injured mitochondria.[26] Only in time could we presume to understand why, but at this point, in late June and early July 1993, our goal was to prove it. Electron microscopy provided a key piece to the puzzle. Ultramicrographs of liver sections from our patients showed large, reduplicated mitochondria lacking their normal internal scaffolding of membranes on which the oxidative, energy-producing enzymatic machinery are assembled.[21] Surrounding these vacant mitochondria were droplets of lipid, clusters of larger and smaller ones, like soap bubbles. Similar collections of fat droplets were seen in muscle fibers and nerve axons. The pancreas did not show clear abnormalities of this type. Once the pancreas is injured, all of the digestive enzymes bottled within it are unleashed, and the tissue autodestructs.

Subsequent work showed that FIAU—which is, after all, an analog of the molecules that are stitched together to make nucleic acids—was being incorporated into cellular and mitochondrial DNA as these molecules were being synthesized.[27,28] Although the normal cellular enzyme that is responsible for synthesis of nuclear DNA, DNA polymerase α, did not utilize FIAU efficiently as a substrate, the mitochondrial enzyme did. Mitochondria contain a different enzyme known as DNA polymerase γ. This enzyme mistook FIAU for being a normal thymidine molecule and inserted a molecule of FIAU in its place. Mitochondrial DNA full of aberrant nucleotides cannot serve as proper templates for the RNA and proteins they are designed to encode. Protein synthesis stops.[29]

Why did the toxicity of FIAU appear in a delayed fashion, weeks or months after the drug was first administered? Our best guess today is based on the life cycle of a mitochondrion. These subcellular organelles have a defined life span of only weeks to months. Assuming that the cell is replete with normal mitochondria at the time FIAU treatment begins, only those mitochondria that are newly formed in the presence of FIAU will be damaged. At first, all of the original mitochondria are in the cell and functioning normally. In time, these mitochondria are replaced, one after the other, with mitochondria containing FIAU-damaged DNA. Eventually, few normal mitochondria remain, and the cell's oxidative machinery disappears. Direct measurement of the mitochondrial enzyme content of cells grown in culture for some time in the presence of FIAU, and of liver from our patients, showed extremely reduced levels of oxidative capacity.

8. REASSESSING THE PRECLINICAL STUDIES

Whatever one thinks of the propriety of animal experimentation, it remains an irreplaceable and underappreciated component of the drug development process. One takes no pleasure in subjecting animals to drug studies, but they provide invaluable proof that a treatment might work in humans. They help point to toxicities that could not be predicted from *in vitro* studies, and they help us decide the dose levels that will be needed for beneficial effects in humans. Without these data, human studies would be much more hazardous.

The development of FIAC and FIAU depended on animal studies. Many studies were done before these drugs were ever given to humans, and even more studies were done once it was inescapable that they are toxic, in an effort to understand that toxicity and to develop means of testing subsequent drugs for similar potential.

Before human studies of FIAC and/or FIAU, multiple studies in mice, rats, dogs, monkeys, and one brief study in woodchucks were done. These tests showed that doses hundreds of times those planned for people were required before any toxicity was apparent, and the toxicities seen in the animals predominantly involved the bone marrow and heart. Hepatic and pancreatic toxicities were not seen.

After the deaths of our patients, consulting toxicologists reviewed all of the prior animal studies, and many were repeated, with the specific goal of seeking mitochondrial injury. Mitochondrial toxicity was virtually unheard of before these trials, and formal tests of new drugs had never been directed at the question. Now we knew what to look for. Curiously, we still never found it in any of the new studies of animals typically used to test new drugs.

The woodchuck, however, proved to emulate what happened to our patients.[30] Recall that woodchuck hepatitis infection is similar to chronic human HBV infection. An early study had shown that four weeks of FIAC suppresses woodchuck hepatitis.[14] A 12-week trial of FIAU was undertaken in woodchucks by Bud Tennant of Cornell University.[31] During the initial eight weeks, the treatment caused a dramatic lowering of the virus levels, but in the final weeks the woodchucks began to weaken and lose weight. Microscopic fat droplets began to appear in their livers. Today, all new hepatitis drugs undergo prolonged testing in woodchucks, and the potential of these drugs for inflicting damage to mitochondria is sought.

Through the course of these studies, it became clear that certain toxicities already recognized in AIDS patients treated with antiretroviral drugs, including AZT (zidovudine), DDI (didanosine), and zalcitabine (DDC), were due to mitochondrial injury.[32–34] It had been difficult to appreciate the scattered reports of a few dozen cases of hepatic failure, acidosis, pancreatitis, or myositis among the many thousands of very complex patients treated for advanced AIDS.[35–40] These drugs, too, caused mitochondrial toxicity, but far less often and obvious than that associated with FIAU.

The two years after the death of the first patient in our six-month trial was marked by more than just intense scientific inquiry that led to an understanding of the cause of FIAU toxicity and development of *in vitro* and animal models for it. It was also a period of public and institutional investigations, some of which seemed to have been designed solely to assign blame for multiple research deaths. None of us who conducted the FIAU studies had imagined the personal and professional risks these investigations would pose.

9. RESEARCH OVERSIGHT

Clinical studies are carefully orchestrated processes that require preparation and oversight. Every aspect of every one of the FIAU studies—their scientific bases, the preclinical data, the choice of study subjects, all of the dose modifications, the decision to repeat treatments and to extend their durations, the criteria for dose modifications according to adverse reactions, the definition of the adverse reactions, the consent forms, and much more—were all subject to prior review and approval. The procedures to obtain approval to conduct clinical studies are formal and sometimes formidable ones that defer casual inquiry, but we rely on them heavily because they provide us an independent assessment of our proposals and the legal basis to pursue them.

Every institution in the United States that engages in clinical research and that receives and expends federal dollars is subject to an elaborate Code of Federal Regulations.[6] The institutions provide written assurance that their scientists will conduct clinical research according to these guidelines and that formal mechanisms will be in place for initial and continuing review of every research project.

Those of us who conducted FIAU studies were subject to oversight by senior colleagues who reviewed the protocols and approved the resources needed to support them, by our IRBs, by our quality assurance committees, and at the NIH by its Office of Human Subjects Research and ultimately by the NIH Office for the Protection of Research Risks. Staff and consultants of the drug sponsors reviewed and monitored the studies as well. The trials done under the auspices of the NIH ACTG were approved by NIH extramural staff and ACTG study committees.

The FDA also played a crucial and active role in our studies. FDA medical officers are assigned to review all studies of experimental drugs and biologies. Before the first tier of such studies can proceed, they examine the existing data about the substance, its action, toxicity, and manufacture. Normally, the FDA uses a fairly passive process for acting on proposed trials of new drugs. Proposals are submitted to the FDA to receive what is known as an investigational new drug exemption, meaning that there is permission to use an experimental substance in a specific context. If FDA reviewers report no objection to the study within 30 days of filing, the study may proceed. The FDA does not actually approve a study; it simply might choose to not disapprove it.

For the FIAC and FIAU studies, the lead investigators met with FDA staff before the first dose of drug was given to a patient. Thereafter, we met every time we planned to modify or extend our studies. From the outset, we knew that we would be exploring a new class of antiviral drugs that could be toxic. We assumed that the potential toxicities would be justifiable, first in the context of CMV disease and the existing treatment for it in AIDS patients, and then later as an alternative to interferon injections for chronic HBV infection. We sought and received an almost unprecedented degree of involvement of the FDA medical officers who helped advise us on our drug development plans and reviewed our study progress in real time. The medical officers who met with us were experienced and eager to see studies done that would bring new therapeutic options as quickly as possible. There was considerable public pressure on the FDA at this time to accelerate drug development, particularly for AIDS, and these medical officers committed their energies to make it happen.

Together, we investigators and FDA reviewers developed a new mechanism for tracking the progress and problems in a research protocol. We established a set of flow sheets that tabulated the data on every patient enrolled in the studies and all of their key laboratory results and symptoms. These flow sheets were faxed to each study center and to the FDA every week during the studies. In all, hundreds of these documents would circle the country before the studies had ended.

10. THE INVESTIGATIONS BEGIN

When serious adverse reactions occur in research patients, many people need to be notified about it quickly. With the hospitalization of our study participant in late June 1993, we contacted all of the other patients and impressed on them the need to stop taking FIAU. The same day, we informed investigators in Boston and Galveston, who had begun other FIAU

studies, and called the drug sponsors. We called our clinical directors, our IRBs, and the FDA. During the next few days, we issued written reports to the FDA, to senior NIH staff, and to the Secretary of Health and Human Services.

During the subsequent few months, every one of the NIH offices charged with clinical research oversight investigated our FIAU studies. Each investigative body had its own concerns. Our IRBs and human subjects research staff sought to verify that all patients had fulfilled the protocol criteria and had signed consent documents. Quality assurance officials wanted to know whether our clinical charts had documented all patient visits fully and more. Over time, formal reports from each of these groups declared that we investigators had followed every procedure appropriately in terms of protocol submission, clinical records, consent forms, and reports. Except for one dosage error that we had reported in the first study, all drug administration and dose modifications were appropriate. Our charts were cited as being above the desired standard in all regards. We knew that our teams of fellows and research nurses had done a great job, but it was reassuring to learn that others thought so as well, particularly as we were doubting our own quality and motivations, having wrought trials that killed several people.

Clinical research, however, is a highly visible and public enterprise, and the pressure to investigate the deaths of several of our research subjects spread quickly beyond the NIH. The flurry of press reports that appeared after the deaths of our patients fanned public interest and inquiry. Seemingly everyone had an opinion on why things went wrong. Some could only imagine that a tragedy of this kind must have stemmed from investigator misconduct. Without such misconduct, the usual layers of IRB and FDA oversight would have succeeded in protecting research subjects from injury and death.

11. SCIENTIFIC MISCONDUCT

As the events of the FIAU trials were unfolding, nationwide attention was already focused on the alleged misconduct of several prominent U.S. scientists. We feared that similar attention would be drawn to us.

The Vietnam War and the Watergate hearings provoked widespread distrust of government and spawned the emergence of the investigative reporter. No longer considered to be muckrakers, these journalists assumed the license to reveal the sinister underbelly of our previously trusted public institutions.

It was not long before the scientific establishment became the focus of investigative reporting as well. In 1983, the influential science journalists William Broad and Nicholas Wade suggested, in a book titled *Betrayers of the Truth*,[41] that fraud is endemic in contemporary science. They highlighted the then recent cases of John Long of the Massachusetts General Hospital, who acknowledged faking laboratory tests of cancer cells, and Vijay Soman of Yale, who resigned after it was revealed that he had misrepresented data in an article on anorexia nervosa. Unlike the great historian of science Thomas Kuhn, who concluded that observer bias is inherent in normal science,[42] Broad and Wade argued that scientists intentionally misrepresent data because the competitive arena of science drives them to do so.

The revelation in 1983 that a promising young cardiologist at Harvard, John Darsee, also faked experimental data and was stripped by the NIH of eligibility for further research grants only supported Broad and Wade's cynical thesis. Congress investigated these incidents, and the NIH felt the pressure to police science rather than wait for outside agencies to do so for them. In 1989, the Department of Health and Human Services established within the NIH the Office of Scientific Integrity (OSI). In 1992, the responsibilities of the OSI were extended as it was removed from the NIH to the office of the Assistant Secretary for Health and renamed the Office of Research Integrity (ORI).

In the first years of its mandate, the OSI investigated Robert Gallo, the codiscoverer of HIV. In 1989, John Crewdson of the *Chicago Tribune* wrote a 50,000-word article asserting that Gallo had stolen his HIV isolate from Luc Montagnier of the Pasteur Institute in Paris. In late 1992, the newly constituted ORI found Gallo and his associate Mikulas Popovic guilty of scientific misconduct. However, in November 1993, on appeal, the ORI verdict was reversed.

As the Gallo investigations were concluding, the ORI was occupied with another case of alleged scientific misconduct. As reviewed by Daniel J. Kevles in *The Baltimore Case*,[43] the charges in this case stemmed from a 1986 article in the journal *Cell* by Thereza Imanishi-Kari of Tufts University and the Nobel Laureate David Baltimore of MIT. Soon after its publication, Imanishi-Kari's postdoctoral fellow Margot O'Toole accused her of faking some of its data. Baltimore defended Imanishi-Kari as having made no meaningful or willful errors in her article. He was rebuked by some leading scientists, by the press, and by Congressman John Dingell in highly publicized hearings for his failure to distance himself from his colleague. Both Baltimore and Imanishi-Kari were tarred with the very same brush. The charges against Imanishi-Kari were not dismissed until 1996 when the ORI verdict of fraud was reversed finally on appeal, but to this day Baltimore stands criticized for his apparent hubris in defending her.

The investigations of Gallo and Baltimore and their colleagues by the ORI and Congress in the mid-1990s fueled public sentiment that science is rife with enormous egos and a penchant for misconduct. Such was the backdrop to the FIAU study deaths. The stakes for clinical research were very high because this was not an episode of faking data in mouse or antibody experiments—human lives were lost. The FDA was compelled to investigate us.

12. THE FDA

Throughout our preparations for the FIAU clinical trials and during them, we interacted frequently and very productively with FDA medical officers. With the deaths of our patients, though, we began to interact with an entirely separate arm of the FDA: the Office of Compliance. Its staff initiated a series of audits and reviews of our studies. FDA inspectors reviewed all of our study records and presented to me FDA Form 483, a Notice of Inspectional Observations. Through this and subsequent communications, I came to know more about FDA procedures than most investigators ever learn in a lifetime of conducting clinical research. Although FDA audits are common and even routine, they must be taken very seriously because they can result in removal of one's privileges to use investigational agents and, in the worst cases, they can have legal consequences.

This initial report was more benign than I had feared. Upon review of all of my records, the FDA investigators issued a one-sentence finding that "the adverse event regarding the hospitalization of Subject 409, although reported by telephone to the Sponsor/Monitor, was not followed by a written report required by the protocol." There is a formal requirement in clinical trials that any "unexpected or serious adverse event" such as the hospitalization of a study subject for any reason must be reported promptly to the drug sponsor and FDA and followed within three working days by a written summary of the event. This affords the FDA the opportunity to temporarily or permanently stop a study before more subjects develop the same reactions. I had failed to follow my telephone notice of the hospitalization of a patient with a written report in a timely manner. Similar, rather concise and procedure-oriented reports were issued to all of the other FIAU investigators. In addition, however, we were each issued lengthy letters detailing how we had failed to understand the true nature of chronic HBV infection and to properly monitor its treatment, how we had misinterpreted all the prior FIAU study data to ignore obvious signs of toxicity, and more.

In November 1993, FDA investigators and consultants issued a 90-page "Report of an FDA Task Force, Fialuridine: Hepatic and Pancreatic Toxicity" that severely criticized our judgment and actions in the studies. There were two major criticisms: that our consent forms failed to disclose all of the potential toxicities of FIAU, and that we had seriously misinterpreted reactions to the drug. As to the first criticism, the protocol consent forms were lengthy and did indicate that there could be bone marrow, pulmonary, gastrointestinal, muscular, renal, or neurologic toxicities. We indicated that FIAU was a new drug, all of whose acute or chronic toxicities were not known. We had not suggested that the drug could injure the liver or prove fatal.

The second criticism was based on a fundamental difference in how expert hepatologists and the FDA viewed changes in liver chemistries observed during our FIAU treatments for hepatitis. The literature, and our prior experience, showed that liver enzyme levels can rise during treatment for HBV, and that these rises correlated with loss of HBV DNA and antigens from the blood, through what was postulated to be immunologically mediated mechanisms that destroy infected hepatocytes.[44,45] The FDA reviewers felt strongly that such enzyme changes must have reflected liver toxicity, and our failure to acknowledge them as such prevented us from predicting that longer courses of FIAU could induce fatal hepatic failure.

Despite the lengthy rebuttals we wrote to these conclusions, the FDA issued to all of the FIAU principal investigators in May 1994 official letters of reprimand. The FDA again enumerated our many "protocol violations," the inadequacy of our consent forms, and the errors in our clinical judgment. We felt quite powerless before the vast regulatory authority of the FDA. It was our speculation that the FDA chose to criticize us, regardless of the scientific merits of its findings, in part to protect itself from claims that it had allowed studies of a toxic drug to proceed. Moreover, throughout the 1980s the FDA had been under tremendous pressure to simplify its reviews so that new drugs would be available more quickly for dying AIDS patients. The FIAU episode could be exploited to prove that a weakening of the FDA would come at the cost of public safety.

13. THE NATIONAL INSTITUTES OF HEALTH

Of the many reviews conducted at the NIH—by the IRBs, the Office of Quality Assurance, the Office of Human Subjects Research, and the Office for the Protection of Research Risks—we investigators thought that none was more welcome than the one commissioned by the NIH director in the fall of 1993. A subcommittee of the Director's Advisory Committee was formed of leading professors of medicine, pharmacology, and nursing and a practicing gastroenterologist. This committee undertook a review of every protocol, all of the IRB minutes, and all of the patient charts. The committee interviewed every physician and nurse involved in our FIAU studies and every patient and the immediate family of those who died.

The advisory committee concluded in June 1994 that "appropriate clinical judgment had been exercised in each of these cases and that patient safety was not compromised." The committee's mandate was different, however, from that of the FDA. It did not concern itself with regulatory requirements about the timeliness or completeness of reporting adverse reactions to the FDA. Nonetheless, the committee stated that

> given the nature of the syndrome of delayed or late toxicity which appeared in these studies, it is unlikely that any of those reporting events were relevant to or could have prevented the tragic outcomes even were they significant, which is under dispute.

We felt a rush of vindication, but only temporarily. The conflicting conclusions of investigations conducted "in-house" at the FDA and the NIH required reconciliation through a more "impartial" investigative body.

14. THE INSTITUTE OF MEDICINE

The Secretary of Health and Human Services, Dr. Donna Shalala, commissioned an independent study by the Institute of Medicine (IOM) of the National Academy of Sciences. This committee was composed of experts in infectious diseases, hepatology, epidemiology, clinical trials, pharmacology, and ethics. It was charged to determine "whether investigators, sponsors, FDA and NIH acted appropriately in all phases of the clinical trials of FIAU and FIAC" and whether "the rules or procedures governing the clinical trial process need to be changed to address the problems, if any, identified in the FDA and NIH reports, or problems identified independently by the committee."

Despite the seemingly endless cycle of investigations to this point, we appreciated that this investigation would prove the pivotal one. This committee was charged to do more than to just review us. It would comment on the entire clinical research process and address our largely unheeded protests that the FDA had been abusive.

The committee was provided access to all records and all prior reviews. It interviewed drug sponsors, all investigators, and 19 of the study patients before issuing its 296-page report in March 1995. In its executive summary, the report stated:

> The overall impression of the IOM committee is that the entire series of trials reviewed was an ethically sound clinical research project designed and carried out by highly competent investigators who frequently went beyond the requirements dictated by regulations or imposed by IRBs to respond to the desires and needs of their patient–subjects.

The IOM committee concurred with the findings by the NIH director's committee and disagreed with many of the FDA's assertions regarding our studies. Specifically, they found "that the [FDA] compliance audit was not as informed or balanced as it should have been." The committee was "troubled as well by a system of communication in regard to warning letters that makes them available to the media and others before their receipt by the parties being cited."

The IOM invested much of its report with scholarly recommendations on clinical trials and the drug review process in general as well as on the excessive attention to mechanics rather than substance. It urged that the drug development system be revised "cautiously," that there be a system of "no-fault compensation for research injury," and, very relevant to this textbook, that

> all clinical investigators engaged in trials should be exposed to explicit training not only in the design and conduct of clinical trials and their ethical obligations to patients but also on their legal and regulatory obligations to both the sponsor and the FDA.

This ended the cycle of formal investigations of the FIAU debacle, but paralleling them had been a series of more public, political, and legal inquiries. The media reported extensively on the investigations, a congressional committee demanded answers, and there were several lawsuits.

15. THE MEDIA

On July 1, 1993, several days before the first patient died, the *Wall Street Journal* cited an Eli Lilly Company press release announcing that it had suspended its studies of a promising new hepatitis drug because of adverse events.[46] With this news, Lilly stock closed 25 cents a share lower for the day. Newspaper reports of the first FIAU study deaths were straightforward summaries of the study and its goals.

By August, however, the journalistic focus evolved from reporting the events to criticizing the research. Something went wrong. Why? Was it the process of clinical research, or had the investigators ignored obvious signs that FIAU would be a deadly poison?

Marlene Cimons concluded in the *Los Angeles Times*[47] that clinical research poses "deadly risks," citing Arthur Caplan, the president of the American Association of Bioethics:

> Over the years, people have tended to mash together research and therapy—when average people hear the term "clinical research" they think, "latest, state-of-the-art therapy." The reality is that "clinical trial" should mean: "Possibly dangerous substance. Beware. Could be fatal."

However, the idea that clinical research could be so dangerous was unsettling. Wholesale acceptance of the idea could make it impossible to conduct the studies that might yield the cure to cancer. Perhaps all clinical research is not this risky.

Lawrence K. Altman of the *New York Times* reported that the deaths of patients in the FIAU trials have "focused attention on the process by which patients come to participate in studies testing the safety and efficacy of new therapies."[48] The process of informed consent is inherently flawed, he argued. Ill patients are too desperate to read the consent forms carefully and ask all of the questions they need to ask, whereas investigators fail to provide balanced descriptions of the experimental process because they "may have vested interests in persuading patients to participate in studies." This harsh conclusion did presage the later episode of gene therapy for ornithine transcarbamylase deficiency.[1–4]

If the process of clinical research is inherently risky, greater safeguards are needed. In his preliminary report on the FDA's review of the FIAU deaths, Commissioner David A. Kessler concluded that scientists need additional oversight because they were "too optimistic about the possibilities for a cure and failed to think skeptically about the data they were collecting."[49]

Of course, investigators would not undertake a trial about which they are not optimistic. So the very enthusiasm and ambition that drive us to undertake novel studies were seen as serious flaws, justifying ever more oversight. It seemed absurd to us that we should be faulted for optimism or that even more intense regulatory oversight could be of any positive value.

More difficult were the ad hominum articles that concluded that we had specifically ignored obvious clues to FIAU's toxicity[50] and that eventually suggested that we were guilty of grave violations of federal regulations in conducting studies of FIAU.[51]

When in June 1994 the NIH advisory committee concluded the FIAU deaths to be an "unavoidable accident," in conflict with the FDA's conclusions, the *Washington Post* reported that Congressman Adolphus Towns of New York had called the NIH report "a whitewash" that showed that "NIH is simply not sufficiently removed from culpability to evaluate impartially the tragic events that occurred."[52]

Press coverage of the FIAU tragedy did not end until IOM issued its report in March 1995. Philip J. Hilts of the *New York Times* quoted the report as concluding, "On review of the FIAU trials, the committee finds no evidence of negligence or carelessness on the part of the investigators or sponsors." He closed his piece, though, by once again quoting the FDA commissioner, who still believed, despite IOM's findings, that the deaths had occurred because "the scientists were too optimistic."[53]

16. THE CONGRESS

Press reports during the summer of 1993 that the FDA had begun investigations of NIH scientists caught Congress' attention. Congress possesses broad statutory oversight of clinical research through the laws it passes and through its appropriations to the NIH and FDA. Congressman Towns of the House Committee on Government Operations requested that the NIH turn over for his staff's review copies of every document in its possession pertaining to FIAU. The NIH agreed to comply, but we investigators had concerns about doing so. There were thousands of pages of patient records, and these all bore personal identifiers. Consider for a moment that both HTV and HBV infections are highly prevalent in gay men, that some of our patients were prominent Washingtonians, and that Congress wanted their medical records. We opposed the release to Congress of sensitive patient records.

After the deadline for release passed, Congressman Towns again chastened the NIH to release all documents "to avoid the appearance of covering up information critical to a resolution of an important public health issue that cost the lives of at least five patients." The Office of the General Counsel, the legal advisors to the NIH, argued that the Code of Federal Regulations allows release of sensitive documents to Congress. Ultimately, with the advice of ethicists, a compromise was reached that permitted us to release only redacted medical records.

In the interim, we were interviewed by congressional staff who seemed to relish the possibility of uncovering evidence of scientific misconduct. The tribulations that Gallo and Baltimore had faced in their interactions with Congress were well-known, and we sensed our professional vulnerability as well.

Throughout the summer of 1993, a disgruntled FIAU recipient, who had been highlighted in a front-page article in the *Washington Post*, penned a series of vitriolic letters to Jay Hoofnagle, his physician for several years. No longer able to maintain the usual physician–patient dialogue with this man, Hoofnagle wrote to him, as many physicians in practice would, to suggest that he seek help elsewhere. The patient promptly copied the letter from Hoofnagle to the major media, which then attacked him for seeking to squelch criticism. Congressman Towns expressed his "outrage" that the NIH would "retaliate" in this way against its patients. At his urging, Hoofnagle was officially reprimanded, and the NIH agreed to convene the Advisory Committee to the Director to review the FIAU deaths. When the FDA and NIH report on FIAU arrived at diametrically opposing conclusions, Congressman Towns wrote to the Secretary of Health and Human Services stating,

> If the FDA findings are correct, then the NIH report appears to be a whitewash of medical negligence and patient mistreatment. . . . On the other hand, if the NIH findings are correct, then the FDA warning letters appear to be an overreach of regulatory action.

It was at his insistence that the IOM investigation was undertaken.

17. THE LAW

The legal implications of the severe and fatal reactions to FIAU did not escape our concerns. During the first weeks after we terminated the FIAU studies, virtually all of our energies were devoted to salvaging the remaining patients. By mid-August, the crisis had stabilized, the newspaper reports were becoming more irritating, and the investigations began. With this background, I contacted the Office of the General Counsel at the NIH to learn it could not represent us. It would, however, defend the U.S. government should lawsuits be filed.

Realizing that I had no personal legal counsel, I interviewed partners in two of Washington's largest and most experienced firms. They both predicted large costs to represent me. Fortunately, members of my family reassured me that it would commit any and all resources needed to defend me. I foresaw that major changes could be forced on my home, and my career in research seemed over. I had learned that as long as my decisions in the FIAU trials were made in the course of my assigned duties, the Justice Department would defend any and all lawsuits against the government, whether or not I was grossly negligent. If, however, my actions were so egregious as to be beyond anyone's definition of my assigned duties, I would be on my own. I held a firm and perhaps naive belief that I had done nothing wrong and that someone would eventually concur and release me from culpability. The private legal option just did not seem right to me, nor was it one that I could afford to pursue.

In November 1993, the first lawsuit against the U.S. government was filed: Two similar suits were filed in 1994. After our providing countless documents and educating many lawyers in the subtleties of viral hepatitis, clinical research, pharmacology, and toxicology, the drug sponsors reached out-of-court settlements with the plaintiffs and with others who threatened to sue. The tide of sentiments in the families ebbed and flowed, understandably, with each newspaper report and with each sequential investigation. One day, they were sympathetic and able to acknowledge that tragic accidents can befall research subjects. When the FDA reported that we had violated research regulations, they were far less understanding.

18. EPILOGUE

With the IOM report and settlement of the lawsuits, it was finally over. Caring for desperately ill patients, the countless meetings to elucidate the nature of FIAU's mitochondrial toxicity, the reviews of the studies, the conferences with lawyers, media inquiries, and more consumed more than one-third of my time during a two-year period. It can be argued that the FDA and congressional investigations focused on procedural details, whereas the NIH and IOM reviews sought the root causes of the episode. Whatever one's view of the process, my career and those of several valued friends and colleagues nearly ended. However, the FIAU deaths and their investigations had additional consequences about which every audience I have addressed on this matter has inquired. They fall into three major categories: how those events changed hepatitis drug development, how they affected changes in the conduct of clinical research, and how they have altered my own decisions and actions.

18.1. Drug Development

With the FIAU deaths, hepatitis drug development ceased for two years. It was revived by the recognition that lamivudine, a drug that inhibits the HIV retrovirus, is also a potent inhibitor of HBV.[54] Extensive studies and licensure of lamivudine for HIV/AIDS proved its overall safety. In late 1998, licensure of lamivudine for the prolonged treatment of hepatitis was recommended by an FDA advisory committee. These successes restored the optimism that safe and effective hepatitis drugs can be developed. Demonstration that combining lamivudine with pegylated interferon-α 2b is superior to lamivudine alone in subsets of patients with chronic hepatitis B infection was another important finding.[55] A number of exciting new drugs, such as tenofovir and adefovir, also have entered human studies.[56,57] These drugs are now being tested in woodchucks, the only animal model that proved to predict FIAU's fatal toxicity.[58] Moreover, all related drugs are now being subjected to *in vitro* assays for mitochondrial injury—assays not available in the early 1990s. Some vital lessons had been learned through the deaths of five FIAU recipients.

The FIAU tragedy affected more than just the development of drugs for hepatitis. It caused a rethinking of the entire drug development process and clinical research in general. To provide some sense of the scope of the deliberations that followed the FIAU study deaths, they are reviewed here according to the broad general questions raised by the episode.

18.1.1. Is Preclinical Testing of New Drugs a Reliable Predictor of Toxicity?

Those of us who administer a new drug to a human for the first time appreciate both the excitement and the tension inherent to the process. On the one hand, we realize the opportunity to do something truly novel; on the other hand, we do it without adequate knowledge of what may ensue.

Our decision to perform these first human studies depended both on the perceived need for the drug and on our projection of how safe it will be. As physicians confronted by sick patients, the need is fairly easy to ascertain. The problem comes in assessing drug safety.

Drugs that inhibit normal cellular pathways, that are toxic to cells in culture, and that injure animals may prove toxic in people as well; however, drugs that appear to have none of these *in vitro* and *in vivo* actions can still prove toxic for humans. No cell culture system or animal model completely emulates the distribution, metabolism, or effects of a drug on a living person. There will never be a substitute for doing studies on people.

On very careful review, it became abundantly clear that none of the preclinical data on FIAU had predicted the nature or the severity of its toxicity. The development of new *in vitro* assays for mitochondrial injury and adaptation of the woodchuck model for prolonged testing of new hepatitis drugs left us better prepared for the future, yet unpredictable outcomes in human trials will continue to be a harsh reality of drug development. All we can do at any point is to ensure that all reasonable steps are taken in preclinical testing to reduce the possibility of bad outcomes.

In addition, we have to acknowledge that the market potential of a new drug, the enthusiasm of investigators, and the clamor of desperate patients can blind us to the potential of a new product to do harm. Other than our own good sense, the best defense against this is the FDA.

18.1.2. Are Patients in Drug Trials Monitored Carefully and Objectively Enough?

One criticism of our conduct as investigators in the FIAU trials was that we had dismissed patient reports of fatigue and nausea, the early signs of metabolic injury. At the time, we rationalized these symptoms as being no worse than those provoked frequently by the then standard treatment for chronic hepatitis, recombinant interferon-α.

The issue, though, is that investigators would benefit from impartial oversight of their work. The question is how to do it. From their reviews of our studies, the FDA proposed sweeping new regulations in which investigators would be required to notify the FDA more frequently and more completely about adverse reactions. On the face of it, this could only benefit clinical research, yet the proposal met numerous objections, and the regulations were never approved.

Industry objected, of course, because all such regulations add to the already high costs of bringing a new product to market. We investigators objected because we already feel overburdened by the paperwork of clinical research.

In my opinion, the most insightful and balanced comments on these regulations were generated by the NIH Director's Advisory Committee and by IOM in their reviews of our studies. Both groups found merit in the intent of the proposed regulations but concluded that they would severely stifle drug development. In its extensive commentary on the matter, IOM proposed[59] that clinical trials need real-time monitoring. Traditional case record forms, they concluded, should be abandoned because they are not reviewed for months to years later. Electronic data entry would permit a more timely review of salient study events.

Although not named as such, IOM also proposed a broader use of data and safety monitoring boards (DSMBs) for routine oversight of studies that carry substantive risk. DSMBs are frequently created today for large cooperative trials. Their universal use, however, would strain the personnel and financial resources of institutions that host clinical trials. Nonetheless, in part as a result of the FIAU episode, all NIH-funded clinical studies are now required to develop and implement a monitoring plan to identify and act promptly upon evidence of unexpected or excessive adverse events.

A second, and important, issue addressed by IOM concerned patient follow-up. It is common in early phases of drug development to monitor study participants for only two to four weeks after they complete the treatment course. The FDA had proposed that all participants be monitored for three months after the treatment ended and that successive trials could not proceed until it was certain that there were no unexpected, delayed toxicities such as those seen with FIAU. IOM endorsed prolonged follow-up but cautioned how difficult it can be to interpret late events in small, early phase trials. The NIH Director's Advisory Committee went further on this issue, concluding that drug development would be slowed unacceptably if one had to complete a prolonged period of follow-up before the subsequent clinical trials could proceed. It would seem that a fair compromise would be to allow sequential trials to proceed while patients in the earlier studies are still under observation.

Finally, the FDA proposed changes to the requirements about reporting adverse reactions. Currently, investigators must report in a timely fashion all *unexpected* or *serious* adverse reactions. An unexpected reaction is one not predicted by the existing information on the drug; a serious reaction is life threatening, permanently disabling, requires instant hospitalization, or is a congenital anomaly, cancer, or overdose.[60]

The FDA proposed redefining the *serious adverse drug experience* as one that is fatal or life threatening, results in persistent or significant disability (incapacity; requires or prolongs hospitalization; necessitates medical or surgical intervention to preclude permanent impairment of a body function or permanent damage to a body structure; or is a congenital anomaly).[61]

Clearly, the FDA intent here is meritorious in requiring a full appraisal of a new drug's potential to do harm. IOM, however, appreciated that this new definition would result in a vast increase in safety reporting. It was "skeptical that the benefits from such added efforts will outweigh the risks."[62]

What is our goal in conducting clinical research? It is to advance medicine with the least risk to subjects. Any requirement for additional paperwork could so distract us as to cause us to neglect our subjects. These requirements would have a chilling effect on research, increasing the costs and exposing investigators to greater risks of noncompliance with intractable regulations.

It must be possible, though, to establish a mechanism by which independent clinicians could actively monitor ongoing studies by scrutinizing the evolving database. The current system of DSMBs is a step in this direction, but it requires periodic submission of cumulative study reports. It would be far better if designated monitors could access an otherwise secure computer file at will and render an opinion on the safety and progress of an ongoing study.

18.2. Clinical Research Training

Tragic outcomes of studies, like those with FIAU, have taught us that clinical research can no longer be considered a cottage industry of well-intended investigators who learn the craft at the sides of experienced senior mentors. It is a formidable undertaking that requires careful training. This book is evidence of institutional commitment to training clinical investigators. In addition, the NIH now funds a whole tier of grant support mechanisms for training and career development and loan repayment programs for clinical investigators (http://grants1.nih.gov./training/careerdevelopment awards.htm).

18.3. Personal Perspectives

Although these onerous FDA proposals were defeated, clinical investigators can be assured of progressively greater oversight and, despite this oversight, clinical research will remain a risky but rewarding undertaking both for the subject and for the investigator. I have continued to conduct clinical research in many disease areas. I am even more obsessive, though, about data management, if that is possible. My consent forms were already and remain still too long and defensive in my efforts to tell prospective subjects everything that could happen to them, and more. I also find myself to be far more cautious about drugs. In retrospect, I realize that I have not undertaken a single phase I drug study since 1991.

I consider myself lucky to have weathered the FIAU investigations, to have learned from them, and to remain vitally engaged in clinical research. The visibility of the NIH and the remarkable collaboration I had with truly distinguished colleagues caused the stakes in this tragedy to be very high for clinical research. I fear that, had I done these studies alone, things might have turned out differently. I find myself collaborating more now than ever before, being willing to sacrifice some independence for greater productivity and security.

Acknowledgments

I thank Drs. John I. Gallin and Ezekiel Emanuel for critical review of the manuscript and Ms. Brenda Rae Marshall for editorial assistance.

References

1. Raper SE, et al. A pilot study of in vivo liver-directed gene transfer with an adenoviral vector in partial ornithine transcarbamylase deficiency. Hum Gene Ther 2002;13:163–175.
2. Raper SE, et al. Fatal systemic inflammatory response syndrome in a ornithine transcarbamylased deficient patient following adenoviral gene transfer. Mol Genet Metab 2003;80:148–158.
3. Savulescu J. Harm, ethics committees and the gene therapy death. J Med Ethics 2001;27:148–150.
4. Citizens for Responsible Care and Research website, www.circare.org/foia3/ihgtdocs.htm.
5. Hacein-Bey-Abina S, et al. LMO2-associated clonal T cell proliferation in two patients after gene therapy for SCID-X1. Science 2003;302:415–419.
6. Federal Regulations for the Protection of Human Subjects. Title 45, Code of Federal Regulations, Part 46. Washington, DC, Office of the Federal Register, National Archives and Records Administration. Available at www.fda.gov.
7. Lopez C, Watamabe KA, Fox JJ. 2'-Fluoro-5-iodo-aracytosine, a potent and selective anti-herpes virus agent. Antimicrob Agents Chemother 1980;7:803.
8. Young CW, et al. Phase 1 evaluation of 2'-fluoro-5-iodo-1-p-D-arabinofuranosylcytosine in immunosuppressed patients with herpesvirus infection. Cancer Res 1983;43:5006.
9. Leyland-Jones B, et al. 2'-Fluoro-5-iodo-arabinosylcytosine, a new potent antiviral agent: Efficacy in immunosuppressed individuals with herpes zoster. J Infect Dis 1986;154:430.
10. Whitley RJ, Gnann JW. Acyclovir: A decade later. N Engl J Med 1993;308:1448–1453.
11. Holland GN, Gottlieb MS, Yee RD, Schanker HM, Pettit TH. Ocular disorders associated with a new severe acquired cellular immunodeficiency syndrome. Am J Ophthalmol 1982;93:393–402.
12. Crumpacker CS. Ganciclovir. N Engl J Med 1996;335:721–729.
13. Hantz O, Allaudeen HS, Ooka T, deClerq E, Trepo C. Inhibition of human and woodchuck hepatitis virus DNA polymerase by the triphosphates of acyclovir, 1-(2'deoxy-2'fluoro—β-D-arabinofuranosyl)-5-iodocytosine and E-5(2-bromovinyl)-2'deoxynridine. Antivir Res 1984;4:187.
14. Fourel I, et al. Inhibitory effects of 2'fluorinated arabinosylpyrimidine nucleosides on woodchuck hepatitis virus replication in chronically infected woodchucks. Antimicrob Agents Chemother 1990;4:473–475.
15. Dusheiko G, Hoofnagle JH. Hepatitis B. In McIntyre N, Benhamou JP, Bircher J, Rizzetto M, Rodes J (eds.) Oxford Textbook of Clinical Hepatology. New York, Oxford University Press, 1991.
16. Hoofnagle JH, et al. Randomized controlled trial of recombinant human interferon-α in patients with chronic hepatitis B. Gastroenterology 1988;95:1318–1325.
17. Perillo RP, et al. A randomized, controlled trial of interferon-α2b alone and after prednisone withdrawal for the treatment of chronic hepatitis B. N Engl J Med 1990;323:295–301.

18. Paar DP, *et al.* The effect of FIAU on chronic hepatitis B virus (HBV) infection in HTV-infected subjects (ACTG 122b). In *Program and Abstracts of the 32nd Interscience Conference on Antimicrobial Agents and Chemotherapy, Anaheim, Calif., October 10–14, 1992.* Washington, DC, American Society for Microbiology, 1992.

19. Fried MW, *et al.* FIAU, a new oral antiviral agent, profoundly inhibits HBV DNA in patients with chronic hepatitis B. *Hepatology* 1992;16:127A.

20. Committee to Review the Fialuridine (FIAU/FIAC) Clinical Trials, Division of Health Sciences Policy, Institute of Medicine. *Review of the Fialuridine (FIAU) Clinical Trials* (Manning FJ, Swartz M, eds.). Washington, DC, National Academy Press, 1995.

21. McKenzie R, *et al.* Hepatic failure and lactic acidosis due to fialuridine (FIAU), an investigational nucleoside analogue for chronic hepatitis B. *N Engl J Med* 1995;33:1099–1105.

22. Grem JL, King SA, Sorensen JM, Christian MC. Clinical use of thymidine as a rescue agent from methotrexate toxicity. *Invest New Drugs* 1991;9:281–290.

23. Seiter K, *et al.* Undine allows dose escalation of 5-fluorouracil when given with N-phosphoacetyl-L-aspartate, methotrexate, and leucovorin. *Cancer* 1993;71:1875–1881.

24. Sussman NL, *et al.* Reversal of fulminant hepatic failure using an extracorporeal liver assist device. *Hepatology* 1992;16:60–65.

25. Stevenson W, *et al.* Clinical course of four patients receiving the experimental antiviral agent fialuridine for the treatment of chronic hepatitis B infection. *Transplant Proc* 1995;27:1219–1221.

26. Cui L, Yoon S, Schinazi RF, Sommadossi J-P. Cellular and molecular events leading to mitochondrial toxicity of l-(2-deoxy-2-fluoro-l-*p*-D-arabinofuranosyl)-5-iodouracil in human liver cells. *J Clin Invest* 1995;95:555–563.

27. Klecker RW, Katki AG, Collins JM. Toxicity, metabolism, DNA incorporation with lack of repair, and lactate production for 1-(2'-fluoro-2'deoxy-β-D-arabinofuranosyl)-5-iodouracil in U-937 and MOLT-4 cells. *Mol Pharmacol* 1994;46:1204–1209.

28. Richardson FC, Engelhardt JA, Bowsher RR. Failuridine accumulates in DNA of dogs, monkeys, and rats following long-term oral administration. *Proc Natl Acad Sci USA* 1994;91:12003–12007.

29. Lewis W, Meyer RR, Simpson JF, Colacino JM, Perrino FW. Mammalian DNA polymerases a, p, and ft incorporate fialuridine (FIAU) monophosphate into DNA and are inhibited competitively by FIAU triphosphate. *Biochemistry* 1994;33:14620–1464.

30. Tennant BC, Gerin JL. The woodchuck model of hepatitis B virus infection. In Arias IM, *et al.* (eds.) *The Liver: Biology and Pathobiology,* 3rd ed. New York, Raven Press, 1994.

31. Tennant BC, *et al.* Antiviral activity and toxicity of fialuridine in the woodchuck model of hepatitis B virus infection. *Hepatology* 1998;28:179–191.

32. Chen CH, Vazquez-Padua M, Cheng Y-C. Effect of anti-human immunodeficiency virus nucleotide analogs on mitochondrial DNA and its implication for delayed toxicity. *Mol Pharmacol* 1991;9:625–628.

33. Lewis LD, Hamzeh FM, Lietman PS. Ultrastructural changes associated with reduced mitochondrial DNA and impaired mitochondrial function in presence of 2',3'-dideoxycytidine. *Antimicrob Agents Chemother* 1992;36:2061–2065.

34. Lewis W, Gonzalez B, Chomyn A, Papoian T. Zidovudine induces molecular, biochemical, and ultrastructural changes in rat skeletal muscle mitochondria. *J Clin Invest* 1992;89:1354–1360.

35. Arnaudo E, *et al.* Depletion of muscle mitochondrial DNA in AIDS patients with zidovudine-induced myopathy. *Lancet* 1991;337:508–510.

36. Freiman JP, Helfert KE, Hamrell MR, Stein DS. Hepatomegaly with severe steatosis in HTV-seropositive patients. *AIDS* 1993;7:379–385.

37. Gopinath R, Hutcheon M, Cheema-Dhadli S, Halperin M. Chronic lactic acidosis in a patient with acquired immunodeficiency syndrome and mitochondrial myopathy: Biochemical studies. *J Am Soc Nephrol* 1992;3:1212–1219.

38. Lai KK, Gang DL, Zawacki JK, Cooley TP. Fulminant hepatic failure associated with 2',3'-dideoxyinosine (ddl). *Ann Intern Med* 1991;115:283–284.

39. Bissuel F, *et al.* Fulminant hepatitis with severe lactate acidosis in HTV-infected patients on didanosine therapy. *J Intern Med* 1994;235:367–371.

40. Maxson CJ, Greenfield SM, Tuner JL. Acute pancreatitis as a common complication of 2',3'-dideoxyinosine therapy in the acquired immunodeficiency syndrome. *Am J Gastroenterol* 1992;87:708–713.

41. Broad W, Wade N. *Betrayers of the Truth.* New York, Simon & Schuster, 1982.

42. Kuhn TS. *The Structure of Scientific Revolutions,* 3rd ed. Chicago, University of Chicago Press, 1996.

43. Kevles DJ. *The Baltimore Case: A Trial of Politics, Science and Character.* New York, Norton, 1998.

44. Liaw Y-F, *et al.* Clinical and histological events preceding hepatitis B antigen seroconversion in chronic type B hepatitis. *Gastroenterology* 1983;84:216–219.

45. Weller FVD, *et al.* Randomised controlled trial of adenine arabinoside 5'-monophosphate (ARA-AMP) in chronic hepatitis B virus infection. *Gut* 1985;26:745–751.

46. Eli Lilly Company. Lilly suspends trials of hepatitis B drug; "Adverse" events cited [Press release]. *Wall Street Journal,* July 1 1993.

47. Cimons M. The deadly risks of research. *Los Angeles Times,* August 25, 1993.

48. Altman LK. Fatal drug trials raise question about "informed consent." *New York Times,* October 5, 1993.

49. Hilts PJ. After deaths, PDA is proposing suffer rules on drug experiments. *New York Times,* November 16, 1993, p. A1.

50. Schwartz J. And then the patients suddenly started dying: How NIH missed warning signs in drug test. *Washington Post,* September 7, 1993, p. A1.

51. Hilts PJ. PDA says 4 scientists broke rules in drug tests. *New York Times,* May 14, 1994, p. A8.

52. Schwartz J. Researchers cleared in drug trial deaths: NIH Advisory Panel's report on hepatitis studies at odds with finding by FDA. *Washington Post,* June 3, 1994.

53. Hilts PJ. Panel clears researchers in 5 deaths. *New York Times,* March 18, 1995, p. 6.

54. Lai C-L, *et al.* A one-year trial of lamivudine for chronic hepatitis B. *N Engl J Med* 1998;39:61–68.

55. Chan HL-Y, *et al.* A randomized, controlled trial of combination therapy for chronic hepatitis B: Comparing pegylated interferon-alpha 2b and lamivudine with lamivudine alone. *Ann Intern Med* 2005;142:240–250.

56. van Bömmel F, *et al.* Comparison of adefovir and tenofovir in the treatment of lamivudine-resistant hepatitis B virus infection. *Hepatology* 2004;40:1421–1425.

57. Karayiannis P. Hepatitis B virus: Old, new and future approaches to antiviral treatment. *J Antimicrob Chemother* 2003;51:761–785.

58. Jacob JR, *et al.* Suppression of lamivudine-resistant B-domain mutants by adefovir dipivoxil in the woodchuck hepatitis virus model. *Antiviral Res* 2004;63:115–121.

59. Committee to Review the Fialuridine (FIAU/FIAC) Clinical Trials, Division of Health Sciences Policy, Institute of Medicine.

Review of the Fialuridine (FIAU) Clinical Trials (Manning FJ, Swartz M, eds.). Washington, DC, National Academy Press, 1995.

60. Code of Federal Regulations 21, Parts 300 to 499. Washington, DC, Office of the Federal Register, National Archives and Records Administration, revised April 1, 1992.

61. Food and Drug Administration. Adverse experience reporting requirements for human drug and licensed biological products:

1995. Proposed rules; 21 CFR, Part 20. *Fed Reg* 1994;59: 54046–54064.

62. Committee to Review the Fialuridine (FIAU/FIAC) Clinical Trials, Division of Health Sciences Policy, Institute of Medicine. *Review of the Fialuridine (FIAU) Clinical Trials* (Manning FJ, Swartz M, eds.). Washington, DC, National Academy Press, 1995.

The Regulation of Drugs and Biological Products by the Food and Drug Administration

KATHRYN C. ZOON* AND ROBERT A. YETTER**

*Division of Intramural Research, National Institute of Allergy and Infectious Diseases, National Institutes of Health, Bethesda, Maryland

**Center for Biologics Evaluation and Research, Food and Drug Administration, Rockville, Maryland

1. INTRODUCTION

The quality and safety of medical products have been of major importance to the United States since the mid-1880s. It was then that the U.S. Congress passed the Drug Importation Act, which required for the first time the inspection and prevention of entry of adultered medicines from abroad. In 1902 and 1906, two laws were passed that form the foundation of the Food and Drug Administration (FDA)—the Biologics Control Act and the Food and Drug Act, respectively. Since that time, Congress has passed additional legislation enhancing FDA's ability to protect the public health. This chapter provides an overview of the FDA and the regulation of human drug and biological products.

2. BACKGROUND

Congress originally enacted the statutes that provide the authority for regulation of drugs and biologics to address significant public health problems. In 1901, during a diphtheria outbreak, several children were given a diphtheria antitoxin made in horses, the best treatment available at the time. Unfortunately, one of the horses used for production of the serum was infected with tetanus. Seven children who received that antitoxin died. The next year saw the passage of the Biologics Control Act of 1902 (Virus, Serum,

Antitoxin Act), which was designed to ensure the purity, potency, and safety of these and other biological products. In 1906, Upton Sinclair published *The Jungle*, an indictment of the meat packing industry. At the same time, Dr. Harvey Wiley, the chief chemist at the U.S. Department of Agriculture, was pointing out that toxic adulterants could be found in foods and medicines. This led President Theodore Roosevelt to sign the Food and Drug Act of 1906, which prohibited interstate commerce of adulterated foods, drinks, and drugs.

By 1933, the FDA had been established and recommended a complete revision of the now obsolete Food and Drug Act of 1906. The first bill was introduced into the Senate and a 5-year debate ensued. It is not clear how long that debate might have continued had it not been for the elixir sulfanilamide tragedy. Sulfanilamide was the most recent advance in medicine, able to destroy a variety of infectious agents. In an effort to make the drug easier to take, one company decided to create a liquid formulation, an elixir. Sulfanilamide, however, was not very soluble in water. Another solvent was found, a raspberry flavor was added and taste tested, and the new elixir sulfanilamide was put on the market in 1937. The new solvent, ethylene glycol, was toxic. Elixir sulfanilamide killed 107 people, mostly children. This led to the passage of the Federal Food Drug and Cosmetic Act (FFDCA) by Congress in 1938. The new FFDCA extended control from food and drugs to cosmetics and devices. It also required that new drugs be shown to be safe before they could be marketed and authorized inspections of factories engaged in the manufacture of regulated products.

Another tragedy, narrowly averted in the United States, led to further food and drug legislation. In 1962, it was found that a new sleeping pill, thalidomide, was responsible for severe birth defects in thousands of babies born in Western Europe. This finding, and reports of the role of Dr. Frances Kelsey, an FDA medical officer, in keeping the drug off the market in the United States created public support for stronger drug regulations. The result was the passage of the Kefauver–Harris amendments in that year to strengthen the drug approval process. For the first time, drug manufacturers were required to prove the effectiveness of a product before it could be marketed.

In 1971, the Bureau of Radiological Health was transferred to the FDA. Its mission was to protect the public from unnecessary radiation from electronic products in the home and the healing arts. In the same year, the National Center for Toxicological Research was established to examine the biological effects of chemicals in the environment. The next year, the Division of Biological Standards, which was responsible for the regulation of biological products, was transferred from the National Institutes of Health to the FDA to become the Bureau of Biologics. The FDA as we know it today was taking shape. Most recently, the Food and Drug Administration Modernization Act of 1997 was signed into law. This law reauthorized the user fee program and codified a number of FDA initiatives intended to speed the availability of new drugs for serious and life-threatening diseases.

3. MISSION, ORGANIZATION, AND TERMINOLOGY

The mission of the FDA is to protect and enhance the public health through the regulation of medical products and food. The scope of its mission is outlined in Table 9-1. The structure of the FDA is shown in Table 9-2. The commissioner of the FDA is nominated by the President and confirmed by the Senate. There are six product-specific centers, the Office of Regulatory Affairs, and a number of smaller offices (e.g., the Office of Orphan Products Development and the Office of Combination Products). In some instances, there is an overlap in the definition of a drug, biological product, or device. In other situations, a product may be a combination product—for example a drug and a biologic, a biologic and a device, or a device and a drug. In each of these cases, the regulation of the combination is

TABLE 9-1   FDA's Mission

1. To promote the public health by promptly and efficiently reviewing clinical research and taking appropriate action on the marketing of regulated products in a timely manner;
2. With respect to such products, protect the public health by ensuring that foods are safe, wholesome, sanitary, and properly labeled; human and veterinary drugs are safe and effective; there is reasonable assurance of the safety and effectiveness of devices intended for human use; cosmetics are safe and properly labeled, and public health and safety are protected from electronic product radiation;
3. Participate through appropriate processes with representatives of other countries to reduce the burden of regulation, harmonize regulatory requirements, and achieve appropriate reciprocal arrangements; and
4. As determined to be appropriate by the Secretary, carry out paragraphs (1) through (3) in consultation with experts in science, medicine, and public health, and in cooperation with consumers, users, manufacturers, importers, packers, distributors, and retailers of regulated products.

From the FDA Modernization Act of 1997 (PL105-115).

TABLE 9-2   Structure of the Food and Drug Administration

Office of the Commissioner
Center for Biologics Evaluation and Research (CBER)
Center for Devices and Radiological Health (CDRH)
Center for Drug Evaluation and Research (CDER)
Center for Food Safety and Applied Nutrition (CFSAN)
Center for Veterinary Medicine (CVM)
National Center for Toxicological Research (NCTR)
Office of Orphan Products Development (OOPD)
Office of Combination Products (OCP)
Office of Regulatory Affairs (ORA)

clarified by FDA intercenter agreements. For products that do not fall clearly under the jurisdiction of one center or another by definition or agreement, there is an FDA process to determine the appropriate regulatory approach for the product that is generally based on primary mode of action.

The regulation of drug and biological products is based on sound science, law, and public health impact. The FDA is composed of scientists of many disciplines, including physicians, biologists, chemists, pharmacologists, microbiologists, statisticians, consumer safety officers, and epidemiologists. The FDA is responsible for the review of regulatory submissions (e.g., applications for clinical research and marketing, and labeling), the development and implementation of regulatory policy, research and scientific exchange, product surveillance (e.g., adverse event reporting and product testing), and compliance (e.g., education, inspections, and enforcement actions). As a science-based institution, the FDA strives to facilitate the development of new safe and effective medical products while ensuring the safety of the products and their uses.

The primary set of laws that governs the regulation of drug and biological products is shown in Table 9-3. Some important regulations for drugs, biologics, and medical devices in Title 21, Code of Federal Regulations (CFR), are shown in Table 9-4. These laws and regulations are intended to protect the public health. One of the FDA's primary functions is to ensure compliance with these laws and regulations. The definitions and explanations of some of the terms used in this chapter's discussion of the FDA's regulation of drugs and biological products are provided in Table 9-5.

Another important role of the FDA is communication. This information often focuses on the quality, safety, and efficacy of medical products. The FDA is one of several entities in a broader risk-management network designed to provide accurate information to health care professionals and the public on product quality, effectiveness, and safety (predominantly in the form of accurate labeling and promotion/advertising and compliance with good manufacturing practice). The FDA website (www.fda.gov) is an extremely

TABLE 9-3   Statutory Authorities

	Drugs	Biologics
Federal Food Drug and Cosmetic Act	✓	✓
Public Health Service Act		✓
Interstate Commerce	✓	✓
Foreign Commerce		✓
Component Jurisdiction	✓	✓
Generic Equivalence	✓	
Prescription Drug User Fee Act	✓	✓
Prescription Drug Marketing Act	✓	✓
FDA Modernization Act of 1997	✓	✓

TABLE 9-4   Principal Regulations for Biological Products and Drugs: Title 21, Code of Federal Regulations

Part 312	Investigational New Drugs
Part 3	Definition of Primary Mode of Action of a Combination Product
Part 50	Protection of Human Subjects
Part 56	Institutional Review Boards
Part 58	Good Laboratory Practices for Non-Clinical Laboratory Studies
Part 314	New Drug Applications
Parts 600–680	Biologics
Part 54	Financial Disclosure by Clinical Investigators
Part 25	Environmental Impact Considerations
Parts 201 & 202	Labeling and Advertising
Parts 210 & 211	Current Good Manufacturing Practices
Parts 800–861	Devices and *In Vitro* Diagnostics
Parts 1270 & 1271	Human Tissues

TABLE 9-5   Definitions and Terms

Law	A statute. An act of Congress that outlines binding conduct or practice in the community.
Regulation	A rule issued by an agency under a law administered by the agency. A regulation interprets a law and has the force of law.
Code of Federal Regulations (CFR)	The compilation of all effective government regulations published annually by the U.S. Printing Office. FDA's regulations are found in Title 21 of the CFR.
Guidance	FDA documents prepared for FDA staff, applicants/sponsors, and the public that describe the agency's interpretation of, or policy on, a regulatory issue. Guidance documents are not legally binding.
Biological	A virus, therapeutic serum, toxin, antitoxin, vaccine, blood, blood component or allergenic product, or analogous product, or arsphenamine or derivative of arsphenamine applicable to the prevention, treatment, or cure of a disease or condition of human beings. This includes immunoglobulins, cytokines, and a variety of other biotechnology-derived products.
Drug	An article intended for use in the diagnosis, cure, mitigation, treatment, or prevention of disease in man or other animals; an article recognized in the United States Pharmacopoeia, the official Homeopathic Pharmacopoeia, or the official National Formulary and their supplements; an article (other than food) intended to affect the structure or any function of the body of man or other animals.
Device	An instrument, apparatus, implement, machine, contrivance, implant, *in vitro* reagent, which is intended for use in the diagnosis of disease or other conditions, or in the cure, mitigation, treatment, or prevention of disease in man or other animals; or is intended to affect the structure or any function of the body of man or other animals; and does not achieve its primary intended purpose through chemical action within or on the body of man or other animals and is not dependent on being metabolized for the achievement of its primary intended purpose.
Investigational New Drug Application	A request for investigational exemption from the approval requirements for new drugs and biologics.
Accelerated Approval	An FDA approval based on a surrogate end point that is reasonably likely to predict clinical benefit or clinical effects that are not the desired ultimate benefit but are reasonably likely to predict such benefit.

valuable tool to access information. Among the documents available on the website are regulations, guidelines, and guidance documents. Guidance documents represent the agency's current thinking regarding a particular issue or product. These documents also greatly facilitate the understanding of laws, administrative directives, and the FDA regulations and policies. Guidances are not binding and are updated regularly to provide accurate and timely information.

The FDA also performs research regarding the products it regulates. Some examples of this research include the establishment of standards and methods, toxicology, product safety, and basic mechanisms of actions or pathogenesis. This research is important for quality review of submissions, development of new policy and guidance, providing advice on product development, and product safety.

4. DRUG AND BIOLOGIC LIFE CYCLE

The life cycle for new drug and biologic products is shown in Figure 9-1. The process is divided into four stages: discovery/preclinical investigation, clinical trials, marketing application, and post approval.

4.1. Preinvestigational New Drug Studies

The earliest stage of product development involves the discovery and initial evaluation of the product. This process can take from 1 to 3 years. In this period, the product is discovered; a production process is established that yields a consistent quality, clinical-grade material; and the product is adequately characterized. Tests and assays to characterize the product should be under development in this stage since they will be necessary to link the product to the outcome of animal or human clinical trials. In addition, at this time the sponsor conducts animal safety studies to determine an appropriate starting dose in humans and to establish the toxicity profile of the product. These studies will assist in designing the clinical trial to ensure that the human participants are properly monitored for potential adverse events. This is the stage in which the biological rationale for the use of the product is proposed. If an animal efficacy model exists, studies in that model should also be performed to support the use of the product in humans. Often, sponsors will call or meet with the agency at the end of this development stage to discuss their data and their future plans prior to submission of their investigational new drug (IND) application. This meeting is referred to as a pre-IND

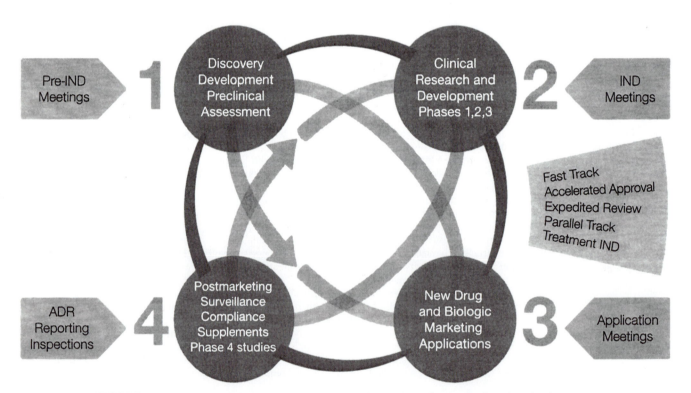

FIGURE 9-1 Biological and drug product life cycle. This figure shows the phases of product development, review and approval, and postmarketing activities. The policies listed in the diagram between 2 and 3 indicate mechanisms available to expedite product development.

meeting and in general is very important in facilitating a successful IND. The FDA has developed a number of guidance documents on considerations in product development and preclinical animal studies to help sponsors develop the necessary information.

4.2. Investigational New Drug Studies (IND)

The FFDCA and the Public Health Service Act require that a new drug or biological product be approved before it can enter interstate commerce. Under its rulemaking authority, the FDA issued regulations found in 21 CFR Part 312 allowing an exemption from the approval requirement for INDs and biologics for which an IND is in effect. These regulations allow investigational products to be legally shipped in order to conduct clinical investigations. The regulations in Part 312 are intended to address two aims: the protection of human subjects from unreasonable research risks and the development of reliable data to support the approval of the product. The duration of this phase of product development is generally from two to ten years, depending on the nature of the product and the intended clinical use. In addition, a number of guidance documents are applicable to the conduct of clinical trials. *The International Conference on Harmonisation (ICH) Harmonised Tripartite Guideline: Guideline for Good Clinical Practice* is a particularly useful reference document on investigational clinical trials that is accepted by the United States, the European Union, and Japan. The term *good clinical practice* (GCP) refers to the design, conduct, recording, evaluating, monitoring, and reporting of clinical trials. The principles of GCP are provided in Table 9-6.

During the clinical development of a product under an IND, additional product process development and testing/validation are performed. Also, additional preclinical information is obtained regarding the safety and efficacy of the product. If certain changes are made to the product, the preclinical studies, or the clinical protocols, FDA regulations require the sponsors to submit an amendment to the IND. These include changes that affect the safety, scope, and scientific quality of the clinical protocol, including its data and analyses, or the addition of a new protocol.

Regarding the clinical development of a product, there are generally three phases of premarketing clinical research to examine the safety and efficacy of a drug or biological product. The first phase (phase I studies) consists of small, dose escalation studies that can include either patients with a particular disease or condition or normal volunteers with the primary goal to assess safety of the product using a particular route of administration. In addition, some phase I studies may examine pharmacokinetics and drug metabolism. It is possible to request an end of phase I meeting to discuss data and the drug development plan. Phase II

TABLE 9-6 Principles of Good Clinical Practice

- Clinical trials should be conducted in accordance with the ethical principles that have their origin in the Declaration of Helsinki, and that are consistent with good clinical practice and the applicable regulatory requirement(s).
- Before a trial is initiated, foreseeable risks and inconveniences should be weighed against the anticipated benefit for the individual trial subject and society. A trial should be initiated and continued only if the anticipated benefits justify the risks.
- The rights, safety, and well-being of the trial subjects are the most important considerations and should prevail over the interests of science and society.
- The available nonclinical and clinical information on an investigational product should be adequate to support the proposed clinical trial.
- Clinical trials should be scientifically sound and described in a clear, detailed protocol.
- A trial should be conducted in compliance with the protocol that has received prior institutional review board/independent ethics committee approval/favorable opinion.
- The medical care given to, and medical decisions made on behalf of, subjects should always be the responsibility of a qualified physician or, when appropriate, of a qualified dentist.
- Each individual involved in conducting a trial should be qualified by education, training, and experience to perform his or her respective tasks.
- Freely given informed consent should be obtained from every subject prior to clinical trial participation.
- All clinical trial information should be recorded, handled, and stored in a way that allows its accurate reporting, interpretation, and verification.
- The confidentiality of records that could identify subjects should be protected, respecting the privacy and confidentiality rules in accordance with the applicable regulatory requirements.
- Investigational products should be manufactured, handled, and stored in accordance with applicable good manufacturing practice. They should be used in accordance with the approved protocol.
- Systems with procedures that assure the quality of every aspect of the trial should be implemented.

From *ICH Harmonised Tripartite Guideline: Guideline for Good Clinical Practice*, Step 4, Secretariat c/o IFPMA, Geneva, Switzerland, 1996.

studies consist of one or more moderate size clinical studies for a particular patient population. The primary goal of these studies is to provide preliminary evidence of efficacy and dosing and supplementary data on safety. Generally, the sponsor meets with the FDA at the end of phase II studies to discuss the outcomes of the studies and the design and analysis plan for the final phase of clinical development. Trials in the last phase of clinical development, phase III, are generally larger studies and are designed to evaluate the risk and benefit of a product in a particular patient population for a defined clinical indication. The safety and efficacy data from these studies are generated to support marketing approval and to provide information to write the instructions for the use of the product for a particular indication. Some key issues for the design, conduct, and analysis of clinical trials include end points, study population, randomization, stratification, blinding, sample size, participant adherence, and study analysis. Information gathered during the conduct of these clinical trials may affect product production and specifications, raise additional preclinical issues, and sometimes warrant additional clinical studies. Following completion of the phase III or pivotal studies, the sponsor again meets with the FDA to discuss a marketing application submission. At any stage in the clinical development of the product, issues or changes may arise that require additional product development work, preclinical studies, or additional clinical data (see Fig. 9-1).

The content and format of the IND application is specified in 21 CFR 312.23. The IND application should include, for example, a table of contents; introductory statement including the biological rationale and general investigational plan; chemistry, manufacturing, and control (CMC) information; pharmacology and toxicology information; previous human experience and other relevant information protocols; and investigator's brochure.

Once the original IND is submitted, the FDA has 30 days to review and notify the submitter or sponsor whether or not the trial has been placed on clinical hold. The initial review is aimed primarily at an evaluation of the safety of the product for human clinical trials. During those 30 days, the sponsor may not initiate the clinical trial. In this time frame, if the agency has no safety concerns regarding the study or does not hear from the FDA, then the IND is allowed to proceed. However, if the FDA has concerns about the IND, it may be placed on clinical hold. A clinical hold notice is issued to notify the sponsor that the clinical trial(s) may not begin until certain stated deficiencies are resolved. Phase I studies may be placed on clinical hold for any of five reasons:

Human subjects would be exposed to unreasonable risk of illness or injury.

There is insufficient information to assess the risk to subjects.

The investigator's brochure is inadequate.

The clinical investigators are not qualified to carry out the study.

The study of a life-threatening disease excludes men and women with reproductive potential (see Chapter 12).

Phase II and III studies may be put on hold for any of the previously discussed reasons. They may also be placed on hold if the study design is inadequate to achieve the stated purpose of the study. If the IND is placed on clinical hold, the sponsor is notified immediately by telephone. This notification is followed with a letter that specifically states the deficiencies. Advice is available on appropriate corrective actions. It is then up to the sponsor to correct the deficiencies and notify the FDA of the corrections in a clinical hold response letter. Once the sponsor submits a complete response to the clinical hold, the FDA then has another 30 days to review the information in the clinical hold response letter. There is no automatic release from clinical hold. In this case, if the sponsor does not hear from the FDA in 30 days the clinical trial may not start. When the review is finished, the sponsor is notified that the trial(s) may proceed or that there are continuing deficiencies.

In addition, FDA regulations require the sponsor to file an IND amendment if major changes are made to the product or the clinical protocol. These include changes in product formulation and changes in protocol that affect safety, scientific quality, or scope of the clinical trials. The sponsor must also file an annual report that includes all changes in and results of the study.

Several mechanisms are available that accelerate the drug development process, such as expedited review for severe and life-threatening illnesses (21 CFR 312 Subpart E); accelerated approval (21 CFR 314.510 and 601.41), and fast track development programs. Subpart E describes procedures to expedite the development, evaluation, and marketing of new therapies intended to treat persons with life-threatening and severely debilitating illnesses who do not have acceptable alternatives. These provisions include early consultations, submissions of treatment protocols, and risk–benefit analysis considerations for review of marketing applications. These provisions have been included and expanded in the fast track program. The Fast Track Guidance was originally developed in 1998 and was revised in 2004. The purpose of this program is to facilitate the development of new drug and biological projects and to expedite the review of new drugs and biologics that are intended to treat serious and life-threatening conditions and that demonstrate the potential to address unmet medical needs. The guidance describes the qualifications for serious and life-threatening conditions and the potential to address unmet needs, the process of designation, and the programs for expediting the development and review of new drugs and biologics. Accelerated approval (21 CFR 314.510 and 601.41) is an FDA approval based on a surrogate end point that is reasonably likely to predict clinical benefit or clinical effects that are not the desired benefit but are reasonably likely to predict such benefit. If a product is approved by accelerated approval and made commercially available the sponsor must conduct a phase IV (postmarketing) study(s) to show clinical benefit (i.e., validate the surrogate end point).

There are also a number of expanded access programs that are available, when appropriate, under IND, including "parallel track" and treatment IND (21 CFR 312.34 and 312.35). The parallel track policy developed by the U.S. Public Health Service was in response to the AIDS epidemic to permit wider availability of experimental agents. Under this policy, patients with AIDS whose condition prevents them from participating in controlled clinical trials can receive investigational drugs shown in preliminary studies to be potentially useful. It can also be used for other clinical conditions when appropriate. The treatment IND (21 CFR 312.34 and 312.35) is for a drug or biologic that is not approved for marketing but is made available for clinical investigation for a serious or immediately life-threatening disease condition in patients for whom no comparable or satisfactory alternative therapy is available. It is generally made available under a treatment protocol when the drug or biologic is being studied in phase III investigation or all clinical trials have been completed; however, it can be made available earlier if appropriate.

4.2.1. Responsibilities and Documentation

4.2.1.1. Sponsors

Several groups have responsibilities in clinical research, including the sponsors, investigators, institutional review boards (IRBs), and the FDA, that are described in the regulations and guidances. The responsibilities of the sponsor are found in subpart D of Title 21 CFR Part 312. The sponsor, generally the developer of the product, is the person or entity who submits the IND. The sponsor is responsible for selecting qualified investigators and providing them the necessary information to conduct the study properly.

The sponsor is also responsible for the trial design, the trial management, data handling and record keeping, allocation of responsibilities, compensation to subjects and investigators, financing, and notification/submission to regulatory authorities (e.g., protocol submission). In addition, the sponsor is required to ensure that there is proper monitoring of the study and that it is in accordance with the general investigational plan. The sponsor must maintain an effective IND and ensure that all participating investigators and the FDA are promptly informed of significant adverse events or risks associated with the product. The sponsor is also responsible for the quality assurance and quality control of the trial. Finally, the sponsor is accountable for maintaining and making available, as necessary, the information on the investigational product, including the manufacture of the product, supplying and handling the investigational product, record access, and safety information. A sponsor may use a contract research organization to conduct some of the activities; however, the sponsor is ultimately responsible for the quality and integrity of the trial.

4.2.1.2. Investigators

Investigators must be appropriately qualified by training and experience to conduct clinical research. They have multiple responsibilities, including following the protocol for the study and complying with all applicable regulations. It is their responsibility to protect the rights, safety, and welfare of subjects in their care. As part of the responsibility for protection of human subjects, an investigator must not involve a human being as a subject in research unless the investigator has obtained the subject's legally effective informed consent. In doing so, the investigator must assure that there is sufficient opportunity for the subject to consider whether or not to participate. The explanation of the study must be in language that the subject can understand and presented in a manner that minimizes the possibility of coercion or undue influence. The consent form must not contain exculpatory language or statements intended to waive the subject's legal rights. The investigators must retain control of the investigational product and maintain records of the disposition of the product, records, and reports (e.g., progress and final reports, safety reports), case histories of the subjects, and termination or suspension of the trial. They are also required to report adverse events observed to the sponsor and the IRB. Investigators must also arrange for review of the IND protocols by the IRB and other communications with the IRB. In addition, because of concerns of potential bias, they are required to supply sponsors with sufficient accurate financial information to allow the sponsor to report on financial interest to the FDA. It is important to realize that repeated or willful failure to comply with the regulations could result in disqualification of the investigator. In that case, the investigator may no longer receive investigational products. The FDA reviews any marketing application that relies on data from studies performed by the disqualified investigator. If the FDA determines that the data submitted by that investigator are unreliable and crucial to the approval, the approval for that product may be delayed pending resolution of the concerns or withdrawn.

If the sponsor and the investigator are the same individual or entity, then all of the responsibilities of the sponsor and investigator must be carried out by that individual or entity with appropriate safeguards or contracting arrangements to ensure the integrity of the trial and human subject safety.

4.2.2. Clinical Protocol

The clinical trial protocol and its amendments are critical elements of clinical research. The protocol should include general information, such as title, number, names of sponsors, medical experts, and investigators, and background information. The background information should include the name and description of the investigational product, nonclinical studies that impact on the clinical trial, the population to be studied, known or possible risks and benefits to human subjects, and administrative information. The protocol should state the objectives and purpose of the trial, the trial design, the selection and withdrawal of subjects, the treatment of subjects, the assessment of efficacy/activity (where appropriate) and safety, and the statistical evaluation plan (where appropriate). It should also address the plan for quality control, monitoring and assurance, data handling, record keeping, and ethical considerations. A more detailed treatment of this subject may be found in Chapter 24.

4.2.3. IRB

The constitution and responsibilities of the IRB are covered by the regulations in Part 56 of Title 21 of the CFR. The IRB is charged with reviewing and approving protocols that are to be carried out in the organization(s) that it serves. It is the IRB's function to ensure that in each protocol the risks to human subjects are minimized and reasonable. IRBs must assure that the selection of subjects is equitable and that informed consent is sought and adequately documented. The regulations specify that the IRB have at least five members with varying backgrounds to promote complete review of research activities at the institution(s). The IRB must have at least one member

whose primary concerns are scientific, another whose primary concerns are nonscientific, and at least one member not otherwise affiliated with the institution. Chapter 5 provides a detailed explanation of the structure and function of the IRB.

4.2.4. FDA

The FDA reviews all INDs and their amendments to determine whether they are in compliance with the appropriate laws and regulations. The regulations establish time frames for the performance for certain reviews and lay out the responsibilities of the FDA in communicating with the sponsors. The primary purpose of the review of the original IND submission and early amendments is to help assure that human subjects are not exposed to unreasonable risk. In the later phases of the IND process, involving studies to support efficacy, FDA review also focuses on whether the studies are constructed and carried out in a way that will yield valid data that can be considered for marketing approval. The FDA also interacts with sponsors through meetings and conference calls, starting at the pre-IND stage and continuing throughout the entire IND process, to address important product development, clinical study design and analysis, and premarket submission issues.

4.2.5. Investigator's Brochure

If the sponsor is not the investigator, there must be an investigator's brochure (IB). It is the sponsor's responsibility to maintain and update the IB and give it to the investigators who are conducting the trial. This document generally includes information regarding the clinical and nonclinical data on the investigational product that are relevant to the use of the product in human subjects.

4.2.6. IND Safety Reports

Sponsors should submit IND safety reports to the FDA as described in 21 CFR 312.32 and 312.33. The reporting requirements for adverse events include expedited reports that consist of written reports and telephone or facsimile reports and annual reports or information amendments. The regulations governing written reports are found in CFR 312.32(c). These include any adverse event associated with the use of the study drug/biologic that is both serious and unexpected or any findings from tests in laboratory animals that suggest a significant risk for human subjects, including reports of mutagenicity, teratogenicity, or carcinogenicity. A serious adverse drug experience is one that results in any of the following outcomes:

death, a life-threatening adverse drug experience, inpatient hospitalization or prolongation of existing hospitalization, a persistent or significant disability/incapacity, or a congenital anomaly/birth defect. A life-threatening adverse drug experience is one that places the subject, in the view of the investigator, at immediate risk of death from the reaction as it occurred. The sponsor must notify the FDA and all participating investigators as soon as possible, but no later than 15 calendar days, after receipt of the information for serious adverse drug experiences. The sponsor shall also notify the FDA by telephone or facsimile of any unexpected fatal or life-threatening adverse experience associated with the use of the drug as soon as possible but in no event later than seven calendar days after the sponsor's initial receipt of the information.

4.3. Marketing Approval/Licensure

Section 351 of the Public Health Service Act requires that a biologics license be in effect for any biological product that is to be introduced into interstate commerce. The FFDCA requires approval of a marketing application [New Drug Application (NDA)] for new drugs. The provisions of the IND regulations allow interstate transportation of drugs, including biologics, for clinical investigations. These investigations are intended to provide data to support a Biologics License Application (BLA) or an NDA. Although there are some slight differences in the way these two types of marketing applications are handled, they are similar enough that we will use the BLA as the example for the development and submission and review of a marketing application for a drug or biologic. In either case, the marketing application process actually begins during the IND phase. The review(s) and response(s) phase of a marketing application ranges from two months to three years.

4.3.1. Pre-submission

Although the IND phase is primarily directed at the collection of clinical data, during this time much of the CMC information needed for a marketing application is also being developed. The formulation to be marketed should be identified and used for the pivotal clinical trials. The product must be adequately characterized and its stability demonstrated. Consistency of manufacture must also be proven. Although the specific approaches to the development of these data vary with the product area, there are a number of guidance documents available that provide insight into what information is important and how the information might be generated.

During the pre-IND and IND stages, it is important that the potential applicant remain in contact with the FDA. It is far easier to address concerns, including both clinical trial and CMC issues, before the clinical protocol is under way. It is in the best interest of both the FDA and the sponsor to work out these details so that when the time comes for a marketing application to be submitted, there are no unexpected problems.

After the sponsor compiles sufficient information, the sponsor will begin to plan the submission of the BLA. The FDA urges the sponsor to have a pre-BLA meeting well in advance of any planned BLA submission. This meeting provides a forum for discussing the content, format, and timing of the proposed submission. This discussion is particularly important for electronic submissions. Through proper communications, most of the problems associated with BLA filings can be avoided.

While the sponsor is preparing to submit a BLA or NDA, the FDA is preparing to review it. A review committee is formed and preliminary decisions concerning the handling of the submission are made. One of the first decisions is whether the review of the product should be handled under a standard schedule or as a priority. The standard and priority review schedules are based on goals agreed to and in conjunction with the Prescription Drug User Fee Act. Currently, the standard schedule requires a complete review in ten months, whereas a priority review is to be completed in six months from the receipt of the BLA or NDA. The review schedule decision is based on the use of the product (for severe or life-threatening illnesses) and whether it fills an unmet medical need.

At this time, the committee will also decide which clinical study sites should be inspected and requests a bioresearch monitoring inspection. This inspection is focused on the verification of the data that are submitted to the FDA. The field investigators will help determine whether the studies were carried out according to regulations and appropriate informed consent was obtained. They also review the record keeping for adequacy and to determine whether protocols were followed. The report of the bioresearch monitoring inspection is a key piece of the review of a BLA or NDA.

4.3.2. The Application

The regulations prescribing the content of a BLA may be found in 21 CFR 601.2 and those for the NDA in 21 CFR 314. The BLA/NDA must contain a signed cover sheet, the Form FDA 356h, which provides information that enables the center to identify the type of submission, the applicant, and the reason for the submission. Because the FDA routinely receives thousands of submissions annually, this form is extremely important. The bulk of the BLA/NDA submission generally consists of preclinical and clinical study reports that the applicant believes provide data supporting the safety and efficacy of the product. The applicant must also submit the proposed labeling for the product, which must be supported by the data.

The BLA/NDA also must contain adequate CMC information to ensure that the product meets standards of purity and potency/efficacy. These data will include information on characterization, stability, the manufacturing process, and the facility (in a BLA) in which the manufacturing is carried out. There are a number of guidance documents available that outline the types of information that are needed for specific product areas. Although most of these documents focus on what to submit, they also provide guidance on how to develop the information needed. In some cases, there are also documents that identify key concerns associated with product classes or manufacturing processes. Although they are not submitted with the original application, the BLA often includes samples of the product for testing by the Center for Biologics Evaluation and Research (CBER). CBER will request these samples during the review process when the battery of product release tests has been determined. The applicant will submit the requested samples of the product and the results of their release tests for confirmation by CBER.

In the BLA/NDA, applicants must include a statement that the nonclinical studies used to support the application were conducted in compliance with regulations on good laboratory practice for nonclinical laboratory studies (Part 58 of Title 21 CFR). If the studies were not conducted according to good laboratory practices, the applicant must explain why they were not. The applicant must certify that all clinical studies were conducted in accordance with the regulations in Parts 50 and 56 of Title 21 of the CFR, which cover informed consent and IRBs. In addition, Part 54 of Title 21 of the CFR requires the submission of a financial certification or disclosure statement or both for clinical investigators who conducted clinical studies submitted in the application.

Every BLA/NDA also must include either a claim of categorical exclusion or an environmental assessment. Under current regulations, most drug and biologic marketing applications are categorically excluded from the need to supply an environmental assessment. However, there are certain categories of products and processes that still require such an assessment. Sponsors should become aware of the need for an assessment during the IND process.

4.3.3. The Review

The receipt of the BLA/NDA at the FDA starts the "review clock." The applications division in the office with product responsibility logs the submission in and routes it to the review committee. The review committee consists of the experts necessary to conduct a review of the submission. Generally, the committee contains specialists in clinical and preclinical data review, product area specialists, specialists in good manufacturing processes, biostatisticians, and a regulatory project manager. Reviewers in other specialty areas are added to the review team as necessary.

The initial review of the BLA/NDA focuses on the suitability of the application for filing. If the application is significantly deficient—that is, it lacks information necessary to conduct a substantive review—the committee may refuse to file it. A "refuse to file" action terminates the review of that application. Although an applicant may elect to file over protest, the refuse to file action indicates a severely deficient submission that is unlikely to lead to an approval in the first review cycle. If the BLA/NDA is complete, the committee files it and the substantive review of the application begins in earnest.

It is not uncommon for questions to arise during the review. If these questions are likely to be readily resolved, CBER/the Center for Drug Evaluation and Research (CDER) raise them in an "information request." The information request may be made by telephone conversation or letter. The responses to these are expected to be short and to facilitate the review. As each discipline finishes its particular review, it prepares a review memo documenting what has been reviewed and any deficiencies that have been found.

Inspections are part of the complete review of a BLA/NDA. One of these is the bioresearch monitoring inspection mentioned previously. This inspection helps provide assurance that the review committee can rely on the clinical data submitted to support the safety and efficacy of the product. The other inspection is a facility inspection in which product specialists and specialists in good manufacturing practice visit the manufacturing facilities. This inspection is aimed at assessing whether the product is made under appropriate conditions and the process for manufacture has been validated and is being followed. All aspects of the manufacture of the product are investigated during this inspection. The applicant is made aware of any significant observations at the end of the inspection. The inspectors complete an inspection report, which becomes part of the review of the application.

CBER/CDER often present issues raised in the review of the application to an advisory committee.

The use of an advisory committee allows the review committee to bring specific questions or concerns to a broader forum of experts. For specific questions, experts in a particular area of concern may be appointed to the committee to provide the best scientific advice available. Not all BLAs/NDAs are presented at an advisory committee. A BLA/NDA may be presented if the product is novel, perhaps the first in its class, or if the review committee has identified particular issues on which they need expert input.

The review of the proposed labeling for the product is a critical part of the review process. Every statement made in the labeling has to be supported by data. The ultimate goal of the review of the proposed labeling is to determine that it clearly identifies the product and provides adequate information to allow the safe and appropriate use of the product. The package insert must include all of the necessary information that will allow the clinician to make the correct decision on the use of the product. Patient labeling, when included, must be both clear and accurate so that the patient will understand how to use it properly. The review committee will work with the applicant to obtain accurate and informative labeling.

After the inspection reports are received, the reviews completed, and any advisory committee advice is considered, the review team makes a recommendation on the BLA/NDA and the center decides on the appropriate action. If the application is approved, the FDA issues a letter that serves as a license (BLA) or an approval (NDA), allowing the applicant to introduce the product into interstate commerce. If the review, including the inspections, has resulted in questions or concerns, the FDA issues a "complete response letter." This letter explains that the application cannot be approved and identifies all of the deficiencies that must be addressed to put the application in condition for approval. When the applicant responds to this letter, the review clock and the review begin again.

In summary, the approval of biological or drug product is based on its purity, potency, safety, and efficacy. In addition, the applicant must be in compliance with good manufacturing practice.

4.4. Postmarketing Surveillance

Following marketing approval, the FDA is responsible for the review of changes to the NDA or BLA, including manufacturing changes and new clinical indications for the lifetime of the product. These changes must be submitted as supplements to the BLA or NDA. Supplements are reviewed and approved (or not) according to the timelines described by the Prescription Drug User Fee Program (www.fda.gov).

In addition, applicants often make a number of commitments in the approval process, such as phase IV clinical studies, pregnancy registries, and additional validation studies. These studies provide additional data to the FDA on a variety of outstanding issues regarding product safety and efficacy. For example, a product that was approved by accelerated approval in which a surrogate end point was evaluated must be studied to obtain additional clinical outcome data in a phase IV study.

Adverse events must be reported according to 21 CFR 600.80 for biological products and 21 CFR 314.80 for drug products. Postmarketing 15-day "alert reports" shall be submitted for adverse events that are both serious and unexpected within 15 calendar days of receipt of information. These are generally reported through Medwatch for drugs and nonvaccine biological products (www.fda.gov/medwatch/what.htm) or the Vaccine Adverse Events Reporting System for vaccines (www.fda.gov/cber/vaers/vaers.htm).

4.5. Compliance

Following the approval of a product, the FDA performs biennial inspections to assess the firm's compliance with current good manufacturing practice (21 CFR 210, 211 and 600–680). If the inspectors observe deviations, they will present the firm with a list of observations (FDA Form 483). The FDA evaluates the observations and determines whether further regulatory action is needed. If the deficiencies are severe, the FDA can issue an "intent to revoke" action and if appropriate corrections are not made, the FDA can revoke the license (BLA). The FDA can also fine the sponsor or seek an injunction to stop the marketing of a product.

5. SUMMARY

The FDA regulates medical products throughout their life cycle to help ensure the quality of the product, the protection of human subjects in clinical trials, and the safety and effectiveness of medical products that are marketed. The FDA regulates these products based on sound science, law, and public health.

Legal Issues

PATRICIA A. KVOCHAK*

NIH Legal Advisor's Office, Office of the General Counsel, U.S. Department of Health and Human Services, Bethesda, Maryland

A topic such as legal issues in clinical research is expansive and cannot be fully explored in a few hours or even a few days. An attempt has been made in writing this chapter to focus on issues commonly encountered by, or of concern to, investigators in the clinical research environment. The issues include those related to (1) informed consent for standard and research care, (2) types of advance directives and other surrogate decision-making requirements, (3) the involvement of children in research, (4) maintenance of adequate medical records, (5) protection of confidentiality, (6) liability of clinical researchers, (7) conflicts of interest, and (8) authorship and rights in data. Although some of the discussion focuses on laws, regulations, and/or policies applicable at federal research institutions (e.g., the National Institutes of Health Clinical Center), the chapter should be useful to investigators in other working environments.

1. LEGAL ISSUES RELATED TO INFORMED CONSENT FOR CLINICAL AND RESEARCH CARE

Probably the predominant legal liability issue in clinical research relates to the presence or absence of adequate informed consent. Although today many of us in the United States take for granted the notion that a patient has the right to be reasonably informed and participate in decisions regarding his or her health care, this was not always the case. The patient's right

*The opinions expressed herein are personal and do not reflect the position of the National Institutes of Health or the Department of Health and Human Services.

to be adequately informed and to provide consent is not universal. For example, there are countries, cultures, and ethnic groups in which the physician and/or family serve as the primary decision maker, and the patient may or may not participate in the decision-making process.[1]

The foundation for the informed consent requirements in the United States grew from the common law action in battery (i.e., the right of an individual to be protected from nonconsensual touching). Damages were based on the occurrence of the touching without consent, whether or not harm had resulted. Beginning in the 1950s, courts began to treat cases based on failure to obtain "proper consent" as negligence actions rather than actions in battery. Use of the elements of common law battery was strained as the focus of the courts switched from whether there was consent to the quality of the consent and compensation for actual injury.

Negligence is different from battery in that there are several elements that have to be proved: (1) the health care provider owed a duty to the patient, (2) the duty was breached, (3) damages occurred, and (4) the damages were caused by the breach of duty. Courts focus on the "quality" of consent, finding no legally effective consent unless the patient understands the procedures/interventions and the risks associated with that treatment or intervention. All information relevant to the patient's decision should be disclosed.

Generally, the legal standard for consent to research is not distinct from that required for consent to standard care. The items required to be disclosed in a legally effective consent to standard care are similar to those elements required in a consent for participation in research under Section 46.116 of Part 46 of Title 45 of the Code of Federal

Regulations (CFR), the Department of Health and Human Services (DHHS) human subject protection regulations, and under section 50.25 of Part 50 of Title 21 of the CFR, the Food and Drug Administration (FDA) protection of human subjects regulations discussed in greater detail herein.

What are the elements of informed consent? The law has developed generally to provide for disclosure of the following items, as applicable, when informed consent is obtained in the standard care setting:

1. Diagnosis (patient's condition or problem)
2. Nature and purpose of the proposed treatment
3. Risks and consequences of the proposed treatment
4. Probability of success
5. Feasible alternatives
6. Prognosis if the proposed treatment is not given.

By comparison, Section 46.116 of the DHHS human subject protection regulations in 45 CFR Part 46 and section 50.25 of the FDA human subject protection regulations in 21 CFR Part 50 provide that the basic elements of informed consent in the research setting include the following:

1. A statement that the study involves research, an explanation of the purposes of the research and the expected duration of the subject's participation, a description of the procedures to be followed, and identification of any procedures that are experimental

2. A description of any reasonably foreseeable risks or discomforts

3. A description of any reasonably expected benefits to the subject or others

4. A disclosure of appropriate alternatives, if any, that might be advantageous to the subject

5. A statement describing the extent, if any, to which confidentiality of personally identifying records will be maintained (and for studies to which the FDA regulations apply, the statement must note the possibility that the FDA may inspect the records)

6. For research involving more than minimal risk, an explanation as to whether any compensation or treatment is available if injury occurs and where further information may be obtained

7. A contact for questions and if research-related injury occurs

8. A statement that participation is voluntary, that refusal to participate will not result in a penalty or loss of benefits to which the subject is otherwise entitled, and that the subject may withdraw at any time without penalty or loss of benefits.

Irrespective of the elements of a standard care or research consent, how much information needs to be given regarding each and how should it be communicated? Increasingly, concerns have been raised in the research setting about the length of consent forms, the necessity of discussing every risk or consequence, the underlying science or pharmacologic action, and so forth. Plato recognized the dilemma of the sick man who, when speaking with his physician, only wants to get well rather than be made a doctor.[2] Clearly, some education is needed and any communication needs to be understandable to the patient.

The most important principle is that any information material to the decision and any reasonably foreseen risks need to be disclosed. In the past, the standard for material information was material to the physician. In recent years, the law has evolved and information that would be material to a reasonable patient's decision making needs to be disclosed. Practitioners need to give careful consideration when developing informed consent language. Often, information that would be viewed as "obvious" to a practitioner may not be so for the patient.

There are various approaches to the discussion of risk in informed consent documents, particularly the likelihood of an adverse event occurring. Some consent forms use terms such as *slight*, *minimal*, or *small* risk. Such terms, however, are subjective and may have different meanings to different people. Although risks may not always be numerically quantifiable (e.g., 1/100 or 4/1000), statements of risk in approximate numerical terms are less capable of misinterpretation, in my opinion. Of course, the subject's ability to relate such numerical risk to his or her own situation may not result in a common understanding either.

In the research setting, it is important to think of consent as a process rather than just the initial explanation and execution of the consent document. Before enrollment in research, it is helpful to send the informed consent document out in advance to give the prospective subject time to consult with others and develop questions. Once a subject has enrolled in research, there is often an ongoing need to make information available and to assess, if not document, the willingness of the subject to continue in the research.

Who can execute a consent? A competent adult or a legally authorized representative is required. State laws authorize the provision of emergency care if a patient is incompetent. The provision of other care or interventions, particularly if research, may be legally problematic, leading into the next topic.

2. ADVANCE DIRECTIVES/ SUBSTITUTE CONSENT

Today, it seems implicit that if an individual has the right to consent to treatment, that same individual has the right to refuse certain procedures or treatments.

This view was not commonly accepted until the early 20th century. Justice Benjamin Cardozo, in a 1914 decision upholding a patient's right to refuse surgery (a leg amputation), held, "Every human being of adult years and sound mind has a right to determine what shall be done with his own body."[3] However, what happens when the individual lacks the capacity to consent (or refuse). Questions arise as to who can and will make decisions and how the individual patient's wishes will be taken into account. Many are familiar with the cases of Karen Ann Quinlan,[4] Nancy Cruzan,[5] and Terri Schiavo,[6] young women who ended up in persistent vegetative states without written advance directives. Since the *Quinlan* decision in 1976, a number of courts have upheld the right of a previously competent patient to refuse treatment even if it is life sustaining, and courts or state laws have authorized family members or other designated individuals to act as surrogate decision makers for mentally incapacitated adults. However, the U.S. Supreme Court in *Cruzan* upheld Missouri's requirement that clear and convincing evidence of the patient's wishes was necessary for a surrogate to forego or withdraw life-sustaining treatment.[7] Because of the case law and related statutory developments as well as the public accounts of the interfamily and other disagreements in the Schiavo case, attention has been focused on the use of advance directives, documents clearly reflecting the patient's wishes and/or designating a substitute decision maker.

Congress recognized the problem in enacting the Patient Self-Determination Act in 1990. This act requires health care institutions that receive Medicare or Medicaid funding to inform patients of their rights under state law to make decisions concerning medical care, including the right to refuse or accept care and the right to formulate advance directives. It also requires the health care institutions to document in a patient's medical record the existence or absence of an advance directive and to provide education.

There are two general types of legally recognized advance directives—the living will and the durable power of attorney for health care. A hybrid form combining both types of advance directives is occasionally seen.

The living will is a document that permits an individual to direct in writing that certain life-sustaining measures be withheld or withdrawn if the individual is in a "terminal condition" and does not have the capacity to make decisions. What is a terminal condition varies from state to state. It generally means a condition from which there can be no recovery and in which death is imminent (i.e., within six months with a reasonable degree of medical certainty). Increasingly, states permit the application of a living will if an individual is in a persistent vegetative state.

A durable power of attorney for health care (also known as a DPA or "health care proxy") is a document in which an individual appoints a surrogate to make decisions in the event he or she becomes incapable. The DPA may or may not contain statements of the patient's wishes to guide the surrogate. Some states only allow surrogates to make decisions that are consistent with the patient's wishes. Others, such as Maryland, permit a surrogate decision maker to make decisions in the best interest of the patient, if the wishes of the patient are unknown.

States set their own requirements regarding the execution and implementation of advance directives: how many witnesses, who they may be, when the instrument becomes operative (e.g., the procedure for certifying an individual as terminal or incapacitated), how the instrument may be revoked or altered, what happens if the individual is pregnant, the need for notarization, etc.

There are some common general requirements. For example, the advance directive must be voluntarily executed in writing and witnessed. Regarding who may witness an advance directive, questions will not usually be raised if the witness is not the person appointed substitute decision maker, related by blood or marriage, a creditor, entitled to inherit in the event of the individual's death, or financially responsible for the medical care or an employee of such. Where such requirements exist, they have been adapted to ensure that the individual executing an advance directive is not subject to any coercion or duress. Lastly, states generally recognize validly executed documents from other states where the individuals reside.

The benefits of a DPA as opposed to those of a living will, particularly in the research setting, are obvious. It is operative for a mentally incapacitated individual at any time, not just when the individual is "terminal." It permits a surrogate to make decisions about any matter, not just with regard to end-of-life decisions. A number of individuals have argued for specific language in the DPA regarding participation in research. At the National Institutes of Health, the Clinical Center (CC) has its own DPA form, which provides for the appointment of a surrogate who is authorized to provide informed consent for participation in research and medical care while the individual is at the NIH. (A copy of the form is provided in the Appendix.) The NIH CC recognizes other advance directives validly executed by the patient. A DPA naming a surrogate to make medical decisions for the patient has been viewed as including medical research decisions. The imposition of a legal requirement that a DPA must specify the writer's agreement to research participation in order for a surrogate to agree to participation will leave many without access to research, having never contemplated the possibility of research participation or having failed to address it.[8] Some have questioned the need for such a requirement, arguing that if a DPA authorizes a surrogate to make life and death decisions for the patient (e.g., withholding or withdrawing treatment), many of which have not been specifically contemplated or anticipated by the patient, why is that any different than decisions about medical research participation? Ample mechanisms exist to protect individuals from unscrupulous surrogates, in my opinion.

Researchers and institutional review boards (IRBs) should consider, depending on the nature of the research or the progression of the disease being studied, whether execution of a DPA should be a requirement for participation in research or, at a minimum, discussed. If it is likely that a subject will lose the mental capacity to provide ongoing consent during research participation, execution of a DPA should be considered. Some examples of studies in which such a requirement has been directed by IRBs include those in patients with early Alzheimer's disease and studies examining neurologic sequelae of HIV/AIDS infection. Consideration can also be given when possible severe neurologic/psychiatric side effects may result from a drug's introduction or withdrawal.

Health practitioners are often concerned with the legal liability associated with the implementation of advance directives by health care providers. State statutes generally provide for no criminal or civil liability for a health practitioner who follows an advance directive in good faith pursuant to reasonable medical standards. A practitioner who fails to follow a patient's wishes in a validly executed document could be sued and damages possibly awarded.

What happens when a patient has not executed an advance directive, or the advance directive is not applicable and the patient becomes mentally incapacitated? If the individual previously expressed his or her wishes regarding standard care, those wishes may be followed. It is helpful if such wishes have been previously documented in the medical record and/or witnessed. In addition, in the absence of a DPA or judicially appointed guardian, certain individuals are authorized by state law to give "substituted consent" for the furnishing (as opposed to withdrawal) of medical and dental care. These individuals may or may not be able to give consent to research participation. In those instances in which an individual may give substituted consent to research, it may only be for certain types of research (e.g., research involving no

more than minimal risk or having the prospect of direct benefit).

3. CHILDREN IN RESEARCH

Although children may be required to assent to their participation in research, they cannot provide legally valid consent. Except as permitted in the human subject protection regulations, contained in Subpart D of 45 CFR Part 46, a child requires the permission of both parents or his or her legal guardian to participate in research. A child is defined as a person who has not attained the legal age for consent to treatments or procedures involved in the research, under the applicable law of the jurisdiction in which the research is conducted. At the NIH CC, anyone younger than age 18 years cannot provide legally effective consent for participation in research, unless he or she is a parent or married.

In general, every effort should be made to obtain the permission of both parents. Section 46.408(b) of 45 CFR provides that unless waived by the IRB, both parents must give permission unless one parent is deceased, unknown, incompetent, or not reasonably available, or when only one parent has legal responsibility for the care and custody of the child. An IRB may find that the permission of one parent is sufficient for research that is no greater than minimal risk or that presents the prospect of direct benefit. Even when the permission of only one parent is determined sufficient, it is often helpful to have the cooperation and participation of the other parent, or an alternative caregiver, in the event the parent who provided permission is temporarily or permanently unable to accompany or care for the child. In such an instance, a temporary guardianship could be granted by the parent to the alternative caregiver.

Determining who has the legal authority to provide consent in the case of a child in foster care calls for careful investigation. States differ as to whether a foster child may participate in research and as to who may provide consent for a foster child to participate in research. Generally, the state agency responsible for the placement of the child in foster care, or a judge, will need to be involved in the research consent process. Some states permit the biologic parent to provide consent if parental rights have not been formally terminated. Rarely does a foster parent have the authority to enter the child in research.

It is important that the individual or entity having legal authority to provide permission for the foster child's participation in research be identified before the child's screening visit or protocol enrollment. If the legally authorized person will not accompany the foster child, procedures need to be in place to procure necessary consents.

What happens if parents or a guardian are unavailable or refuse consent? If the parents or guardian are unavailable and have not previously consented, emergency care may be provided to the child. If parents/guardian refuse permission for a child to participate in research, that decision governs. If parents/guardian refuse clinically accepted care in which the benefits outweigh the risks (e.g., refusal of consent for life-sustaining blood transfusion based on religious grounds), states will ordinarily assume their *parens patriae* role and mandate treatment. This may be done either through the involvement of child welfare authorities or by court order.

There are instances in which minors may provide their own consent for participation in research or for routine medical care. Pursuant to section 406.408(c) of 45 CFR, an IRB may determine that a research protocol is designed for conditions or for a subject population for which parental or guardian permission is not a reasonable requirement (e.g., neglected or abused children) and may waive parental consent, provided an alternative mechanism for protecting the subjects is substituted. Use of the waiver must be consistent with applicable law.

Pursuant to state law, some minors can consent to testing and treatment without parental consent.[9] The statutes generally specify the age at which a minor may consent and for what conditions. Lastly, a court may hold a hearing and determine a minor to be a "mature minor" and capable of making his or her own decision regarding medical or research care. Such a determination ordinarily requires clear and convincing evidence that the minor fully understands the risks, the nature of the treatment, and appreciates the consequences of his or her actions.

4. MEDICAL/RESEARCH RECORDS

Complete and accurate medical or research records are not only necessary to provide quality care to patients and ensure scientific integrity and verification but also become the most essential evidence in the event of subsequent litigation, review, audit, or other inquiry. For example, most litigation takes place two to five years after an event. To accurately reconstruct what took place, the record is most important given that memories fade, personnel change, and so forth.

There are three basic rules of medical documentation. First, documentation should be complete. The documentation should account for all treatment/interventions and observations. Failure to do this could result in the level of care being misinterpreted. For example, a patient postsurgery requires vital sign measurements every 15 minutes. Rather than accounting for all vital sign measurements taken, a nurse documents only abnormal findings, although the signs were taken every 15 minutes. A jury could infer no other observations were made and find that the standard of care was violated. The second basic rule of medical documentation is that documentation should be accurate. If discrepancies are found in a subsequent audit, the correctness of other entries, even if accurate, could be called into question. The last basic rule of documentation is that entries should be timely (i.e., made at the time treatment is given or observation is made or as close in time as possible). Late entries are subject to question and may raise concern about accuracy or reason for delay.

If an error is noted in documentation, how should it be corrected? Entries should never be obliterated or removed. If correction is needed, a line should be drawn through the incorrect entry and the correct information should be entered, initialed, and dated. If this is not possible, the incorrect entry should be lined out, and an explanation of the change should be written as close as possible to the original entry, signed and dated. Corrections should only be made by the original author; if that is not possible, a correction to the medical or research records should only be made by a supervisor or as otherwise provided by institutional policy.

5. CONFIDENTIALITY

5.1. Federal Privacy Act

The medical record is a confidential document and should be treated as such. It is the responsibility of health care professionals to safeguard patient confidentiality and patient records. Under the Privacy Act, 5 USC 552a, disclosure of any personally identified information from a patient's medical record in a federal facility, except to another employee who has a need to know the information in order to perform his or her job, may not be made without the patient's consent, unless one of the exceptions to the Privacy Act applies. These exceptions to the Privacy Act are limited and rarely apply to personally identified information. Some exceptions in which release may be considered include to an individual pursuant to a showing of compelling circumstances affecting the health or safety of any individual, pursuant to a court order signed by a court of competent jurisdiction, or to another government agency for a law enforcement activity if the activity is authorized by law and the head of the agency submits a written request specifying the record desired and the law enforcement activity for which the record is sought.[10]

The Privacy Act applies to all federal government records, not just medical records, that contain information on individuals and that are filed so that the records are retrieved by use of the person's name or some other personal identifier. The Privacy Act applies to personal information stored in computers as well as paper files. Violations of the Privacy Act, such as improper disclosures or maintenance of a system of records without proper notice, can carry both civil and criminal penalties.

The Privacy Act is a federal statute, but many states have adopted similar laws that govern the records of state agencies (including state university records).

5.2. Privacy Rule (Health Insurance Portability and Accountability Act)

The federal government published regulations, commonly referred to as the Privacy Rule, to protect the privacy of individually identifiable health information, known as protected health information (PHI), held or disclosed by a covered entity (i.e., health plans, health care clearinghouses, and those health care providers that conduct certain financial and administrative transactions electronically).[11] Although not all researchers will have to comply with the Privacy Rule, because the Privacy Rule regulates the use and disclosure of PHI by covered entities, it could affect certain aspects of research. In general, the Privacy Rule allows covered entities to disclose PHI with an individual's written permission called an "authorization." PHI may be used and disclosed for research without an authorization in very limited circumstances—that is, with a waiver issued by the IRB or privacy board,[12] when the subject of the research is deceased; when the data are disclosed in a limited data set[13] and an agreement is entered into between the covered entity and the researcher regarding the ways the information will be used and how it will be protected; or for "reviews preparatory to research" when the researcher assures that disclosure is solely to prepare a research protocol, no PHI will be removed from the facility, and the PHI is necessary to the research.

The Privacy Rule establishes minimum federal standards for protecting the privacy of PHI held by covered entities. Covered entities that fail to comply with the Privacy Rule may be subject to civil monetary penalties, criminal monetary penalties, and imprisonment. Whether a researcher must comply with the Privacy Rule is fact sensitive and, therefore, necessitates an individualized determination. Consultation with appropriate institutional officials is recommended.

5.3. Certificates of Confidentiality

Section 301(d) of the Public Health Service Act, 42 USC 241(d), provides that the Secretary of the U.S. Department of Health and Human Services may authorize persons engaged in biomedical, behavioral, or other research to protect the privacy of individuals who are research subjects by withholding from all persons not connected with the conduct of such research the names or other identifying characteristics of such individuals.[14] Researchers so authorized may not be compelled in any federal, state, or local civil, criminal, administrative, legislative, or other proceedings. Although the certificate may be helpful in initially refusing to provide information in response to law enforcement inquiry or subpoena, it appears that because voluntary disclosure by the researcher/research entity is not precluded and research subjects may provide consent to release of the information and the researcher even knowing the subject was under some pressure to consent to release is required to disclose the information, the protection afforded by certificates of confidentiality may be overstated.

In addition to the confidentiality protections described previously, there are also federal regulations governing the confidentiality of alcohol and drug abuse patient records that are maintained in connection with any federally assisted drug abuse or alcohol program.[15] Human subject protection regulations also require the research consent to contain language as to what efforts will be made to protect patients' confidentiality.[16] In addition, a subject should be informed of, and consent to, any possible access that will be given to his or her personally identified information (e.g., sharing of data with outside collaborators or drug company sponsors and FDA audits). The State and licensing or accreditation bodies (e.g., Joint Commission on Accreditation of Healthcare Organizations) may specify requirements regarding the confidentiality of medical records, which need to be observed.

What is a researcher to do? How does he or she decide what privacy protections to apply? In general, applicable federal law overrides state laws governing the privacy of information. However, state laws that offer more protection to PHI than the Privacy Rule and state or local laws that offer more protection to human subjects' confidentiality will continue to apply.

6. LEGAL LIABILITY

It is a general principle that the federal government may not be sued unless it has consented to be sued. This principle is known as the doctrine of sovereign immunity. The Federal Tort Claims Act

(FTCA) (28 USC 2671 et seq.) largely eliminated the federal government's immunity from tort liability and established the conditions for suits against the U.S. government. Pursuant to the FTCA, the United States is liable for certain torts (civil wrongs) in the same manner and to the same extent as a private individual under like circumstances, although the government is not liable for punitive damages. Actions for damages for alleged negligent or wrongful acts or omissions of federal employees done while performing their official duties are within the provisions of the FTCA. Section 224 of the Public Health Service Act, 42 USC 233, generally provides that the FTCA is the exclusive remedy available to an individual injured as the result of negligence by an officer or employee of the Public Health Service while providing health care (including the conduct of clinical studies) within the scope of his or her employment. These provisions operate to limit the naming of individuals who work at NIH specifically as defendants in lawsuits in their personal capacities and require that the United States be substituted as a party.

To file suit against the government, an individual must first exhaust administrative remedies. A claim, involving an NIH employee, must be filed with the PHS Claims Office within two years of when the incident occurred and specify the amount of the damages sought. An individual may not file a case in federal court until the claim has been denied administratively.[17]

The federal government self-insures. Professional liability insurance, therefore, is not maintained for federal employees. Clinical researchers at NIH are subject to actions for negligence or malpractice with less frequency than health professionals not involved in research. The types of claims filed most commonly involve allegations of error or mistake in treatment or diagnosis or defects in informed consent. Health professionals, who are not federal employees or volunteers, who practice at the NIH CC are required to be insured and to maintain professional liability insurance with designated coverage amounts. This is not unlike what investigators not covered by institutional liability policies must do.

Drug and technology development companies, as well as other entities, often ask investigators interested in receiving materials or doing collaborative studies with them to provide an assurance that the government will indemnify them for any costs in the event something goes wrong. Absent express statutory authority, the federal government (or its employees on its behalf) may not enter into an agreement to indemnify where the amount of the government's liability is indefinite, indeterminate, or potentially unlimited. Similar restrictions apply to investigators at a num-

ber of state universities and other governmental agencies.

Investigators may consider purchasing project casualty or liability insurance to cover the costs associated with any clinical trial mishap. Because of the possible latency of adverse events, professional liability "tail coverage" is also advisable if a researcher changes insurance carriers and does not have protection for prior acts. It is possible that drug companies will agree to offer coverage to research subjects who are injured owing to participation in clinical trials of their products, particularly in early phases of study. Researchers should consider the options to maximize their own protection and that of their subjects.

7. CONFLICT OF INTEREST

This subject has become increasingly important to the research enterprise to protect the integrity of research results and to eliminate questions of bias because of financial conflicts of interest. Pursuant to criminal statutes and implementing regulations, federal employees are prohibited from participating in an official capacity in matters affecting their own financial interests or the financial interests of other specified persons (spouse and dependent children) or organizations (trusts and partnerships).[18] If the interest is disclosed and it is determined to be not so substantial as to be deemed likely to affect the integrity of the services provided by the employee, a waiver may be granted.[19] If not, disqualification may be required. In a limited number of cases, divestiture of the financial interest may be required.

Conflicts of interest can raise concerns in a variety of circumstances. For example, in the procurement/acquisition situation, a requester may not have a financial interest in a manufacturer or vendor if purchasing products from that vendor or manufacturer. If an individual is evaluating competing products, the individual should not have a financial interest in any product under consideration.

In addition to the possible conflict of interest caused by remuneration from an outside activity (e.g., consulting), an NIH employee should keep a few other things in mind. Any professional outside activity needs to be approved and cannot involve the use of government time or resources. In evaluating whether an outside activity should be approved, the reviewer must consider whether the activity will interfere with NIH responsibilities. The reviewer must also consider whether the activity will result in the employee taking a position contrary to the government or result in the representation of the organization to the government.

These circumstances often arise when the activity involves serving as an officer of an organization/entity or on its board of directors. Representation of an organization by a government employee back to the government is prohibited by 18 USC 203 and 205.

If engaged in an outside clinical practice, NIH employees must arrange for patient coverage during their NIH hours and may neither refer their outside practice patients to NIH nor refer NIH patients to their outside practices. The latter is prohibited by one of the standards of conduct applicable to federal employees (i.e., a government employee may not use his or her public office for private gain).[20]

To avoid a conflict of interest, the standards of conduct for federal employees set strict limits on the receipt of gifts. An NIH employee may not solicit a gift. He or she may not receive a gift valued at more than $20 (market value) per occasion nor receive gifts valued at more than $50 per year from a "prohibited source."[21] A prohibited source is considered to be any individual or entity having official dealings or seeking official action with the employee's agency. To avoid the appearance of a conflict of interest, the receipt of gifts is discouraged regardless of their value.

Investigators not employed at NIH are subject to the conflict of interest policies of their employing institutions. Historically, institutions that received PHS funds were required to establish safeguards to prevent employees or consultants from using their positions for purposes that gave the appearance of being or were motivated by a desire for financial gain for themselves or others such as those with whom they have family, business, or other ties. Section 493A of the Public Health Service Act, added by Public Law 103-43, mandated regulations defining and setting standards for the management of financial interests that will, or may be reasonably expected to, bias a clinical research project to evaluate the safety or effectiveness of a drug, medical device, or treatment. Regulations were developed for PHS grantees with the following goals in mind: (1) to ensure the objectivity of research, (2) to meet the statutory requirements, (3) to minimize burdens on institutions, and (4) to avoid unnecessary restrictions on technology.

The regulations can be found in 42 CFR Part 50, Subpart F, for grants and cooperative agreements and in 45 CFR Part 94 for research contracts. The National Science Foundation has similar provisions.[22] The regulations apply to all applicants for PHS research funding. The regulations require that a grantee institution, before any expenditure of grant funds, certify that no conflicting interests exist or that conflicts have been resolved that could directly and significantly affect the design, conduct, or reporting of proposed PHS-funded research. Principal investigators and any other persons responsi-

ble for the design, conduct, or reporting of research must disclose significant financial interests (including those of the spouse and dependent children) to the designated institution official by the time an application is submitted to PHS. Significant financial interest is defined in the regulations in 42 CFR 50.603 and 45 CFR 94.3. Institutions may resolve conflicts of interest in a variety of ways, including, but not limited to, (1) public disclosure, (2) monitoring of research by independent reviewers, (3) disqualifying the investigator, (4) modifying the research plan, or (5) requiring the investigator to sever the relationships creating the actual or potential conflict.

Although much of the focus on conflict of interest has centered on the possibility of personal or family financial gain, other circumstances, such as personal relationships, academic rivalries, and the need for professional advancement, may pose conflicts of interest. Drug company sponsors may wish to control the public's access to information to enhance a company's position. To preserve public trust and confidence in research, researchers and research institutions must be vigilant in avoiding conflicts or the appearance of conflicts of interest.[23]

8. AUTHORSHIP/RIGHTS IN DATA

Authorship questions are ordinarily resolved by the primary author and the research group. Although there are no specific legal requirements governing who may or may not claim authorship of a scientific article, professional standards, such as those established by the International Committee of Medical Journal Editors, require that the designation of authorship should be based on a substantial contribution to (1) the conceptualization, design, analysis, and/or interpretation of the research study; (2) drafting or critically revising the article; and (3) final approval of the version to be published.[24] Authors must also be willing to take responsibility for the content and defense of the study.[25] Lesser contributions should be handled through acknowledgments.

Data management, including the decision to publish, is the responsibility of the principal investigator. Research data and supporting materials, such as unique reagents, of NIH investigators/employees belong to NIH and should be maintained in the laboratory in which they are developed. Ownership of data, in this case by NIH, generally carries with it the right to decide when and how to disclose it and how to control its use. Departing NIH investigators, with approval, may take copies of laboratory notebooks and other materials for further work. Certain restrictions related to patient privacy, prepublication review, and intellectual property may apply to the copying and sharing of clinical and other research data.

Other institutions may have similar or distinct rules. It is critical that investigators understand the policies of their institutions and that collaborators discuss any issues in advance and during the project.

NIH investigators may receive requests for research data or records pursuant to the Freedom of Information Act (FOIA)[26] or legal process such as subpoena or court order. FOIA operates generally to make government records available to the public subject to a number of exceptions. Nongovernmental trade secrets or proprietary information and personal private information in government records are ordinarily protected from public release.[27] If the information is contained in a Privacy Act system of records (i.e., retrieved by personal identifier such as a subject's name), the person whose file it is may authorize release of the information. Ordinarily, records with personal identifiers are not released without the subject's consent. NIH investigators receiving requests for data or records should consult with the appropriate records officials before any release.

If NIH sponsors extramural research, who owns the data? Ownership of data depends on the funding mechanism and the terms of the award. Generally, for grants, the grantee owns the data in the absence of a specific grant condition to the contrary. In the case of contracts and cooperative agreements, ownership of data is dependent on the terms of the award. Ownership of data does not preclude access to the data by NIH.[28] In addition, whether an extramural investigator has an ownership interest in the data depends on the policy of his or her employing institution.

References and Notes

1. See, for example, Blackball LJ, Murphy ST, Frank G, Michel V, Azen S. Ethnicity and attitudes toward patient autonomy. *JAMA* 1995;274(10):820–825.
2. Levine R. *Ethics and Regulation of Clinical Research*, 2nd ed. New Haven, CT, Yale University Press, 1988.
3. *Schloendorff v. Society of N.Y. Hasp.*, 105 N.E. 92 (N.Y. 1914).
4. In re Quinlan, 355 A.2d 647 (N.J. 1976).
5. *Cruzan v. Director, Missouri Dept. of Health*, 497 U.S. Ill, 110 S.Ct. 2841 (1990).
6. *Schindler v. Schiavo* (In re Schiavo), 780 So. 2d 176 (2001); subsequent proceedings at 789 So. 2d 551; 800 So. 2d 640; 851 So. 2d 182; *Bush v. Schiavo*, 861 So. 2d 506; 866 So. 2d 136; 885 So. 2d 321; Schiavo ex rel. *Schindler v. Schiavo*, 358 F. Supp. 2d 1161 (M.D. Fla.), aff'd 402 F.3d 1289 (11th Cir.) (denying injunction), reh'g en banc denied, 404 F. 3d 1270 (11th Cir.) reh'g denied, 404 F. 3d 1282 (11th Cir.) stay denied, 125 S.Ct. 1722 (2005).
7. *Cruzan*, supra note 5, at 2854.
8. For an interesting discussion of the issue, see Muthappan G., Forster H., Wendler D. Research advance directives: Protection or obstacle? *Am J Psychiatry* 2005;162:2389–2391.
9. See, for example, Md. Code Ann., Health-Gen. 20-102.
10. See 45 CFR Part 5b, Privacy Act Regulations of the Department of Health and Human Services.
11. See 45 CFR Parts 160 and 164.
12. The IRB or privacy board may waive or alter the authorization, in whole or in part, if it determines (1) the use or disclosure of the PHI involves no more than minimal risk to the privacy of individuals based on (a) an adequate plan to protect the PHI from improper use or disclosure, (b) an adequate plan to destroy identifiers at the earliest opportunity, and (c) adequate written assurance that the PHI will not be reused or otherwise disclosed; (2) the research could not be practicably carried out without the waiver or alteration; and (3) the research could not practicably be conducted without access to and use of the PHI.
13. A limited data set excludes 16 categories of direct identifiers but permits the retention of ages, date of birth and death, date of admission and discharge, zip codes or other geographic subdivisions, and other numbers or codes not listed as direct identifiers.
14. Implementing regulations can be found at 42 CFR Part 2a.
15. See 42 CFR Part 2.
16. 45 CFR 46.116(a)(5).
17. See 45 CFR Part 35, Tort Claims against the Government.
18. 18 USC 208.
19. Pursuant to recent regulations, NIH senior employees who file public financial disclosure reports are prohibited from acquiring or holding financial interests, such as stock, in a substantially affected organization, such as a biotechnology, pharmaceutical, and medical device companies and others involved in the research, development, or manufacture of medical devices, equipment, preparations, treatments, or products above a combined total of $15,000 per company. The regulations further provides that any NIH employee who is a clinical investigator is required to report an existing financial interest and subsequently any acquisition of an interest in a substantially affected organization. 70 *Fed Reg* 51559–51574 (August 31, 2005).
20. 5 CFR 2635.702, Use of Public Office for Private Gain.
21. 5 CFR Part 2635, Subpart B, Gifts from Outside Sources.
22. 60 *Fed Reg* 35820–35823 (July 11,1995).
23. Korn D. Conflicts of interest in biomedical research. *JAMA* 2000;284(17):2234–2236.
24. International Committee of Medical Journal Editors. Uniform requirements for manuscripts submitted to biomedical journals. *Ann Intern Med* 1997;126:36–47.
25. Despite publication of the uniform requirements and institutional policies consistent with them, a number of articles have demonstrated increases in honorary or ghost authors. For example, Flanigan A, *et al*. Prevalence of articles with honorary authors and ghost authors in peer-reviewed journals. *JAMA* 1998;280(3):222–224; Drenth JPH. Multiple authorship: The contribution of senior authors. *JAMA* 1998;280(3):219–221.
26. 5 USC 552 (and implementing regulations at 45 CFR Part 5).
27. 45 CFR 5.65, Exemption Four: Trade Secrets and Confidential Commercial or Financial Information; and 5.67, Exemption Six: Clearly Unwarranted Invasion of Personal Privacy.
28. See 45 CFR 74.53, Retention and Access Requirements for Records.

Further Reading

Hospital Law Manual, New York, Aspen, 2006.
National Institutes of Health. *Guidelines for the Conduct of Research at the NIH, 3rd ed*. 1997. *NIH Durable Power of Attorney for Health Care Decision Making*, 2000. Bethesda, MD, National Institutes of Health.

NIH Advance Directive for Health Care and Medical Research Participation

Instructions

The NIH is committed to respecting your health care and medical research participation wishes. As long as you are able to make decisions for yourself, we will determine what you want by speaking with you. However, it is possible that you may lose the ability to make your own decisions. At that point, it could be difficult for us to determine what kind of care you want. The NIH advance directive addresses this difficulty by allowing you to indicate in *advance* your health care and medical research wishes. This form goes into effect only if you lose the ability to make your own decisions. If you are completing this form, and have a non-NIH advance directive that you would like to remain in effect during your stay at the NIH, a copy of the non-NIH advance directive must be attached to this form.

The NIH advance directive is designed for use at the NIH Clinical Center. In addition, it can provide evidence of your wishes outside the Clinical Center. You can change this form at any time. You may fill out as much or as little as you want. This form must be signed and witnessed. You should keep the gold copy and give the pink copy to the person you name in part 1, if any. You should then give the remaining copies to your nurse or doctor. If you have any questions, or would like additional information, please speak with the members of your medical team, or contact the Department of Clinical Bioethics (301-496-2429).

PART 1: Your Choice for a Substitute Decision Maker: This section is similar to a durable power of attorney (DPA) for health care. It allows you to name someone to make medical research and health care decisions for you if you ever become unable to make these decisions for yourself. To ensure that the person you name can make the decisions you want, you should *discuss your health care and medical research wishes with the person you name.*

PART 2: Your Wishes About Medical Research Participation: This section allows you to indicate any wishes you have about your medical research participation in the event you become unable to make your own decisions. Some issues you may want to consider are listed below. You should *discuss your medical research wishes with your research team.*

PART 3: Your Wishes for Health Care: This section is similar to a *living will.* It allows you to indicate any wishes you have for your health care in the event you become unable to make your own decisions. Some issues you may want to consider are listed below. You should *discuss your health care wishes with the doctor taking care of you.*

Issues for Consideration and Discussion

Think about the things that are most important to you (your core values). Use these core values to decide which treatments you would or wouldn't want, and what types of research, if any, that you would be willing to participate in, if you lost the ability to make your own decisions. For instance, some people value certain abilities (such as the ability to communicate) so much that they would not want to be kept alive if they lost these abilities. In contrast, some people value life itself so much that they would want treatments to keep them alive no matter what their circumstances. Below are some additional issues that you may want to consider in thinking about, and discussing, your preferences with your doctor, substitute decision maker, and family.

Medical Conditions Relevant to End-of-Life Decision Making

Terminal condition: A medical condition from which, in the opinion of the patient's doctors, there is no reasonable chance of recovery and the use of life-sustaining treatments would only prolong the dying process.

Permanent coma: A complete loss of consciousness that the patient's doctors believe is not reversible.

Loss of the capacity for communication: The inability to communicate and interact with others.

Loss of the capacity for self care: The inability to perform the activities of daily living, such as bathing, eating, and dressing, without substantial assistance from others.

Intractable pain: Persistent and significant pain that continues despite maximum pain relief efforts.

Treatment Options

Emergency resuscitation: The attempt to restart a person's breathing and/or heartbeat. Resuscitation efforts include cardiopulmonary resuscitation (CPR), which involves pushing on the patient's chest or inserting a breathing tube in the patient's throat. Resuscitation efforts may also include the use of drugs or electric shock.

Do Not Resuscitate (DNR) order: When patients do not want emergency resuscitation attempted in the event their breathing or heart stops, instructions are written not to attempt resuscitation. This is called a DNR order.

Ventilatory support: A ventilator is a machine that helps patients breathe when their lungs fail. Ventilator support involves a breathing tube being placed in the patient's throat.

Artificial nutrition and hydration: Nourishment and fluids provided by tubes into the stomach or veins or by other artificial means.

Comfort measures: Treatments, such as pain killers, that are intended to keep patients comfortable.

Kinds of Research

Research with the potential for direct medical benefit: Research that offers the chance of improving the subject's medical condition.

Research with no potential for direct medical benefit: Research that does *not* offer the chance of improving the subject's medical condition, but will help doctors learn more about the disease under study and thus may help others with that disease.

In general, clinical research is divided into two categories of risk: minimal risk and greater than minimal risk of harm. Minimal risk means that the likelihood and degree of harm that you might experience in the research are no greater than those encountered in everyday life such as routine physical examinations and blood tests.

MEDICAL RECORD   **NIH Advance Directive   for   Health   Care and Medical Research Participation**

PART 1: Your Choice for a Substitute Decision Maker

I authorize the person(s) named below to make decisions for me concerning my health care and participation in medical research in the event that I become unable to make these decisions for myself:

Primary Substitute Decision Maker	Alternate (Used if Primary Substitute Decision Maker is Unavailable)
Name:	Name:
Address:	Address:
Telephone #	Telephone #

PART 2: Your Wishes About Medical Research Participation

A. If you lose the ability to make decisions, you may continue in your present study or be enrolled in a new study if your substitute decision maker agrees. You may also initial the following statements that reflect your wishes.
   If I lose the ability to make my own decisions:
   __ I do NOT want to participate in any medical research.
   __ I am willing to participate in medical research that might help me.
   __ I am willing to participate in medical research that will not help me medically, but might help others and involves minimal risk of harm to me.
   __ I am willing to participate in medical research that will not help me medically, but might help others and involves greater than minimal risk of harm to me.

B. You can use this space to indicate any values, goals, or limitations you would like to guide your participation in medical research. For more space use the NIH-200-1 Continuation form.

PART 3: Your Wishes for Health Care

A. You may initial the statements below that reflect your wishes. Your doctors can then make medical decisions for you based on your wishes and specific situation. If you have any questions about the situations you might face in the future, please speak with your medical team.
   _ I want all effective treatments for keeping me alive, no matter what my condition.
   OR
   I do NOT want life-sustaining treatments if:
   _ I have a condition that cannot be cured and will soon lead to my death, and life-sustaining treatment will only prolong the process of dying.
   _ I am in a permanent coma.
   _ I am awake, but have permanently lost the ability to communicate and interact with others.

B. You can use this space to indicate any values, goals, or limitations you would like to guide your health care. For more space use the NIH-200-1 Continuation form.

Patient Signature	Witness Signature
Print Name Date	Print Name Date

Patient Identification

NIH Advance Directive for Health Care and Medical
Research Participation
NIH-200 (10-00)
P.A. 09-25-0099
File in Section 4: Advance Directives
WHITE-Medical Record   GOLD-Patient   PINK-Substitute Decision Maker   GREEN-CCBioethics(1C118)

Rules to Prevent Conflict of Interest for Clinical Investigators Conducting Human Subjects Research

ROBERT B. NUSSENBLATT* AND MICHAEL M. GOTTESMAN**

*Laboratory of Immunology, National Eye Institute, National Institutes of Health and Office of Protocol Services, National Institutes of Health Clinical Center, Bethesda, Maryland;
**Office of the Director, National Institutes of Health, Bethesda, Maryland

The personal integrity of the physician is a paramount concern of society that dates back to the beginning of written history. The Hippocratic Oath is one of the earliest examples of an attempt to define community standards of behavior and promote such integrity. Because clinical research involves a somewhat different relationship between investigators (many of whom are not physicians) and patients, it has been necessary to develop a new set of community standards to assure the integrity of the clinical research process. One set of ethical standards relates to the need to protect human subjects involved in clinical research. Dr. Grady's chapter on this subject in this volume very amply and expertly covers this aspect of the ethics of clinical research. Another concern relates to the way in which real or perceived conflicts of interest may affect the integrity of clinical research. This subject has become a very active area of scrutiny in both the public and the private sectors. This chapter addresses the concept broadly and also describes the efforts taken by the National Institutes of Health (NIH) to prevent conflicts of interest by investigators involved in clinical research.

Emanuel[1] has described the three primary interests of a physician as (1) promoting patients' well-being and health, (2) advancing biomedical knowledge through research, and (3) training future physicians and other health care professional. A conflict of interest occurs when other interests that the physician may have undermine, or appears to undermine, his or her conduct in meeting those goals. Such interests may call into question the validity of the research process or put patients at unnecessary risk. The conflicts of interest may result in inappropriate acts of commission. The physician may have a financial interest in a company or other incentive that motivates a study that is costly to society or puts subjects at risk or affects the interpretation or reporting of data. Even the appearance of such a conflict, without intent on the part of the investigator, is corrosive to the integrity of clinical investigation. The investigator may commit acts of omission if it is in his or her interest to do less, such as failing to report adverse events or investigate potential complications occurring in a clinical study. Thus, it is clear that given the vulnerability of human subjects and the fragility of the clinical research enterprise, it is essential for the clinical investigator to avoid conflicts of interest, real or perceived, in protocols in which the investigator is responsible or plays a role.

Although most clinical investigators will deny vehemently that their financial interests would affect their research and clinical activities, studies have shown that interactions with pharmaceutical firms can have an affect on decision making by physicians.[2,3] Those who were receiving remuneration of some kind from pharmaceutical firms were more likely to support the safety of the drugs of those companies, and we can presume that research activities would be similarly affected.

Pharmaceutical and biotechnology companies are frequent supporters of clinical studies. There is no evidence to suggest that the quality of these studies is less rigorous than studies supported by nonprofit organization. However, Bekelman and colleagues[4] reviewed systematically the results of studies reported in the literature in which there was pharmaceutical support. They found that there was a statistically significant association between pharmaceutical industry support and positive results for the agent produced by the industry (odds ratio, 3.60; 95% confidence interval, 2.63–4.91; Fig. 11-1). In another review of 136 randomized studies that focused on the treatment of multiple myeloma, when studies were sponsored by a drug company, positive results for a new treatment were reported in 74% of the studies compared to 47% for those not sponsored by a drug company.[5] It is likely that these data indicate a bias in favoring reports that are positive when the sponsor benefits financially; more negative reports may appear when the sponsor has no financial ties to the study.

Another kind of conflict that may undermine the credibility of clinical researchers has been highlighted in a series of articles in the *New York Times* and elsewhere.[6,7] Agents that give financial advice to investors have been turning increasingly more frequently to physicians and clinical investigators to seek their opinions about the likelihood that new drugs and devices will be marketable. Large financial premiums are paid to hear "thumbs up" or "thumbs down" from informed clinical investigators, raising the possibility (or the appearance) that the integrity of clinical trials information could be violated by scientists wanting to continue to receive large payments for their advice. Clinical

investigators who are actively involved in research protocols are well-advised to avoid such consultations.

Another area of concern is in the reporting and reviewing of results of clinical trials in the literature and in oral presentations. Investigators who have a financial or other interest in companies that may benefit from their published research or written evaluation of drugs or devices must realize that their credibility may be questioned based on conflict of interest. Respectable journals have attempted to manage this problem by reporting that authors have such conflicts, a process known as disclosure, but this does not eliminate the conflict and leaves the critical reader in a quandary as to whether to trust the research results. The Accreditation Council for Continuing Medical Education (ACCME) has taken a tough stand on this issue concerning ACCME-accredited training activities, requiring no conflict rather than just disclosure.[8]

Three general approaches have been advocated for how professional societies and other venues of continuing medical education can approach conflicts of interest. As noted, (1) one may ask investigators to disclose any interest; (2) if conflicts exist, they can be managed in a variety of ways but not eliminated; and (3) they can be prohibited. Many professional organizations have developed guidelines utilizing all three of these approaches. Clearly, the avoidance of conflicts or appearance of conflict is the most straightforward approach. Based on federal government ethics laws and regulations, including a conflict of interest regulation issued in August 2005,[9] the NIH has attempted, insofar as is possible, to eliminate conflicts of interest for clinical researchers. Although this approach may not be ideal for clinical researchers who are not government employees, it sets a high standard that could be emulated by clinical researchers elsewhere who seek to avoid any appearance of conflict.

1. PREVENTING CONFLICT OF INTEREST IN CLINICAL RESEARCH IN THE NIH INTRAMURAL RESEARCH PROGRAM

The bedrock of the new NIH conflict of interest regulation is that no NIH employee may consult for remuneration with a significantly affected organization, including pharmaceutical companies, biotechnology firms, health services organizations, or agents of such organizations (see Appendix). This eliminates the major concern that an NIH investigator will appear to be "serving two masters"—that is, receiving payments from a company whose product is the subject of

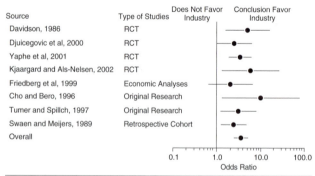

RCT indicates randomized controlled trial. Error bars indicate 95% confidence intervals.

FIGURE 11-1 Relation between industry sponsorship and study outcome in original research studies. RCT, randomized controlled trial. Error bars indicate 95% confidence intervals. From Bekelman JE, Li Y, Gross CP. Scope and impact of financial conflicts of interest in biomedical research. A systematic review. *JAMA* 2003;289:454–465.

government-supported clinical research. In addition, stockholdings or other equity or leadership positions for the investigator or his or her immediate family in such companies are not allowed. Every clinical investigator who is named on the cover sheet for a clinical protocol must report any holdings in significantly affected organizations. If any such organizations have a financial interest in a product under study in a protocol in which the investigator is involved, the investigator is encouraged to divest the holding. Although the government rule sets a *de minimis* of $15,000 for divestiture, NIH guide points out that the investigator with such stock holdings should be aware of the appearance of conflict that results from having such stock.

Every principal investigator on a protocol is responsible for informing all investigators on their protocols about the current intramural "Guide to Preventing Conflicts of Interest in Human Subjects Research at the NIH" (see Appendix). This document defines why this guide is needed and the kinds of conflicts that are of concern to the NIH. It further gives examples of conflicts, financial and otherwise, that are prohibited. It should be emphasized that these specific prohibitions pertain to the research performed under the specific clinical protocol in question and covers both clinical investigators and institutional review board (IRB) members. Once conflicts are eliminated, the protocol can proceed to the IRB. We trust that outside collaborating investigators who are not NIH employees will abide by these same requirements.

Section V of the guide deals with an important issue for intramural scientists at the NIH. It is the intramural investigator's legal responsibility under the Federal Technology Transfer Act to report new discoveries. The NIH may decide to pursue patent protection for these observations. Clinical research projects may very well be based on these observations. Investigators who have made such discoveries are not prohibited from participating in such studies. However, there needs to be full disclosure to the IRB and to the patients who participate that both the NIH and NIH investigators may receive royalties as a result of some kinds of clinical research. Additional oversight, including review of the results by the institute clinical director and a data and safety monitoring committee, is also required.

2. PREVENTING CONFLICT OF INTEREST BY MEMBERS OF INSTITUTIONAL REVIEW BOARDS

Members of IRBs and DSMBs are also expected to avoid conflicts of interest in their deliberations. At the beginning of discussion of each protocol, the chair asks any members who are in conflict to leave the room. To ensure that each member understands the kinds of interests that are considered in conflict, each member is given a copy of the "Guide to Preventing Conflict of Interest in Human Subjects Research at the NIH."

The pursuit of ethical clinical studies is one of the most important tasks for clinical investigators in the NIH intramural program and elsewhere. No one wishes to make this task more difficult than it already is. However, there must be assurance that there is no real or perceived conflict in our endeavors, and both the federal regulations governing conflict of interest and the "Guide to Preventing Conflict of Interest in Human Subjects Research at the NIH" will help to avoid such a situation.

References

1. Ezekiel Emanuel, personal communication.
2. Stelfox HT, Chua G, O'Rourke K, Detsky AS. Conflict of interest in the debate over calcium-channel antagonists. *N Engl J Med* 1998;338(2):101–106.
3. Blumenthal D, Campbell EG, Anderson MS, Causino N, Louis KS. Withholding research results in academic life science. Evidence from a national survey of faculty. *JAMA* 1997;277(15):1224–1228.
4. Bekelman JE, Li Y, Gross CP. Scope and impact of financial conflicts of interest in biomedical research. A systematic review. *JAMA* 2003;289:454–465.
5. Djulbegovic B, Lacevic M, *et al.* The uncertainty principle and industry-sponsored research. *Lancet* 2000;356:635–638.
6. Saul S, Anderson J. Doctors' links with investors raise concerns. *New York Times*, August 16, 2005, p. 1.
7. Steinbrook B. Wall Street and clinical trials. *N Engl J Med* 2005;353(11):1091–1093.
8. Steinbrook B. Commercial support and continuing medical education. *N Engl J Med* 2005;352(6):534–535.
9. *Fed Reg* 701(168):51559–51574, August 31, 2005. Available at www.nih.gov/about/ethics_COI.htm.

A P P E N D I X

A Guide to Preventing Financial and Non-financial Conflicts of Interest in Human Subjects Research at NIH

Avoiding financial and other conflicts of interests is important for NIH, where the trust and protection of research subjects is vital to our mission to improve the public health. The number and complexity of laws and regulations in this area makes it difficult to know when there is a problem and what to do. This guide is intended to assist clinical investigators and NIH IRB members in avoiding real or perceived financial and non-financial conflicts of interest.

I. WHAT ARE A CLINICAL INVESTIGATOR'S POTENTIAL CONFLICTS OF INTEREST?

All clinical investigators have primary obligations. These include obtaining knowledge that will promote health and health care and helping ensure the safety and health of research participants. Clinical investigators may also have other, personal or secondary interests, which could include teaching trainees, supporting a family, and earning income. These secondary interests are not, themselves, unethical, but in some circumstances they have the potential to compromise, or appear to compromise, the judgment of clinical researchers regarding their primary obligations. When these secondary interests compromise judgment, or appear to do so, there is a conflict between the secondary and primary interests.

This guide provides information to prevent financial and other conflict, thereby helping to ensure both the integrity of our research and the safety of participants.

II. TO WHOM DOES THE GUIDE APPLY?

The restrictions discussed in this guide are based on the laws that apply to NIH employees.[1] Thus, all NIH employees who are listed as investigators[2] on the front sheet of a protocol because they substantively participate in the development, conduct, or analysis of clinical research protocols (both diagnostic and therapeutic) must adhere to the rules described below. These rules also apply to NIH employees who serve on NIH Institutional Review Boards (IRBs) and Data Safety and Monitoring Boards (DSMBs). It is expected that non-employees who serve as investigators and IRB and DSMB members will review this guide and adhere to rules set out to the extent practical. These non-employees should be mindful of real and potential conflicts and discuss such conflicts with the protocol's PI.

III. EXAMPLES OF INVESTIGATOR AND IRB AND DSMB MEMBER FINANCIAL CONFLICTS OF INTEREST

As noted when applicable, some of these examples of financial conflicts of interest are prohibited by regu-

---

[1]NIH employees are those NIH staff with an appointment to the federal government pursuant to, for example, Title 5, 38 or 42, or the Commission Corps, and may include some fellows. Some IPA personnel may have federal government appointments as well.

[2]Investigators are those NIH employees who occupy the following positions: Principal Investigator; Co-Principal Investigator; Associate Investigator; Medical Advisory Investigator; and Research Contacts.

lation for NIH employees. We list them, however, as guidance for non-employee investigators and IRB and DSMB members who are reviewing this guide. It should be noted that in addition to his or her own financial interests and outside interests, an NIH employee's financial interests also include the financial interests of others, such as his or her spouse, dependent children, or household members. Examples of such interests are:

- Serving as a director, officer or other decision-maker for a commercial sponsor of the human subjects research (prohibited activity for NIH employees);
- Holding stock or stock options in a commercial sponsor of the human subjects research (unless below the applicable de minimis amount or held within a diversified, independently managed mutual fund);
- Receiving compensation for service as consultant or advisor to a commercial sponsor of the human subjects research (excluding expenses) (prohibited activity for NIH employees);
- Receiving honoraria from a commercial sponsor of the human subjects research (prohibited activity for NIH employees);
- Personally accepting payment from the human subjects research sponsor for non-research travel or other gifts (for NIH employees, government receipt of in-kind, research-related travel is not included and other exceptions may apply);
- Obtaining royalties or being personally named as an inventor on patents (or invention reports) for the product(s) being evaluated in the human subjects research or products that could benefit from the human subjects research (special rules apply in this case when NIH holds the patent — see Section VII below);
- Receiving payments based on the research recruitment or outcomes (prohibited activity for NIH employees);
- Having other personal or outside relationships with the commercial sponsor of the human subjects research (prohibited activity for NIH employees);
- Having financial interest above the applicable de minimis in companies with similar products known to the investigator to be competing with the product under study (prohibited activity for NIH employees); or
- Participating in an IRB or DSMB decision that has the potential to affect your spouse's employer (prohibited activity for NIH employees).

IV. EXAMPLES OF NON-FINANCIAL REAL OR APPARENT CONFLICTS OF INTEREST FOR IRB AND DSMB MEMBERS

- Voting on a protocol when a member of the IRB is the protocol's Principal Investigator, Associate Investigator or study coordinator;
- Voting on a protocol when a member of the IRB or DSMB is a spouse, child, household member or any other individual with whom the protocol's Principal Investigator, Associate Investigator or study coordinator has a close personal relationship[3]; or
- Voting on a protocol when the protocol's Principal Investigator is the IRB member's supervisor (up the chain of command to the Clinical Director).

V. NIH'S SYSTEM TO ASSIST IN IDENTIFYING AND PREVENTING FINANCIAL CONFLICTS FOR INVESTIGATORS IN CLINICAL RESEARCH

The Principal Investigator is responsible for assuring that each investigator listed on the protocol front sheet receives a copy of this guide. The guide should be distributed to any new investigators added to a protocol while the protocol is active.

a. New Protocols

At the earliest point possible, the PI is responsible for providing his or her IC Deputy Ethics Counselor (DEC) with a list of all investigators. The Protocol COI Statement (see Appendix I) or an electronic equivalent should be used to provide this information. This submission date will be noted on the form 1195.

Upon receipt of the Protocol COI Statement, the IC DEC will verify that all investigators who are employees have a form 716/717 on file and that the personal investment information on the form 716/717 is current as of the date on the Protocol COI Statement. The IC DEC will then review file copies of each PI's and AI's 716 or 717 forms that enumerate stock holdings in all organizations that are significantly affected by the NIH (referred to as "SAOs").

---

[3]The IRB or DSMB member determines, in his/her own opinion, whether a close personal relationship with the protocol's Principal Investigator or another member of the research team exists. If such a determination is made, the IRB or DSMB member shall disqualify him or herself from the protocol to avoid any appearance of bias.

For each protocol, the DEC will provide the PI with an anonymous list of AIs' holdings in SAOs reported on these forms so the PI can determine if any pose a conflict of interest for the protocol in question. Any investigator who has a potential conflict will be contacted by his or her DEC to determine how to resolve any actual or apparent conflict. The employee's supervisor and/or the Clinical Director will be consulted as necessary if a conflict exists. The conflicts review will occur in parallel to the IRB submission process.

At the completion of the conflicts review, the IC DEC will return a signed copy of the Protocol COI Statement to the PI. The PI will then note the date of DEC clearance on the Form 1195 and ensure that the Protocol COI Statement is included in the protocol packet.

The DEC clearance form will become part of the protocol packet forwarded to the IRB Chair for final approval. The IRB chair may not provide final approval by signing a protocol until the completed Protocol COI Statement is included in the protocol packet.

b. Continuing Review

A COI analysis will take place at the time of continuing review using the same process as described above. The Protocol COI Statement will be used for this process. For the conflicts analysis, the addition of new investigators, any changes related to the use of commercial products or any change to an IND/IDE will be evaluated by the IC DEC.

c. Amendment

A COI analysis will take place for amendments involving the addition of investigators to a protocol, any changes related to the use of commercial products or any change to an IND/IDE. The Protocol COI Statement will be used for this process following the procedure above.

Although government-wide regulations allow NIH employees to hold de minimis amounts of publicly-traded stock without triggering conflict of interest restrictions, there may be other factors to consider with respect to stock ownership. For example, new NIH policy will require that the informed consent document signed by protocol participants contain a statement that one or more investigators own a de minimis amount of stock in the company that makes the product being tested in the protocol. Also, if a publication should result from the protocol, most journals require the authors to disclose individual financial holdings within the text of the published paper. Such disclosures could raise at least the appearance of the conflict

of interest. Thus, all investigators should consider these outside factors when making personal financial investments.

VI. IRB AND DSMB CLEARANCE FOR COI

- Before beginning protocol review activities, the Chair asks whether any member is aware of any real or apparent conflict of interest. The response of an individual who has a conflict of interest is noted in the minutes. No IRB or DSMB may have a member participate in the initial or continuing review of any project in which the member has a conflicting interest, except to provide information requested by the IRB or DSMB.
- When the Principal Investigator or Associate Investigator is the Institute Director, or Scientific Director, the protocol will be reviewed by an IRB not affiliated with that institute.
- When the Principal Investigator is the Clinical Director (CD) it shall be the prerogative of an IRB either to review such protocols or refer them to another Institute's IRB. IRBs reviewing protocols in which their CD is the PI must have a majority of members who are not employed by the CD's Institute otherwise any alternative plan must have prior approval by the Director, CC, and the Deputy Director for Intramural Research.

VII. NIH INTELLECTUAL PROPERTY AND ROYALTIES

In some instances, NIH clinical research protocols will evaluate or potentially advance product(s) in which NIH (i.e., the government) owns patents or has received invention reports. In such cases:

- An NIH investigator may participate in the clinical trial, even if the investigator is listed on the patent or invention report and/or may receive royalty payments from the NIH for the product(s) being tested.
- When such an investigator participates in a trial, there should be full disclosure of the relationship to the IRB and to the research subjects (i.e., information should appear in the consent form) with review and approval by the IRB.
- In the case of continuing review of current protocols where NIH has an intellectual property interest in the invention, investigators should provide a new human subjects consent form or correspondence

outlining the relationship, for review and approval by the IRB.

- An independent entity, such as a DSMB, must review the results of all such human subjects research.

- These relationships must be reported to the DDIR as part of the quarterly report, without reference to specific individuals, but should not impede the pursuit of the trial.

PROTOCOL CONFLICT OF INTEREST STATEMENT
(Appendix 1)

Date of Memo:

Date of IRB Meeting:

Date Protocol Expires:

To: _____
        I.C. Deputy Ethics Counselor

From: _____
        Principal Investigator
        CC:

Re: Documentation of Discussion of Conflict of Interests with P.I.

Date Received by Ethics Office _____

____ New Protocol (attach **précis**)
____ Continuing Review
____ Amendment

Protocol #:

Type of Protocol:

Title:

Principal Investigator's IC:

Responsible IRB:

Product(s) made by commercial entity that is the subject of the study:
Manufacturer of study product(s) (drug or device):
IND/IDE # (if applicable):
IND/IDE Holder (if applicable):
Do you know of competitors for study drug or device manufacturer(s) for purposes related to this protocol?
Key words as per 1195:

Accountable Investigator:

Medical Advisory Investigator:

Research Contact:

Lead Associate Investigator:

List of Associate Investigators:
        <u>**Name of Investigator**</u>   <u>**NIH Employee's Institute or Non-NIH Affiliation**</u>

____ No conflicts identified   ____ Conflicts if identified are resolved.
                                                            Explain:

_____   _____   _____
Deputy Ethics Counselor for IC of P.I.   Date Signed   Date Returned to P.I.

12

National Institutes of Health Policy on the Inclusion of Women and Minorities as Subjects in Clinical Research

MIRIAM KELTY,* ANGELA BATES,** AND VIVIAN W. PINN**

*National Institute on Aging, National Institutes of Health, Bethesda, Maryland
**Office of Research on Women's Health, Office of the Director, National Institutes of Health, Bethesda, Maryland

Approximately one in two women develop coronary heart disease (CHD)[1] and one in three die from it, accounting for over 250,000 deaths in women per year.[2] Despite the high prevalence of CHD in women, it has traditionally been thought of as a disease of middle-aged men, perhaps because women tend to develop CHD about a decade later in life than men.[3] During the last two decades, multiple important studies have helped define accurate clinical tests, important risk factors, preventive interventions, and effective therapies for CHD. Unfortunately, the majority of these studies have either excluded women entirely or included only limited numbers of women.[4] Thus, much of the evidence that supports contemporary recommendations for testing, prevention, and treatment of coronary disease in women is extrapolated from studies conducted predominantly in middle-aged men. Applying the findings of studies in men to management of CHD in women may not be appropriate since the symptoms of CHD, natural history, and response to therapy in women differ from those in men.[5]   —Grady et al.[6]

The establishment and implementation of policies for the inclusion of women and minorities in clinical research funded by the National Institutes of Health (NIH) have their origins in the women's health movement.

As the last decade of the 20th century began, interest in women's health was increasing throughout the general populace, the scientific community, the media, and the government. There was growing recognition that despite the enormous strides that had been made in biomedical research, there still remained many unanswered questions about women's health: Not only did much need to be learned about the diseases, disorders, and conditions that are unique to or more prevalent in women but also there continued to be gaps in scientific knowledge about disease processes, their underlying mechanisms, and the best way to prevent disease or treat women for diseases that affect both men and women.

In addressing this situation, it became clear that the major reason for the dearth of knowledge about both women's diseases and diseases in women had been the widespread exclusion of women from participation in research, especially clinical trials. The general reluctance of investigators to enroll women as research subjects reflected the prevailing biases of the times within the scientific community and was defended on the grounds of both practical considerations and ethical concerns. There was concern that periodic changes in hormone levels in women of reproductive age might affect therapeutic interventions and necessarily make research designs more complicated. Since the thalidomide tragedy and revelations about carcinogenesis related to intrauterine exposure to diehtylstilbestrol, there were concerns about the risk of adverse outcomes to offspring if a woman were to become pregnant during the course of a clinical trial. The outcome of this approach was that important questions about women's health were not being appropriately addressed because women often were not included in research studies. This outcome was no longer scientifically, socially, or politically acceptable.

In 1990, the Office of Research on Women's Health (ORWH) was established to ensure the inclusion of women and minorities in NIH-funded research. With a record number of women then elected to Congress, the time was ripe for the enactment of landmark legislation. A legislative mandate that women and minorities must be included in all clinical research studies was incorporated into the language of the NIH Revitalization Act of 1993.[7] This mandate was implemented in 1994, when the NIH published its *Guidelines on the Inclusion of Women and Minorities as Subjects in Clinical Research*.[8] Although the Revitalization Act stipulated a number of requirements not present in the earlier NIH policy, its greatest impact was that NIH policy goals assumed the force of law. Thus, the 1994 NIH inclusion guidelines represent another phase of NIH's long-standing commitment that all members of our society share in the benefits and burdens of biomedical and behavioral research. The policy in essence directs NIH-funded biomedical and behavioral research to be designed such that differences or similarities between men and women can be determined; similarly, just as the assumption should not be made that men and women are the same, neither should the assumption be made that all men and all women are also the same. Therefore, the policy directs attention to racial and ethnic characteristics and to determining if there are differences in health or disease characteristics for different racial and ethnic groups.

The revisions made to the NIH inclusion policy in 1994 were challenged by some members of the scientific community. However, the fears of that time that this legislative mandate would impede research did not materialize, and the policy is now fully implemented.[9] The policy provides a valuable tool to assist investigators in answering important questions about the differences and similarities in health and disease between women and men (Figs. 12-1 and 12-2).

An additional and more general reason for studying differences between the sexes is that these differences, like other forms of biological variation, can offer important insights into underlying biological mechanisms.

—Wizemann TM, Pardue M-L. *Exploring the Biological Contributions to Human Health: Does Sex Matter?* Washington, DC, National Academy Press, 2001

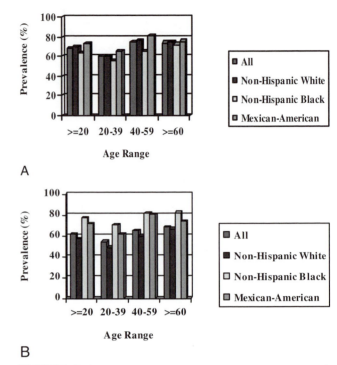

FIGURE 12-1 Overweight and obesity prevalence (1999–2002) for men (BMI ≥25) (**A**) and women (BMI ≥25) (**B**). Based on data from Hadley AA, Odgen CL, Johnson CL, *et al*. Prevalence of overweight and obesity among U.S. children, adolescents and adults, 1999–2002. *JAMA* 2004;291(23):2847–2850.

1. NIH POLICY

The NIH Revitalization Act of 1993 essentially gave force of law to existing NIH policy and added four major requirements. The NIH must:

- Ensure that women and members of minority groups and their subpopulations are included in all human subjects research
- For phase III clinical trials, ensure that women and minorities and their subpopulations must be included such that valid analysis of differences in intervention effect can be accomplished
- Not allow cost as an acceptable reason for excluding these groups
- Initiate programs and support for outreach efforts to recruit these groups into clinical studies.

As a result of these requirements, it is now the policy of NIH that women and members of minority groups and their subpopulations must be included in all NIH-supported biomedical and behavioral research projects involving human subjects, unless a clear and compelling rationale and justification establishes, to the satisfaction of the relevant institute/center director, that

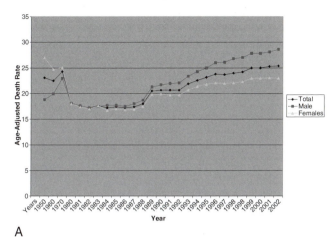

A

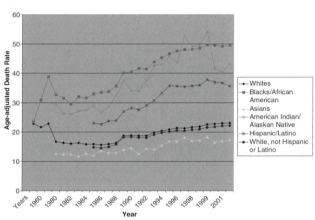

B

FIGURE 12-2 Diabetes: Age-adjusted death rates by sex (per 100,000) from 1960 to 2002 (**A**) and by race from 1960 to 2001 (**B**). Data source: Department of Health and Human Services, Centers for Disease Control and Prevention, National Center for Health Statistics. *Report on Trends in the Health of Americans*, 2005. Available at www. cdc.gov/nchs/data/hus/hus05.pdf.

inclusion is inappropriate with respect to the health of the subjects or the purpose of the research.[10]

In addition, phase III clinical trials, as defined in the 2001 inclusion guidelines,[11] are to be designed and carried out in a manner that will provide for valid analysis of whether the variables being studied affect women or members of minority groups differently from other subjects in the trial.[12]

The inclusion guidelines described some very limited exceptions to policy, as delineated by the law.[13] In all cases, the research study designs are evaluated prospectively by the NIH, and funding is contingent on submission of a research plan that meets all the inclusion requirements.[14–16]

A later report by the U.S. General Accounting Office, *Women's Health: NIH Has Increased Its Efforts to Include Women in Research*,[17] acknowledged that NIH had made significant progress in implementing a strengthened

policy but also concluded that more emphasis was needed in identifying and reporting potential sex/gender differences in phase III trials. The report made two specific recommendations to the director of NIH to ensure:

- That the requirement be implemented that phase III clinical trials be designed and carried out to allow for the valid analysis of differences between women and men; that this requirement is communicated to applicants as well as requiring peer review groups to determine whether each proposed phase III clinical trial is required to have such a study design; and that summary statements document the recommendations of the initial reviewers, and

- That the NIH staff receives ongoing training on the purpose and requirements for data transmission to the NIH population tracking system.

Several actions resulted to clarify the requirement for NIH-defined phase III clinical trials to include women and minority groups, if scientifically appropriate, and for analysis of sex/gender and/or racial/ethnic differences to be planned and conducted by investigators engaged in NIH-funded research. These included the following:

- Updates to the *NIH Policy and Guidelines on the Inclusion of Women and Minorities as Subjects in Clinical Research*. This version (2001) incorporates the definition of clinical research as reported in the *1997 Report of the NIH Director's Panel on Clinical Research*[18] and the Office of Management and Budget (OMB) Directive 15 racial and ethnic categories[19] to be used when reporting population data. It also provides additional guidance on reporting analyses of sex/gender and racial/ethnic differences in intervention effects for NIH-defined phase III clinical trials.

- The *1997 Report of the NIH Director's Panel on Clinical Research* defined clinical research as (1) patient-oriented research. This is research conducted with human subjects (or on material of human origin such as tissues, specimens, and cognitive phenomena) for which an investigator (or colleague) directly interacts with human subjects. Excluded from this definition are *in vitro* studies that utilize human tissues that cannot be linked to a living individual. Patient-oriented research includes mechanisms of human disease, therapeutic interventions, clinical trials, and development of new technologies; (2) epidemiologic and behavioral studies; and (3) outcomes research and health services research.[20]

- The 1997 OMB Directive 15 minimum standards for maintaining, collecting, and reporting data on race and ethnicity were published in the *NIH Guide for Grants and Contracts*. The primary differences from the previous categories were (1) the Hispanic population is considered an ethnicity and reported separately from racial data, (2) there is a separate racial category for Asian population data and for Hawaiian and Pacific Islander population data, and (3) respondents are given the option of selecting more than one race.[21]

- An NIH guide notice was posted on the Internet (http://grants.nih.gov/grants/funding/women_min/women_min.htm) restating that all applications and awards after October 2000 that have NIH-defined phase III clinical trials must include a description of plans to conduct analyses, as appropriate, by sex/gender and/or racial/ethnic groups. The results of subset analyses must be reported to NIH in annual progress reports, competitive renewal applications (or contract renewals/extensions), and in the required final progress report. NIH-defined phase III clinical trials must be designed and conducted to allow for a valid analysis of whether the variables being studied affect women or members of minority groups differently than other subjects.

- Guidelines and instructions for reviewers and scientific review administrators were developed to emphasize and clarify the need to review research proposals that are classified as NIH-defined phase III clinical trials for both inclusion requirements and issues related to analyses by sex/gender and/or race/ethnicity. Summary statements must document adherence to these policies.

2. FOCUS ON SCIENTIFIC CONSIDERATIONS

The 1994 NIH inclusion guidelines emphasize that the policy is intended to address gaps in scientific knowledge and state that "since a primary aim of research is to provide scientific evidence leading to a change in health policy or a standard of care, it is imperative to determine whether the intervention or therapy being studied affects women or men or members of minority groups and their subpopulations differently."[22] A clinical study without appropriate numbers of women or minority subjects may not be able to address unanswered scientific questions for those populations. Therefore, the inclusion of women and minorities as research subjects is considered an issue of scientific merit.

The intent of the NIH inclusion guidelines is to ensure that scientific norms for health, disease, treat-

ments, and other medical interventions are applicable to all populations (men and women, diverse racial/ethnic groups) based on scientific evidence established by studying those populations; that is, are there biological or other differences in effect based on sex/gender or race/ethnicity?

In defining its standards for inclusion, the NIH has consistently focused on scientific questions: "It is not anticipated that every study will include all minority groups and subgroups. The inclusion of minority groups should be determined by the scientific questions under examination and their relevance to racial/ethnic groups." The 2003 *Outreach Notebook for the Inclusion, Recruitment and Retention of Women and Minority Subjects in Clinical Research*[23] and the accompanying *Frequently Asked Questions*[24] document the circumstances in which it may be acceptable to study groups that lack women or minority participants, provided that the justification is compelling and that the scientific objectives of the research are not compromised. The focus on scientific inquiry also was apparent in the broad definition of clinical research in the inclusion guidelines, which recognizes the need to obtain data about minorities and both men and women in phase I and II studies so that pilot and preliminary data can be included in the design of phase III clinical studies.

The NIH policy for inclusion of women and minorities in clinical research allows for single-sex composition of studies when that is justifiable.[25] In addition to sex-specific studies of the reproductive system and menopause, for example, results from studies that have previously been conducted only in men—such as a number of studies related to diagnosis and treatment of cardiovascular disease—must be validated in women.[26] Furthermore, the causes, treatments, and prevention of disparities among those subpopulations of women and men may allow single-sex composition to define biological behavioral factors that may contribute to differences in health status or outcomes.[27,28]

3. ROLE OF THE NIH OFFICE OF RESEARCH ON WOMEN'S HEALTH

ORWH was established in 1990 and has a mandate to

- set an agenda for future directions in women's health;
- increase and fund research projects on women's health and related sex/gender factors;
- ensure that women are appropriately represented in biomedical clinical research studies; and

- develop opportunities for the recruitment, retention, reentry, and advancement of girls and women in biomedical careers and encourage both women and men to pursue women's health research.

Although ORWH was established in response to concerns about the inclusion of women as subjects in clinical research studies, the 1994 NIH inclusion guidelines, policies, and procedures equally encompassed minorities. In 2000, legislation authorized the establishment of the National Center on Minority Health and Health Disparities (NCMHD) within the NIH.[29] That center continues the legacy of the former NIH Office of Research on Minority Health in partnering with the NIH institutes and centers to support programs of health disparities research with a focus on basic and clinical research, training, and the dissemination of health information. In particular, NCMHD serves as the focal point for coordinating and focusing the minority health disparities research and other health disparities research programs at the NIH into a national health research agenda.

In 1992, ORWH commissioned a report by the Institute of Medicine to address some of the ethical and legal issues associated with including women in clinical studies,[30] and it sponsored public hearings and a workshop in 1995 titled "Recruitment and Retention of Women in Clinical Studies" to address barriers to women's participation in research.[31]

Although much progress has been achieved, the retention of women in clinical studies and the recruitment of populations of women who have been difficult to recruit into clinical research can be improved. A workshop titled *Science Meets Reality: Recruitment and Retention of Women in Clinical Studies and the Critical Role of Relevance*[32] examined the critical role of inclusion in increasing knowledge about the contributions of sex differences and/or similarities to the health and disorders of women, men, and minorities and lessons learned concerning the recruitment and retention of women and other participants from clinical prevention and treatment trials. Emerging ethical and policy issues that present both challenges and opportunities for women's health research and for studies that will elucidate sex and gender factors in health and disease were carefully considered.

ORWH continues to monitor implementation of the inclusion guidelines by overseeing the compilation of aggregate, trans-NIH demographic data on subjects enrolled in NIH-supported studies. ORWH cochairs the NIH-wide Tracking and Inclusion Committee, which was established to address policy compliance as well as data collection, reporting, and quality issues. Additional oversight is provided by the advisory councils of each of the NIH institutes and centers[33] and by the Advisory Committee on Research on Women's Health, which is charged by the Revitalization Act to assist in monitoring compliance with the inclusion requirements.[34]

4. ROLE OF PEER REVIEW

NIH inclusion guidelines emphasize that the policy is intended to address gaps in scientific knowledge, and that inclusion is considered an issue of scientific merit. NIH initial review groups and study sections are instructed to assess a project's inclusion plan as part of their overall evaluation of the research design and reflect that assessment in the priority score.

They assign a gender/minority code that indicates whether the proposed study population meets the inclusion standard, including the requirement to design phase III trials in a manner sufficient to provide for valid analysis of differences in intervention effect (Table 12-1).

Reviewers have the flexibility to assess each research study in light of the scientific questions to be addressed. It is possible for a study that does not include women or minorities to receive an acceptable code, if a convincing justification has been provided. Under NIH

TABLE 12-1   Explanation of Gender/Minority Codes Assigned by NIH Initial Review Groups and Study Sections during Scientific Peer Review[a]

G1A	Includes both genders, scientifically acceptable
G2A	Includes only women, scientifically acceptable
G3A	Includes only men, scientifically acceptable
G4A	Gender representation unknown, scientifically acceptable
G1U	Includes both genders, but scientifically unacceptable
G2U	Includes only women, scientifically unacceptable
G3U	Includes only men, scientifically unacceptable
G4U	Gender representation unknown, scientifically unacceptable
M1A	Includes minorities and nonminorities, scientifically acceptable
M2A	Includes only minorities, scientifically acceptable
M3A	Includes only nonminorities, scientifically acceptable
M4A	Minority representation unknown, scientifically acceptable
M1U	Includes minorities and nonminorities, but scientifically unacceptable
M2U	Includes only minorities, scientifically unacceptable
M3U	Includes only nonminorities, scientifically unacceptable
M4U	Minority representation unknown, scientifically unacceptable

[a]When an application receives a "U" (unacceptable) code it automatically receives a bar-to-funding as well. If the bar is removed, the "U" is converted to "R" to designate that change in status.

review procedures, any application or proposal that is deemed unacceptable with regard to inclusion during initial review receives an administrative bar-to-funding, as does one found to be unacceptable with regard to safeguarding the welfare of human subjects and vertebrate laboratory animals. When this happens, the problem must be corrected before an NIH institute or center may make an award. Thus, the initial review groups play an important role in assessing whether research plans meet the inclusion requirements and have scientific merit.

Most applications describing human subject research meet the inclusion standard as submitted.[35] For applications that are barred because of failure to meet the inclusion requirements, the deficiency found at initial review is addressed by obtaining additional information from the applicant.

NIH's administrative procedures give program staff the flexibility to work with an applicant to ensure that the subject composition is in compliance with the policy. Finally, lack of inclusion in an individual study of men and women and/or minority groups may be justified if that same scientific question is addressed elsewhere for those populations so that together the research portfolio adequately addresses the particular research question for women and minorities.

To assist both reviewers and applicants, NIH published *Frequently Asked Questions* to provide policy guidance and address some of the more commonly asked questions about implementation of the inclusion guidelines.[36] An *Outreach Notebook* that outlines key elements in the outreach process offers some practical suggestions and provides references to additional sources of information.[37]

5. ROLE OF THE INSTITUTIONAL REVIEW BOARD

The institutional review board (IRB) plays an important role in protecting human welfare. This includes the right of subjects to participate or not to participate in research studies as well as their right to share in the potential benefits of research.[38] Both must be considered when designing trials and selecting subjects.[39]

The Office for Human Research Protections emphasizes the role of IRBs in implementing the equitable selection of subjects.[40] Institutions have a responsibility to create an environment in which equitable selection of research participants is fostered and to promote effective recruitment strategies and communication mechanisms to ensure policy implementation.

IRBs should continue to examine research protocols for representation of women, men, and minority groups and must recognize the need for appropriate, not just convenient, population samples. As investigators place more emphasis on the recruitment of women and minority subjects, IRBs need to be particularly sensitive to any special vulnerability of participants with regard to education level or socioeconomic status. For example, they should consider whether consent is informed and ensure that any monetary reimbursements do not promote coercion or undue influence. Cultural sensitivity can be promoted by community members and/or ad hoc advisors who understand the perspectives of various populations and by translators who understand the nuances of communication in another language. Finally, by paying attention to the requirements of the NIH guidelines, IRBs can render an additional service to investigators by identifying weaknesses with regard to subject selection.

6. ROLE OF VOLUNTEERS AND THEIR COMMUNITIES

Many questions remain about why there are disparities in disease prevalence, progression, health outcomes, and excessive mortality for a number of populations in the United States. Although limited access to health care is an important contributor to health status, it is not the only factor that influences differential health status and outcomes. In our evidenced-based health care system, it is essential to understand all of the parameters involved in the disparities in health status and outcomes for minorities, from genetic, biologic, and environmental factors to contributions of culture, behavior, health care, and health care policies (Figs. 12-3 and 12-4). Therefore, it is crucial for women as well as men, for members of diverse racial/ethnic groups, and for those who are disadvantaged by socioeconomic status, geographic location, or other factors, to participate in clinical research, both as study volunteers and as full participants in the planning, implementation, and interpretation of such studies.

As barriers to minority participation in clinical trials are examined, the legacy of the Public Health Service (PHS) syphilis study conducted at Tuskegee figures prominently in the fear about participating in research. The resulting mistrust of the research establishment, especially "the government," is manifested in concerns about being used as a "guinea pig." Thus, the attention focused on the inclusion of racial/ethnic minority subjects in clinical trials must be accompanied by a true sensitivity to the legitimate concerns of the people who are being recruited as research subjects. There must also be a firm commitment and adherence to policies

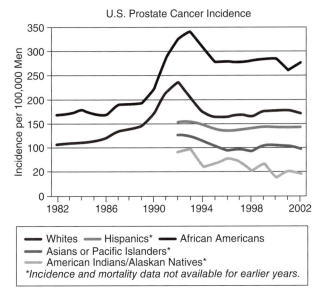

FIGURE 12-3 U.S. prostate cancer incidence, 1982–2002. Data source: Surveillance, Epidemiology, and End Results (SEER) program and the National Center for Health Statistics (www.seer.cancer.gov).

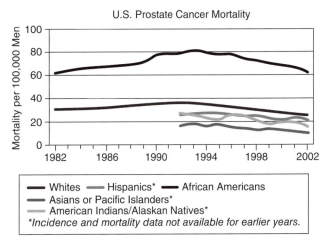

FIGURE 12-4 U.S. prostate cancer mortality, 1982–2002. Data source: Surveillance, Epidemiology, and End Results (SEER) program and the National Center for Health Statistics (www.seer.cancer.gov).

that protect participants in research[41] to allay any lingering fears and mistrust.

The NIH inclusion guidelines ask not just scientists and IRBs but also other groups to ensure that a diversity of study participants are included, that they are protected from harm, and that a mutually beneficial relationship exists between investigators and the populations of interest. The scientific community must address challenges in the recruitment and retention of minority populations and subpopulations and must understand the practices that result in overrepresentation or underrepresentation of subjects. Collaboration with scientists in other communities may be an effective strategy to ensure appropriate representation in clinical studies.

Researchers are encouraged to establish relationships with the community from which participants are recruited in the early stages of study development, incorporate the community's agenda into their research designs, and share their findings with the community. Community representation on IRBs should improve communication, enhance sensitivity to community needs, and foster community involvement in the research process. Effective partnership with the community can foster a sense of mutual responsibility, acknowledge the mutual need for such a partnership, and lead to benefits for both parties. Within the context of investigator–community partnerships, community advocacy can and should exert influence and bring changes in health care standards and policies through

biomedical research that addresses both health and disease issues that may affect the survival and quality of life of a community's population. A true partnership between the scientific enterprise and the broader targeted community can foster biomedical research so that together both can define research questions, determine ways to collect data, inform target populations, and encourage volunteers for studies to elucidate and mitigate the ways in which gender, age, ethnic background, economic status, and lifestyle influence health status.

Together, researchers and the community can develop research initiatives that address the both ethical issues and health needs of the community and thus overcome the negative legacies and memories of the historical events of the PHS syphilis study conducted at Tuskegee.

Communities of potential research participants must be aware, involved, and knowledgeable about the potential risks and benefits of every proposed study. The community and scientific partners should together make efforts to secure and assure trust, participation, informed consent, as well as ensure that the injustices that occurred at Tuskegee will never happen again. Members of the targeted community, local health care institutions, and IRBs share a duty to ensure that risks are minimized, that selection of subjects is equitable, and that the rights and welfare of subjects are maintained. Researchers must become involved in developing culturally respectful community-based research initiatives founded on trust and understanding. This can best be done by including representatives of the community in responsible roles in the planning of research studies, in the recruitment of volunteers, and as members of IRBs.

Both investigators and their community partners must maintain highly ethical research practices with cultural sensitivity throughout the process. Effective recruitment of minorities as volunteers in research protocols can be assisted further by efforts to educate the community about the disease to be studied, its impact on the community, and the need for the information to be derived from the project. Involved community members can help change health and public policy and assure that research and health services are germane to community needs.

Successful research efforts in minority populations depend on collaboration with members of the community at every phase of the research, and should include culturally diverse researchers who can effectively convey research concepts, encourage the inclusion of diverse populations of subjects, and carry out research that is sensitive and appropriate to the health needs of the involved communities. Researchers who are themselves members of the communities being recruited may facilitate the research (Table 12-2).

7. DEMOGRAPHIC DATA ON SUBJECT ENROLLMENT

Research grant and contract applications must include recruitment targets that demonstrate compliance with NIH policies for inclusion of women and minority groups in the study design. When research is in progress, annual progress reports must demonstrate the aggregate accrual to allow tracking and assessment of the extent to which the recruitment plan is successful. These data, along with information about scientific findings and other measures of progress, are evaluated by staff to determine whether stated goals have been met and whether funding should be continued.[42]

When assessing aggregate inclusion data, enrollment figures should not be directly compared with the national census figures. The appropriate numbers of women and minority subjects included in a particular study will depend on the scientific question addressed in the study and the prevalence among women and minorities of the particular disease, disorder, or condition under investigation. The goal of the NIH inclusion policy is not to satisfy any quotas for proportional representation but, instead, to conduct biomedical and behavioral medicine research based on science-driven hypotheses, the results of which will be generalizable to the at-risk or affected populations.

TABLE 12-2   Five Elements of Outreach

Understand the study population. Learn about the people you hope to recruit. Prior beliefs may need to be changed with a goal of valuing, knowing, trusting, and understanding would-be participants. Identify the potential research participants, the medical settings in which they are found, and/or the community in which they reside. Try to learn something about their cultural norms, migration patterns, and reasons for seeking health care.

Establish an explicit outreach plan. Establish specific goals for recruiting and retaining study participants. Where possible, involve formal and informal decision makers from local organizations and institutions, as well as the main communication channels in each medical setting or community. Establish lines of communication to promote continuing awareness of and trust in the project.

Achieve agreement on research plans. Confirm that the investigators, medical staff, and community all agree on the design, methodologies, implementation, and conduct of the study.

Design and conduct evaluations. In cooperation with health care staff, community leaders, and potential participants, pretest and periodically retest the recruitment and retention strategies—including resources, incentives, and problem-solving mechanisms—to ensure that they conform with the needs and values of the research participants and their communities. Monitor subject accrual on a frequent and regular basis and compare results with established goals.

Establish and maintain communication. Keep everyone informed of progress and findings, including research staff, health care providers, participants and their families, and communities. This will increase awareness of the project and demonstrate that the participants and community are valuable partners in the scientific process.

Modified from "Outreach Notebook for the Inclusion, Recruitment, and Retention of Women and Minority Subjects in Clinical Research," NIH Publication No. 03-7036. Bethesda, MD, National Institutes of Health, 2003.

7.1. Tracking System and Aggregate Enrollment Data

A centralized population tracking system was developed to help NIH monitor its performance with respect to the inclusion policy. A trans-NIH committee monitors inclusion data and policy compliance. Table 12-3 demonstrates the data report format utilized for each clinical study.[43]

In addition, the advisory councils of each institute and center review research initiatives and data and certify every two years that their institute or center is in compliance with the NIH inclusion guidelines.[44]

In fiscal year (FY) 2004, there were more than 18 million participants reported for all clinical research conducted at or supported by NIH. Of these, approximately 57.5% were women, 40.9% were men, and 1.5% did not report sex/gender (Table 12-4). Although the number of participants in clinical research significantly increased over the prior years, there was no substantial

TABLE 12-3 Annual Report Form for Each Study Reflecting the 1997 OMB Directive 15 Race and Ethnicity Categories

	American Indian or Alaskan	Asian	Black	Hispanic	White	Race	Other or Unknown	Total	Not Hispanic	Hispanic or Latino	Other or Unknown	Total
Female												
Male												
Unknown												
Total												

The federal Office of Management and Budget (OMB) specifies racial and ethnic categories for data collection and reporting. The NIH collects and reports data according to the 1997 OMB standards for categorizing race and ethnicity. Data collecting format found in the Application for Continuation of a Public Health Service Grant (PHS Form 2590).

TABLE 12-4 Overview of Extramural and Intramural Clinical Research Conducted at or Supported by NIH: FY2004 Enrollment Data

Enrollment Reported	Clinical Studies (Not NIH Defined Phase III)	NIH Defined Phase III Clinical Trials	Total All Clinical Studies
Females enrolled	10,602,296	286,801	10,889,097
%	57.6	55.5	57.5
Males enrolled	7,513,411	228,481	7,741,892
%	40.8	44.2	40.9
Sex of subjects unknown	291,853	1,078	292,931
%	1.6	0.2	1.5
Total subjects enrolled	18,407,560	516,360	18,923,920
%	100.0	100.0	100.0

Source: Spring 2005 comprehensive report: Tracking of human subjects research as reported in fiscal year 2003 and fiscal year 2004. In *Monitoring Adherence to the NIH Inclusion Policy on the Inclusion of Women and Minorities as Subjects in Clinical Research.* Bethesda, MD, U.S. Department of Health and Human Services, National Institutes of Health.

percentage change in the ratio of women and men. However, when sex-specific studies were excluded, the proportions of women and men in all clinical research were proportional to the percentages of the general population.

8. WOMEN OF CHILDBEARING POTENTIAL, PREGNANT WOMEN, AND CHILDREN

For many years, there was a presumption that women of childbearing age should be excluded from clinical studies. Such an approach led to gaps in knowledge about metabolic activity and drug interactions in this group. As a result, drugs have been marketed with undetected side effects, and the benefit of potential treatments may have been delayed for women who did not have access to novel interventions as early as their male or infertile female counterparts.

Historically, the fear that some women may become pregnant contributed to the rationale to exclude all women of childbearing potential from clinical studies. Medical researchers and pharmaceutical manufacturers feared that if a woman participating in research became pregnant and her fetus was harmed, they might be held liable.[45] This fear was often the reason for the exclusion of women from clinical trials, despite a low reported incidence of research injuries and few reported legal cases concerning such injuries.[46] Questions concerning liability risk are difficult to resolve, but there is growing consensus that the exclusion of women from research studies may pose just as much risk of liability as their inclusion. Liability issues were not addressed in the 1994 NIH inclusion guidelines but have been discussed in detail in the Institute of Medicine (IOM) report[47] and in other commentaries.[48]

Investigators are now encouraged to include fertile women earlier in clinical trials. The rationale for this shift in emphasis was described in detail in the IOM report that was commissioned by ORWH.[49] The report emphasized respect for the autonomy of women to make decisions regarding their participation in clinical research studies and recommended that women who participate in research studies should be permitted to select voluntarily the contraceptive method of their choice where there are no relevant study-dependent, scientific reasons for excluding certain contraceptives, such as drug interaction. The IOM report further recommended that federal policy should ensure that neither women nor men of reproductive age should be excluded from participation in clinical studies. Instead, both should have the opportunity to participate in the benefits and burdens of research.

The NIH inclusion guidelines state that "[w]omen of childbearing potential should not be routinely excluded from clinical research"[50] but do not specifically address the participation of pregnant women. In discussing this issue, the IOM report concluded that pregnant women should be presumed eligible for participation in clinical studies.[51] The IOM report further recommended that pregnant woman be excluded only when the IRB finds that there is no prospect of medical benefit to the pregnant woman and there is significant risk of harm to the potential offspring.[52] It is important to note, however, that presuming pregnant women to be eligible is not the equivalent of advocating their active recruitment into every clinical study, because there may be scientifically and medically valid reasons for excluding pregnant women from a particular study.[53]

In moving from a paradigm of exclusion of vulnerable populations to one of inclusion, much still needs to be done to overcome some of the barriers that have prevented women from full participation. Subpart B of the Department of Health and Human Services regulations for the protection of human subjects (45 CFR 46) reflects the presumption that pregnant women are as competent as nonpregnant persons to weigh the risks and benefits of participation in an approved clinical study.[54]

The rationale for requiring the inclusion of women and minorities was based on two important needs: the need for justice in providing access to potential life-saving therapies and the need to obtain information and address gaps in scientific knowledge. Similar arguments could be made regarding the inclusion of the elderly, the disabled, and children. To a large degree, research on the elderly and the disabled is addressed by existing NIH policies and practices. For example, the National Institute on Aging was established at NIH for the specific purpose of examining diseases, disorders, conditions, and natural processes associated with aging. Moreover, previous restrictions on the participation of women of childbearing age often led investigators to select older (postmenopausal) women as research subjects. Likewise, the National

Institute of Child Health and Human Development Center for Medical Rehabilitation Research sponsors projects and training to promote the health, productivity, independence, and quality of life of people with disabilities by conducting research that improves rehabilitation methods, technology, and lifelong care. Respect for autonomy favors voluntary participation of older people and people with disabilities. Both should have the opportunity to participate in the benefits and burdens of research.

A separate policy was developed about the inclusion of children, if scientifically appropriate. Historically, there have been strict safeguards to protect children from the potential risks of participating as research subjects.[55] These safeguards, however, had the effect of presenting obstacles to children's access to the potential benefits of clinical research. In 1998, the NIH implemented guidelines on the inclusion of children in clinical research.[56] Although not binding by law, as is the inclusion of women and minorities, these guidelines require that the inclusion of children as subjects be addressed in all research applications submitted to the NIH, and that the inclusion of children be evaluated during peer review as an issue of scientific merit. This policy was designed to increase our understanding of both diseases of children and disease processes in children.

9. FUTURE QUESTIONS

This chapter has described the evolution of the NIH policy on inclusion of women and minorities in clinical research. Now that the shift to include women in clinical research is being implemented, it is important to remind ourselves that the overarching goal is equitable selection of research participants. With data demonstrating that approximately 59% of participants in clinical research are women, attention now needs to be directed at retaining women as research participants. Sex/gender differences or similarities must be determined through the analysis of results of interventions from studies that include both women and men. Also attention needs to be directed at equitable participation and retention of minorities of both sexes.

The question of race in biology has generated controversy in recent years, with some experts questioning the scientific validity of the very concept of race.[57] In many instances, members of racial/ethnic minorities may display measurable differences in terms of income, education, and employment patterns. In that regard, race/ethnicity may be viewed as a social construct. Socioeconomic status and cultural values both can play an important role in compliance with therapy, access to preventive services, coping strategies, and general outlook on life. At the same time, the possibility of a genetic basis for some differences between populations cannot be overlooked, as clearly illustrated by the potential life-threatening situation in which an individual with the genetically linked glucose-6-phosphate dehydrogenase deficiency is administered certain antimalarials. The health impacts of race and ethnicity are difficult to address, especially in a heterogeneous and mobile society with many confounding factors. Nevertheless, such questions remain in the context of reducing disparities in health status.

Some argue that we should focus on the social and biological influences inherent in racial designations.[58] Burchard et al.[59] hold the view that "evidence does exist, and that recording race allows racially biased health policy and practices to be uncovered." Phimister[60] and others argue that it is important to report race regardless of whether it is a genetic characteristic or a surrogate for one. Phimister et al. state that "the goal of personalized medicine is the prediction of risk and the treatment of disease on the basis of a person's genetic profile, which would render biologic consideration of race obsolete."[60] However, it seems unwise to abandon the practice of recording race when we have barely begun to understand the architecture of the human genome and its implications for new strategies for the identification of gene variants that protect against, or confer susceptibility to, common diseases and modify the effects of drugs.

Observed differences in pharmacologic responses have been reported in racial/ethnic subgroups to a number of therapeutics, such as antidepressants, antipsychotics, antihypertensive medications, and interferon-α treatment of hepatitis C.[61–66] However, isosorbide dinitrate/hydralazine hydrochloride (BiDil), approved in 2005 by the Food and Drug Administration (FDA) for the treatment of heart failure in black patients, was the first drug approved by that agency for race-based clinical application.[67] This has revitalized the discussion and the debate about the role of race in clinical research.

The FDA issued Guidance for Industry: Collection of Race and Ethnicity Data in Clinical Trials in 2006, which provides further discussion of this topic, especially as it relates to drug trials for FDA approval. The NIH policy that requires attention to race and ethnicity in design and analysis of research studies and requires the inclusion of racial and ethnic minorities in clinical research promises to be useful in addressing these issues.

10. CONCLUSIONS

NIH focused on scientific considerations in constructing its guidelines to meet the congressional mandate requiring the inclusion of women and minorities as subjects in clinical research. Implementation of the inclusion guidelines has supported the emerging field of sex-/gender-based medicine. Outreach efforts to communicate the requirements of this policy have been effective and have promoted continued dialog among investigators, IRBs, potential volunteers, and their communities. The policy and procedures for implementing the requirements are effectively monitored and revised as science demands. Studies that do not meet the requirements are being properly identified, and administrative procedures allow for timely resolution of problems. Procedures have been developed to monitor and measure inclusion after research awards are made.[68] Appropriate measures have also been instituted to ensure that the requirement to provide valid analysis of differences in intervention effect for NIH-defined phase III trials is being met and to encourage the publication of this information.

As a result, we have data to track aggregate demographics for study populations on an NIH-wide basis. Substantial numbers of women and minorities are included as research subjects in both phase III trials and other clinical research studies, and NIH institutes and centers are in compliance with legal and policy requirements. When assessing these data, enrollment figures must not be directly compared to the national census figures. The goal of the NIH policy is not to satisfy any quotas but to conduct biomedical and behavioral research in such a manner that the scientific knowledge acquired will be generalizable to the entire population of the United States. The answer to the question, "How many women or minority subjects should be included?" will depend on the scientific question addressed in a particular study and the prevalence of the disease, disorder, or condition under investigation. This answer will vary on a case-by-case basis. The aggregate data provide a measurement of inclusion so that more specific questions about any apparent gaps in enrollment may be formulated and appropriate studies to answer those questions may be designed.

Perhaps the most cogent argument in favor of the NIH inclusion policy is the societal cost of continued gaps in scientific knowledge about important health problems that affect both women and men of diverse racial/ethnic groups. Sex and racial/ethnic differences must be appraised when generalizing results to entire populations because a "one size fits all" standard of care is no longer acceptable.[69]

Acknowledgments

We are particularly indebted to Dr. Eugene Hayunga, who coauthored the original version of this chapter. Special recognition is extended to members of the NIH Tracking and Inclusion Committee for their leadership in ensuring effective implementation of the inclusion policies. The efforts of countless NIH review, program, grants management, and other staff, who continue to monitor compliance, are deeply appreciated.

References and Notes

1. Grady D, Rubin SM, Petitti DB, *et al.* Hormone therapy to prevent disease and prolong life in postmenopausal women. *Ann Intern Med* 1992;117(12):1016–1037.

2. American Heart Association. *Heart Disease and Stroke Statistics—2002 Update.* Dallas, American Heart Association, 2002.

3. Lerner DJ, Kannel WB. Patterns of coronary heart disease morbidity and mortality in the sexes: A 26-year follow-up of the Framingham population. *Am Heart J* 1986;111(2):383–390.

4. Healy B. The Yentl syndrome. *N Engl J Med* 1991;326(2): 274–276.

5. Wenger NK, Speroff L, Packard B. Cardiovascular health and disease in women. *N Engl J Med* 1993;329(4):247–256.

6. Grady D, Chaput L, Kristof M. *Results of Systematic Review of Research on Diagnosis and Treatment of Coronary Heart Disease in Women. Evidence Report/Technology Assessment No. 80,* AHRQ Publication No. 03-0035, p. 11. Rockville, MD, Agency for Healthcare Research and Quality, 2003.

7. National Institutes of Health Revitalization Act of 1993 (Public Law 103-43), 107 Stat. 22 (codified at 42 U.S.C. Subsection 289. a-1), June 10, 1993. Available at http://orwh.od.nih.gov/inclusion/revitalization.pdf

8. U.S. Department of Health and Human Services, National Institutes of Health. *NIH Guidelines on the Inclusion of Women and Minorities as Subjects in Clinical Research.* Bethesda, MD, U.S. Department of Health and Human Services, National Institutes of Health, amended October 2001. Available at http://grants. nih.gov/grants/guide/notice-files/NOT-OD-02-001.html

9. Hayunga EG, Pinn VP. Women and minorities in clinical research: NIH response to researchers' concerns. *Appl Clin Trials* 1996; 5(1):59–64.

10. *Revitalization Act, supra* note 7, at Section 492B(b); *Guidelines, supra* note 8, at Section II.A.

11. The 2001 NIH guidelines define a phase III trial as a broadly based prospective clinical investigation for the purpose of evaluating an experimental intervention in comparison with a standard or control intervention or comparing two or more existing treatments. Often, the aim of such investigations is to provide evidence leading to a scientific basis for consideration of a change in health policy or standard of care. The definition includes pharmacologic, nonpharmacologic, and behavioral interventions given for disease prevention, prophylaxis, diagnosis, or therapy. Community trials and other population-based intervention trials are also included. See *Guidelines, supra* note 8, at IV.B.

12. *Revitalization Act, supra* note 7, at Section 492B(c); see also *Guidelines, supra* note 8, at II.B and IV.C for discussion of valid analysis and specific requirements for phase in clinical trials.

13. The 2001 NIH guidelines (*supra* note 8) repeat the language of PL 103-43 (*supra* note 7) to explain that "exclusions to the requirement for inclusion of women and minorities are stated in the statute as follows: The requirements established regarding women and members of minority groups shall not apply to the project of clinical research if the inclusion, as subjects in the project, of women and members of minority groups, respectively—
a. Is inappropriate with respect to the health of the subjects;
b. Is inappropriate with respect to the purpose of the research; or
c. Is inappropriate under such other circumstances as the director of NIH may designate" 492B(b); and "(B) In the case of a clinical trial, the guidelines may provide that such inclusion in the trial is not required if there is substantial scientific data demonstrating that there is no significant difference between—
  (i) The effects that the variables to be studied in the trial have on women or members of minority groups, respectively; and
  (ii) The effects that variables have on the individuals who would serve as subjects in the trial in the event that such inclusion were not required" 492B(d)(2).

14. LaRosa JH, Seto B, Caban CE, Hayunga EG. Including women and minorities in clinical research. *Appl Clin Trials* 1995;4(5): 31–38.

15. Hayunga EG, Pinn VW. Women and minorities in clinical research: Implementing the 1994 NIH guidelines. *Appl Clin Trials* 1996;5(10):34–40.

16. *Ibid, supra* note 8; LaRosa *et al., supra* note 14.

17. U.S. General Accounting Office. *Women's Health: NIH Has Increased Its Efforts to Include Women in Research,* GAO/HEHS-00-96, May 2000.

18. The *1997 Report of the NIH Director's Panel on Clinical Research Definition of Clinical Research.* Available at http://www.nih.gov/news/crp/97report/execsum.htm.

19. U.S. Office of Management and Budget (OMB) Directive 15, *Revisions to the Standards for the Classification of Federal Data on Race and Ethnicity.* Available at http://www.whitehouse.gov/OMB/fedreg/ombdir15.html.

20. NIH clinical research definition, *supra* note 18.

21. U.S. OMB Directive 15, *supra* note 19.

22. *Guidelines, supra* note 8, at Summary section.

23. U.S. Department of Health and Human Services, National Institutes of Health. *NIH Outreach Notebook for the Inclusion, Recruitment and Retention of Women and Minority Subjects in Clinical Research,* NIH Publication No. 03-7036. Bethesda, MD, Available at http://orwh.od.nih.gov/inclusion/outreach. pdf.

24. U.S. Department of Health and Human Services, National Institutes of Health. *NIH Frequently Asked Questions Concerning the NIH Guidelines on the Inclusion of Women and Minorities Subjects in Clinical Research.* Bethesda, MD. Available at http://orwh. od.nih.gov/inclusion/outreachFAQ.pdf.

25. Pinn VW. Sex and gender factors in medical studies: Implications for health and clinical practice. *JAMA* 2003;289:397–400.

26. Pinn VW. Women's health: Research, progress and priorities. *Women's Health* 2005;1(1):1–4.

27. Pinn VW. From policy to science: Enhancing the inclusion of diverse ethnic groups in aging research. In Curry L, Jackson J (eds.) *Recruiting and Retaining Racial and Ethnic Elders in Health Research,* pp. vii–viii. Washington, DC, Gerontological Society of America, 2003.

28. Pinn VW. Research on women's health: Progress and opportunities. *JAMA* 2005;294(11):1–4.

29. U.S. Department of Health and Human Services, National Institutes of Health, National Center for Minority Health and Health Disparities website: www.ncmhd.nih.gov.

30. Mastroianni AC, Faden R, Federman D (eds.). *Women and Health Research. Ethical and Legal Issues of Including Women in Clinical Studies.* Washington, DC, Institute of Medicine, National Academy Press, 1994.

31. U.S. Department of Health and Human Services, National Institutes of Health. *Recruitment and Retention of Women in Clinical Studies. A Report of the Workshop Sponsored by the Office of Research on Women's Health*, NIH Publication No. 95-3756. Bethesda, MD, U.S. Department of Health and Human Services, National Institutes of Health, 1995.

32. U.S. Department of Health and Human Services, National Institutes of Health. *Science Meets Reality: Recruitment and Retention of Women in Clinical Studies and the Critical Role of Relevance*, Publication No. 03-5403. Bethesda, MD, U.S. Department of Health and Human Services, National Institutes of Health, 2003. Available at http://orwh.od.nih.gov/pubs/SMR_Final.pdf.

33. See *Revitalization Act, supra* note 7, Section 486, Subsection [d][4][D], which states that "the advisory council of each national institute shall prepare biennial reports describing the manner in which the institute has complied with this section."

34. See *Revitalization Act, supra* note 7, Section 486, Subsection [d][4][D], which states that the Advisory Committee on Research on Women's Health shall "assist in monitoring compliance with section 492B regarding the inclusion of women in clinical research."

35. *Implementing the 1994 Guidelines, supra* note 15.

36. *Frequently Asked Questions, supra* note 25.

37. *NIH Outreach Notebook, supra* note 24.

38. McCarthy CR. Historical background of clinical trials involving women and minorities. *Acad Med* 1994;69:695–698.

39. Levine RJ. The impact of HIV infection on society's perception of clinical trials. *Kennedy Institute Ethics J* 1994;4(2):93–98.

40. U.S. Office for Human Research Protections (OHRP) website: www.hhs.gov/ohrp. [See IRB Registrations and Assurances.]

41. Pinn VW. The role of NIH's Office for Research on Women's Health. *Acad Med* 1994;69:698–702; See also *IOM Report, supra* note 32.

42. The Application for a Public Health Service Grant (SF424(R&R)/PHS Form 398) and the progress report for noncompeting continuation (PHS Form 2590) request target enrollment figures be provided in a similar format. Applicants are asked to describe their projected study populations quantitatively and in sufficient detail for full consideration during initial scientific review. Reporting such data facilitates direct comparison of target and actual enrollments during review of the annual progress reports to determine how well an applicant's stated goals have been met.

43. NIH has reported participant data in both the 1977 OMB standards (combined race/ethnicity format) and the 1997 OMB standards (separate race and ethnicity formats). The FY2000 data tables, which report data according to the 1977 OMB categories and standards, are the final tables produced from the old NIH tracking system. Since 2002, investigators have been required to report data using the 1997 OMB categories and standards. As a result of these changes, direct comparisons between the 1977 and the 1997 data formats cannot be accurately combined for a specific race/ethnicity category. For example, following the 1997 OMB standards, "Asian" is a separate race from "Hawaiian/Pacific Islander," and "Hispanic or Latino" is an ethnicity and also reported by racial category.

44. See *Revitalization Act* §486[d][5][A], *supra* note 7.

45. Rothenberg KH. Gender matters: Implications for clinical research and women's health care. *Houston Law Rev* 1996;32(5):1201–1272; see also *IOM Report, supra* note 31, at page 150.

46. As noted in IOM *Report, supra* note 31, at page 12, "It is impossible to quantify the risk of tort liability from the inclusion of women in clinical studies at this time, because: (1) there is no complete compendium of unreported cases involving settlements and (2) pregnant women and women of childbearing age have not been included in some major studies in the past." See also *IOM Report, supra* note 32, at page 151.

47. IOM *report, supra* note 31.

48. For a general overview of liability issues, see Hayunga EG, Rothenberg KH, Pinn VW. Women of childbearing potential in clinical research: Perspectives on NIH policy and liability issues. *Food Drug Cosmetic Medical Device Law Digest* 1996;13(1):7–11. For a more complete review of the topic, see Rothenberg K. 1996. Gender matters, *supra* note 46.

49. IOM *Report, supra* note 31; a succinct summary of these recommendations may be found in Rothenberg KH. *The Institute of Medicine's Report on Women and Health Research: Implications for IRBs, Houston Law Rev* 1996;32(5); see also Applebaum PS. Drug-free research in schizophrenia: An overview of the controversy. *IRB* 1996;18:1–3.

50. *Guidelines, supra* note 8, at Section II.A.

51. IOM *Report, supra* note 31, at page 195.

52. *Ibid.*

53. For example, a pregnant woman could be appropriately excluded from a study of hormone replacement or contraception. See *Rothenberg, supra* note 46, citing IOM *Report, supra* note 31, at page 17.

54. U.S. Department of Health and Human Services. Protection of human subjects: Pregnant women, human fetuses, and newborns as research subjects and pertaining to human *in vitro* fertilization: Proposed rules. *Fed Reg* 1999;63:27794–27804.

55. Department of Health and Human Services, Food and Drug Administration (FDA), 45 CFR 46 Federal Policy for the Protection of Human Subjects: Additional Protections for Children Involved as Subjects in Research, Subpart D. *Fed Reg* 1991;56:29756.

56. NIH Policy and Guidelines on the Inclusion of Children as Participants in Research Involving Human Subjects, Notice 98-024, March 6 1998. Available at http://grants.nih.gov/grants/guide/notice-files/not98-024.html.

57. Freeman HP. The meaning of race in science—Considerations for cancer research. Report of the President's Cancer Panel, National Institutes of Health, National Cancer Institute, Bethesda, MD. *Cancer* 1998;82(1):219–225.

58. Cooper RS, Kaufman JS, Ward R. Sounding board: Race and genomics. *N Engl J Med* 2003;348:1166–1170.

59. Burchard EG, Ziv E, Coyle N, *et al.* The importance of race and ethnic background in biomedical research and clinical practice. *N Engl J Med* 2003;348:1170–1175.

60. Phimister EG. Perspective: Medicine and the racial divide. *N Engl J Med* 2003;348:1081–1082.

61. Xie H, Kim R, Wood A, Stein C. Molecular basis of ethnic differences in drug disposition and response. *Annu Rev Pharmacol Toxicol* 2001;41:815–850.

62. Exner D, Dries D, Donamski M, Cohn J. Lesser response to angiotensin-converting-enzyme inhibitor therapy in black as compared with white patients with left ventricular dysfunction. *N Engl J Med* 2001;344:1351–1357.

63. Yancy C, Fowler W, Colucci E, *et al.* Race and the response to adrenergic blockade with carvedilol in patients with chronic heart failure. *N Engl J Med* 2001;344:1358–1365.

64. McHutchison J, Poynard T, Pianko S, *et al.* The impact of interferon alphas plus ribaviran in response to therapy in black patients with chronic hepatitis C. The International Hepatitis Interventional Therapy Group. *Gastroenterology* 2000;119: 1317– 1323.

65. Reddy KR, Hoofnagle J, Tong WL, *et al.* Racial differences in responses to therapy with interferon in chronic hepatitis C. Consensus Interferon Study Group. *Hepatology* 1999;30: 787–793.

66. U.S. Department of Health and Human Services, Food and Drug Administration, Center for Drug Evaluation and Research (CDER), Center for Biologics Evaluation and Research (CBER), Center for Devices and Radiologic Health (CDRH), and Office of the Commissioner (OC). *Guidance for Industry: Collection of Race and Ethnicity Data in Clinical Trials*, September 2005. Available at www.fda.gov/cder/guidance/5656fnl.htm.

67. Taylor AL, Ziesche RN, Yancy C, *et al.* Combination of isosorbide dinitrate and hydralazine in blacks with heart failure. *N Engl J Med* 2004;351:2049–2057.

68. Hayunga EG, Costello MD, Pinn VW. Women and minorities in clinical research: Demographics of study populations. *Appl Clin Trial* 1997;8(l):41–45.

69. Pinn VW. Equity in biomedical research. *Science* 1995;269:739.

CHAPTER

13

Clinical Research:
A Patient Perspective

SUSAN LOWELL BUTLER

DC Cancer Consortium, Washington, DC

In 1995, I began my personal experience with clinical research when I participated in a clinical trial at the National Cancer Institute at the National Institutes of Health.

Diagnosed with simultaneous breast and ovarian cancers, I was admitted on a compassionate basis to an intense, three-drug protocol aimed at obliterating ovarian cancer; the thinking was that the medications would probably prove efficacious for the breast cancer as well, and further treatment for that cancer could follow the initial assault on my stage III-C ovarian cancer.

Since that time, more than 10 years from this writing, I have become an active survivor, helping to found a national advocacy organization for ovarian cancer survivors and participating in a number of activities with and for the National Cancer Institute (NCI), including service as a peer reviewer for NCI grant proposals and service for 7 years on the NCI Director's Consumer Liaison Group and on the Patient Advisory Group for the NIH Clinical Center.

Through my engagement with NIH, I have had the opportunity and privilege to observe the process of patient participation in clinical trials. In 2003, at the request of Dr. John I. Gallin, Director of the NIH Clinical Center, I became a faculty member for the course "Introduction to the Principles and Practice of Clinical Research," when I was asked to provide my observations on patient participation in a clinical research setting. Now it is my privilege to share my observations with those of you, nationally and internationally, who have an interest in clinical research. In light of the potential for saving lives and advancing knowledge—

and in light of my personal history—I can think of no higher calling.

1. THE PATIENT–SCIENTIST PARTNERSHIP

I believe that a clinical treatment trial is a partnership between patient and researcher that is based on trust and understanding. This is a partnership of peers, with both participants hoping for a favorable outcome that saves lives and advances science. At the heart of this partnership—for patient and scientist alike—is a basic optimism and hope for the future.

Most patients, when asked, have an understanding of why clinical research scientists engage in trials: to make important contributions to their field, to advance science, and, most important, to save lives.

Similarly, researchers should understand why patients choose to participate in trials; understanding should help light the way to designing trials that will successfully attract participants. Research and observation on the subject show an array of patient motivations:

- To survive, to regain health, perhaps to find a "miracle"
- Refusal to surrender to their disease, and the will to fight on
- To take advantage of a new treatment that may be inaccessible any other way
- Because of pressure from family

- Trust in the physicians and other staff who will provide care
- Trust in the physician who referred the patient to the trial
- Because the physicians and staff conducting the clinical trial are considered top experts in the field
- To contribute to the knowledge base of their disease, even if what is learned may not be of personal benefit
- To have the satisfaction of knowing that one has done everything possible to fight the disease
- Because the trial represents a tolerable risk, in that a person may quit the trial at any time without negative consequences to subsequent treatment
- For monetary gain, although for most treatment trials this is a minor motivation, if it is present at all.

1.1. A Good Start

The literature on patient motivation for trial participation, as well as anecdotal conversations with other patients, shows that the manner in which the trial initially is explained is of pivotal importance. Institutions doing clinical research should invest time and thought into evaluating how clinical trials are presented to prospective patients, with particular attention to several key elements:

Make the institution welcoming. Remember that most people receive the majority of their health care in the setting of a physician's office or a relatively small clinic. Large clinical research institutions are strange and new to the majority of patients, and it is important to factor into planning how patients will be welcomed and guided when they first arrive to discuss the clinical trial. A negative impression may begin at the front door.

Hold the discussion about the trial in a private setting, where others cannot overhear the discussion or interrupt. This is not always how presentations are made, and some patients have reported being presented with complex clinical trial information in the middle of a busy clinic.

Have the principal investigator present to meet the patient and family, as well as other staff with whom the patient will interact; this initial contact should neither be hurried nor cursory. Patients report deciding positively or negatively about the trial in some measure at the time of meeting and talking with the principal investigator.

Important materials, such as the informed consent form and information sheet(s), should be presented to the patient slowly and thoroughly, allowing ample time for questions. Even well-educated individuals may be confused by the language of these forms and may need (even if they do not always ask) "translation" as each aspect of the trial is discussed. People with less education probably will have even more difficulty, and this should be anticipated when the consent form and other materials are prepared. In the best of all possible worlds, time also would be taken to develop a low-literacy version of important materials drafted, so no one is excluded simply because the material is inscrutable. In all cases, it would be useful to provide patients (who want them) with CDs, DVDs, and disks containing the most important information, as well as printed materials.

Despite a growing emphasis on creating relatively simple informed consent forms and other materials, it seems that anxious attorneys and others are pushing the other way. This is resulting, in some places, in what are described as "frightening" materials on possible side effects that are scaring prospects away if health care professionals are not there to discuss the reality of side effects and what is likely to happen and what is not.

Do not push for a decision. For many people, deciding to participate in a clinical treatment trial can be life altering. It may not be possible or wise to decide whether or not to participate at the time of the initial conversation about the trial. In general, at least a day or two—longer if need be—should be factored in to give people a chance to discuss the trial with loved ones and think about what will be the best choice. During this period of decision making, make sure that the patient and family have access to contact the principal investigator if there are questions.

Be very clear about how costs relating to the trial are to be handled. The patient and family must know, up front, what costs the institution covers and what costs it does not cover. It is of no benefit to the host institution to enroll people in a clinical trial and have some depart when they later learn there are related costs that they must personally bear. In addition, trial planners must realize that although the well-insured will find that many of the costs patients must bear are covered, those with weak insurance or none are inherently and unfairly disadvantaged. Provision—in the planning stage—should be made to cover additional costs for eligible patients who would otherwise have to decline to participate for this reason.

The randomization ("guinea pig") factor can be very discouraging to patients. Some may perceive the possibility of not being selected for the new agent or treatment as an extreme disadvantage. In general, randomization and the reasons for it are not well understood by the public, and it is worth time and energy in the trial discussion process to address this fully, including the explanation that all participants will receive state-of-the-art care for their cancers or other disorders if they are not randomized to the trial agent, and that should the new agent be a success, the other trial participants will be the first to be receive treatment with it.

Take time to explain the "nuts and bolts." Patients should come away from the initial conversation about the clinical trial fully understanding how every aspect of the process will work: where and how often to come to the research institution, who to ask for, what will happen next, how the testing process works, how the drugs will be administered, how long the process will take, and how he or she will feel once the visit concludes. All of this should be in writing as well. Transportation issues play an important role in deciding about participation for many people, so advice, support, and information about this aspect of the process are also important. Also, patients will need to know whether or not they should plan to bring a caregiver along for the whole process or perhaps just have someone accompany them home. Remember that patients have, in many instances, experienced a complete upending of their lives and routines. If this is kept in mind, providing concrete details meets a need for assurance and comfort that will aid the patient.

Discuss important issues relating to a patient's life. People want to know answers—or at least educated guesses—about the practical aspects of life while undergoing treatment. Will they be able to continue to care for their family? Go to work? Maintain most of their daily schedules? Although it may be impossible to determine how an individual patient will fare, it can be helpful to speak in general terms about what has been observed in other trial participants. For example, people with cancer are actively struggling to adapt to the new reality that has overwhelmed their lives, and the insight and information health care professionals have are valuable.

Make patients aware of the array of services the trial institution has available that can help—social workers, chaplains and other faith-based support, registered dieticians, rehabilitation therapists, and more. In some institutions, there may be facilities, such as a business center, where patients and family members can access the Internet to do work or research their illness.

Remember the primary care physician. Sometimes patients considering clinical trial participation are encouraged to do so by their primary care physician, and sometimes they are not. In any event, it is helpful to both the patient and the research or clinical staff to know about the relationship and attitudes of this physician and to assure the patient that if he or she goes on to the clinical trial, the primary care physician will be kept apprised of progress on a regular basis.

2. WALKING AWAY: WHY PATIENTS REFUSE TO PARTICIPATE IN CLINICAL TRIALS

There is a growing body of research addressing the reasons people choose not to participate in clinical trials that they may be made aware of in the course of treatment of cancer and other serious illnesses, quite apart from potential participants who may be excluded because they do not meet the criteria. Since my focus here is to address the experience of patients who choose to participate in treatment trials, I provide only a snapshot of the reasons people just say no, in the hope that understanding these reasons may spark a commitment to do whatever possible in the clinical setting to forestall them. I also address specifically the reasons minority populations and the elderly reject trial participation.

The following are general reasons cited for rejecting trial participation:

Poor presentation of the clinical trial: The setting for discussion may be relatively public, the individuals presenting the material may be rushed and lack answers to a patient's questions, there is no contact from the principal investigator, the description of side effects is frightening and unclear—there are many things that can make up a poor, confusing introduction to a trial.

The information is too confusing/complicated/frightening: Patients, especially those who have less than a high school education, sometimes struggle with what are often complex written materials. People are sometimes suspicious of what is not readily understood, comfortable, and familiar.

Since the people who have prepared the trial and the materials for patients often are highly educated, it can be helpful to have outside lay people, particularly those skilled in personal communications, review the materials and make recommendations to make the information as clear and simple as possible.

Concern about relationship with the primary care physician: Some physicians discourage trial participation, and patients may value their primary care physician's views highly.

Language barriers: For people with little or no command of English, enrollment in a clinical trial can be particularly difficult. In many instances, trial providers are not able to provide regular and reliable translation support, making it difficult for potential participants to ask questions and gather information. Adding this barrier to the struggle with serious illness can be overwhelming.

Some institutions seeking to accommodate non-English speakers have relied on family and friends to translate for the patient. Although this may be successful in some instances, it may not always work unless the translating individual has some knowledge of medical terminology.

Cost factors: If a trial requires that some costs be borne by the patient and/or the patient's insurance company, and if the trial institution has not set aside funds for those without insurance or those with poor insurance coverage, some patients will withdraw.

Transportation: For a person who is ill, getting to treatment, especially if complex and frequent travel is required, may be impossible. Some patients inevitably will opt to seek care in a more convenient setting.

Objection to randomization: Sometimes patients enroll in clinical trials because they are seeking specific access to a new drug that is being tested. If they are randomized to standard care, they may drop out. Also, others object to randomization on principle or do not understand fully that they will receive state-of-the-art care if they are not randomized to the new treatment.

Fear of the treatment: The process of absorbing the information about the trial protocol can be frightening, and some people may think they cannot physically withstand what is to come.

Conflict with religious or cultural beliefs: Some individuals may find the trial requirements, the examinations, and other procedures in conflict with the tenets of their faith.

Fear/mistrust of research and the institutions that conduct it: Sensational stories about "trial scandals" have impacted some people's minds to the degree where considering participation in a research trial is frightening. Underscoring this for some may be a deep and recent skepticism about so-called

"safe" medications approved by the Food and Drug Administration, promoted by the recent outcry and trials regarding the COX-2 inhibitor Vioxx and the withdrawal of Bextra from the market.

Dislike of the clinical research environment: For people who have been treated only in the setting of a private physician's office, the clinical research environment can be intimidating and unwelcoming. Most individuals overcome this initial reaction, but not all.

2.1. Why African Americans Are Underrepresented in Clinical Trials

Despite the fact that African Americans have the highest age-adjusted cancer incidence and mortality in the nation, they continue to be significantly underrepresented in clinical trials for cancer. This fact continues to undermine research validity in some instances since a significant population has not been appropriately represented in studies. This also denies some African Americans state-of-the-art cancer care.

An overview of the research reveals an array of reasons, in addition to the reasons people in general reject trials listed previously, including the following:

Deep mistrust of research performed on minorities: The searing Tuskegee Institute experiment, in which African Americans with syphilis were untreated for decades, has left a wide, deep, and lingering scar. Mistrust of the medical profession and of clinical research is the result.

Restrictive trial participation criteria: Some research has shown that African American cancer patients may present with numerous comorbidities that frequently bar them from enrolling in trials.

Religious faith: Some African Americans' faith in God is such that they believe that God will decide whether or not they will die from their disease, regardless of treatment approaches.

Barriers presented by transportation difficulties, costs, other demands, and time: These are also common factors for why African Americans do not participate in trials.

Absence of minority health care professionals: Minorities sometimes will note specifically the absence or scarcity of people of color working in the research environment, and this can cause both suspicion and discomfort.

Lack of awareness of clinical trials. Recent research shows that African-Americans are largely unaware of clinical trials and therefore do not consider this option when considering treatment decisions.

In addition to previous suggestions on how to make the recruitment process work better, there are other ideas that may help encourage African Americans to participate. Of course, the institution conducting the trial must begin by making a strong commitment to enrolling minorities proportionate to the area population and back that commitment with concrete steps to improve performance. These steps might include the following:

- Recruiting and utilizing African American health care professionals and recruiters to work in the community and in the institution on recruitment.
- Asking African American patients who have successfully participated in the trial (and are no longer enrolled) to talk directly with potential enrollees about their experiences in the trial.
- Working directly with African American community physicians to make them aware of available clinical trials and how these might benefit their patients.
- In some instances, it might be feasible to reevaluate trial requirements to determine if some might be reconsidered in light of otherwise eligible African American prospective participants.

2.2. Why the Elderly Are Underrepresented in Clinical Trials

Elderly cancer patients are recruited for clinical trial participation less frequently than other populations. The research on this matter suggests a variety of reasons:

- Elderly patients often take an array of prescription and over-the-counter medications that can cause significant drug interactions when combined with treatment in clinical trials. Stopping or reducing dosage of these medications can be complicated if this is necessitated by trial requirements.
- The elderly may have considerably more difficulties in coping with trial logistics, especially costs and travel.
- Some protocols may have unnecessarily strict exclusion requirements that by their nature rule out most of the elderly population. For example, very few elderly people have no preexisting conditions.
- Determining the appropriate dosage of new drugs being tested in the trial may present difficulties when working with older patients.
- There may be the general perception on the part of some health care professionals that elderly people are inherently too frail or even incompetent to participate in trials.

Since cancer is so often a disease of the elderly, the relatively low and slow participation of this population seems a significant disservice to research progress. The following steps might be considered in recruiting older patients:

- Rapport and trust with the clinical trial staff members can be very important, and a special effort should be made with older patients so they feel welcome and comfortable and able to express their needs and concerns freely.
- Enlisting the help and support of an older patient's family and friends to help with travel and other logistical matters.
- Having a consistent person to contact in the trial structure can prove helpful (for all patients, not just the elderly).
- Contact with the patient's primary care doctor can be especially important with elderly patients, who may visit this individual frequently.
- Sensitivity to the length and stress levels of a patient's treatment day at the clinic can be important, and efforts should be made to avoid overtaxing older patients.

3. THE TRIAL BEGINS: UNDERSTANDING THE PATIENT EXPERIENCE

The patient participating in a clinical treatment trial is, in many ways, undergoing a very different set of experiences than that of a patient pursuing standard treatment. To make the patient–scientist partnership work, it is important that the impact of these experiences be understood by trial administrators.

3.1. The Worst News

First, patients choosing a clinical treatment trial are sometimes very ill or have been given a dire or even terminal diagnosis. For individuals who have not experienced serious illness, it is a great stretch to place oneself in the patient's shoes, but it is important to try.

Advanced illness has a profound and generally negative initial impact on an individual's perspective and ability to function, absorb information, cope with family environment, employment, and social relationships—literally everything that makes up the fabric of life. Basic anchors in life—one's health, family, work, friends, and routines—are all disrupted.

A deep, pervasive sense of loss is sometimes present and can be tantamount to grief. Some patients entering a trial may still be "grieving" their diagnosis, even as they begin treatment they hope will help them. This terrible sense of loss that a bad diagnosis delivers can include a perception that life as one knew it has ended forever; that one's body has failed; and that there may be, in reality, little or no hope for the future. This burden of grief should not be underestimated or forgotten because it can sometimes linger throughout the course of treatment. The patient before you is not the individual he or she was before advanced illness became a reality. How each individual handles this major disruption—and it will change over time—will of course vary by personality and other intangible factors, but health care providers dealing with people enrolling in clinical trials should be prepared to understand the situation and the reasons for it.

3.2. A New World

A second significant factor for most patients entering trial participation, at least for those receiving care in a cancer center or other large institution, is adapting to an entirely new environment—at a time when one's ability to cope may be undermined by advanced illness.

Cancer centers, like many large institutions, have a life, language, and rhythm of their own, and this is a world entirely foreign to most patients. Patients report experiencing the following:

A loss of personal and professional identity: One is no longer Jane Smith, wife, mother, attorney, and colleague. Now it is Jane Smith, "the colon cancer," Number 55-089-004H, assigned to Ward 6 on the fourth floor. Now it is Jane Smith, alone in a hospital bed, gazing out into the night, often alone, wondering what lies ahead for her, and whether her life will ever again be whole.

Difficulty coping with a new, sometimes large treatment institution, where everything can be confusing, exhausting, and sometimes difficult to navigate: In large institutions, some patients report struggling in a "maze," shuttling from radiology to chemotherapy treatment rooms to phlebotomy to wards—and around again—either alone or escorted by strangers, treated by strangers, and processed by strangers. For many patients, coping with this new maze, and a sea of strangers, while seriously ill is a daunting challenge.

A struggle to master the "language" of one's disease, one's trial, medications, monitoring needs, and more: This struggle is particularly intense in the days and weeks immediately following initial diagnosis. Even highly educated professionals must struggle unless they come from a health care professional

environment. This struggle is amplified by a very human tendency to cling to every word, every nuance of conversation with health care professionals, to attempt to determine how the treatment is going and whether the disease is responding.

In time, most patients orient themselves, and a setting that was once alien and confusing becomes familiar. However, understanding the sadness, disorientation, and mind-set of patients can help those who plan clinical trials to design systems and structure that provide better support and comfort. Doing so, in turn, can help patients adapt more quickly to their environments and encourage them to remain trial participants.

3.3. The Lay Expert

In today's environment, many patients, especially those who have been struggling with a serious illness for some time, will come to a clinical trial having done considerable research about their disease. Although the Internet offers its fair share of unreliable and inaccurate information, many patients today have learned to find their way to reliable information and often will spend many hours researching and learning and absorbing. It is extremely important that research scientists and other health care professionals understand why patients and their family members do this—and equally important that these efforts to learn and to make informed choices be respected and even encouraged.

Why do patients research their illnesses? There are many reasons—to understand their illness in all of its ramifications, to find the best possible treatment options, and to consider their prognosis and make plans. However, what is underlying this effort in many patients is an intense desire to have some *control* over one's life and one's health. Knowledge truly can be power for a patient with a difficult disease, and for this reason alone, if for no other, clinicians must respect patient efforts to learn. If one loses control, then the next loss may well be hope. And without hope, interest in trial participation, and indeed life, can be lost.

Inevitably, some patients will come to trial participation and treatment with at least some misinformation. How the clinician handles this can be important. A best course to consider would be to respectfully correct the misinformation not only verbally but also perhaps by directing the patient to sources of reliable information, on the Internet and elsewhere.

One unfortunate approach entails directing a patient to "stop researching" and just rely on information provided by the health care team. Not only will this prove unsuccessful but also it will fracture what may already be a fragile partnership because the patient will be offended. In general, treating patients like tall children is ineffective and alienating.

4. UNDERSTANDING THE CAREGIVER

Like patients, caregivers have had their lives utterly disrupted. Their responsibilities have suddenly increased, perhaps dramatically. They are consumed with worry for their family member and concerned about the uncertainties ahead. The realities of this new environment and the demands on caregivers' time may also spark considerable financial concerns. Caregivers, for the most part, want to spend as much time with the patient as possible, and pressures (other family members' needs and employer demands) can make this difficult.

According to a 1996 study by the National Alliance for Caregiving and AARP, 50% of caregivers perform care for at least 8 hours a week, and nearly 20% provide 40 or more hours of care per week. Nearly 64% of caregivers are working full- or part-time, and 40% of those providing more than 50 hours per week of caregiving are also working. This is a portrait of overload.

Caregivers have a profound need for information from the staff, especially if the patient is newly diagnosed and/or acutely ill. In some instances, caregivers may not ask for this information, for a variety of reasons, including not wanting to take time or attention away from the patient and not yet really knowing what they need to know. The clinical trial staff members can be most helpful by taking the initiative in working with the caregiver. Suggestions include the following:

- Make sure the caregiver knows how to reach the trial staff, 24 hours a day, 7 days a week, if he or she has questions. This may mean providing a day number and a night number for questions that just cannot wait until the daytime but may not be full-blown emergencies.
- The caregiver should be given comprehensive information about all of the support services the cancer center or hospital offers and introduced to key individuals who provide those services, such as oncology social workers, registered dieticians, chaplains, and travel and transportation coordinators. Making this personal connection will encourage the caregiver to make use of the information and services available to benefit the patient and the family.

- Trial staff should give the caregiver a reasonable picture of how much hands-on care and support the patient is likely to need. This enables the caregiver to plan effectively and minimizes stress on all concerned. Although it is not always possible to anticipate all the needs for care that may arise, an overview can be invaluable and ensure that the caregiver is not caught by surprise with needs that are difficult to meet. Will the patient be able to take care of most daily life functions, such as bathing, using the toilet, eating, walking, and dressing? Will he or she be able to see normally? Move comfortably? Drive? If the patient lives alone, should the primary caregiver consider seeking some outside support, such as a cleaning service, food delivery, or nursing care, to support the patient? These are the kinds of issues that should be addressed as treatment gets underway.
- If a patient and his or her caregiver are from another state or community, trial staff should provide general information and advice on how to seek out support services in the hometown. Unless one has dealt with advanced illness in the past, familiarity with social services and how to locate them may be very limited or nonexistent, and some basic direction and advice from trial personnel can be very helpful. The kinds of services needed might include help with meals and food preparation, transportation, home modification, house cleaning and yard care, as well as health care support from aides and nurses.
- Caregivers need instruction in the classic warning signs that might indicate the onset of a significant medical problem in the patient, such as protracted nausea and vomiting, persistent shortness of breath, temperature above 100.5 degrees, and inability to eat or drink. In addition, they need to know how to get help at the institution where the clinical trial is being conducted. They will also need to know the difference between the need for immediate attention and an emergency so they can reach out in a timely manner for the appropriate kind of help.
- Depression is a not uncommon symptom in the course of cancer treatment, and it can affect both the patient and the caregiver. Caregivers should be instructed in the warning signs of depression and encouraged to make use of available supportive services for both themselves and the patient.

5. THE ROLE OF PALLIATIVE CARE

In the past, "palliative care" was largely thought to mean care provided to patients at the end of life. Today, most practitioners view this discipline as defined by the World Health Organization: palliative care improves the quality of life of patients and families who face life-threatening illness, by providing pain and symptom relief, spiritual and psychosocial support from diagnosis to the end of life and bereavement.

Palliation—relief of symptoms, both physical and mental—is an important element of care and, as such, should be addressed by those working in clinical treatment trials and planned for, from the start of treatment to its conclusion. Although symptom management has always been considered an important part of care for trial participants, providing true palliative care is a larger universe. The addition of attention to the psychosocial and spiritual needs of patients is significant and important. The 2001 Institute of Medicine report, *Improving Palliative Care for Cancer*, clearly recommends that more attention be given to the emotional, spiritual, and practical needs of patients and their families.

"I'm fine!" It is important to understand that some patients in clinical trials may be very reluctant to complain about side effects they may be experiencing. There are a number of possible motivations behind this, including a reluctance to distract the physicians and other care providers from the disease being treated, a fear that the presence of side effects may mean that the disease is worsening, and a fear that complaining about side effects may increase the odds that the patient might be asked to leave the clinical trial. Although these concerns may be puzzling to the trial team, they become more understandable if one takes the perspective of a person in a clinical treatment trial that offers perhaps the last best hope for successful treatment, and perhaps life itself.

How to manage these concerns? Taking the initiative—explaining, perhaps repeatedly, that side effects are to be expected, that they do not necessarily mean the disease is worsening, and that they probably will not jeopardize participation in the trial—is probably the best approach. Patient literature on the trial also should emphasize these points. At the least, when patients are participating in a treatment trial in which side effects are common, regular and thorough interviews with each patient throughout the course of the clinical trial can be used to "hone in" on the realities of how the patient is managing.

6. MANAGING DIFFICULT NEWS

Sometimes circumstances in a patient's course of treatment on a clinical trial may mean that continued participation is no longer possible. Some causes include clear indications that the treatment is not effective, severe and debilitating side effects become overwhelming, or the patient is not complying with treatment requirements. Whatever the cause, having to inform a patient that he or she cannot continue on the trial any longer can be devastating. Although there is no way to really soften this news or put a good face on it, there are steps that physicians and other trial providers can take that may help patients:

- Deliver the bad news in an appropriate setting and allow plenty of time—meaning not in a hurried conversation in the hall or in the middle of a busy clinic. A private setting should be selected so there is ample time for the patient and family to ask all the necessary questions and get as much information as possible at that time. If possible, the principal investigator should take responsibility for this conversation.
- Make sure the patient is not alone, if possible. Coping with bad news can be very difficult if you are alone and potentially facing a long trip home. If necessary, call and ask the patient's caregiver to attend this meeting, if the caregiver does not routinely accompany the patient.
- Avoid language such as "There is nothing more that can be done" or other sweeping phraseology that connotes hopelessness and abandonment. Although it may be true that "nothing more can be done" by participating on this trial, there may be other care options available. Even if it is likely that no further treatment can help the patient, other options, such as hospice care, are important and should be presented if appropriate. Ideally, trial staff will have communicated with the patient's referring physician about the need to end trial participation in advance of the conversation with the patient. This should be communicated to the patient so the patient knows that his or her physician will be available and ready to help with the next steps.
- Have supportive care available. If your patient needs additional help, support, and information to cope with present realities and the next steps, make appropriate staff available for that purpose, such as psychologists, social workers, and spiritual counselors. Have this information at the ready; now is not the time to make the patient cast about for help and information.
- Make yourself available later. In the face of bad news, people do not often ask all of the questions that may come to mind. Therefore, the principal investigator and other staff with whom the patient and family have interacted should be available for further conversations in a mutually convenient way.

7. EFFECTIVE PATIENT COMMUNICATIONS: RECOMMENDATIONS AND CONSIDERATIONS

People enrolled in clinical trials and their caregivers generally have consistently complained about behaviors that can interfere with establishing and maintaining an effective patient–clinical investigator partnership. These include the following:

- The trial staff, particularly physicians, talk down to patients and their caregivers, addressing them by first names, communicating the impression of more important things to do than speak with the patient, and talking on the fly, in halls, and while standing.
- Clinical trial staff, particularly physicians, communicates in medical jargon, often without explanation or an attempt to translate terms.
- Chronic lateness—in clinic, in physician appointments, and in the start of treatment and procedures.
- Rudeness or indifference by some staff. Although reports of this kind of treatment have centered on support staff, in some instances others working on the trial or providing services such as phlebotomy, scans, and radiation can behave rudely or indifferently as well.
- Unreturned, or very slowly returned, phone calls from patients and caregivers.
- Hassles getting copies of medical records.
- Some caregivers complain that despite a continuing presence with the patient, they are largely ignored and unacknowledged as contributing to patient care.
- Some patients complain that if their caregiver is present, much of the conversation from trial staff is directed to the caregiver.
- Chronic interruptions when physicians are meeting with patients and caregivers; pagers vibrate, cell phones ring, and people interrupt.

At the heart of all of these issues is communications failure. Although no one suggests that seriously ill patients, worried and overworked caregivers, and hard-pressed clinical trial staff will have an easy time of communicating effectively all of the time, it is nonetheless possible to consider some steps and strategies that can help.

To work effectively, a partnership, including a partnership between a research institution and its patients, must be a relationship of peers, not that of suppliant and superior. The following are approaches, behaviors, and tactics that can balance the equation toward success:

When physicians and other staff are talking with patients and caregivers, they should sit down. Sitting down, at least on those occasions when the dialogue is going to last more than a minute or so, communicates attention. By sitting, instead of standing over a patient, the staff member conveys a sense that this is a conversation of peers and of importance to both parties. Hovering in a doorway and talking, or standing over a patient in bed, communicates haste and even disinterest in the conversation.

Respect patient privacy. Never hold important conversations in a hallway, in the midst of a busy clinic, or in any other setting where others can overhear it.

Explain, upfront, the teaching function present in many clinical trial environments. This should be done early in a person's consideration of participating in a clinical trial, and people should be made aware that some care will be provided by "residents" or "fellows," who these people are, what training they have had, and any other information necessary. Most patients are accepting and even enthusiastically welcome the opportunity to help young physician trainees and others learn, but not everyone—and it is far better to find that out before a person enrolls in the trial than after the fact.

Make eye contact with patients and caregivers when having a conversation. This communicates attention, engagement, and focus.

If possible, when speaking with a patient, do not write while the patient talks. Instead, listen attentively. The time for notes can come when the conversation ends.

Use the patient's name (and the caregiver's, if possible) in conversation. If a staff member uses a patient's or caregiver's first name, then it should be expected, and accepted, that the staff member's first name should be used as well. In short, if you are Dr. Jones, then I am Mrs. Butler. If I am Susan, then you are Joe.

Talk to patients, whenever possible, when they are fully dressed. In clinical settings, this is not always going to be possible, but it should be encouraged. People in street clothes feel much less vulnerable in the world of white coats than people in backless hospital gowns.

Work hard on time management. Nothing offends patients more than physicians and other health care professionals who begin a conversation by saying, "Well, I only have a minute . . ." followed by a furtive (and even not so furtive) glance at a watch or clock. When a physician says he or she must rush off, we wonder who it is whose care must matter so much more than our care. Patients and caregivers alike are justifiably alienated and insulted by this conduct.

Have the clinic schedule reflect consideration for the patients' and caregivers' time. People in clinic often complain of waiting, sometimes for hours, for their scheduled appointment, without any real sense of why the delay is occurring and even that any of the clinic staff are particularly concerned. First and foremost, clinic scheduled appointments should be real and meaningful, and if this is not happening, then the staff has an obligation to work with problems until they are resolved. It is the height of disrespect to keep sick people sitting and sitting, trapped in the clinic, without attention or consideration, and all necessary steps must be taken to stop this practice. Some clinics have adopted the technique restaurants use to manage the time of people waiting for tables—flashing pagers that enable people to go to other areas of the hospital, go for a meal, or even go for a walk—when there are delays in honoring appointments.

Avoid offensive language. One chronic complaint cancer survivors have is the phrase physicians and others use: "You've failed your chemotherapy (or other treatment)." By any reasonable measure, the failure is that of the treatment, not the patient, and hearing a phrase like this contributes mightily to making bad news even more difficult to take. This pejorative kind of language should be forbidden.

Be accessible. Patients and caregivers have a real need to call between clinic visits, and there should be a reliable mechanism for this. Patients should have phone numbers that can yield very quick information and answers when needed, and patients should not be expected to wait for hours before receiving a response.

Be honest. Although trial staff should work to have a sense of just how much information, in how much detail, patients want to receive and respond accordingly, for the most part, patients want a factual understanding of what is happening, good or bad. Knowing the truth allows patients and caregivers to plan the next steps in their lives.

8. THE ASSERTIVE PATIENT: ALLY IN SCIENTIFIC RESEARCH

For several reasons, I believe that strong, self-advocating patients are an advantage to clinical trials:

- The assertive patient is more likely to monitor treatment details, symptoms, side effects, etc.—reporting these in a timely manner, working to keep compliant with the trial protocol, and generally working to provide good self-care. Of course, this means that the patient is knowingly optimizing his or her chances for success.
- The assertive patient is likely to be happier with the treatment received than others, and recent research has shown that "patient activation" was the major determinant of patient satisfaction with care. *Activation* is defined as patients asking questions, initiating discussions about what is on their minds, and insistence on discussing topics of importance.

It is entirely possible to encourage patient assertiveness in the setting of a clinical treatment trial. These steps can help:

Encourage patients and caregivers to be open, honest, and direct with trial staff all the time. Make sure they understand that comments and information are welcomed, not merely tolerated.

Allot time for meaningful conversation with patients, rather than cursory or rushed visits.

Generally, talk "up" to patients rather than "down." Also, check to make sure—if you are doubtful at all—that the information provided makes sense and is understood. Some patients might say, "Uh huh, uh huh, uh huh . . ." and not truly understand but not want to interrupt. Encourage interruptions and check on clarity.

Be welcoming to patients' caregivers. These people are often essential to the patient's well-being and treating them respectfully and as a full partner in the patient's care makes a difference in the patient attitude toward the trial staff.

Encourage note taking and tape recording. Patients want to do this when they are receiving a lot of new information. This can be very helpful to patients once they have left the clinical setting and reinforce the important information they need.

Encourage patients to make full use of the hospital's supportive care services. These range from institution to institution, but in general most trial environments offer spiritual counseling, social work services, nutrition advice and support, and more. By encouraging this use, the message communicated to the patient is that the trial staff is interested in and concerned with the whole patient, not just the body part under scrutiny in the trial.

Make patients aware of credible advocacy organizations specializing in their disease or condition. Although some patients will not be interested, many more will be—if not during the treatment phase, then thereafter.

Ask, on a regular basis, how patients feel about the quality of care they are receiving and how the trial experience is going for them. Make sure they understand that candor is welcomed because the clinical trial staff wants to make the experience as effective and supportive as possible.

Be willing to accommodate special human needs of patients. Sometimes this may translate into deferring a chemotherapy treatment for a few days to better celebrate a birthday or other special event. Rigidity can be an important aspect of a patient's care, but so can human compassion and understanding without a lecture.

Consider creating a patient advisory committee. If a research institution or hospital consistently offers numerous clinical trials, it may be useful to imitate the National Institutes of Health Clinical Center and create a patient advisory body to meet regularly to advise the staff on what is and is not working in clinical trial operations.

9. CONCLUSION

There is an old expression that says, "If you want to know how to do a job right, ask the person who has to do it." I believe that this applies to the opportunity I have had in this chapter to discuss the patient perspective on the clinical trials process. My goal has been to communicate ideas and techniques that, from the patient perspective, may encourage our great scientific research institutions to take steps that will make the clinical trial process more welcoming to more people, and that it will improve the quality of care patients in trials receive.

Further Reading

Adams-Campbell LL, Ahaghotu C, Gaskins M, *et al.* Enrollment of African Americans onto clinical treatment trials: Study design barriers. *J Clin Oncol* 2004;22:730–734.

Advani AS, Atkeson B, Brown CL, *et al.* Barriers to the participation of African American patients with cancer in clinical trials: A pilot study. *Cancer* 2003;97:1499–1506.

Brawley OW, Freeman HP. Race and outcomes: Is this the end of the beginning for minority health research? *J Natl Cancer Inst* 1999;91:1908–1909.

Brown RF, Butow PN, Butt DG, Moore AR, Tattersall MHM. Developing ethical strategies to assist oncologists in seeking informed consent to cancer clinical trials. *Soc Sci Med* 2004;58:379–390.

Cassidy EL, Baird E, Sheikh JI. Recruitment and retention of elderly patients in clinical trials: Issues and strategies. *Am J Geriatr Psychiatry* 2001;9:136–140.

Cohen GI. Clinical research by community oncologists. *CA Cancer J Clin* 2003;53:73–81.

Cox K. Informed consent and decision making: Patients' experiences of the process of recruitment to phases 1 and 2 anti-cancer drug trials. *Patient Education Counseling* 2002;46:31–38.

Cox K, McGarry J. Why patients don't take part in cancer clinical trials: An overview of the literature. *Eur J Cancer Care* 2003;12:114–122.

Ehrenberger HE, Breeden JR, Donovan ME. A demonstration project to increase the awareness of cancer clinical trials among community-dwelling seniors. *Oncol Nurs Forum* 2003;30:E80–E83.

Ellis PM. Attitudes towards and participation in randomized clinical trials in oncology: A review of the literature. *Ann Oncol* 2000; 11:939–945.

Ellis PM, Butow PN, Tattersall MH, Dunn SM, Houssami N. Randomized clinical trials in oncology: Understanding and attitudes predict willingness to participate. *J Clin Oncol* 2001;19: 3554–3561.

Fleming TR, Ellenberg S, DeMets DL. Monitoring clinical trials: Issues and controversies regarding confidentiality. *Statistics Med* 2002;21:2843–2851.

Goodwin PJ, Black JT, Bordeleau LJ, Ganz PA. Health-related quality-of-life measurement in randomized clinical trials in breast cancer—Taking stock. *J Natl Cancer Inst* 2003;95:263–281.

Hayes MA, Smedley BC. *The Unequal Burden of Cancer: Committee on Cancer Research among Minorities and the Medically Underserved.* Washington, DC, Institute of Medicine, National Academy Press, 1999.

Hussain-Gambles M, Atkin K, Leese B. Why ethnic minority groups are under-represented in clinical trials: A review of the literature. *Health Soc Care Community* 2004;12(5):382–388.

Lengacher CA, Gonzales LL, Giulano R, Bennett MP, Cox CE, Reintgen DS. The process of clinical trials: A model for successful clinical trial participation. *Oncol Nurs Forum* 2001;28:1115–1120.

Markman M. Ethical conflict in providing informed consent for clinical trials: A problematic example from the gynecologic cancer research community. *Oncologist* 2004;9:3–7.

Pace C, Miller FG, Danis M. Enrolling the uninsured in clinical trials: An ethical perspective. *Crit Care Med* 2003;31:S121–S125.

Raich PC, Plomer KD, Coyne CA. Literacy, comprehension and informed consent in clinical research. *Cancer Invest* 2001;19: 437–445.

Ruffin MT, Baron J. Recruiting subjects in cancer prevention and control studies. *J Cell Biochem Suppl* 2000;34:80–83.

Silverman HJ, Luce JM, Lanken PN, *et al.* Recommendations for informed consent forms for critical care clinical trials. *Crit Care Med* 2005;33:867–882.

Wiklund I. Assessment of patient-reported outcomes in clinical trials: The example of health-related quality of life. *Fundam Clin Pharmacol* 2004;18:351–363.

14

The Clinical Researcher and the Media

JOHN BURKLOW

Office of Communications and Public Liaison, National Institutes of Health, Bethesda, Maryland

If you are a scientist conducting clinical research, chances are that you will interact with the media at some point in your career. Dealing effectively with reporters is like any other skill—to succeed, you need to learn the basics, be open to coaching, and practice. Also like any other skill, it may look easy, but actually doing it can be another matter. The purpose of this chapter is to review what makes news and why, and to discuss how to handle media inquiries, steps to take when you get a call or do a media interview, and what to say and how to say it. Your goal is to be astute, caring, absolutely correct, and to get your message across while avoiding common pitfalls.

Since health and medicine are two of the most popular topics among the public, they are covered daily by the media. Consequently, reporters are constantly searching for new stories, new angles. Whether it is good news or bad news in medical research, the media wants your story. If your research shows results that could lead to a promising treatment, people want to know about it. The more impact a disease has on society, the more the public wants to know. And they might want to know before you have all the answers to provide them.

Conversely, if something bad happens during your research, the media will be on your doorstep too. If a patient dies because of an adverse reaction to the investigational therapy, a patient suffers because of a protocol that did not comply with regulations or guidelines, or an investigative reporter gets wind of an allegation of conflict of interest, you can count on a call—or many calls—from the media. This applies in many circumstances, not only if the situation directly involves you. There is a good chance that you will be called if the event happens at an outside research insti-

tution as well. Reporters will want to get your reaction or perspective. In addition, if it was a National Institutes of Health (NIH)-sponsored clinical trial, reporters frequently assume NIH staff is involved in some way.

If you are a clinical investigator funded by the NIH, you work for an agency supported by public dollars. The public has a right to know the good news and the bad. That means that you have to be ready to deal with media inquiries, whether they are about positive or negative stories.

The media disseminates information that often brings about needed changes in human subject protection. Many of the major changes, rules, and regulations that govern modern clinical practice have occurred as a result of missteps and abuses in history. From the revelation of Tuskegee to the tragic death of a young man in 1999 who was participating in a gene therapy clinical trial, the media has played an important role in society's perception of the ethics of clinical research and the necessity of changes.

Your research with human subjects could end up on the front page of the major newspapers or on the evening news. A good example was the announcement in July 2002 that the Women's Health Initiative revealed unexpected results regarding long-term hormone therapy and its increased risk for several diseases. Until that point, hormone therapy was thought to have had a preventative effect. The media coverage was extensive—print, radio, television, and Web. The public reaction was strong and immediate. NIH scientists, both intramural and extramural, were bombarded with questions from the media and other health professionals. Although it was a science advance with important information, the initial public reaction was anger and confusion because it contradicted conventional wisdom and did not provide clear alternatives. NIH responded by convening a scientific workshop on the state of the science and what health care providers should tell their patients. The workshop generated a new round of media inquiries, but this time they resulted in positive stories because the scientists conveyed a sense that they understood the public wanted practical advice.

1. WHAT MAKES NEWS IN SCIENCE AND MEDICINE?

Most of the time, media coverage of your clinical trials is desirable, and you may wonder why a particular study attracts media attention while others do not. The following categories describe what draws reporters to cover science and medicine. Keep in mind, however, that large clinical studies will get more attention than basic laboratory findings and phase I and II clinical studies.

1.1. Published Science: The Media's Bread and Butter

Scientific studies and research advances that have been published in peer-reviewed journals get the most newsprint and television air time by a huge margin. This constitutes the major source of news in science and medicine. Journals such as *Science*, *Nature*, *New England Journal of Medicine* (*NEJM*), and the *Journal of the American Medical Association* (*JAMA*) dictate, by and large, what is covered by science and health reporters from week to week. If you have a paper accepted at a major journal, begin thinking now about how you will handle media inquiries and what should be your "core" message. Later in this chapter, there is a more in-depth discussion of what to do once you get the media's attention.

1.2. Novelty

As with all news stories, the "unusual" in science and medicine gets the attention of the general public. Even people with little knowledge or interest in science want to hear about cloning sheep, face transplants, or new therapies for weight loss. In addition, since NIH studies hundreds of common and rare conditions and diseases, there is a reasonable chance that an area of research being studied at NIH will generate media attention, as hard news or perhaps a human-interest story.

1.3. The Unexpected

The results from the Women's Health Initiative are an example of the unexpected. It is safe to presume that the majority of women—and their health care providers—expected positive results that would confirm the belief that hormones were a wonder drug, a claim that had been encouraged for many years. In early 2004, reports of the increased risk of cardiovascular disease from taking certain over-the-counter pain relievers created a media firestorm. The public expects drugs to be safe and effective, particularly if they are available without a prescription.

When they hear unexpected news, often it is when something previously considered safe poses risks. As researchers, expect the public reaction to not only be negative but also demanding. Most people will want to learn how to assess their own risk and about safe alternatives. It will not be sufficient to announce research results and merely recommend that the public talk to their health care provider. With immediate access to health information through the Internet, many patients have either the same or different information about a certain topic than their provider. If there is an information vacuum, both the patient and his or her provider are frustrated. Sometimes there is no safe alternative. In that case, you need to convey through the media that you understand the difficult situation and that you feel compassion for patients and the public, and that the research will continue to seek answers.

1.4. Celebrity

Think about the impact that Christopher Reeve's paralysis had on the national attention to spinal cord injury and research and, later, on the debate over human embryonic stem cell research. When Michael J. Fox, the actor, announced that he had Parkinson's disease, public awareness and interest in the disease were heightened. Former First Lady Betty Ford raised public awareness about breast cancer and addictions by coming out and talking candidly to the media. The public wants to know about celebrities and their medical problems.

You may treat a celebrity in the clinic, or a celebrity may develop a disease that happens to be in your area of expertise. Often, this generates a great deal of media interest, and you may be called on to comment on the celebrity's condition. It is a general policy that clinicians do not comment on their patients or acknowledge that the individual is under their care. Furthermore, it is best not to comment to the media about a certain disease, such as AIDS or stroke, if a famous person has it. If you do, there is a good chance that you will be perceived as commenting on the celebrity, even if you are speaking about the disease.

1.5. Tragedy and Controversy

One of the most explosive research controversies in recent years is the safety of human gene transfer. The debate was intensified by the tragic death of a young man in a gene therapy trial at the University of Pennsylvania, a highly respected institution. The fallout centered not just on gene therapy but also on patient safety in clinical research throughout the country. It also put NIH's policies and procedures under intense scrutiny. After the initial media reports, Congress called for hearings, which were heavily covered by the press. As reporters continued to write more stories, more hearings took place. In addition, NIH's and the Food and Drug Administration's responses to the patient safety issue also triggered more media attention. The clinical researchers and administrators involved had to deal with an onslaught of media calls and interviews. For some, it was their first experience speaking with national reporters. In addition, it was their first experience testifying before Congress. Sometimes the two activities occurred simultaneously.

There is often interplay among NIH, Congress, the Administration, and the media. A prime example in recent years has been the debate over federal funding of human embryonic stem cell (HESC) research. For the past 5 years, no other science and medicine story has dominated the news like stem cell research. Although much of the focus has been on HESCs, the press has reported on all types of stem cell research. This has involved researchers from institutions throughout the world. Since it is such a politically and emotionally charged issue, it requires a great deal of judgment, tact, and preparation on the part of the researcher speaking to the press. Also, one must keep in mind the various audiences who are reading or listening to one's words.

1.6. Impact

Research that has a major, immediate impact on people receives a great deal of media coverage. For example, findings in the late 1990s that tamoxifen reduced the risk of developing breast cancer by 50% drew more than 100 reporters to the press conference that took place in the Health and Human Services building in Washington, DC. The hormone therapy results in 2002 had direct implications for more than 14 million women, and extensive media coverage reflected this impact. The announcement of the Cancer Genome Atlas—to study genes associated with cancer—in December 2005 made headlines even though no cures or treatments were announced. The topic drove the media coverage. On a smaller scale, some diseases may be rare and not affect large numbers of people, but their impact on people's lives is devastating, and this also will draw media attention.

2. WHY TALK TO REPORTERS?

Although some scientists would prefer to go through their entire career without speaking to a reporter, the media lets the public know that research is always moving forward and is helping to bring advances to human health. No other human endeavor, except perhaps sports, generates this kind of automatic news interest. Since the majority of the public receives much of its health and medical information from the media, it is a logical extension that scientists should be facile in dealing with reporters and work at becoming effective public communicators. Specifically, since NIH conducts and supports a great deal of the clinical research conducted in the United States, it is part of NIH's mission to let people know about medical research progress.

3. WHY REPORTERS WANT TO TALK TO YOU

Reporters tenaciously seek out quotes from experts for many reasons, including the following:

Credibility: Quotes from experts and the people directly involved make the story more credible. In reports on HESC research, for example, reporters always try to include a quote from the NIH director, the head of the NIH Stem Cell Task Force, or a well-known researcher in the field. If they are covering a clinical trial result, reporters will want to talk to the lead scientists and their patients. Most institutions, including the NIH, go to great lengths to protect patient privacy, and that includes shielding patients from the media. Sometimes, a patient wants to talk to reporters, and that is his or her prerogative. The patient's clinician, however, must ensure that the patient does not feel obligated or that there is an implicit expectation that he or she speak to the press.

Clarity and lively flavor: Quotes are what make a news story different from an editorial or an essay. A good quote is usually more interesting than the same information written in a reporter's words and quotes often make the story easier to read.

Tension: Reporters want interviews because they often find hints of controversy in what you say, and controversy heightens the interest of the public. This fact exemplifies the need for you to think through your words carefully, get advice when needed, and be well prepared before you speak to a reporter. If a reporter asks you to comment on another scientist's comments, defer and transition to your core message. This is discussed later in the chapter.

Limited time: News reporters are in a hurry. In contrast, scientists are deliberate and meticulous in their process. Sometimes these worlds clash. Scientists will complain that the reporter waited until the last minute to request an interview and did not allow enough preparation time. The reality is that the news cycle is short and quick, and reporters often only have a few hours to put a story together, and quotes save them from the time it takes to uncover the facts from other sources. This does not mean that you should feel rushed into an interview. Normally, there is at least a window of time between the reporter's first call and his or her deadline. Allow yourself enough time to gather your thoughts and adequately prepare for the interview.

4. WHY YOU SHOULD TALK TO REPORTERS

Occasionally, a principal investigator does not want to talk to reporters. He or she would rather have an official institutional spokesperson talk to reporters about the research, or send a written statement through e-mail. This is almost never satisfying to a reporter because the spokesperson is simply whoever has been designated to speak on the subject. The spokesperson is not viewed as credible as the principal investigator because he or she is not the expert. The clinical researcher—you—are the expert.

NIH spokespeople do speak for the NIH in a crisis or in a story involving a sensitive issue. They also help their scientists prepare to talk with reporters. The NIH firmly believes that one should talk to reporters, unless an individual believes that he or she is not the appropriate person to be interviewed or he or she has evidence that the reporter is not operating in good faith. There are clear benefits to speaking to reporters, and they usually outweigh the risks of being misquoted or having your comments taken out of context. The following are examples of benefits:

You can improve the accuracy of the story. Many science reporters are smart and very experienced, but they do not know everything about every scientific or medical subject. Even the best science reporters need guidance on emphasis, nuance, or help in understanding methodology.

For those investigators at NIH, you help create a favorable climate for NIH. Your input improves public understanding of the importance of medical research and its relevance to people's everyday lives. In turn, it helps maintain public support for NIH and medical research in general.

You owe it to the American taxpayers. Since the taxpayers support NIH, NIH-funded researchers owe it to them to explain their work. The best way to do this is through the mass media.

Clinical investigators especially need to help people understand what an advance in medical research could mean or not mean to their lives. Your participation in the story provides an appreciated context for the American people.

Your Words Have Impact: A True Story

When the discovery of the *BRCA1* gene was about to be announced in *Science* in 1997, the embargo was broken a couple of days early. NIH knew that the impact of the announcement would be enormous for millions of women who were concerned about breast cancer.

The NIH quickly put together a press conference and involved scientists from three different NIH institutes to answer the inevitable questions that would put the discovery into everyday meaning for the public: What does the discovery mean to women? Is there a screening test? When will there be one? Should every woman get the test? Does it relate to all breast cancer?

There were more than 100 reporters at the press conference, and the outcome was highly informative and successful. The scientists provided consistent and accurate messages, saving time and preventing confusion for everyone concerned.

---

5. ENGAGING THE MEDIA: THE PROCESS

Reporters may call you directly, without going through your institution's communications office. There are specific steps you can take to handle such a call. Even if the reporter is friendly and wants to do a positive story, you can find yourself in an unfortunate position if you forget to ask a few key questions. For NIH intramural investigators, although you do not have to get official permission from NIH to talk to a reporter, it is considered best practice to inform your communications office and seek its advice before you agree to an interview. This is especially important when called by major media outlets such as the *Washington Post*, *New York Times*, *Wall Street Journal*, or any of the radio or television networks. Also, your laboratory, institution, NIH institute, etc. might have a rule requiring clearance. Make sure you know your local policy before you talk to the media.

6. A WORD ABOUT E-MAIL AND THE INTERNET

Traditionally, the major media outlets generate the most attention. Although that is still the case, generally speaking, e-mail and the Internet have leveled the playing field somewhat. It is often difficult for the casual reader to distinguish between a major news story and an obscure one. Therefore, keep in mind that your interview with a relatively unknown Web-based news outlet may reach millions. In addition, e-mail enables broad redistribution of interviews with all types of media, including industry and advocacy organization newsletters. Also, if you are in a newswire service story, such as the Associated Press (AP) or Reuters, your name will appear in Google News and other search engines each time the story appears, which could be hundreds of citations. Therefore, it is wise to be as careful about what you say to a newsletter reporter as what you would say to the *Washington Post* or an AP reporter.

If you receive a call from a reporter requesting an interview, there are several things you should routinely do. First, take the reins confidently and ask a few questions of your own. You need the answers to the following questions from the reporter before you agree to do anything and be sure to write the answers down:

1. What is your name and phone number?
2. What publication/network/station are you with?
3. What is your deadline? (This gives you an idea of how much time you have to think about your answer.)
4. What is the angle or story line?
5. Who else are you talking to?
6. What information are you looking for from me? (Do they just want a background discussion about T

cells? Do they want to feature you or just get a quick quote on someone else's work?)

Alternatively, you can ask your press officer in your communications office to gather this information and report back to you. To allow yourself some time, you can say, "I would be happy to talk with you. Could you coordinate this with our communications office? Here's who you should call, and here's their number." You should always have the name and phone number of your press officer within reach—at your desk and in your wallet or pocketbook.

Your next step should be to consult the communications office of your institution or NIH institute. Your press officer can help you with the following types of questions:

*   Are you the right person to talk to the reporter or is this an issue that should be handled by the communications office, the NIH spokesperson, or someone else?
*   Should someone outside NIH field the questions?
*   Is there an NIH position on the subject in question (for those at NIH)?
*   What information about the reporter would be helpful to you, such as the line of questioning he or she might take?
*   What experience has your press officer had with this reporter, if any?
*   How do you decide to say "No" gracefully, if that is what you decide?

If you decide not to do the interview, decline truthfully and firmly. The following are three common answers, but the first one is not recommended:

"I would like to talk to you, but I've been told not to." This is a fairly common, but inappropriate, answer. It will only entice the reporter to pursue you and it will become its own story.

"I'm not the best person to talk to you about this. Why don't you call _____?" This is an appropriate answer if it is true. Be sure to give the person you named the courtesy of telling him or her of your referral.

"It's really too early in the research to have anything firm to say about it." Again, this is a good answer if it is true. Give the reporter a projected date when he or she could call back for better information.

7. THE INTERVIEW

If you decide to do the interview, keep in mind that reporters may be friendly, but they are not your pals. They are not cheerleaders for science or your point of view. They are in the business of reporting what they think is a news story. The overwhelming majority of science and health reporters, however, strive for accuracy and context. You can help them achieve their goals while you convey important health and science information to their readers, listeners, or viewers.

If you have not had any experience speaking with reporters, it is a good idea to get some training. It could be given by your communications office or by a specialist who conducts media training for a living. Usually, the sessions last a half day and include mock on-camera interviews as well as role-playing. Even if you have media experience, it is helpful to take a brief training session to refine your skills.

If you do not have time to get training before you are faced with a media interview, here are a few tips to keep in mind:

1. Although your training and orientation may lead you to provide detailed, comprehensive answers to questions, with many qualifiers and caveats, your answers to reporters should be concise, directed to the question asked, and void of any "filler." You should prepare in your mind—and on paper—your core message. On an index card, write the message—the exact words—that you hope will appear in the story. This is the message that you will keep coming back to, or "bridge to," in all of your answers. The point is to increase your chances that what you want reported actually gets in the story. All of your answers should support your core message. Do not add qualifiers or caveats. They will increase your chances of not getting your message across. When you are speaking, think of holding a tape measure that is being pulled out. As you talk, think of the inches as words. You hope the reporter uses every inch. Instead, the reporter uses 1–8 and 16–23, and it is out of context. One of the most common complaints from scientists about news stories is that the stories did not include what they told the reporter. Say only the words you would like to see in print, and keep it focused. Interviews should not be free-flowing discussions. Everything you say is potentially fair game to the reporter.

2. Your core message must encapsulate not only the facts but also your perspective, your institution's perspective, the NIH perspective, and, if applicable, a human dimension. It is important to take time to think about your message, write it, ask others to review it, and edit it. It should be brief, direct, compelling, and interesting. Always speak in plain language, avoid technical jargon, and never use acronyms, even if you think everyone knows what they stand for.

3. Beware of a few terms when you are talking with a reporter. Terms of the journalistic trade may not mean what you think they mean. Keep the following definitions in mind, in case you hear the reporter use them:

On the record: This means a reporter can quote you directly, using your name and title.

Not for attribution and on background: This means that the information you give, including direct quotes, can be used by the reporter but you are not to be named. You may be identified as an institution's or NIH official or source. It is rare to have science and medical sources speak under a condition of anonymity. It is recommended that you stay on the record at all times.

Off the record: This means the reporter cannot use your information in a story as coming from you; however, he or she can use it in other ways—for example, to get another source to respond to your comment.

Work out the ground rules with the reporter regarding how he or she plans to use your comments before your interview begins. You cannot take it back after you have said it. Despite these informal rules, the strong recommendation is to always speak as if you are on the record.

4. Keep the interview relatively short. Make it clear in the beginning that you have 10–15 minutes to talk. Ideally, your press officer has called the reporter and relayed to you the types of questions he or she intends to ask. That will help you prepare for the interview. If you allow the interview to go on too long, you become fatigued or the reporter moves on to other topics, and the chances for unintended results increase.

5. Be cordial but not too casual, cavalier, or humorous. Humor is important; however, in a media interview, it can easily be taken out of context or misinterpreted by the reader. It is best to keep it to a minimum.

6. Decide ahead of time if you prefer to do the interview over the telephone or in person. You have more control over your time if it is over the phone. You may, however, want to invite a reporter to your office if you are beginning to establish a long-term working relationship with him or her.

7. If you are doing a television interview, try to do it in person, as opposed to a remote location. It is difficult to concentrate on a camera lens and easy for your eyes to shift around in a remote format.

8. Have someone (preferably a communications expert) in the room with you when you do the interview. If it is by phone, tell the reporter who else is in the room and if you are tape-recording the interview.

9. Keep focused on the outcome. Some of the reporter's questions may irritate you or surprise you. Sometimes that is a technique used to evoke an emotional response, which makes for an interesting quote, but at your expense. Remain calm and think about what you want to read in the story. If the reporter asks a negative question, do not use the reporter's words to answer. For example, the reporter may ask, "Were you involved in the scandal?" and your answer is, "I had nothing to do with that scandal." Now there is a quote from you mentioning a scandal. That is what readers will remember. Do not take the bait. Rather, you answer with "No" and move on to your core message.

10. Practice. Like everything else, practice improves your chances for success. Rehearse with your press officer or colleague before the interview, particularly if it is about a complex or controversial issue.

8. WHAT IF YOU ARE MISQUOTED?

Misquotes happen. Even if you follow all of this advice and more, there is a chance your information will be reported in a different light than you anticipated. If the health message is incorrect and may have an effect on patients or the public, it is important to get the mistake corrected. Call the reporter immediately with the correct information.

If the health message is accurate but you feel misrepresented, you can call the reporter or the editor or write a letter to the editor. Talk to your communications office and he or she will help you decide on a case-by-case basis. As counterintuitive as it may seem, sometimes the best thing to do is just let it be. If you pursue a story correction, you may inadvertently keep the story in the news longer than it would have been if you had done nothing about it.

9. WHAT THE PUBLIC DOES NOT KNOW ABOUT SCIENCE

Surveys show that approximately 70% of Americans say they get their health information from the mass media. That means your words have a great deal of impact.

There are some good basic guidelines to keep in mind when you do an interview with the mass media. The first is that the public—your audience—does not know much about how science works. What you say and what they hear might not be the same thing:

What you know: Research yields new knowledge and raises new questions.

What the public perceives: A piece of published science is "The Truth." For example, you might view a study on high-fiber diets and cancer as raising more questions than it answers. The public might view it as a definite cancer prevention method.

What you know: Legitimate scientific differences of opinion exist.

What the public perceives: They view differences as confusion. They want the final answer. The challenge is not to overstep what is known and at the same time avoid waffling on the issue.

10. UNEXPECTED QUESTIONS

Another possibility that you should be prepared for are questions that will come at you unexpectedly. These questions will not necessarily be about your research or even science. The following are examples of unexpected questions that were asked at the *BRCA1* gene press conference mentioned previously:

Who holds the patent?
What will the test cost the country, and what will it save?
Will insurance companies cover the cost of the test once it is developed?
Did you have a financial interest in this discovery?

People care about these issues today. Even if you do not feel competent to answer them, anticipate that they will arise. Work with your communications office to be prepared for the toughest questions. The rule of thumb is to be responsive to the question but move immediately to the main purpose of the interview or press conference.

11. WHEN THE NEWS IS NOT GOOD

Clinical research has had a bumpy road in the press in the past several years. You can assume that if you or your research encounter certain types of problems, including ethical questions, you will have to deal with media attention. Some types of issues are guaranteed to attract the attention of the mass media. An unexplained death, or deaths, in a study will draw immediate media attention. Other bad news includes scientific misconduct and allegations of conflict of interest, both of which can be extremely painful.

From December 2003 to December 2006, NIH endured a long series of newspaper stories, television programs, and congressional hearings regarding alleged conflicts of interests among NIH scientists and the pharmaceutical industry. Some of the allegations were found to be groundless, whereas others are still being reviewed. Regardless of the ultimate outcome of this process, the researchers profiled in the news stories have had their professional and personal lives greatly disrupted. At the same time, they have had to field questions from the press or have been quoted testifying before Congress. In these cases, it is especially important for NIH scientists involved, or not involved, to think through what they are going to say to reporters, if they choose to speak, and to seek advice from their communications staff. NIH scientists should also become very familiar with the recently revised NIH ethics rules.

NIH, as an institution, cannot legally comment at all when legal proceedings or investigations are under way. The intention is not to "hang [the accused] out to dry." NIH simply cannot legally comment in these types of situations. This may frustrate some reporters and they may write stories about the fact that NIH is not commenting on the matter. There also may be a perception among the scientists that NIH is not defending its staff. Again, NIH must abide by the rules regarding investigations, and NIH scientists need to be aware that commenting to the press in any way could put them at risk.

12. A WORD ABOUT INVESTIGATIVE REPORTERS

Before moving away from the whole idea of doing or not doing interviews, take some time to consider a special type of reporter: investigative reporters. They are likely to be the reporters who uncover the bad news. Keep the following in mind about these tenacious reporters:

1. They have more time than the average reporter, who has a daily deadline to file.
2. They are most interested in irregularities, violations, and/or misconduct and seek to confirm or not confirm allegations.
3. They try to cultivate "whistle-blowers" or unconventional sources.
4. They will use the Freedom of Information Act to obtain documents, whereas a daily reporter would not generally do this. They may collect thousands of pages over many years.
5. From their painstaking research, they become extremely knowledgeable about your organization and gather a great deal of information, sometimes from people with an axe to grind.
6. Keep all of the previously mentioned facts in mind when considering how to engage them if they ask to interview you.

If you determine that the reporter interested in interviewing you is an investigative reporter, contact your institution's communications office. It, in turn, will call the reporter and decide how to proceed.

13. THE FREEDOM OF INFORMATION ACT

The Freedom of Information Act (FOIA) makes documents available to anyone, whether or not they are a citizen and whether or not we think they have a need to know. You cannot withhold documents because they make us look bad or because they could be misinterpreted by the public. There are nine exemptions in the FOIA that permit NIH to withhold documents. The following two exemptions are most often used:

Invasion of personal privacy, such as release of medical records
Commercial or financial information

The following are documents that are available to anyone at any time under FOIA:

Approved research protocols
Minutes of NIH institution review boards, with some possible deletions
Your e-mail messages
Your computer files
Document drafts

Under FOIA, it does not matter if you stamp a document "Confidential" or not. Each request is considered anew. Each NIH institute has a FOIA officer to help you with requests. You will be involved in the process, but only one person at NIH—an attorney in the Office of Communications—has the authority to deny documents under FOIA.

Take precautions. How would your documentation look if it were released to someone who wants to sue you or an investigative reporter who thinks you are the villain of the story?

14. EMBARGOES

An embargo is an agreement between a scientific journal and reporters. It designates the time frame in which a story may be released. In other words, embargoes are dates established by scientific journals that prevent the release of stories before a certain date.

For example, the December issue of *NEJM* hits the newsstands on the 14th. Copies are sent to the press and NIH several days ahead of that date, but we cannot release any information on the stories until the evening of December 13. Likewise, if you are an author of an embargoed story, you must remind reporters that you are speaking under an embargo.

In addition, refrain from talking to nonjournalists about an embargoed article because they are not held to the same restraints as reporters. There is a danger in science and medicine that talking about a pending study result could start "insider trading" and stimulate an investigation by the Securities and Exchange Commission; for example, if the stock market showed unusual movement. In fact, if a research result is considered by NIH to be market sensitive, the press release is not made available until the stock market closes that day.

Sometimes an ambitious reporter will jump the gun and break a story before the embargo. What happens then?

The journal may lift the embargo.
News stories may run ahead of schedule. You may be permitted to proceed immediately with interviews.

Whatever happens, you should contact your communications office for instructions. It is important to note that the future of embargoes is uncertain. With more clinical papers having many authors, with fierce competition from reporters, with people posting their data on the Internet, and with the economic importance of clinical advances, embargoes are unlikely to stand over time.

15. THE INGELFINGER RULE

The Ingelfinger rule, named after a former editor of *NEJM*, was levied in the 1960's to control early release of *NEJM* article information. The Ingelfinger rule has succeeded in intimidating some scientists to the point that they feel uncomfortable giving media interviews —even delivering abstracts at a medical meeting—for fear they will not get published.

The rules have since been clarified by *NEJM* and *JAMA* and are more liberal. You can talk freely at meetings and still get published. You can talk to reporters about what you presented at meetings, but it is probably not a good idea to go beyond what you presented in public sessions or to discuss the details of your data before publication. Also, never give out your manuscript to a reporter.

Events do infrequently occur that make both embargoes and the Ingelfinger rule moot.

16. CLINICAL ALERTS

In the 1990s, journals moderated their views about releasing details concerning certain studies prior to publication, in part persuaded by NIH and in recognition of the public's need to know. These cases, which remain rare, are wrapped up in the term *clinical alerts*.

Some prominent journals now allow agencies such as NIH and the Centers for Disease Control and Prevention to hold press conferences prior to publication when the data are very compelling and have a very immediate impact on public health. In short, if lives can be saved by disseminating the information immediately, then it is allowed without jeopardizing publication.

One example would be when a data safety monitoring board, in examining a study, sees a clear advantage or disadvantage in one arm of the study and recommends to the funding agency that the study be discontinued for ethical reasons. Then NIH, as an institution, may find that it cannot ethically keep the information from the wider public.

17. WHEN TO CONTACT YOUR COMMUNICATIONS OFFICE

Throughout this chapter, several circumstances have been discussed that call for assistance from your institution's communications office. The following list condenses those circumstances in which it would be appropriate for you to contact the communications office:

- When you receive a request from a reporter for an interview. It is the policy for most NIH institutes and other institutions and a good idea for your own comfort level.
- Any time you are doing an interview with a major newspaper, magazine, or television network.
- To receive help regarding how to phrase answers for the public, in the interest of plain language.
- To do a dry run for a television or radio interview.
- To learn what (NIH) policy is on a matter.
- When you are concerned about an investigative reporter who wants to talk to you.
- If you have a question about embargoes.

18. CONCLUSION

As a clinical researcher, you may find yourself in demand for media interviews. Remembering these key points will help you successfully deal with the media:

- People want to hear science and medical news.
- For those at government institutions, you are obligated to inform the public about your work.
- Understand that bad news or an ethically questionable problem draw media attention.
- NIH encourages its scientists and researchers to talk with reporters. It adds credibility and reflects well on NIH and the biomedical research community.
- If you receive a call from a reporter, get the information you need before agreeing to the interview.

- Use plain language in explaining your work for the general public.
- Develop a core message and various ways to convey it to increase the chances that you will be accurately quoted.
- Assume everything is on the record and only say what you would want to read in the story.
- Be aware of media issues, such as embargoes, that are unique to scientists.
- Contact your communications office for assistance.

An Introduction to Biostatistics: Randomization, Hypothesis Testing, and Sample Size Estimation

LAURA LEE JOHNSON,* CRAIG B. BORKOWF,[†,**] AND PAUL S. ALBERT[‡]

*Office of Clinical and Regulatory Affairs, National Center for Complementary and Alternative Medicine, National Institutes of Health, Bethesda, Maryland

[†]Cancer Prevention Studies Branch, Center for Cancer Research, National Cancer Institute, National Institutes of Health, Bethesda, Maryland

[‡]Biometric Research Branch, Division of Cancer Treatment and Diagnosis, National Cancer Institute, National Institutes of Health, Bethesda, Maryland

1. INTRODUCTION

The goal of this chapter is to provide an intuitive explanation of the statistical principles used in medical research. This chapter provides basic information so that the reader will be able to understand the results of analyses presented in medical research papers. In particular, this chapter discusses the principles of biostatistics related to the types of studies conducted at the National Institutes of Health (NIH). These principles are illustrated by three motivating studies that are typical of the studies conducted within NIH's intramural and extramural programs.

This chapter contains three major sections. In the first section, we discuss issues in randomization and study design. We review the reasons for randomization and describe several types of randomized study designs. In addition, we compare randomized experimental studies to nonrandomized observational studies and discuss the difficulties inherent in nonrandomized experimental studies. Furthermore, we illustrate the underlying theory and mechanisms of randomization. The next section introduces the principles of hypothesis testing from a conceptual perspec-

tive. We define and discuss commonly used terms, such as p-values, power, and type I and type II errors. We discuss commonly used statistical tests for comparing two groups of measurements and illustrate the use of these tests with analyses of the three motivating studies. In the third section, we discuss the intuition behind power and sample size calculations. We present the sample size formulas for common statistical tests for comparing two groups of measurements and illustrate these methods by designing new studies based on the motivating examples.

Statistical analyses are used to answer scientific questions. The scientific question of interest in a study should dictate both the design of the study and the analysis of the collected data. A set of well-defined questions will guide all of the statistical aspects of a study. Although some statistical plans may be routine for simple studies, statistical plans for other studies may require substantial thought and planning. Statistical practice is based on consensus across many organizations and schools of thought. Some organizations, such as the Food and Drug Administration (FDA) (see Chapter 9)[1] and counterparts worldwide,[2] are continuing to develop guidelines, especially related to the conduct and analysis of data from clinical studies and studies using newer technologies. Even with such guidance documents, there may be multiple valid

*Current address: Centers for Disease Control and Prevention, Atlanta, Georgia.

ways to approach a problem and the challenge lies with the investigator and governing bodies of interest [e.g., institutional review boards (see Chapter 5)[3] and FDA (see Chapter 9)[1]] to decide what assumptions or biases are acceptable. It is wise to speak with a statistician early in study planning.

In the remainder of this introduction, we describe the main features of three motivating examples. Additional details about these examples are presented and developed throughout this chapter.

1.1. Three Motivating Examples

Here, we describe three examples of actual studies that will be used to illustrate the concepts and methods presented in this chapter. The first two studies are from the NIH intramural program, and the third is similar to studies conducted through the NIH extramural program. Each of these studies provides different degrees of evidence for the effectiveness of the therapeutic agents under investigation. Each study also has a different design and consequently a different sample size.

1.1.1. A Study of Beta-Interferon on Disease Activity in Multiple Sclerosis

The intramural research program of the National Institute of Neurological Disorders and Stroke (NINDS) conducted a series of studies to evaluate images of contrast-enhanced lesions as a measure of disease activity in early relapsing-remitting multiple sclerosis (RRMS). The contrasting agent gadolinium causes areas of blood–brain barrier breakdown to appear on magnetic resonance images (MRIs) as bright spots or lesions. Traditional clinical measures of disease activity, such as those based on assessing physical or mental disability, are known to be very insensitive during the early phase of the disease. By comparison, it is thought that the number and area of these lesions as measured by serial monthly MRIs may be a more sensitive measure of disease activity during this phase.[4,5] A series of phase II (safety/efficacy) studies were conducted at NINDS to screen new agents, including beta-interferon, for effectiveness. We consider a study to examine the effect of beta-interferon on lesion activity during the early phase of RRMS.[6,7]

The beta-interferon study was designed to have 14 patients followed for 13 months. Patients remained untreated during the first 7 months (seven serial MRI measurements) and then were treated with beta-interferon during the last 6 months (six serial MRI measurements). The primary outcome or response in this study was the average monthly number of lesions on treatment minus the corresponding average number during the untreated baseline period. The study results showed that beta-interferon significantly reduced the number of lesions compared to baseline. This study is a nonrandomized study in which all patients were switched over to the investigational treatment after 6 months. This type of nonrandomized design is often used to screen for new therapeutic agents.

1.1.2. A Clinical Trial of Felbamate Monotherapy for the Treatment of Intractable Partial Epilepsy

The intramural research program of NINDS also conducted a clinical trial to study the efficacy of felbamate monotherapy for the treatment of intractable partial epilepsy.[8] The patients in this study had partial and secondarily generalized seizures and were undergoing presurgical monitoring. The effectiveness of felbamate monotherapy was compared to that of a placebo (an inert, dummy pill). Forty patients were randomized to either felbamate ($n = 19$) or placebo ($n = 21$) and followed in the clinical center for two weeks. The patients' numbers and types of seizures and were recorded daily for two weeks. The primary outcome of this study was daily seizure rates for patients on treatment or placebo. The study results showed that felbamate monotherapy significantly reduced the number of seizures compared to the placebo. This type of randomized design is often used to test promising new treatments for efficacy.

1.1.3. The ISIS-4 Trial: Drug Treatment on Survival after a Heart Attack

A multinational collaborative group designed the Fourth International Study of Infarct Survival (ISIS-4) clinical trial.[9] This study was designed as a large multicenter randomized trial to assess early oral captopril, oral mononitrate, and intravenous magnesium sulphate treatments in patients with suspected acute myocardial infarction (MI). Approximately 58,000 patients were randomized to one of eight treatment groups, which correspond to all $2 \times 2 \times 2 = 8$ combinations of the three study treatments. The primary outcome was whether each patient was alive or dead at 35 days after randomization (35-day mortality). The study results showed that oral captopril significantly reduced 35-day mortality, whereas oral mononitrate (marginally worse) and intravenous magnesium sulphate (marginally better) did not. This type of randomized design is often used to test whether at least one of several treatments has a small but clinically meaningful effect on mortality or survival.

The diverse study designs of these examples illustrate fundamental issues. The first study, on the effect of beta-interferon on lesions detectable by MRI, is a nonrandomized study, whereas the felbamate monotherapy trial and the ISIS-4 trials are randomized clinical trials. How does one design a study to test whether a treatment is effective? How does one construct two or more groups of patients that are comparable on both known and unknown characteristics so that the results of a study will reflect the efficacy of the treatment rather than group differences? These issues are addressed in the section on randomization and study design.

Also, in each study the investigators wished to determine whether a given treatment or treatments were effective in the care of patients with a specific disease or medical risk. Each study was conducted because the investigators wanted to be able to treat not only the patients in that study but also all patients with similar characteristics, diseases, and medical risks. How does one use the results of a study to test whether a given treatment is effective? These essential questions are addressed in the section on hypothesis testing.

Furthermore, these studies had diverse numbers of patients enrolled in them. The beta-interferon and MRI study had 14 patients, the felbamate monotherapy study had 40 patients, and the ISIS-4 trial had approximately 58,000 patients. How many subjects does one need to enroll in a trial to have a "good chance" of determining that a clinically significant effect exists when in reality such an effect does indeed exist? Conversely, how many subjects does one need to enroll in a trial to have a "good chance" of determining that no clinically meaningful effect exists when in reality no such effect exists? These parallel questions are addressed in the section on sample size calculation.

2. ISSUES IN RANDOMIZATION

2.1. Reasons for Randomization

It is useful to consider first what is meant by random allocation. Altman[10] provided a useful definition: "By random allocation we mean that each patient has a known chance, usually equal chance, of being given each treatment, but the treatment to be given cannot be predicted." The idea of randomness accords with our intuitive ideas of chance and probability, but it is distinct from those of haphazard or arbitrary allocation.

Randomization ideally helps researchers compare the effects of the treatments among two or more groups of participants with comparable baseline characteristics, such as the distribution of age or the proportion of high-risk participants in each group, as well as other known and unknown covariates. Randomization also

ensures researchers do not deliberately or inadvertently create groups that differ from each other in any systematic way. For example, proper randomization techniques help to eliminate the problem of selection bias, where investigators give the healthiest participants the treatment and the sickest participants the placebo, which would result in a biased assessment of treatment effect. Similarly, proper randomization techniques help to eliminate investigator bias, where researchers subjectively perceive study participants on treatment as performing better, and for patient response bias or the so-called placebo effect, where study participants respond to treatment just because they know that they are on an "active" treatment. Without randomization and a placebo comparison group, we may misinterpret these potential biases as the true effect. Optimally, studies will be double-blind, in that neither the investigators nor the study participants should know which participants received which treatments. Furthermore, proper randomization helps ensure that statistically significant differences between treatment groups are indeed due to the actual treatment effect, rather than known or unknown external factors. Finally, randomization guarantees that statistical tests based on the principle of random allocation will be valid.

2.2. Types of Randomized Studies

We now consider various types of randomized studies. These study designs require different amounts of resources and provide different degrees of evidence for treatment effectiveness. In addition, there are different inherent assumptions about the nature of the treatments as well as about characteristics of the study outcome in some of these studies.

2.2.1. Parallel Groups Designs

In parallel groups designs, participants are randomized to one of k treatments. Interest focuses on comparing the effects of the k treatments on a common response or outcome. One of these groups may be a placebo group (a group assigned to a dummy or inert pill) or a control group (a group assigned to a standard or an alternative treatment). The effect on the response could be adjusted for baseline measurements of patient characteristics. The felbamate monotherapy clinical trial was conducted with a parallel groups design, with $k =$ two groups. The response in this trial was the average daily seizure frequency over the two-week follow-up period. Furthermore, the double-blind randomized parallel groups design is the "gold standard" to which all other designs should be compared. It is the ideal study to arrive at a definitive answer to a clinical question and is often the design of choice for large-scale definitive clinical trials. The challenge of this design is that it often requires large sample sizes and thus requires large amounts of resources.

In *sequential* trials, the *k* parallel groups are studied not for a fixed period of time but, rather, until either a clear benefit from one treatment group appears or it becomes highly unlikely that any difference will emerge. These trials tend to be shorter than fixed-length trials when one treatment is much more effective than the other treatments. In *group sequential* trials,[11] the data are analyzed after a certain proportion of the observations have been collected, perhaps after one-fourth, one-half, and three-fourths of the expected total number of participants or events, and once more at the end of the study. Data analyses of the primary outcome variables during a study are called *interim analyses*. Group sequential trials are easier than sequential trials to plan regarding duration and resources, and they can also be stopped early if one treatment is much more effective than the others. All trials must have a mechanism for stopping early if evidence of harm due to the treatment emerges. Trials also may be stopped for futility, where futility is defined as the unlikelihood that a positive treatment effect will emerge as the result at the end of the trial. Talk to a statistician about how to plan interim analyses for efficacy or futility.

2.2.2. Randomized Crossover Designs

In the two-period crossover design, participants are randomized to one of two sequences, with each sequence having a probability of 50% of selection. These sequences are (1) placebo → washout period → treatment or (2) treatment → washout period → placebo. The advantage of a crossover design is that each person receives both the placebo and the treatment, and thus each person serves as his or her own control, and in turn the sample sizes for crossover designs can be substantially smaller than those for comparable parallel group designs. The required sample size depends on the amount of between-subject variability relative to the amount of within-subject variability. For disease processes such as epilepsy or multiple sclerosis, where the between-subject variation is relatively large, there could be a substantial reduction in sample size in using a crossover design. The major problem with crossover designs is the potential for carryover effects. Carryover effects exist when the treatment received in period 1 affects the response in period 2. For example, the pharmacokinetic and systematic effects of a new drug may last a few weeks after the last dose, and thus study participants will still show residual effects when they receive placebo if there is not a sufficiently long period before treatment periods. The time between treatment periods is called the *washout period*, and it needs to be sufficiently long to avoid carryover effects. A challenge with crossover designs is that the required length of the washout period for a particular therapeutic is usually unknown before the study is conducted. In addition, certain therapies change the natural history of the disease process, which creates a carryover effect no matter how long the washout period. This reason explains why a two-period randomized crossover design was not considered for either the beta-interferon/MRI study or the felbamate monotherapy trial.

There are statistical tests to examine for the existence of carryover effects. A major challenge is to design crossover trials with sufficiently large sample sizes so as to be able to detect meaningful carryover effects with a good statistical chance or high power.[12] Crossover designs may have more than two periods as long as carryover effects and study retention are not problematic.

2.2.3. Factorial Designs

In a factorial design, each level of a factor (treatment or condition) occurs in combinations with every level of every other factor. Experimental units are assigned randomly to treatment combinations rather than individual treatments. ISIS-4 was designed as a factorial study with three treatments: oral captopril, oral mononitrate, and intravenous magnesium sulphate. Each of the three treatments could be delivered at one of two levels (e.g., placebo, standard dosage). Therefore, for this study there are eight ($2 \times 2 \times 2 = 8$) possible treatment combinations. Each patient is randomized to one of the eight combinations with a probability of 1/8 (12.5%). Because every treatment combination is tested on a different group of participants, we are able to estimate the interactions or synergistic effects between various treatments on the response (e.g., 35-day mortality). A major challenge of factorial designs is to choose a sufficiently large sample size to be able to detect meaningful interactions with high power or a good statistical chance of seeing an interaction if it in truth is present. The main reason factorial designs are used is to examine multiple hypotheses with a single study. For example, the ISIS-4 study was designed to simultaneously examine the role of three treatments in reducing 35-day mortality in treating acute MIs. Designing a factorial study saved resources compared to designing three separate parallel group studies for each of the experimental treatments. Note that if some particular treatment combinations are not of interest, a partial factorial design that omits the less interesting combinations may be used.

2.3. Alternatives to Randomized Studies

Not all studies in science and medicine involve randomization. We discuss the use of observational studies in epidemiology, the use of nonrandomized historical controls in early drug development trials, and nonrandom methods of allocating participants to treatment or control groups. These studies are useful in many circumstances, but they may lead to difficulties in inference and interpretation.

2.3.1. Epidemiologic or Observational Studies

One can classify common study designs used in medicine and epidemiology according to three major criteria.[10] First, studies can be either *experimental* or *observational*. In experimental studies, researchers deliberately influence the course of events and investigate the effects of an intervention or treatment on a carefully selected population of subjects. Experimental studies conducted on humans are called *clinical trials*, with attendant ethical considerations such as informed consent. In observational studies, researchers merely collect data on characteristics of interest but do not influence the events. Second, studies can be *prospective* or *retrospective*. In prospective studies, researchers collect data forward in time from the start of the study, whereas in retrospective studies researchers collect data on past events from existing sources, such as hospital records or statistical abstracts, or from interviews. Third, studies can be *longitudinal* or *cross-sectional*. In longitudinal studies, researchers investigate changes over time, like a movie film, possibly in relation to an intervention or treatment, whereas in cross-sectional studies researchers observe the subjects at a single point in time, like a snapshot. Note that experiments and clinical trials are both prospective by their very nature. Agresti[13] and Fleiss[12] provide excellent discussions of observational studies with many examples.

Consideration of observational studies as alternatives to randomized clinical trials provides insight into the advantages and disadvantages of the latter from a scientific perspective. Suppose that one wishes to study the effects of oral contraceptives (the "pill") on the risk of breast cancer over 30 years in women who began to use the pill in their early 20s. From the scientific perspective, the ideal way to address this question would be through a clinical trial. The researchers would randomly assign women in the trial to either treatment or placebo groups and then follow them

prospectively for 30 years and observe which group experiences more cases of breast cancer. Such a study would present many ethical challenges, because reproductive rights are a core human right, and also would prove impractical since it would be impossible to blind the subjects and researchers as to the treatment assignment, at least after the first pregnancy.

From the ethical and practical perspectives, the best way to address these questions would be through an observational study. In a *cohort* study, women would choose whether or not to use the pill in their early 20s, and the researchers would merely follow them prospectively and longitudinally over 30 years to observe who develops breast cancer. One would need to consider whether the group of women who chose to use the pill differed systematically from the women who chose not to do so. In a *case–control* study, researchers would construct groups of women in their 50s who had or had not developed breast cancer and then retrospectively look into the past to determine which women had used oral contraceptives and determine which other life events may have influenced the risk of breast cancer. One would need to consider how well the women selected for this study reflect the original population of women who began to use the pill in their early 20s. One would also need to consider if other risk factors were well understood and data reliably collected on all risk factors on all women in the study. In a *cross-sectional* study, researchers would collect a sample of women in their 50s and then simultaneously classify them on contraceptive use and breast cancer. One again would need to consider how well one could use this sample to make inference back to the population of women who began to use the pill in their early 20s.

Regardless of how well an observational study is constructed, questions may arise about applicability and unknown risk factors. Currently, different oral contraceptive pills and doses are used compared to 30 years ago, so can the previous study shed light on today's women in their early 20s and their future health? Are there unknown or unmeasureable risk factors that might be playing a role in the study results? In observational studies, we can only control for known and measured variables.

Moving beyond known and measured factors, in the scientific perspective, experimental studies and clinical trials can establish *causation* of a response by a treatment, whereas observational studies can merely show an *association* between a risk factor and a response. The fact that observational studies can only find a weaker degree of connection reflects the fact that they are subject to *confounding*. Two or more variables, whether known or unknown to the researchers, are confounded when their effects on a common response variable or outcome are mixed together.

To illustrate, consider the following examples. A mother's genome has a causal relationship to a daughter's height because the mother gives part of her genes that influence height to her daughter. However, a son's genome is merely associated with a daughter's height because both children receive part of their mother's genes, but the son does not give any of his genes to the daughter (except in Greek tragedies). Finally, a mother's genome and a nutrition program are confounded because their effects on a daughter's height are mixed together.

To understand the importance of randomized clinical trials to establish causation, consider the following example. The Physicians' Health Study Research Group (Harvard Medical School) conducted a clinical trial to test whether aspirin reduced mortality from cardiovascular disease.[14] Every other day, the physicians in this study took either one aspirin tablet or a placebo, and the physicians were blinded as to which tablet they were taking. Of the 11,034 physicians randomized to the placebo group, 18 had fatal MIs and 171 had nonfatal MIs. Of the 11,037 physicians randomized to the aspirin group, 5 had fatal MIs and 99 had nonfatal MIs. The results of this trial are highly statistically significant. Are the researchers justified in concluding that taking one aspirin tablet every other day prevents fatal and nonfatal MIs? Yes, because this study is a clinical trial, one can posit causation. Note that one could not have conducted an observational study to definitively answer this question. For example, the results for a specific population naturally taking such high levels of aspirin, such as patients with arthritis, cannot easily be generalized to the population at large.

By contrast, to illustrate the weakness of observational studies in this regard, consider the following hypothetical example. Suppose a team of researchers designs a cohort study to address the question of whether smoking causes premature death. They may construct two groups of middle-aged men (50–55 years old) who are smokers and nonsmokers, with 2500 subjects in each group. The subjects may be examined at baseline, followed prospectively and longitudinally, and their age at death recorded. Suppose the median time to death is 8 years earlier for smokers than for nonsmokers, and that this difference is statistically significant. Are the researchers justified in concluding from this study that smoking causes premature death? No. The tobacco companies can respond that smokers are inherently different from nonsmokers. Perhaps there are some genetic, socioeconomic, or behavioral factors that cause (or predispose) people to smoke and that also cause them to die at an earlier age. Are the

researchers, nevertheless, justified in concluding from this study that smoking is associated with premature death? Yes, that is the precise function of observational studies—to propose associations.

A large set of observational studies led to a set of ambitious clinical trials in women's health. The Women's Health Initiative (WHI) was launched in 1991 and consisted of a set of trials in postmenopausal women motivated by several prevention hypotheses.[15] The hormone replacement therapy (HRT) hypothesis assumed women assigned to estrogen replacement therapy would have lower rates of coronary heart disease (CHD) and osteoporosis-related fractures. Progestin and estrogen were to be used in women with a uterus, and breast and endometrial cancers would be monitored. The hypothesized cardioprotective effects of HRT in postmenopausal women could not be proven in observational studies but had evolved over time due to the adverse affects of menopause on the lipid profile. Epidemiologic evidence, the majority of 30 observational studies, reported a benefit among estrogen users in age-adjusted all-cause mortality. Questions remained about the demographic profile associated with the observational studies' participants being both rather healthy and younger with little pertinent data on women beginning hormones after age 60 years; the use of the combination treatment estrogen plus progestin instead of unopposed estrogen, which had been the focus of most studies; and the overall risk and benefit trade-off. Observational studies had noted a modest elevation in the risk of breast cancer with long-term estrogen use; however, adverse event data on progestin were inconsistent at the time. At the inception of the WHI, it was to be the study with the longest follow-up. The questions addressed in the clinical trials were posed based on epidemiological evidence. When approaching an randomized control trial (RCT) from a base of cohort studies, several important points must be addressed. If the motivation for a cohort study is to evaluate risks associated with a treatment or an exposure, then the study needs not only long-term users but also sufficient number of newly exposed participants to assess short- and long-term intervention effects. Time variation must also be taken into account and exposure effects may need to be evaluated over defined exposure periods.[16] The estrogen plus progestin portion of WHI stopped early after finding estrogen plus progestin did not confer cardiac protection and could increase the risk of CHD, especially during the first year after the initiation of hormone use.[17] The take-home messages were not simple yes/no answers and women were advised to talk with their doctors about their personal health and family history. Whereas some believed the WHI hormone therapy results were surprising, others did not. What the experience has taught us is to pay close attention to observational study design and analysis. Hypothesis development, particularly in prevention, which many times is based on cohort data, is vital and far more difficult than imagined at first glance. As researchers, we must always think ahead while planning our current study to the next several trials and studies that may result from our anticipated (and unanticipated) findings.

In summary, observational studies are valuable alternatives when ethical considerations, costs, resources, or time prohibit one from designing a clinical trial. They may also be useful in providing preliminary evidence of an effect, which, ethics and reality permitting, can subsequently be studied with a well-designed randomized clinical trial. It should be recognized that observational studies have the scientific weakness that they can be used only to find associations between risk factors and responses, but they cannot establish causation.

2.3.2. Phase I Study Designs

Phase I studies, often called dose-finding studies, traditionally are nonrandomized. The fundamental goal of these studies is to find appropriate dose levels and to detect potential toxicities due to the investigational drug. A dose limiting toxicity (DLT) threshold or physical event must be defined in order to create a stopping rule. Usually, the definition of a DLT is based on criteria such as a certain grade of toxicity, as defined by the National Cancer Institute Common Toxicity Criteria (NCI CTC), although for some newer interventions a traditional DLT may not be the appropriate stopping rule. Different studies and substances will require different stopping rules, especially since some treatments are nontoxic at all dose levels.[18,19]

Traditionally, a few dose levels or categories (e.g., four) are decided on and at least three participants are treated at each dose level, escalating through the dose levels in the following manner. Three participants are enrolled at the lowest dose level in the protocol, and if none of the three (0/3) develop a DLT then the study escalates to the next dose. If a DLT is observed in one of the three (1/3) participants, then three additional participants are enrolled, so now six people will receive the current dose. If none of the three additional participants (0/3) develop a DLT, then the study escalates to the next dose. Participants are not entered at a new dose level until all participants in the previous levels remain free of toxicities for a specified period of time.

In addition, the maximum tolerated dose (MTD) is the dose level immediately below the level at which two or more participants experienced a DLT. Usually, the study aims to find a safe dose defined as the MTD or the study finishes at the maximum dose that is pre-specified in the protocol. As the use of this methodology has expanded from phase I oncology studies to broader contexts, the definition of MTD has also expanded to include nontoxic but nevertheless undesirable events.

There is increased interest in the phase I/II format for examining trade-offs between toxicity and efficacy.[20] Statisticians have developed special study designs that examine the potential plateaus in drug responses while carefully balancing the risk of over- and underdosing.[21] Some investigators believe that because of the small sample sizes at each level, trials should involve both dose escalation and de-escalation to find the optimal thresholds.[20] Some believe that the traditional numbers of participants, usually three, at each dose are too small and want to increase this number. Currently, statisticians debate how much of this work belongs in phase I (dose-finding) and how much in phase II (safety/efficacy) studies. Regardless of the blends in design, the design of any study should fit the purpose of the study, the problems in the specific medical area, and the guidelines of any agency that needs to review the study (e.g., the FDA for an investigational new drug).

2.3.3. Historical Control Studies and Other Phase II Study Designs

The screening of new therapies is often conducted with single arm or nonrandomized studies. In cancer treatment, for example, phase II studies are often conducted in which patients are treated and their responses (often whether or not a clinical response is obtained) are observed. The purpose of these studies is not to prove that a new therapy is efficacious on the ultimate outcome of interest (e.g., survival) but only that it has sufficient activity (e.g., tumor reduction) to now be tested in a randomized study. These designs often require a small number of patients; when the evidence shows that the benefits of the new therapy are relatively small, the designs prevent large numbers of patients from being exposed to potentially toxic treatment. The beta-interferon/MRI trial is an example of a screening trial. The major advantage of this type of trial design was that only 14 patients were needed to conduct this study of beta-interferon. There are serious disadvantages of this trial, however, such as the problems of the placebo effect and investigator bias because all patients were treated in an unblinded fashion. In addition, in this example, there is the problem of potential regression to the mean. For example, patients with MS are screened for moderate MRI activity and then followed longitudinally. Because the natural process is relapsing–remitting, there is the potential that patients will be screened when in a relapse and naturally move into a remitting phase over time. Thus, one may see a reduction in disease activity over time, even if the experimental therapy is ineffective.

Another type of design is a historical control study. These studies have commonly been used in cancer research.[22] Here, instead of creating a randomized comparison group, a single group of patients is treated, and their responses are compared with controls from previous studies. These studies have the advantage of using only half the number of patients, none of whom receive a placebo. They also have serious disadvantages. In addition to the other problems with a nonrandomized study, these controls often do not provide a good comparison with the new treatment patients. For example, controls are often taken from studies conducted years ago. Also, treatments, technology, and patient care often change over time. In addition, the patient population characteristics may change. These changes, which are often not recognized or reported, can result in serious biases for assessing treatment efficacy.

Two other methods for phase II studies that are used, especially in oncology, are optimal two-stage designs[23] and randomized phase II clinical trials.[24,25] In optimal two-stage designs, a certain number of participants are enrolled in the first stage, which is followed by an interim analysis. During this analysis, the study may be terminated for insufficient biological or clinical activity, but it cannot be stopped early for efficacy. If the trial is not terminated, additional participants are enrolled in a second stage. At the end of the second stage, participant data from both stages are used in an analysis of the biological or clinical activity end point. Optimal two-stage designs have the advantage that they require on average (under the null hypothesis of insufficient activity) the fewest number of participants to detect a specified effect or to conclude that no such effect exists.

Randomized phase II clinical trials can avoid the problems of the nonrandomized studies described previously. They try to select the superior treatment arm from k arms and use a binary outcome describing failure or success along with statistical selection theory to determine sample size. The goal is to have a low probability of choosing an inferior arm out of the k arms. Essentially, a single sample phase II study is imbedded into each study arm. In the randomized phase II design, an interim analysis to stop the study early for sufficient biological or clinical activity is possible, allowing the agent to be moved forward more quickly to another, larger study. In contrast, using

other phase II designs, such as the optimal stage-2 design, interim analyses may stop early for insufficient activity, so it is important to choose the design that is best for the study question.[26]

2.3.4. Uncontrolled Studies and Nonrandom Methods of Allocation

The designs discussed in this section have several major weaknesses and should not be used except as a last resort when ethical or practical considerations prevent one from designing a randomized placebo-controlled clinical trial. In *uncontrolled* (or open label) studies, researchers follow a group of participants at baseline and then on treatment. These studies have no placebo or control group (e.g., a standard therapy). Theoretically, measurements from participants at baseline serve as the controls, and the measurements from participants on treatment serve as the treatment group. However, such studies are unreliable because participants tend to do better simply by receiving some treatment (placebo effect), researchers and participants tend to interpret findings in favor of their new treatment (investigator or patient bias), and it is impossible to distinguish between the effect of treatment and the effect of time on the response (confounding between treatment and time).

In studies with *systematic allocation*, researchers allocate treatments to participants according to a system, for example, by each participant's date of birth, date of enrollment, terminal digit of hospital ID or Social Security number, or simply alternately into different treatment groups according to whim. Such studies are unreliable because they are not blinded and hence have the danger of bias or outright manipulation. For instance, if a receptionist knew that participants who arrived at the hospital on Tuesdays received the placebo and those who arrived on Thursdays received the treatment, and his favorite aunt had the particular disease under study, on which day would he recommend that she enter the trial?

In studies with *nonrandom concurrent controls*, researchers choose groups for participants nonrandomly or participants get to choose their group. Such studies are unreliable because it is impossible to establish that the treatment groups are comparable, volunteers for the treatment tend to do better than nonvolunteers (volunteer bias), and ineligible persons tend to be assigned to the control group (ineligibility bias). For example, suppose that researchers wish to compare weight gain in infants who are given either natural fruit juice or sweetened, colored water. A mother is asked to enroll her infant, but she knows that people in her family have a pattern of childhood obesity. In which group does she choose to enroll her infant?

In short, trials with nonrandom allocation tend to be a waste of time and resources and a source of grief and controversy for researchers. Altman[10] described a nonrandomized study of the possible benefit of vitamin supplements at the time of conception in women at high risk of having an infant with a neural tube defect (NTD). The researchers found that the vitamin group subsequently had fewer NTD infants than the placebo group. He noted, however, that because the study was not randomized, the findings were not widely accepted, and the Medical Research Council (UK) subsequently ran a large randomized trial to try to get a proper answer to the question. Indeed, in the nonrandomized study, the control group included women ineligible for the trial as well as women who refused to participate.

2.4. Types of Randomization

We are all familiar with randomization in our daily lives, such as the coin toss before a football game, dice thrown in a board game, and names drawn out of a hat to determine the order of several contestants, not to mention gambling and lotteries. The goal of randomization in these cases is to choose an outcome or set of outcomes in a fair but unpredictable manner. Likewise, this section discusses the conceptual types of randomization that one can employ to construct treatment groups in a clinical trial in a fair and reliable manner.

2.4.1. Simple Randomization

In simple randomization, each patient is randomized to a treatment or a treatment sequence with a known probability. For a two-group trial such as the felbamate monotherapy trial, the results of randomization would correspond to flipping a coin and giving treatment when the coin toss resulted in heads (H) and giving the patient placebo when the toss resulted in tails (T). Such a random sequence of treatment assignments might look like

H H T H T T H.

Although on average there will be approximately equal numbers in the two groups, in small samples there is the potential for severe imbalance. For example, the sequence of 18 tosses,

T H T T H T T T T H T T T T H T T T,

contains 4 heads and 14 tails, a severe imbalance. Based on binomial probability calculations, the probability of

imbalance as extreme or more extreme than this is 0.03. Thus, with simple randomization in a small study (n = 18), there is a small but significant chance of having this degree of severe imbalance. Other randomization procedures have been developed to reduce or eliminate the possibility of this type of imbalance and to avoid the occurrence of trends in the order of heads and tails.

2.4.2. Block or Restricted Randomization

Block or restricted randomization is a type of randomization used to keep the number of subjects in the different groups closely balanced at all times. For example, with a block size of 4, balance will be achieved for sample sizes that are multiples of 4. For example, the sequence

HHTT THTH THTH HTTH HHTT THHT

is a block randomized sequence of 24 treatment assignments randomized with a block size of 4. Thus, every four treatment assignments results in equal numbers in the two groups. In addition, even if the study stops early, the most severe imbalance that can occur is being off by 2. This randomization scheme also eliminates the possibility of trends in the order of heads and tails. The felbamate monotherapy trial was designed with a block size of 4. In randomized clinical trials, randomization is often done with a variable block size (permuted block randomization). This provides an additional layer of blindness to the investigators.

2.4.3. Stratified Randomization

Simple or block randomization may result (by chance) in the differences of important characteristics between treatment groups. For example, these methods may result in equal numbers of participants in the treatment and placebo arms of a trial but with very different distributions of gender in the two treatment arms. Stratified randomization is used to achieve balance on important characteristics. In this approach, we perform separate block randomizations for each subgroup or stratum. Common strata are clinical center, age, sex, and medical history. An example of a stratified randomization by center is as follows:

Center 1: HHTT HTHT THHT

Center 2: THHT TTHH HTHT.

Within each center, we have balance between the treatment groups at the end of the 24-subject study.

Although permuted block randomization stratified by center is a preferable method of randomization, sometimes stratification may be extended too far. In stratified randomization, each stratum preferably has several people in it, and if multiple strata are used, the implementation may become difficult. Thus, the only variables that should be considered for stratification are potential confounders that may be strongly associated with the outcome of interest.

Expanding the previous example, suppose in addition to two treatment groups and two centers, we also want to stratify on sex (male/female), body mass index (BMI) classification (underweight/normal/overweight/obese I/obese II), and several age groups (18–24.9, 25–34.9, 35–44.9, and 45–55 years) but maintain the same sample size. Each of the combinations of stratification factors may be called a cell. As we add more stratification variables, it becomes obvious we may have sparse or empty cells. Two centers × 2 sexes × 5 BMI categories × 4 age groups alone provide 80 cells. Within each of these cells/strata, subjects are assigned to the two treatment groups. Although this example may seem to be an exaggeration, in some studies there are 30-plus different strata as investigators add various "important" variables. For this and other reasons, minimization or dynamic allocation/stratification was developed. This will be discussed further in the special considerations section. It is important to note that although our examples use two group study designs, everything may be extended to studies with three or more treatment groups.

2.5. Mechanisms of Randomization

In practice, random numbers can be generated with a computer or random number tables. Random digit tables are presented in many textbooks.[10,27] This table presents random numbers as pairs of digits from 00 to 99 in a large matrix form. Reading across any row or column results in a random sequence of these numbers. We can use either tables or software packages to generate a randomization scheme for a parallel groups design with two groups (groups A and B) with a total sample size of 20. Starting with a typical row in a random digit table and reading down, we have the numbers 56, 74, 62, 99, 76, 40, 66, 75, 63, 60, etc. We can assign participants to group A if the digit is even and group B if the digit is odd. Corresponding to the digits, 5 6 7 4 6 2 9 9 7 6 4 0 6 6 7 5 6 3 6 0, we have the following treatment assignments: B A B A A A B B B A A A A A B B A B A A. Thus, the randomization resulted in 12 participants being assigned to group A and 8 to group B.

An alternative to simple randomization is to assign short sets or blocks of treatment assignments to sets of participants using block randomization. Suppose we decide to perform block randomization with block sizes of size 4. There are six equally likely combinations that will achieve balance in each block after every fourth assignment:

1. A A B B
2. A B A B
3. A B B A
4. B B A A
5. B A B A
6. B A A B

We can generate these sequences by assigning treatment sequence 1 to 6 with equal probabilities of 1/6. We can generate random numbers from 1 to 6 by reading across either a row or a column of a random digits table and assigning the sequence corresponding to numbers obtained between 1 and 6, and ignoring 0, 7–9. For example, with the sequence, 56, 74, 62, 99, 76, 40, 66, 75, 63, 60, 29, 40, . . . , we get a listing of numbers of 5 6 4 6 2. This would construct the sequence

BABA BAAB BBAA BAAB ABAB.

Suppose that we plan to have two centers in a study, with one center contributing 8 participants and the other 12 participants, and we wish to perform a randomization stratified by center. This stratified randomization could be accomplished by performing block randomization in each center separately. Specifically, we could randomize the participants in center 1 using the first two blocks of the previous sequence (BABA BAAB) and the participants in center 2 using the remaining three blocks of 4 participants (BBAA BAAB ABAB).

There are many extensions of the basic principles of randomization presented here. For example, one can design methods to randomize participants to three or more treatments. It pays to work with a statistician to develop practical and efficient methods for randomizing subjects in a particular study.

3. OVERVIEW OF HYPOTHESIS TESTING

We begin this section by presenting the basic ideas of statistical hypothesis testing. We then introduce the concepts of type I and II errors, *p*-values, and, finally, power. We next summarize one-sample and two-sample tests for dichotomous (binomial) and continuous (modeled by the normal distribution, also known as the bell curve or Gaussian distribution) data that are commonly found in clinical trials, and illustrate the use of these tests with the analysis of data from the motivating examples.

3.1. The Goals of Statistical Inference

Statistical inference is the procedure through which inferences about a population are made based on certain characteristics calculated from a sample of data drawn from that population. In statistical inference, we wish to make statements not merely about the particular subjects observed in a study but also, more importantly, about the larger population of subjects from which the study participants were drawn. In the beta-interferon/MRI study, we wish to make statements about the effects of beta-interferon, not only in the 14 participants observed in this study but also in all patients with RRMS. Similarly, in the felbamate monotherapy study, we want to make a decision about the effectiveness of felbamate for all patients with intractable partial epilepsy. Statistical inference can be contrasted with exploratory data analysis, where the purpose is to describe relationships in a particular data set without broader inference. Inferential techniques attempt to describe the corresponding characteristics of the population from which the sample data were drawn.

To develop a conceptual view of hypothesis testing, we first need to define some terminology. A *statistic* is a descriptive measure computed from data of a sample. For example, the sample mean (average), median (middle value), or sample standard deviation (a measure of typical deviation) are all statistics. A *parameter* is a descriptive measure of interest computed from the population. Examples include population means, population medians, and population standard deviations. The distribution of all possible values that can be assumed by a particular statistic, computed from random samples of a certain size repeatedly drawn from the same population, is called the *sampling distribution* of that statistic. The goal in statistical inference is to use probability theory to make inferences about population parameters of interest. For example, for the felbamate monotherapy trial, the parameter of interest is the change in daily seizure rates due to felbamate treatment. The statistic is the mean number of seizures per day for participants in the placebo arm minus the mean for participants randomized to the felbamate arm of this trial. Although we cannot observe the population and hence the sampling distribution directly, we can model them based on our understanding of the biological system and the sample that we are studying.

There are two broad areas of statistical inference: statistical estimation and statistical hypothesis testing. Statistical estimation is concerned with best estimating a value or range of values for a particular population parameter, and hypothesis testing is concerned with deciding whether the study data are consistent at some level of agreement with a particular population parameter. We briefly describe statistical estimation and then devote the remainder of this section to providing a conceptual overview of hypothesis testing.

There are two types of statistical estimation. The first type is point estimation, which addresses what particular value of a parameter is most consistent with the data. For example, how do we obtain the best estimate of treatment effect for the beta-interferon/MRI data? Is the best estimate obtained by taking the mean or median reduction in the number of monthly lesions? Depending on the skewness of the data and the exact question of interest, one estimate may be preferable to the other; this is another time to talk with a statistician about the best way to evaluate the effect of interest. The second type of statistical estimation is interval estimation. Interval estimation is concerned with quantifying the uncertainty or variability associated with the estimate. This approach supplements point estimation because it gives important information about the variability (or confidence) in the point estimate. An example would be the statement of the 95% confidence interval for the mean effect of felbamate in the epilepsy clinical trial. This interval gives us an idea of the variability of the treatment effect as well as its size. One can interpret these confidence intervals in a frequentist fashion; in the long term, 95% of similarly constructed confidence intervals will contain the true mean effect. However, one cannot determine whether a particular interval does or does not contain the true mean effect. More loosely one might discuss being 95% confident that the true treatment effect occurs between two stated values, with the caveat of understanding this in a frequentist fashion and not exactly as stated.

Hypothesis testing has a complementary perspective. The framework addresses whether a particular value (often called the null hypothesis) of the parameter is consistent with the sample data. We then address how much evidence we have to reject (or fail to reject) the null hypothesis. For example, is there sufficient evidence in the epilepsy trial to state that felbamate reduces seizures in the population of intractable partial epilepsy patients?

3.2. Basic Concepts in Hypothesis Testing

The purpose of hypothesis testing is to make decisions about a population parameter by examining a sample of data from that population. For the MS study, we wish to test whether beta-interferon has any effect on disease activity. For the felbamate trial, we wish to test whether felbamate lowers the propensity to have seizures in patients with intractable partial epilepsy. A key question in the ISIS-4 trial is whether magnesium sulphate administered soon after an MI lowers the risk of 35-day mortality.

A hypothesis test involves specifying both a *null hypothesis* and an *alternative hypothesis*. The null hypothesis is often stated as the negation of the research question. In this section, we focus on the most common hypothesis test in medical statistics, the two-group comparison of population means. In this instance, the null hypothesis is that the two means, denoted μ_1 and μ_2, are equal. This hypothesis is stated as H_0: $\mu_1 = \mu_2$. The alternative hypothesis is that the null hypothesis is not true. There are a few formulations for the alternative hypothesis for two-group comparisons. One possibility is that we are interested in any difference between the means. This hypothesis can be stated as H_A: $\mu_1 \neq \mu_2$. Tests with this type of alternative hypothesis are referred to as *two-sided tests*. There may be occasions in which interest only focuses on detecting differences between the population means in one direction. We may, for example, be interested in testing whether the mean in treatment group 1 is larger than the mean in treatment group 2 and would not at all be interested in the result if the opposite were true. This is called a *one-sided test* and the alternative hypothesis is stated as H_A: $\mu_1 > \mu_2$.

When should one conduct a one-sided versus a two-sided test? In a trial in which the interest is in determining that the treatment effect is different in one group compared to another group, but there is no *a priori* reason to suggest which group should have the stronger effect, one should conduct a two-sided test. In the very limited situation in which there is specific interest in demonstrating that the treatment effect is larger in one group than in the other group, and the reverse situation would be scientifically or clinically uninteresting, then a one-sided test might be appropriate. The decision to conduct either a one-sided or a two-sided test must be made before the study begins.

Having set up these hypotheses, we conduct an experiment (collect a sample of data) and calculate a *test statistic*, a value that can be compared with the known distribution of what we expect when the null hypothesis is true. This reference distribution depends on the statistical model for the data, and its formulation requires assumptions about the distribution of the outcome variable. Thus, different test statistics and reference distributions are formulated for many situations that are common in medical statistics. We review some of these in the next subsection.

Test statistics considered here have the form

$$\text{Test statistic} = \frac{\text{point estimate of } \mu - \text{target value of } \mu}{\text{known value or point estimate of } \sigma},$$

$$(15.1)$$

where σ denotes the population standard deviation. For a two-sided test, we reject the null hypothesis H_0 when the test statistic is in the upper or lower $100 \times \alpha/2\%$ of the reference distribution (i.e., the so-called tails of the distribution). Typically, α is chosen to be equal to or smaller than 0.05. Thus, we reject H_0 when it is very unlikely (< 5% by chance) that we would have observed a test statistic as large or larger in magnitude as the one we did if the null hypothesis were true.

There are two ways that one can err in a hypothesis test. We can reject the null hypothesis when the null hypothesis is "in truth" true, or conversely, we can fail to reject the null hypothesis when the alternative hypothesis is in truth true. The first of these errors is called the *type I error* and the second is called the *type II error*. There is a trade-off between minimizing the rates of these two types of errors in a hypothesis test. Statistical hypothesis testing is based on ensuring that the probability of a type I error is very small (often chosen to be equal to or smaller than 0.05 in many trials). This chosen rate is called the *significance level* of the test and is denoted as α [i.e., $\alpha = P$ (type I error) or the probability of a type I error]. The *power* of a test is the probability of rejecting the null hypothesis (H_0) when the alternative hypothesis (H_A) is true. Alternatively, power equals one minus the probability of a type II error [i.e., power $= 1 - \beta = 1 - P$ (type II error)]. Power can be influenced by various factors. Power depends on the actual value of the alternative hypothesis, the sample size of the study, and the chosen significance level of the test. Good studies are designed to have sufficient power (at least 80–90%) to detect scientifically or clinically meaningful effects.[28] We discuss how to design studies with high power in the next section.

Often, statisticians are asked which is more important, significance level or power. The answer, of course, depends on the research questions the study needs to address. It is important to think about the implications on the individual and population levels for health outcomes and costs. Often, it is easier to think in terms of the implications of a false positive or a false negative, and in screening studies to look at the positive and negative predictive values of various laboratory tests. This is briefly discussed at the end of this chapter, but other books[29] describe these topics in detail.

A commonly reported value for a hypothesis test is the *p*-value (probability value). The *p*-value is the probability of observing a test statistic as extreme or more extreme than that observed if the null hypothesis is true. A *p*-value for a one-sided test only involves computing the probability in the direction of the alternative hypothesis. The *p*-value for the two-sided test involves computing this probability in either direction.

The p-value measures how unlikely the value of the test statistic is under the null hypothesis.

3.3. The Formulation of Statistical Hypotheses in the Motivating Examples

We now illustrate the development of hypothesis tests using the three motivating examples.

3.3.1. Hypotheses for the Beta-Interferon/MRI Study

The outcome variable for the beta-interferon/MRI study is the change in the mean number of gadolinium-enhanced lesions per month during the seven-month baseline period minus the mean number of lesions per month during the six-month treatment follow-up period. Data are paired in that each subject is observed before and during treatment. Exploratory data analysis (histograms) suggests that the difference in the means is approximately normally distributed on the log scale. In addition, observations between subjects are independent from each other; this is an important assumption for most standard statistical methods. Scientific interest focuses on detecting either a significant increase or decrease in lesion frequency. Hence, the proposed test will be two-sided. We can write the hypothesis test as

$$H_0: \mu_{\text{treatment–baseline}} = 0 \text{ vs. } H_A: \mu_{\text{treatment–baseline}} \neq 0. \quad (15.2)$$

3.3.2. Hypotheses for the Felbamate Monotherapy Trial

The felbamate monotherapy trial was designed as a parallel groups design. The unit of analysis is mean daily seizure rates during the 14-day period. Each subject has only a single outcome, so the data are unpaired. Exploratory data analysis (histograms) suggests that the mean daily seizure rates are approximately normally distributed on the square root scale (often, statisticians will examine data on a variety of scales, such as the natural logarithmic or square root scales). The scientific interest focuses on detecting either a significant increase or decrease in seizure frequency due to felbamate. Because we would be interested in reporting the result even if we found that seizure frequency was significantly larger on treatment than on placebo, we conducted a two-tailed test. We can write the hypothesis test for this example as

$$H_0: \mu_{\text{treatment}} = \mu_{\text{placebo}} \text{ vs. } H_A: \mu_{\text{treatment}} \neq \mu_{\text{placebo}}. \quad (15.3)$$

3.3.3. Hypotheses for the ISIS-4 Trial: Comparing the Magnesium and No Magnesium Arms

The ISIS-4 trial was a factorial design with three treatments. We focus on comparing the magnesium and no magnesium groups. The outcome is whether a randomized subject was dead 35 days after randomization. Because these data have only two possible outcomes, they are binary or dichotomous. In addition, these binary outcomes are independent of each other. Interest focuses on detecting either a positive or a negative effect of magnesium on mortality. We can write the hypothesis test for this example as

$$H_0: p_{\text{treatment}} = p_{\text{placebo}} \text{ vs. } H_A: p_{\text{treatment}} \neq p_{\text{placebo}}. \quad (15.4)$$

3.4. One-Sample Hypothesis Tests with Applications to Clinical Research

We begin with a discussion of hypothesis tests for the one-sample problem. In the next section, we extend these basic principles to the two-sample problem.

3.4.1. Tests for Normal Continuous Data

Suppose we collect conceptually continuous measurements on a sample of n individuals, $x_1, x_2, \ldots, x_n$. We may represent the sample mean or average of these measurements by $\bar{x}$ and the sample standard deviation by s_x, which estimate the true or population mean μ_x and standard deviation σ_x, respectively.

We can use these statistics calculated from the original data to test several hypotheses. If we want to test whether μ_x equals some target value μ_0, we can denote the null hypothesis (H_0) of equality and the alternative hypothesis (H_A) of inequality of this two-sided test as

$$H_0: \mu_x = \mu_0 \text{ vs. } H_A: \mu_x \neq \mu_0. \quad (15.5)$$

If we want to test whether μ_x is strictly less than (greater than) some target value, we can denote the null and alternative hypotheses of this one-sided test as

$$H_0: \mu_x \geq \mu_0 \text{ vs. } H_A: \mu_x < \mu_0 \quad (15.6)$$
$$(H_0: \mu_x \leq \mu_0 \text{ vs. } H_A: \mu_x > \mu_0).$$

Note that the null hypothesis always contains the case of equality, and the alternative hypothesis contains the research question. When σ_x is known, the test statistic takes the form

$$Z = (\bar{x} - \mu_0)/(\sigma_x / \sqrt{n}), \quad (15.7)$$

whereas when σ_x is unknown, the test statistic takes the form

$$T = \frac{\bar{x} - \mu_0}{s_x / \sqrt{n}}, \text{ where } s_x = \sqrt{\frac{1}{n-1} \sum_{i=1}^{n} (x_i - \bar{x})^2}. \quad (15.8)$$

These tests are referred to as z tests and t tests. Under the null hypothesis, Z has the standard normal distribution and T has the Student's t distribution with $n - 1$ degrees of freedom (df). The df parameter describes the shape of the reference distribution whereby the distribution becomes closer to normal as the df becomes large.

3.4.2. Determining Statistical Significance

3.4.2.1. Critical Values

Critical values for these distributions are available in any introductory statistics textbook.[10,30–32] The expression z_p or t_p gives the cut point or percentile of the normal (z) or t distribution, respectively, such that $100p\%$ of the probability lies left of that cut point. Commonly used cut points of the normal distribution are $z_{0.8} = 0.841$, $z_{0.9} = 1.282$, $z_{0.95} = 1.645$, and $z_{0.975} = 1.960$. Note also that by symmetry, $z_p = -z_{1-p}$.

To determine whether a test statistic provides significant evidence against the null hypothesis, we compare the observed values to the critical values or percentiles of the appropriate distribution. If the test statistic has an extreme value compared to the reference distribution, we reject the null hypothesis. For example, in a two-sided test at the $\alpha = 5\%$ significance level, we compare the observed z test statistic value to the critical values -1.960 and 1.960 (from the normal distribution), and if it falls either above or below these values we reject the null hypothesis. If the z test statistic value falls between the two critical values, then we do not reject the null hypothesis. Similarly, in a one-sided test of H_0: $\mu_x \geq \mu_0$ versus H_A: $\mu_x < \mu_0$ at the $\alpha = 5\%$ significance level, we compare the observed z test statistic value to the critical value 1.645, and if it exceeds this value we reject the null hypothesis.

3.4.2.2. Confidence Intervals

Another way to evaluate the evidence is by using a *confidence interval* (CI). A $100 \times (1 - \alpha)\%$ CI for a population parameter is formed around the point estimate of interest. The most basic CI is that for the mean, μ. If variance is known, the CI has the following formula:

$$\left(\bar{x} - \frac{z_{1-\alpha/2}\sigma}{\sqrt{n}}, \bar{x} + \frac{z_{1-\alpha/2}\sigma}{\sqrt{n}} \right). \tag{15.9}$$

By contrast, if the variance is unknown, then s_x is used instead of the standard deviation σ and the T critical value is used instead of the corresponding Z value. There is an important parallelism between hypothesis testing and CI construction for normal and Clopper-Pearson CIs. Specifically, if the hypothesized population parameter falls within the CI, we do not reject the null hypothesis. For example, for a 95% CI this is similar to performing a test at the $\alpha = 5\%$ significance level.

3.4.2.3. z Tests or t Tests

The choice between t and z is an important one. Although some people will switch to the normal as soon as sample size looks slightly large (e.g., $n > 30$), it can be problematic. Looking at the 0.975 cutoff for the upper end of a 95% CI, at df = 30 the T cut point is 4% larger than the normal cut point. At df = 120, there is still a 1% difference between the t distribution and the normal. Often, this will not matter; a test is highly significant or nonsignificant, but in general it is best to use a Student's t distribution if indeed that is what the test and data warrant.

Suppose an investigator wishes to determine whether pediatric anesthesiologists have unusually high serum Erp 58 protein levels. This protein is associated with industrial halothane hepatitis. Suppose that she collects $n = 9$ blood samples, and the sample mean and standard deviation of the protein levels are $\bar{x} = 0.35$ and $s_x = 0.12$ (optical density units), respectively. If the mean protein level is over 0.28, it will suggest that further study is needed. She chooses the $\alpha = 5\%$ significance level. This hypothesis test corresponds to H_0: $\mu_x \leq 0.28$ versus H_A: $\mu_x > 0.28$. Using the previous formula, one can calculate

$$T = \frac{0.35 - 0.28}{0.12 / \sqrt{9}} = 1.75, \tag{15.10}$$

which is less than the 95% percentile, $t_{1-\alpha,8} = 1.860$, of Student's t distribution with $n - 1 = 8$ degrees of freedom (p-value = 0.06). Thus, she does not reject the null hypothesis, although she may wish to collect a larger sample to explore this question further. In practice, one would collect a larger sample and use a more advanced method such as multivariate regression to adjust this hypothesis test for covariates such as age, gender, work experience, body mass, and medical history.[33]

3.4.3. Binary Data

Just as we can perform hypothesis tests on continuous data, we can perform them on proportions, with due alteration in details. Binary or dichotomous data have two possible outcomes, such as success or failure, presence or absence of a disease, or survival or death. Thus, a proportion is simply the average of dichotomous data, where each observation is scored as a 1 (success) or a 0 (failure).

3.4.3.1. Developing a Test

There are a variety of different tests that can be used with binary data, including the z test, continuity corrections to the z test, and exact tests. Let p_1 denote a population proportion, let $\hat{p}_1$ denote a sample estimate of that proportion, and let p_0 denote that proportion's value under the null hypothesis. To test the two-sided hypothesis $H_0: p_1 = p_0$ versus $H_A: p_1 \neq p_0$ (or a corresponding one-sided hypothesis), one can use the test statistic

$$Z = \frac{\hat{p}_1 - p_0}{\sqrt{p_0(1 - p_0)/n}}, \quad (15.11)$$

which, by the central limit theorem, has approximately the standard normal distribution for large enough sample sizes ($n > 25$, $np_0 > 5$, and $n(1 - p_0) > 5$).

If this test statistic falls in the extreme percentiles of the standard normal distribution (or beyond the appropriate lower or upper percentiles for a one-sided test), one can reject the null hypothesis. One can improve this test statistic by adding a small sample continuity correction or by performing an exact test, refinements that we briefly describe here (see Altman[10] for details).

3.4.3.2. Continuity Correction

Since the binomial distribution is discrete whereas the normal distribution is continuous, the normal approximation methods for binomial hypothesis tests are approximate. One way in which these normal approximation methods can be improved is by adding a *continuity correction* when calculating the sample proportion. This method places a band of unit width around each outcome, half above and half below. Let $p_1 = (x - 1/2)/n$ and $p_2 = (x + 1/2)/n$. Then, for a two-sided test of the hypotheses $H_0: p = p_0$ versus $H_A: p \neq p_0$, we reject the null hypothesis when either

$$Z_1 = \frac{(\hat{p}_1 - p_0)}{\sqrt{p_0(1 - p_0)/n}} < -z_{1-\alpha/2} \quad \text{or}$$

$$Z_2 = \frac{(\hat{p}_2 - p_0)}{\sqrt{p_0(1 - p_0)/n}} > z_{1-\alpha/2}. \quad (15.12)$$

3.4.3.3. Exact Tests

Alternatively, with sufficient computing power, we can perform an *exact* binomial test. In an exact binomial test, we enumerate the true binomial probabilities for each of the possible numbers of events ($0, 1, 2, \ldots, n$) and then reject the null hypothesis when the sum of the probabilities for values as extreme or more extreme than the observed value is less than the significance level. For example, suppose that the null hypothesis is $H_0: p = 0.35$ and $n = 6$. Under this hypothesis, the true binomial probabilities of observing exactly 0, 1, 2, 3, 4, 5, and 6 events out of six trials are 0.075, 0.244, 0.328, 0.235, 0.095, 0.020, and 0.002, respectively. Thus, if the alternative hypothesis is $H_A: p < 0.35$ and we observe five events, the one-sided p-value is 0.020 +

0.002 = 0.022. By comparison, if the alternative hypothesis is $H_A: p \neq 0.35$, the two-sided p-value is double the smaller one-sided p-value, namely 0.044. In both of these examples, we would reject the null hypothesis at the 5% significance level but not at the 1% significance level.

3.4.3.4. Confidence Intervals

Similar to constructing the test statistic, binomial confidence interval construction may follow the normal approximation but improvements can be made. Indeed, statistical research has shown that the normal approximation methods for binomial confidence interval construction tend to produce confidence intervals that are too small on average and thus have lower coverage rates than the specified confidence levels. One classical approach for obtaining better binomial confidence intervals is the Clopper–Pearson method,[34] which gives confidence intervals with guaranteed nominal coverage for all proportion parameters and sample sizes. The Clopper–Pearson confidence intervals consist of all proportion parameters that are consistent with the observed binomial data at a particular significance level using the exact binomial test with a two-sided hypothesis. Most statistical software can easily provide the Clopper–Pearson exact confidence bounds for proportions.

An alternative approach is to use the normal approximation method for binomial confidence interval construction given in Eq. (15.12) but with an adjusted sample proportion. For example, for $100 \times 1-\alpha\%$ confidence intervals, the popular Agresti–Coull method[35] symmetrically adds $z_{1-\alpha/2}^2/2$ imaginary failures (non-events) and $z_{1-\alpha/2}^2/2$ imaginary successes (events) to the original binomial data. Thus, if x events are observed, the adjusted sample size is $\tilde{n} = n + z_{1-\alpha/2}^2$ and the adjusted proportion is $\tilde{p} = (x + z_{1-\alpha/2}^2/2)/\tilde{n}$. In particular, for 95% CIs, the Agresti–Coull method is approximately equivalent to adding two imaginary failure and two imaginary successes to the original binomial data. This method gives the correct coverage rate on average, but individual confidence intervals can still be somewhat too small.

Another alternative is the single augmentation with an imaginary failure or success (SAIFS) method,[36] which approximates the Clopper–Pearson confidence interval. In the SAIFS method, we asymmetrically add a single imaginary failure (success) to the observed binomial data to obtain an adjusted proportion with which to compute the lower (upper) confidence bound using the standard formula. Thus, we compute the lower and upper confidence bounds with the adjusted proportions $\hat{p}_{\text{lower}} = (x + 0)/(n + 1)$ and $\hat{p}_{\text{upper}} = (x + 1)/(n + 1)$, respectively. The SAIFS method gives approximately the correct coverage for all confidence intervals for most underlying proportion parameters and sample sizes.

For example, with $n = 60$ trials and $x = 15$ successes or events, the (a) normal approximation, (b) Clopper–Pearson, (c) Agresti–Coull, and (d) SAIFS methods give 95% CIs of (a) (0.140, 0.360), (b) (0.147, 0.379), (c) (0.157, 0.373), and (d) (0.137, 0.374), respectively. A statistician can help you implement these improved methods for binomial confidence interval construction, which essentially build on the conceptual framework presented in this chapter. Although these improved methods may seem to create a little extra work, they can be crucial when the binomial test statistic lies near the boundary of significance.

3.4.4. Example

Suppose that in response to complaints about allergies, a large hospital changes the standard brand of rubber gloves that it supplies to a new but more expensive brand. An administrator wishes to know what proportion of nurses in that hospital prefer the new gloves, p_1, and if that proportion is at least $p_0 = 40\%$, she will consider the change worthwhile. She chooses a one-sided significance level of $\alpha = 5\%$. This hypothesis test corresponds to H_0: $p_1 \leq 0.4$ versus H_A: $p_1 > 0.4$. She finds that out of a sample of 30 nurses, 18 prefer the new brand and the rest are indifferent. Hence $n = 30$, $\hat{p}_1 = 18/30 = 0.6$, and using the previous formula,

$$Z = \frac{0.6 - 0.4}{\sqrt{(0.6 \times 0.4)/30}} = 2.24, \quad (15.13)$$

which exceeds the 95% percentile, $Z_{1-\alpha} = 1.645$, of the standard normal distribution (p-value = 0.01). By comparison, using an exact binomial test the p-value is 0.02. Thus, she rejects the null hypothesis and decides to adopt the new brand of gloves. Indeed, examining two-sided 95% CIs she finds similar results using the normal approximation or Wald (0.425, 0.775), Clopper–Pearson (0.406, 0.773), Agresti–Coull (0.423, 0.754), and SAIFS (0.404, 0.787) methods.

3.5. Two-Sample Hypothesis Tests with Applications to Clinical Research

Here, we develop hypothesis tests for comparing the means of two normal populations in both paired and unpaired analyses. We also discuss hypothesis tests for comparing two population proportions. These tests will then be used to analyze the data from the motivating examples in the next section.

3.5.1. Tests for Comparing the Means of Two Normal Populations

3.5.1.1. Paired Data

We first consider the paired analysis. This analysis corresponds to the beta-interferon/MRI trial, in which measurements on each patient are observed both before and during treatment. In this situation, we have two observations on every patient, from which we can compute the difference $d_i = x_i - y_i$. The data consist of n differences: $d_1, d_2, \ldots, d_n$, where n is the number of subjects in the study. In the beta-interferon/MRI study, $n = 14$. The observations x_i and y_i correspond to suitably transformed individual mean monthly lesion counts on baseline and on treatment for the ith subject. The hypothesis we will be testing is

$$H_0: \mu_d = 0 \text{ vs. } H_A: \mu_d \neq 0. \quad (15.14)$$

We need to make modeling assumptions to set up a hypothesis test. This will allow us to develop a sampling distribution for the test statistic under the null hypothesis that the mean difference is 0 (i.e., there is no effect of beta-interferon). We assume that the differences for each patient are independent and are normally distributed from a population with mean μ_d and variance σ^2, where $\bar{d}$ is the mean of the differences on all the n subjects. When σ^2 is known, the test statistic

$$Z = \frac{\bar{d}}{\sigma/\sqrt{n}} \quad (15.15)$$

has the standard normal distribution under the null hypothesis. When σ^2 is unknown (as is common in most situations in medical statistics), we need to estimate the variance σ^2. When σ^2 is unknown, the test statistic is

$$T = \frac{\bar{d}}{s/\sqrt{n}}, \text{ where } s = \sqrt{\frac{1}{n-1}\sum_{i=1}^{n}(d_i - \bar{d})^2}. \quad (15.16)$$

This test statistic has Student's t distribution with $n - 1$ degrees of freedom under the null hypothesis. Before we begin the study, we choose a significance level (the amount of evidence we need to reject the null hypothesis). If the Z or T test statistic's value is in the upper or lower $100 \times \alpha/2\%$ percentiles of the reference distribution (standard normal or Student's t distribution, respectively), we reject the null hypothesis and conclude that the means in the two groups are not equal. If the test statistic is not in the extreme tails of the distribution, we conclude that we fail to reject the null hypothesis, and hence that there is insufficient evidence to conclude that the means in the two groups are different.

The p-value is the probability of observing a Z or T test statistic value larger (in magnitude or absolute value) than what one observed. Suppose the observed value is Z_{obs}, and let Z denote a random normal variable. Then the p-value is $P(z < -Z_{obs}) + P(z > Z_{obs})$ for a two-sided test. The p-value for a one-sided test with alternative hypothesis $H_A: \mu_d > 0$ is $P(z > Z_{obs})$.

Tests based on Z and T test statistics values are called *paired z-tests* and *paired t-tests*, respectively. A paired z test is used when σ^2 is known, and a paired t-test is used when σ^2 needs to be estimated from the data.

3.5.1.2. Unpaired Data

We next consider tests of two normal population means for unpaired data. We discuss the cases of equal variances and different variances separately. We begin with a discussion of the equal variance case. The example that corresponds to this test is the felbamate monotherapy trial, and it is similar to many other parallel groups designs. We assume that we have observations from two groups of subjects, with sample sizes n and m. We assume that the observations $x_1, x_2, x_3, \ldots,$ x_n and $y_1, y_2, y_3, \ldots, y_m$ come from two independent normal distributions with a common variance σ^2 and means μ_1 and μ_2, respectively. The hypothesis test for this situation is

$$H_0: \mu_1 = \mu_2 \text{ vs. } H_A: \mu_1 \neq \mu_2. \qquad (15.17)$$

When σ is known, the test statistic

$$Z = \frac{\bar{x} - \bar{y}}{\sigma \sqrt{1/n + 1/m}} \qquad (15.18)$$

has the standard normal distribution. When σ^2 needs to be estimated from the data,

$$T = \frac{\bar{x} - \bar{y}}{s\sqrt{1/n + 1/m}}, \text{ where}$$

$$s = \sqrt{\frac{\sum_{i=1}^{n} (x_i - \bar{x})^2 + \sum_{i=1}^{m} (y_i - \bar{y})^2}{n + m - 2}}, \qquad (15.19)$$

has the Student's t distribution with $n + m - 2$ degrees of freedom under the null hypothesis. The preceding estimate of σ is the pooled sample standard deviation and is based on the assumption of equal variances in the two groups. As in the previous hypothesis test, if the Z and T test statistics values are in the upper or lower $100 \times \alpha/2\%$ percentiles of this reference distribution, we reject the null hypothesis. Tests based on the Z and T test statistics values are called two-sample z-tests and two-sample t tests, respectively. Two-sample z-tests are used when σ^2 is known, and two-sample t-tests are used when σ^2 needs to be estimated from the data.

In many situations, the assumption of a constant variance in the two treatment groups is not a good assumption. Since treatments may be effective in only a fraction of subjects, often the variability of the outcome in the treatment group is larger than that of the placebo group. The test statistic to use in this situation is

$$Z = \frac{\bar{x} - \bar{y}}{\sqrt{\sigma_x^2/n + \sigma_y^2/m}}, \qquad (15.20)$$

where both the sample sizes in each group are large or when σ_1^2 and σ_2^2 are known. The Z test statistic has the standard normal distribution when the null hypothesis is true.

When the variance estimates are unknown and need to be estimated using the data, the test statistic is

$$T = \frac{\bar{x} - \bar{y}}{\sqrt{s_x^2/n + s_y^2/m}}. \qquad (15.21)$$

Under the null hypothesis that the means in the two groups are equal, the preceding test statistic has a distribution that is approximately the Student's t distribution with ω degrees of freedom (determined by Satterthwaite's formula), where

$$\omega = \frac{(s_x^2/n + s_y^2/m)^2}{\dfrac{(s_x^2/n)^2}{n-1} + \dfrac{(s_y^2/m)^2}{m-1}}. \qquad (15.22)$$

Because this result may not be an integer, ω should be conservatively rounded downward. As with the other hypothesis tests we discussed, if the Z and T test statistics values are in the upper or lower $100 \times \alpha/2\%$ percentiles of the reference distribution, we reject the null hypothesis and conclude that the means in the two groups are unequal. The t-test with unequal variances is often called Welch's t-test.

3.5.2. Tests for Comparing Two Population Proportions

Binary or dichotomous outcomes are common in medical research. The binary responses are often no or yes responses, such as death or survival, presence or absence of disease, and reaction or insensitivity to a particular diagnostic test. In the ISIS-4 study, the primary outcome was a binary variable signifying whether a randomized patient was alive or dead at 35 days after randomization. Our interest focuses on comparing participants who were randomized to receive magnesium and those not randomized to magnesium. The typical data structure for this two-sample problem involves the number of positive responses in n subjects from group 1 and the number of positive responses in m subjects from group 2. The hypothesis for this test is

$$H_0: p_1 = p_2 \text{ vs. } H_A: p_1 \neq p_2. \qquad (15.23)$$

The assumptions for the test are that (1) the data are binary, (2) observations are independent, and (3) there is a common probability of a "yes" response for each of the two groups. For large sample sizes [n and m both greater than 25, and np_1, $n(1 - p_1)$, mp_2, and $m(1 - p_2)$ each greater than 5], we can use a two-sample z-test for comparing the two population proportions. The test statistic is

$$Z = \frac{\hat{p}_1 - \hat{p}_2}{\sqrt{\dfrac{\hat{p}_1(1 - \hat{p}_1)}{n} + \dfrac{\hat{p}_2(1 - \hat{p}_2)}{m}}} \qquad (15.24)$$

which, for large sample sizes, has approximately the standard normal distribution under the null hypothesis that the population proportions are equal in the two groups. Other tests have been developed for small samples. For example, the Fisher's exact test is a valid test for any values of the proportions and sample sizes, no matter how small.[13]

3.6. Hypothesis Tests for the Motivating Examples

We now conduct hypothesis tests to analyze the data from the three motivating examples.

3.6.1. Hypothesis Tests for the Beta-Interferon/MRI Study

The beta-interferon/MRI study consisted of 14 patients followed for 13 months—7 months on baseline and 6 months on treatment. The outcome was the average number of monthly contrast-enhanced lesions on treatment minus the corresponding average number during baseline.

Table 15-1 summarizes the data from the trial. A total of 13 of 14 patients had decreased lesion frequency on treatment compared with their baseline frequency. This result suggests that beta-interferon lowers disease activity in early RRMS. The inferential question is this: Do the data provide enough evidence to make a statement about the population of all RRMS patients? The hypothesis test is used to address this question. We conducted a two-tailed test of whether there is a difference between lesion frequency during baseline and lesion frequency after treatment. We chose a significance level of 0.05 before the study began. First, note that the structure of the data suggests that a two-sample paired t-test may be appropriate. Data are paired since observations on different patients are independent, and the variance of the difference in lesion activity for each subject is unknown. In addi-

TABLE 15-1   Beta-Interferon and MRI Study

Patient No.	Baseline (Mean Lesions/Month)	6-Month Treatment (Mean Lesions/Month)
1	2.43	0
2	1.71	0.67
3	3.14	1.00
4	1.29	0.33
5	0.57	1.67
6	2.00	0
7	6.00	0.33
8	0.43	0
9	12.86	0.17
10	6.42	0.67
11	0.57	0
12	0.71	0
13	1.57	0.17
14	3.17	1.67

tion, the data transformed to the log scale appeared to be approximately normally distributed. The data were transformed so that $d_i = \log[(\text{7-month baseline mean}) + 0.5] - \log[(\text{6-month treatment mean}) + 0.5]$. The constant 0.5 was added to all numbers since the log of 0 is undefined. We use a paired t-test with a test statistic computed as

$$T = \frac{\bar{d}}{s/\sqrt{n}} = -4.8. \qquad (15.25)$$

The test statistic has a t distribution with $14 - 1 = 13$ degrees of freedom when the null hypothesis is true. The $100 \times \alpha/2\%$ (2.5%) lower and upper percentiles of the reference distribution are -2.16 and 2.16, respectively. Since -4.8 is less than -2.16, we reject H_0 and conclude that there is a difference between lesion frequency during baseline and lesion frequency on beta-interferon. The p-value for the two-sided test can be computed as $P(t_{13} < -4.8) + P(t_{13} > 4.8) = 0.0004$, where T_{13} denotes a random variable with the t distribution on 13 df. This means that if the null hypothesis of no effect were true, there would only be a 1 in 2500 chance of observing a test statistic as large (in absolute value) as the one we observed.

How would a one-tailed test of change from baseline in mean number of lesions after treatment be conducted? Would the results be different than with the two-tailed test? We would calculate the same test statistic. Our criterion for rejection, however, would be different. We would reject the null hypothesis if the test statistic was smaller than the $100 \times \alpha$ lower (5%) percentile of the reference distribution (t_{13}). This value is -1.77. Since -4.8 is less than -1.77, we would reject the null hypothesis and conclude that lesion frequency was reduced on beta-interferon. The p-value for this one-sided test is $P(t_{13} < -4.8) = 0.0002$. Note that the p value for this one-sided test is smaller than the p-value for the corresponding two-sided test. In general, for the same significance level (α), any test that would reject with a two-sided test would always reject with a one-sided test. However, a test may reject with a one-sided test and yet fail to reject with a two-sided test. This fact is one reason why investigators are often eager to perform one-sided tests; less evidence is required to reject the null hypothesis and conclude a significant result. Nevertheless, investigators should be cautious about using one-sided tests, which are only appropriate when there is interest in detecting a beneficial effect from treatment and there would be no interest in detecting a harmful effect.

3.6.2. Hypothesis Tests for the Felbamate Monotherapy Trial

The felbamate monotherapy trial was designed as a parallel groups design with 19 patients randomized to the felbamate arm and 21 patients randomized to the placebo arm. Seizure frequency was monitored during the 2-week follow-up period in the hospital or until a patient dropped out of the study. The outcome was daily seizure rates over follow-up period. The test was a two-tailed test of whether there is a difference in seizure frequency between the felbamate and placebo arms. We chose a significance level of 0.05 before the study began. The hypothesis is

$$H_0: \mu_{\text{treatment}} = \mu_{\text{placebo}} \text{ vs. } H_A: \mu_{\text{treatment}} \neq \mu_{\text{placebo}}. \quad (15.26)$$

The appropriate test is an unpaired t test. The data are independent and approximately normally distributed on the square root scale (by taking square roots of the mean daily seizure counts on all patients). On the square root scale, the mean seizure rates are $\bar{x} = 1.42$ in the placebo group and $\bar{y} = 0.42$ in the treatment group. The sample standard deviations were $s_x = 1.3$ and $s_y = 1.0$, suggesting that there are higher amounts of variation in the placebo arm. We begin by performing a test under an assumption of equal variances in the two groups. Using formulas (15.16) and (15.19), we find the common variance $s = 1.17$.

The test statistic assuming that both populations have a common variance is

$$T = \frac{\bar{x} - \bar{y}}{s\sqrt{(1/n) + (1/m)}} = 2.71. \quad (15.27)$$

When the null hypothesis is true, the test statistic has a t distribution with $n + m - 2$ (38) degrees of freedom. The $100 \times \alpha/2\%$ (2.5%) lower and upper percentiles of the t distribution with 38 df are -2.02 and 2.02, respectively. Because 2.71 is greater than 2.02, we

reject the null hypothesis and conclude that there is a difference in seizure frequency in the placebo and felbamate arms. The p-value equals $P(t_{38} > 2.71) + P(t_{38} < -2.71) = 0.01$, which means that the chance is approximately 1 in 100 of getting a test-statistic this large (either positive or negative) if the null hypothesis is true. Thus, we can reasonably reject the null hypothesis at this significance level.

By comparison, a Welch's t-test, which does not assume an equal variance for the two populations, was conducted. The test was done on the square root scale and resulted in $T = 2.74$, df $= 37.09$ rounded down to 37, and a p-value of 0.009, which is similar to the result from the test assuming a common population variance.

3.6.3. Hypothesis Tests for the ISIS-4 Trial: Comparing the Magnesium and No Magnesium Arms

The ISIS-4 study was a factorial design of three treatments. We focus on comparing participants receiving magnesium to those not receiving magnesium. A total of 58,050 MI patients were randomized: 29,011 received magnesium and 29,039 did not receive magnesium. The inferential question was whether the proportion of participants dying during the first 35 days after an MI differed between the two groups. The hypothesis is

$$H_0: p_{\text{Mg+}} = p_{\text{Mg-}} \text{ vs. } H_A: p_{\text{Mg+}} \neq p_{\text{Mg-}}. \quad (15.28)$$

The test was two-sided and conducted at the 0.05 significance level. We assume that individual binary outcomes are independent with a common probability of dying in each group, and we note that the sample sizes are large, so we can test this hypothesis with a two-sample z-test. The data from the study can be presented in the following 2×2 table:

	Mg+	Mg−
Dead	2,216	2,103
Alive	26,795	26,936
Total	29,011	29,039

The proportion dead at 35 days after randomization (35-day mortality) can be estimated as $\hat{p}_{\text{Mg+}} = 2216/29011 = 0.0764$ and $\hat{p}_{\text{Mg-}} = 2103/29039 = 0.0724$. The mortality rate is slightly larger in the magnesium arm. We can formulate the hypothesis test with the test statistic,

$$Z = \frac{\hat{p}_{\text{Mg-}} - \hat{p}_{\text{Mg+}}}{\sqrt{\dfrac{\hat{p}_{\text{Mg-}}(1 - \hat{p}_{\text{Mg-}})}{n} + \dfrac{\hat{p}_{\text{Mg+}}(1 - \hat{p}_{mg+})}{m}}} = -1.82. \quad (15.29)$$

The test statistic, at least approximately, has the standard normal distribution when the null hypothesis is true. The 2.5% lower and upper percentiles of the z distribution are -1.960 and 1.960, respectively. Since -1.82 falls between -1.960 and 1.960, we do not reject the null hypothesis and we do not have enough evidence to conclude that the population proportions are unequal. The p-value is $P(Z < -1.82) + P(Z > 1.82) = 0.07$.

3.7. Common Mistakes in Hypothesis Testing

As the previous section shows, hypothesis testing requires one to make appropriate assumptions about the structure and distribution of a data set, especially the relationships between the sample observations. There are a number of mistakes that researchers commonly make in hypothesis testing due to ignoring the structure of the sample data or failing to check the assumptions of the hypothesis test. Some of these common mistakes, illustrated in the context of the t-test, are the following:

1. Ignoring the pairing between observations within subjects. Testing paired continuous data with a two-sample unpaired t-test.

2. Incorrectly assuming a paired structure between two independent samples. Testing unpaired continuous data with a paired t-test.

3. Ignoring the dependence that occurs when multiple observations are made on each subject. For example, if there are five subjects and 3, 2, 1, 2, and 2 measurements are made on these subjects, respectively, there are not 10 independent observations. In this case, more complicated methods, such as mixed models regression, must be used to analyze the data.

4. Ignoring the apparent sample distribution of observations, especially features such as skewness, outliers or extreme values, and lower or upper limits on measurement accuracy. Performing a t-test on highly skewed data without appropriate adjustments.

5. Assuming equal variances in two groups without examining the data, either graphically or numerically. Performing a pooled t-test instead of a Welch's t-test for two samples with very different variances.

Mistake 1 is often committed by careless researchers. Although the t-test remains valid (correct type I error rate), there could be a substantial loss of power or efficiency. By contrast, mistake 2 is more serious and could lead to the wrong inference. Mistake 3 is both very common and serious because observations on the same subject tend to be more similar (positively correlated) than those on different subjects. Use of a one-sample t-test will tend to give p-values that are too small compared to the correct values, which in turn will lead one to conclude that the data provide more evidence against the null hypothesis than they actually do. Finally, the t-test is generally robust against mistakes 4 and 5, such as ignoring moderate amounts of skewness for sufficiently large samples. Indeed, regarding mistake 5, the felbamate monotherapy example showed that the two-sample t-test is robust to ignoring the differences between the variances of the two samples.

3.8. Misstatements and Misconceptions

The following are some of the major misstatements and misconceptions that arise when performing hypothesis tests and reporting the results:

1. Using a small p-value to conclude that two sample means ($\bar{x}$ and $\bar{y}$) are significantly different from each other. This approach is incorrect because the p-value is a statistical tool for making inferences about the true population means.

2. Failing to reject the null hypothesis (H_0) means that it is true. On the contrary, failing to reject the null hypothesis may merely indicate that there is not enough evidence to state that it is false at a particular significance level. The null hypothesis may be true or it may be false, but we do not have the evidence to reject it.

3. Focusing on the significance of an effect (its p value) but ignoring its magnitude or size. In a study with multiple explanatory variables, there will often be several variables that appear to be related to the outcome of interest. A small p-value demonstrates significant evidence that the effect of the variable on the outcome is nonzero, whereas the point estimate and confidence intervals for the magnitude of the effect demonstrate how much of an impact that variable has on the magnitude of the response.

4. Confusing statistical significance with clinical significance. In the ISIS-4 trial, the participants who received intravenous magnesium sulfide had a 35-day unadjusted mortality rate of 7.64%, whereas those who did not receive that treatment had a corresponding mortality rate of 7.24%. If the two-sided p-value had been equal to 0.007 (it was actually $p = 0.07$), we would need to ask ourselves, even though the p-value was quite significant at 0.007, was the increase in mortality of 0.40% on the treatment clinically troubling? Possibly, it is troubling if 0.4% is equal to many lives per year. Possibly, such a small difference is not troubling in some studies. Just because a finding is statistically significant does not make it clinically significant.

3.9. Additional Topics

Most of this section was devoted to establishing a conceptual framework for statistical hypothesis testing. We focused primarily on tests for comparing two populations because these tests are the most common types of tests used in clinical research. Here, we briefly describe other methodology that is commonly used in analyzing the data from medical studies. More details on all these subjects can be found in the references.

3.9.1. Comparing More Than Two Groups: One-Way Analysis of Variance

The analysis of variance (ANOVA) framework extends the methodology for comparing the means of two populations to more than two populations. This method may be applicable in multiarm clinical trials in which interest focuses on detecting any differences among the various treatments. The hypotheses for comparing k population means with ANOVA can be written as

$$H_0: \mu_1 = \mu_2 = \ldots = \mu_k \text{ vs. } H_A: \text{Some } \mu_i \neq \mu_j. \quad (15.30)$$

The assumptions for this test are that the data are normally distributed with a constant population variance across the k groups. In addition, it is assumed that the data for each of the subjects are statistically independent. The test statistic used is the ratio of the between-subject variance to the within-subject variance. Under the null hypothesis of equal population proportions, the test statistic has an F distribution, and one can obtain a p-value to assess the significance of this test (see Altman[10] for more details).

3.9.2. Simple and Multiple Linear Regression

Simple linear regression is a technique used to examine the strength of a linear relationship in a set of bivariate or paired data, where one variable acts as the predictor and the other as the response. For example, one may be interested in examining whether there is a linear increase in blood pressure-with-age over a particular range of ages. The model for simple linear regression is

$$y_i = \beta_0 + \beta_1 x_i + \varepsilon_i, \quad (15.31)$$

where β_0 and β_1 are the intercept and slope for the regression line, respectively. In addition, ε_i is an error term (normally distributed with mean $= 0$ and variance $= \sigma^2$) that characterizes the scatter around the regression line. The intercept (β_0) and slope (β_1) parameters are estimated using least squares fitting. Least squares fitting involves choosing the line that minimizes the sum of the squared vertical differences between the responses and the points predicted by the fitted line at values of the predictor variable. Hypothesis testing also plays an important role in regression. We often wish to test whether there is a significant increase of one variable with each unit increase in a second variable, not only with the data we observed in the sample but also in the population from which the sample data were drawn. The hypotheses for linear regression can be stated as

$$H_0: \beta_1 = 0 \text{ vs. } H_A: \beta_1 \neq 0. \quad (15.32)$$

The assumptions for this test are that response observations are independent and normally distributed (with constant variance) around the regression line. The test statistic for a significant linear relationship is the ratio of the variance of the data points around the average y value ($\bar{y}$) relative to the variance around the regression line. A large test statistic of this type reflects either a steep slope or small variability around a slope. This test statistic has an F distribution under the null hypothesis that the slope is zero (i.e., a horizontal line), and one can obtain a p-value to assess the significance of this test.

Multiple or *multivariate regression* is an extension of simple linear regression to more than one independent or predictor variables. We may be interested in examining for a linear increase in blood pressure with age (x_i) after adjusting for weight (z_i). The multiple regression model can be written as

$$y_i = \beta_0 + \beta_1 x_i + \beta_2 z_i + \varepsilon_i. \quad (15.33)$$

The hypotheses, one for each β, for multiple regression are formulated in a similar way as for simple linear regression.

3.9.3. Multiple Comparisons

When making many statistical comparisons, a certain fraction of statistical tests will be statistically significant even when the null hypothesis is true. In general, when a series of tests are performed at the α significance level, approximately $100 \times \alpha\%$ of tests will be significant at the α level even when the null hypothesis for each test is true. For example, even if the null hypotheses are true for all tests, when conducting many hypothesis tests at the 0.05 significance level, on average (in the long term) 5 of 100 tests will be significant. Issues of multiple comparisons arise in various situations, such as in clinical trials with multiple end points and multiple looks at the data. Pairwise comparison among the sample means of several groups is an area in which issues of multiple comparisons may be of concern. For k groups, there are $k(k-1)/2$ pairwise comparisons, and just by chance some may reach significance. Our last example is with multiple regression analysis in which many candidate predictor variables are tested and entered into the model. Some of these variables may result in a significant result just by chance. With an ongoing study and many interim analyses or inspections of the data, we have a high chance of rejecting the null hypothesis at some time point even when the null hypothesis is true.

There are various approaches to the multiple comparisons problem. One rather informal approach is to choose a significance level α lower than the traditional 0.05 level (e.g., 0.01) to prevent many false-positive conclusions or to "control the false discovery rate (FDR)." The number of comparisons should be made explicit in the article. More formal approaches to control the "experiment-wise" type I error using corrections for multiple comparisons have been proposed. An example is the Bonferroni correction, in which the type I error rate is taken as α/n, where n is the number of comparisons made. Interim analysis methods are available for various study designs.[11]

It is best to address the issue of multiple comparisons during the design stage of a study. One should determine how many comparisons will be made and then explicitly state these comparisons. Studies should be designed to minimize the number of statistical tests at the end of the study. Ad hoc solutions to the multiple comparisons problem may be done for exploratory or epidemiologic studies. Multiple comparison adjustments should be made for the primary analyses of definitive studies (such as phase III confirmatory studies). Studies that focus on a single primary outcome and data analyzed at the end of study avoid the issue of multiple comparisons.

3.9.4. *Nonparametric versus Parametric Tests*

Inferential methods that make assumptions about the underlying distributions from which the data derive are called *parametric methods*, whereas those that make no distributional assumptions are called *nonparametric methods*. Nonparametric methods are often used when data do not meet the distributional assumptions of parametric methods, such as asymmetric distributions or unusual numbers of extreme values. Nonparametric methods are usually based on the ranks of observations as opposed to their actual values, which lessens the impact of skewness and extreme outliers in the raw data. Hypotheses are usually stated in terms of medians instead of means. Corresponding to the two-sample hypothesis tests of means discussed in this chapter are the following nonparametric analogs:

- Paired *t*-test: Wilcoxon signed rank test or the sign test

- Two-sample *t*-test: Wilcoxon rank sum test
- Analysis of variance: Kruskal–Wallis test

In general, nonparametric tests have somewhat lower power than their parametric counterparts. This is the price one pays for making fewer assumptions about the underlying distribution of the data. Fewer assumptions, however, does not necessarily mean no distributional assumptions. For large sample sizes, parametric and nonparametric tests generally lead to the same inferences. More information about nonparametric approaches can be found in van Belle *et al.*[37]

4. SAMPLE SIZE AND POWER

This section introduces concepts in sample size and power estimation. A definition of power is given, and why it is important is discussed. Sample size calculations for the one-sample and two-sample problems are summarized. In addition, we discuss how to design new studies based on the motivating examples.

4.1. Basic Concepts

Power is the probability of rejecting the null hypothesis when a particular alternative hypothesis is true. Power equals one minus the probability of making a type II error. When designing studies, it is essential to consider the power because it indicates the chance of finding a significant difference when the truth is that a difference of a certain magnitude exists. A study with low power is likely to produce nonsignificant results even when meaningful differences do indeed exist. Low power to detect important differences usually results from a situation in which the study was designed with too small a sample size. Studies with low power are a waste of resources since they do not adequately address their scientific questions.

There are various approaches to sample size and power estimation. First, one often calculates power for a fixed sample size. The following is a typical question: What is the power of a study to detect a 20% reduction in the average response due to treatment when we randomize 30 participants to either a placebo or treatment group? Second, one often wishes to estimate a required sample size for a fixed power. The following is a typical question for this approach: What sample size (in each of the two groups) is required to have 80% power to detect a 20% reduction in the average response due to treatment using a randomized parallel groups design? The focus of this section is on the latter approach, namely, estimating the required sample size for a fixed power.

Sample size and power calculations are specific for a particular hypothesis test. One needs to specify a model for the data and propose a particular hypothesis test to compute power and estimate sample size. For continuous outcomes, one needs to specify the standard deviation of the outcome, the significance level of the test, and whether the test is one-sided or two-sided. Power and sample size depend on these other design factors. For example, power changes as a function of the following parameters:

1. Sample size (n): Power increases as the sample size increases.

2. Variation in outcome (σ^2): Power increases as variation in outcome decreases.

3. Difference (effect) to be detected δ: Power increases as this difference increases.

4. Significance level α: Power increases as the significance level increases.

5. One-tailed versus two-tailed tests: Power is greater in one-tailed tests than in comparable two-tailed tests.

By comparison, sample size changes as a function of the following parameters:

1. Power ($1 - \beta$): Sample size increase as the power increases.

2. Variation in outcome (σ^2): Sample size increases as variation in outcome increases.

3. Difference (effect) to be detected δ: Sample size increases as this difference decreases.

4. Significance level α: Sample size increases as the significance level decreases.

5. One-tailed versus two-tailed tests: Sample size is smaller in one-tailed tests than in comparable two-tailed tests.

4.2. Sample Size Calculations for the One-Sample Problem

We begin with a discussion of sample size calculations for the one-sample problem. In the next section, we extend these basic principles to the two-sample problem. To calculate a sample size, we need to specify the significance level α, the power ($1 - \beta$), the scientifically or clinically meaningful difference δ, and the standard deviation σ. First, for a two-sided hypothesis test involving the mean of continuous data, the sample size formula is

$$n = \frac{(Z_{1-\alpha/2} + Z_{1-\beta})^2 \sigma^2}{\delta^2}. \qquad (15.34)$$

If n is not an integer, it should be rounded up. For a one-sided test, replace $Z_{1-\alpha/2}$ by $Z_{1-\alpha}$. For example, patients with hypertrophic cardiomyopathy (HCM) have enlarged left ventricles (mean, 300 g) compared to the general population (mean, 120 g). A cardiologist

studying a particular genetic mutation that causes HCM wishes to estimate the mean left ventricular mass of patients with this particular mutation within $\delta = 10$ g and compare it to the mean for other patients with HCM. If previous laboratory measurements suggest that $\sigma = 30$ g, and he chooses a significance level of $\alpha = 5\%$ and a power of 90% ($\beta = 0.1$), what sample size does he need?

This hypothesis is two-sided, so $Z_{1-\alpha/2} = 1.960$ and $Z_{1-\beta} = 1.282$. Using the previous formula, one calculates

$$n = \frac{(1.960 + 1.282)^2 \times (30)^2}{(10)^2} = 94.6 \approx 95. \qquad (15.35)$$

Thus, the required sample size is $n = 95$ in this study. In practice, the sample size calculations for such a study could be more complicated. For example, these calculations could take into account age, gender, body mass, hormone levels, and other patient characteristics.[38]

Second, for a two-sided hypothesis involving a proportion, the sample size formula is

$$n = \frac{(Z_{1-\alpha/2} + Z_{1-\beta})^2 p_0(1 - p_0)}{\delta^2}. \qquad (15.36)$$

As before, if n is not an integer, it should be rounded up. For a one-sided test, replace $Z_{1-\alpha/2}$ with $Z_{1-\alpha}$. For example, suppose that one wishes to conduct a phase II (safety/efficacy) clinical trial to test a new cancer drug that one has recently developed. If only 20% of patients will benefit from this drug, one does not wish to continue to study it because there already are drugs with comparable efficacy available. Conversely, if at least 40% of patients will benefit from this drug, one wants to detect this effect with 80% power ($\beta = 0.2$). The significance level is $\alpha = 5\%$. How many participants should one enroll in the clinical trial?

This hypothesis is one-sided, so $Z_{1-\alpha} = 1.645$ and $Z_{1-\beta} = 0.841$. The null proportion is $p_0 = 0.2$ and the difference is $\delta_0 = 0.2$. Using the previous formula, one calculates

$$n = \frac{(1.645 + 0.841)^2 (0.2 \times 0.8)}{(0.2)^2} = 24.7 \approx 25. \qquad (15.37)$$

Thus, the required sample size is $n = 25$ in this clinical trial. By comparison, with 90% power ($\beta = 0.1$, $Z_{1-\beta} = 1.282$), the required sample size is $n = 35$ in the clinical trial.

It is important to recognize that there are many other approaches to sample size calculation, most of which are beyond the scope of this introductory chapter, and it is wise to consult with a statistician to determine which method is best for one's particular research problem. For example, for the preceding problem, one could consider a *two-stage* design.[23,28] Two-stage designs are optimal in the sense that they have the smallest expected or average sample size under the null hypothesis. With 80% power and a significance level of $\alpha = 0.05$, in the first stage one would enroll $n_1 = 13$ participants, and if $r_1 = 3$ or fewer study participants respond positively to the drug, one should terminate the trial and abandon the drug. In the second stage, one would enroll up to 30 additional participants sequentially, for a maximum of $n_2 = 43$; if $r_2 = 12$ or fewer study participants out of 43 respond, one should abandon the drug, whereas if 13 or more participants respond the drug should be considered for further study. If the null hypothesis is true ($p_1 = 0.2$) and with 80% power, one will need to enroll on average 21 participants in the trial to conclude that the drug should be abandoned. By comparison, with 90% power ($n_1 = 19$, $r_1 = 4$, $n_2 = 54$, $r_2 = 15$), if the null hypothesis is true, one will need to enroll on average 30 participants in the trial to conclude that the drug should be abandoned.

Finally, it is important to recognize that the sample size formulas presented in this chapter are approximate (and based on the more tractable standard normal rather than the t distribution). Thus, there is a tendency for these calculations to result in slight underestimates of sample size in small samples, and the adding of a few extra subjects to small sample sizes is therefore recommended.[39] It also would be wise to calculate several different sample sizes under various assumptions.

4.3. Sample Size Calculations for the Two-Sample Problem

As with the discussion of hypothesis testing, we discuss sample size estimation for both testing the differences in population means between two groups for continuous data and testing the difference in population proportions for two group comparisons.

4.3.1. Sample Size Calculations for the Comparison of the Means of Two Normally Distributed Populations

We begin with a discussion of sample size for the paired analysis. As in the beta-interferon/MRI trial, we compute the difference of the two observations on each subject $d_i = x_i - y_i$. Assumptions are that the differences are normally distributed with a variance σ^2. The hypothesis of interest are

$$H_0: \mu_d = 0 \text{ vs. } H_A: \mu_d \neq 0. \quad (15.38)$$

The required sample size can be computed with the following formula:

$$n = \frac{(Z_{1-\alpha/2} + Z_{1-\beta})^2 \sigma^2}{\delta^2}, \quad (15.39)$$

where

- δ is the paired difference one wishes to detect. It represents a scientifically or clinically meaningful effect on the scale of the outcome.
- σ^2 is the variance of the difference in the paired observations.
- α is the significance level of the test and $1 - \beta$ is the specified power.
- $Z_{1-\alpha/2}$ and $Z_{1-\beta}$ corresponding to the upper $100 \times \alpha/2\%$ and $100 \times \beta\%$ percentiles of the standard normal distribution.
- for $\alpha = 0.05$ and a power of $1 - \beta = 0.8$, we have that $Z_{1-\alpha/2} = 1.960$ and $Z_{1-\beta} = 0.841$.

We now give a hypothetical example to illustrate how these calculations can be performed. Suppose an investigator wishes to design a pilot study to investigate the effect of a new pharmacologic agent on diastolic blood pressure. He plans to take two measurements on each subject, one on no medications followed by the other on the new agent. Suppose the investigator wishes to test whether there is a change in average blood pressure on the new agent with a two-sided hypothesis test with a 0.05 significance level. How many subjects should the investigator enroll to have a 90% chance of detecting an average drop of 5 mmHg units in blood pressure on treatment? Is any additional information needed to make this calculation? The standard deviation in the difference of the measurements needs to be specified. Say that the standard deviation is 25 mmHg. The required sample size can be computed as

$$n = \frac{(Z_{1-\alpha/2} + Z_{1-\beta})^2 \sigma^2}{\delta^2}. \quad (15.40)$$

Thus, the required sample size is

$$n = \frac{(1.960 + 1.282)^2 (25)^2}{5^2} = 262.7 \approx 263. \quad (15.41)$$

How could the investigator reduce this sample size? Taking three repeated observations both on and off therapy results in a standard deviation of the difference in the sets of measurements of 15 mmHg. Thus, the sample size is now computed as

$$n = \frac{(1.960 + 1.282)^2 (15)^2}{5^2} = 94.6 \approx 95. \quad (15.42)$$

Thus, the required sample size is substantially smaller (threefold smaller) by taking these additional observations on each subject. Suppose that 95 participants are still too many subjects for a study at this institution. We could specify a lower power than 90% to detect an average drop in blood pressure of 5 mmHg. If we specify a power of 80%, the required sample size is

$$n = \frac{(1.960 + 0.841)^2 (15)^2}{5^2} = 70.6 \approx 71. \qquad (15.43)$$

We could reduce the sample size further by specifying a larger minimum detectable difference. Suppose that the investigator now states that he is only interested in this therapy if the reduction in blood pressure is more than 10 mmHg. In other words, a reduction less than 10 mmHg is not clinically meaningful. The required sample size is now

$$n = \frac{(1.960 + 0.841)^2 (15)^2}{10^2} = 17.7 \approx 18. \qquad (15.44)$$

It is clear then that the required sample size is highly dependent on the particular design parameters we choose. As mentioned previously, this design is a non-randomized baseline versus treatment design, which is subject to various problems, such as regression to the mean, bias if there is a time trend, placebo effect, and investigator bias. An alternative design is a parallel groups design with or without baseline measurements.

We now discuss sample size estimation for the unpaired two-group comparisons. We consider the case of equal variance and sample size in the two groups first. The assumptions are the same as those for the unpaired *t* test, namely that outcomes are from two normal populations with means μ_1 and μ_2 and common variance σ^2. The required sample size for *each* of the two groups is

$$n = \frac{2(Z_{1-\alpha/2} + Z_{1-\beta})^2 \sigma^2}{\delta^2}, \qquad (15.45)$$

where δ is the meaningful difference in population means $(\mu_1 - \mu_2)$ we wish to detect, σ^2 is the variance of the observations in each group, and $Z_{1-\alpha/2}$ and $Z_{1-\beta}$ are percentiles of the standard normal defined previously.

We now return to the hypothetical example to illustrate how this formula could be used in study design. An investigator wishes to design a study to investigate the effect of a new pharmacologic agent on diastolic blood pressure using a parallel groups design. He plans to randomize study participants either to a placebo or treatment arm and collect one blood pressure measurement at baseline and another follow-up measurement. Suppose the investigator wishes to test whether the average blood pressure in the treatment arm is different from that in the placebo arm with a two-tailed hypothesis test at the 0.05 significance level. How many subjects would the investigator need to enroll to have 90% power to detect an average decrease of 5 mmHg units in blood pressure on treatment? As before, the standard deviation for the difference between the follow-up and baseline blood pressure measurement is assumed to be 25 mmHg in both the placebo and treatment groups. The required sample size (in *each* of the two groups) can be calculated as

$$n = \frac{2(1.960 + 1.282)^2 25^2}{5^2} = 525.5 \approx 526. \qquad (15.46)$$

Thus, more than 1000 participants would be required to perform the best designed study. How could the investigator reduce this sample size? Taking the average of three repeated blood pressure measurements at baseline and at follow-up evaluation reduces the standard deviation of the difference from before treatment to after the initiation of treatment to 15 mmHg. Thus, the *per* arm required sample size becomes

$$n = \frac{2(1.960 + 1.282)^2 15^2}{5^2} = 189.1 \approx 190. \qquad (15.47)$$

Specifying a lower power of 80% results in the *per* arm following calculation:

$$n = \frac{2(1.960 + 0.841)^2 15^2}{5^2} = 141.3 \approx 142. \qquad (15.48)$$

Finally, specifying a larger minimum detectable difference of 10 mmHg results in

$$n = \frac{2(1.960 + 0.841)^2 15^2}{10^2} = 35.3 \approx 36. \qquad (15.49)$$

Thus, even this last calculation demonstrates that we need at least 72 participants to test this new investigational drug with a parallel groups design. This is compared with a total of 18 participants with comparable design parameters to test the drug with a baseline versus treatment design. The low number of participants for the baseline versus treatment design is often the motivation for this type of study. In particular, when one is screening many potential toxic treatments (as is often done in cancer research) performing many screening studies with nonrandomized designs, identifying potentially active treatments and bringing these to more definitive testing with parallel groups designs may optimize limited resources.

We can also estimate sample sizes for testing differences in population means when the variances in the two groups are unequal. When the variances are not equal, we can compute the sample size in *each* group as

$$n = \frac{(Z_{1-\alpha/2} + Z_{1-\beta})^2 (\sigma_1^2 + \sigma_2^2)}{\delta^2}, \qquad (15.50)$$

where σ_1^2 and σ_2^2 are the variances in groups 1 and 2, respectively.

There may be situations in which one may want to design trials with different numbers of participants in the two groups. For example, in placebo-controlled trials, one may want to give a higher proportion of participants the treatment. One may want to randomize two participants to the treatment arm for every patient randomized to the placebo arm. We need to specify the ratio of n_2 to n_1, namely $\lambda = n_2/n_1$. Then

$$n_1 = \frac{(Z_{1-\alpha/2} + Z_{1-\beta})^2 (\sigma_1^2 + \sigma_2^2/\lambda)}{\delta^2}, \qquad (15.51)$$

and, in turn, $n_2 = \lambda n_1$.

4.3.2. Sample Size Calculations for the Comparison of Two Population Proportions

The assumption for the statistical test on which the sample size calculations are based is that the binary observations are independent with common probability of p_1 in group 1 and p_2 in group 2. The required sample size in *each* of the two groups is

$$n = \frac{(Z_{1-\alpha/2} + Z_{1-\beta})^2 (p_1(1-p_1) + p_2(1-p_2))}{(p_1 - p_2)^2} \qquad (15.52)$$

where p_1 and p_2 are estimates of the proportions in each of the two groups. We can illustrate this calculation with further discussion about the hypothetical example. Suppose the investigator wishes to consider additional designs for the study of the investigational drug for hypertension. The design he is considering is a parallel groups design in which hypertensive participants are randomized to either treatment or placebo and the outcome is whether the proportion of participants who are still hypertensive (defined as diastolic blood pressure > 100 mmHg) is different in the two groups. Suppose the investigator wishes to use a two-tailed test with a 0.05 significance level. How many subjects would be required to have 90% power detect a difference in the proportions of 0.05? Is any other information needed to make this calculation? Do we need to have an estimate of the proportion of participants still hypertensive in the placebo group? Suppose that from other studies we know that this proportion is 0.9. Thus, we have that $p_1 = 0.9$ and $p_2 = 0.85$. The required sample size in *each* of the two groups is

$$n = \frac{(1.960 + 1.282)^2 (0.9 \times 0.1 + 0.85 \times 0.15)}{(0.05)^2} = 914.4 \approx 915.$$

$$(15.53)$$

Thus, more than 1800 hypertensive participants would need to be enrolled in this parallel groups study. If the power was reduced to 80%, then *per arm*

$$n = \frac{(1.960 + 0.841)^2 (0.9 \times 0.1 + 0.85 \times 0.15)}{(0.05)^2} = 682.9 \approx 683.$$

$$(15.54)$$

The *per* arm sample size required to have 80% power to detect a difference of 0.25 in the proportions is

$$n = \frac{(1.960 + 0.841)^2 (0.9 \times 0.1 + 0.65 \times 0.35)}{(0.25)^2} = 39.9 \approx 40.$$

$$(15.55)$$

These calculations demonstrate that we need a sample size of approximately 80 subjects to detect a very large effect with 80% power. The choice of outcome has a large effect on required sample size. Using a continuous variable as an outcome, if sensible from a scientific perspective, results in a more efficient design than categorizing a continuous variable.

4.4. Designing New Studies Based on the Motivating Studies

We illustrate these sample size calculations by redesigning studies similar to the three motivating examples.

4.4.1. Sample Sizes Based on the Beta-Interferon/MRI Study

For the beta-interferon/MRI study, 14 participants were followed for 7 months on baseline and 6 months on treatment. Based on the results of this study, how many participants would be required to conduct a similar study and be able to detect a similar size effect to what we observed in the beta-interferon trial? Suppose that the trial results will be analyzed with a paired t test with a 0.05 significance level. In addition, the test will be two-sided. Noting that the mean difference in average lesion counts was 1.12 on the log scale and that the variance in the difference between baseline and treatment counts was 0.770, we would need a sample size of

$$n = \frac{(1.960 + 1.282)^2 (0.770)}{(1.12)^2} = 6.5 \approx 7 \qquad (15.56)$$

in order to be able to detect this size reduction with a power of 0.9. (Often, when a small sample size is found using formulae based on the standard normal distribution, we use the t distribution in an iterative process to find a more accurate sample size estimate; one should consult a statistician for this tricky calculation.) The baseline versus treatment design was chosen over a parallel groups design because of the smaller number of participants required to screen for new therapies. Limitations on numbers of participants were due to the desire not to subject a large group of participants to a potentially toxic agent and the difficulties of recruiting and monitoring (with monthly serial MRI) a large cohort of participants in a single center. How many participants would we need to conduct a parallel groups study with the same design parameters? The required sample size in *each* of the two groups is

$$n = \frac{2(1.960 + 1.282)^2 (0.770)}{(1.12)^2} = 12.9 \approx 13. \qquad (15.57)$$

Thus, we would need approximately 26 participants with a parallel groups design to have high power to detect the very large effects found in the beta-interferon trial. The study sample size increases fourfold for the total study using the parallel groups design compared with the single group baseline versus treatment design. This is the price of increased credibility.

4.4.2. Sample Sizes for a New Felbamate Trial

In the felbamate monotherapy trial, 40 participants were monitored in a placebo-controlled trial on either felbamate ($n = 19$) or placebo ($n = 21$). The outcome was the average number of daily seizures during the 2-week follow-up period. We will use the data from this trial to help us design a new trial testing the effect of another antiseizure medication. How many participants would be required in a parallel groups design to detect a similar reduction in seizure frequency with a power of 0.8? The analysis will be based on a two-sided test of average daily seizure counts using a t test with a 0.05 significance level.

Noting that on the square root scale the average daily seizure counts were 1.42 and 0.42 in the placebo and treatment groups and the variances were 1.69 and 1.00 in these groups, respectively, we compute the sample size in *each* group as

$$n = \frac{(1.960 + 0.841)^2 (1.69 + 1.00)}{(1.42 - 0.42)^2} = 21.1 \approx 22. \qquad (15.58)$$

Thus, the required sample size would be 44 participants, which is close to the original design for the fel-

bamate trial. Now suppose that instead of a placebo-controlled study, an investigator wishes to design an "add-on" trial in which participants are randomized either to carbamazepine or to carbamazepine and felbamate. The scientific interest here is whether felbamate has an additional antiseizure effect over carbamazepine alone. Estimates of the mean and variance in the carbamazepine-alone group are 0.42 and 1.00, respectively; the latter value is assumed to be the same as that for the felbamate-alone arm in the felbamate trial. We also assume that the carbamazepine plus felbamate combination has a variance of 0.8. If we want to be able to detect a 50% reduction in seizure frequency with a power of 0.8, the required sample size in *each* group is

$$n = \frac{(1.960 + 0.841)^2 (1 + 0.8)}{(0.42 - 0.21)^2} = 320.2 \approx 321. \qquad (15.59)$$

This calculation demonstrates the major reason why the original felbamate trial was designed as a placebo-controlled trial compared to an add-on trial. It was impossible to conduct a trial with more than 600 participants at the NIH Clinical Center.

4.4.3. Sample Sizes Based on the ISIS-4 Trial Findings

We now illustrate the design of a study to examine the effect of magnesium on MI fatality. It is postulated that one of the major reasons why ISIS-4 found no effect of magnesium on 35-day mortality was that participants were not given magnesium early enough after experiencing chest pain. A new randomized clinical trial was designed to further examine the effect of magnesium on MI mortality.[40] Assuming a two-tailed test of population proportions at the 0.05 significance level as the hypothesis test, we demonstrate how sample size can be estimated. Using a 35-day mortality of 15%, how many participants would be required in each of two groups to detect a 20% reduction in mortality in the magnesium arm with a power of 0.8? The required sample size in *each* of the two groups can be computed as

$$n = \frac{(1.960 + 0.841)^2 (0.15 \times 0.85 + 0.12 \times 0.88)}{(0.15 - 0.12)^2} \approx 2033. $$

$$(15.60)$$

The 35-day mortality in the placebo arm may be closer to 0.1 than 0.15. We examined the sensitivity of the sample size estimate to reducing this placebo event rate. Using an event rate of 0.1 in the placebo arm, the required sample size in each arm is

$$n = \frac{(1.960 + 0.841)^2 (0.10 \times 0.90 + 0.08 \times 0.92)}{(0.10 - 0.08)^2} \approx 3211.$$

(15.61)

Thus, more than 6000 participants will be needed to study this question adequately.

In general, other factors may increase the required sample sizes, such as drop-in (participants randomized to the placebo who start taking the treatment), drop-out (participants randomized to the treatment who stop taking their medication), participants who switch treatment arms, and partial or noncompliance with treatments.[41] Investigators must responsibly plan for these and other likely complications in designing their studies.

We stress the importance of designing studies with sufficient power to detect meaningful differences. Sample size calculations are vital to ensuring that studies are not doomed from the start because of low power. As a general rule, sample sizes should be calculated for a number of different design parameters to examine the sensitivity of the final sample size to these parameters. In confirmatory trials, the most conservative design parameters should be used to ensure a study with high power. Note also that in this chapter we have only discussed sample sizes for one-sample and two-sample comparisons because these are the most common calculations in clinical medicine. These calculations can be done using a handheld calculator. For other statistical analyses, sample size estimation is much more complicated and there are software packages devoted to computing sample sizes in these cases. One example in which sample size calculation is more complicated is in survival analysis, the topic of a subsequent chapter.

5. SPECIAL CONSIDERATIONS

The following topics extend the concepts presented in the previous sections. They represent ideas that arise frequently in consultations with clinicians, and further details can be found in the references.

5.1. Maintaining Balance in Randomization: Dynamic/Adaptive Allocation or Minimization

In addition to block randomization, another way to achieve some degree of balance on important characteristics is to maintain balance "on the margins" or totals for each variable. That is, instead of trying to fill

each possible cell in a multiway stratification design, we use special software to assign subjects to treatments in a way that keeps the totals balanced for particular characteristics summed over all other stratification categories. Similar cautions about the choice of stratification variables are true for dynamic allocation. Some investigators believe that dynamic allocation is the best way to avoid having any predictor variables assigned unequally across treatment groups in a randomized study. Logrank tests, used in survival analysis, may be biased by imbalances. Although there are statistical methods to deal with unbalanced totals among particular covariates, some journal readers will look at the preliminary tables in a paper and dismiss a study because of imbalance in the baseline characteristics, although other journal readers will carefully consider the magnitude of imbalance and potential impact of the imbalances and how they were handled in the analyses. The randomization methods described previously minimize the chance of this happening in a study, although some degree of imbalance may always remain.

Although dynamic or adaptive allocation has been used in cancer research since the 1970s, there remains a lack of consensus in the field of clinical trials about the value of this form of randomization. For example, the International Conference on Harmonization (ICH) E9 Statistical Guidance document[42] withholds judgment on the topic, but the European Medicines Agency's (EMEA) Committee for Proprietary Medicinal Products (CPMP) 2003 document[43] on baseline covariate adjustment strongly discourages the use of dynamic allocation. The release of the new CPMP guideline caused a flurry of articles and letters in support of dynamic randomization. Researchers need to be prudent and weigh the value of using dynamic allocation in their research and the general acceptability of this method in their particular fields.

Response Adaptive Allocation is yet another set of methods that utilize interim outcome data from an ongoing trial to stop a study early, influence or unbalance the allocation probabilities for group assignment, or perform sample size reassessment. This topic is broad and an active area of work and discussion.

5.2. A Trick for Confidence Interval Estimation When No Events Occur

One trick related to estimating the upper confidence interval limit is particularly useful in small laboratory experiments with a binomial outcome. The "rule of three"[44] states that the 95% upper bound for a binomial proportion when no events occur is approximated by $3/n$. The improved "rule of three"[45] is $3/(n + 1)$ and is a uniformly better approximation than $3/n$. For example, if we conduct 25 rodent experiments and have no fatal outcomes, then the 95% upper confidence bound on the true rate of mortality is approximately equal to 3/25 or 12% using the older rule. Using the improved rule of three, the upper bound may be better approximated by 3/26 or 11.5%. The exact calculation for the upper bound in this case is 11.3%, slightly less than the two quick approximations.

5.3. Tricks for Common Sample Size Problems

In general, it is easiest to estimate sample size for balanced designs in which the number of individuals assigned to each treatment group is the same. Unfortunately, for practical reasons, we cannot always obtain the sample size we desire for one of the groups of interest, especially in observational studies. For example, in a case–control study there may be a fixed number of cases or devices that are available, and yet the standard sample size calculation may indicate that more observations are required. If we want to obtain a specific value of power with a fixed number of cases, we may use the following formula to determine the ratio of controls to cases, namely,

$$k = \frac{n}{2n_0 - n}, \qquad (15.62)$$

where n is the number of subjects in each group required for the given power under the balanced design, n_0 is the fixed number of cases, and, in turn, kn_0 will be the number of controls. For example, if we have $n_0 = 11$ cases, then $k = 1.44$, and thus we need $kn_0 = 16$ controls and $n_0 = 11$ cases to achieve the same power as a balanced design with $n = 13$ controls and $n = 13$ cases.[44]

In the sample size and power calculations presented previously, we have assumed particular values for the significant differences of interest and the true variances. Often, these values are chosen based on previous experience, but sometime there exists much uncertainty about appropriate choices for these values. In this case, one should construct a table showing the sample size or power values using a wide variety of possible assumptions about parameters, such as the differences of interest and the true variances. If possible, one should aim to preserve sufficiently large values of sample sizes and power in a study for a wide variety of possible parameter values, rather than merely have satisfactory power for a narrow set of assumptions.

5.4. Data Dependencies

5.4.1. Correlation

Correlation coefficients are measures of agreement between paired variables (x_i, y_i), where there is one independent pair of observations for each subject. The general formula for the sample (Pearson) correlation is

$$r = \frac{\sum_{i=1}^{n}(x_i - \bar{x})(y_i - \bar{y})}{\sqrt{\sum_{i=1}^{n}(x_i - \bar{x})^2 \sum_{i=1}^{n}(y_i - \bar{y})^2}}. \qquad (15.63)$$

The sample correlation r lies between the values -1 and 1, which correspond to perfect negative and positive linear relationships, respectively. Values of $r = 0$ correspond to no linear relationship, but other nonlinear associations may exist. Also, the statistic r^2 describes the proportion of variation about the mean in one variable that is explained by the second variable. One may compute p-values for the hypothesis of zero correlation, although the strength of the correlation is often more important than the fact that it is nonzero. In addition, if one replaces the raw data for each variable by the respective ranks of that data, one obtains Spearman's rank correlation. For further details, see Altman.[10]

5.4.2. Relationships in Organization, Space, and Time

It is important to recognize the various structures and relationships that may exist among the data in organization, space, and time. In some studies, there may be hierarchical relationships among subjects. For example, in a communitywide observational study on the health of school-aged children, we may have children nested within classrooms, within schools, within school districts, etc. Similarly, in a study with a geographical or spatial component, measurements on locations that are closer together may tend to be more similar, and those nearer large cities may tend to have different traits than more rural locations. Furthermore, in a longitudinal study in which repeated measurements are made on each subject, the measurements made closer together in time may be more highly related to each other than those made at more distant times. It is important to recognize these various structures and relationships because they need to be considered appropriately in the statistical design of studies and the analysis of data.

5.5. Essential Issues in Microarrays, fMRI, and Other Applications with Massive Data Sets

Structural relationships in data sets are important for all studies, but they are especially important for the massive data sets we find in microarray and functional MRI (fMRI) experiments. The basic rules and assumptions described previously pertain to these data sets, and slight variations from those assumptions may result in large changes in inference, such as the difference between the t and z distributions at a given cut point for a certain sample size. When trying to draw inference to a population we need to look at numerous independent samples. Looking at 100,000 pieces of information for five people means we know those five people very well but does not mean we can extrapolate that knowledge to the entire human race. The structure inherent in a data set may make many of the observations correlated or otherwise related. Furthermore, multiple comparisons issues abound; data mining, hunting for patterns and useful information in massive data sets, is common; and many other issues arise not only in analysis but also in computing a reasonable sample size estimate for a study.

The simple hypothesis testing methods discussed previously relied on having independent samples and measurements. The only exception we saw was for the paired t-test, but there we looked at independent differences within each pair of subjects. If we take multiple measures over time in the same person, as in a longitudinal study, these measures are not independent, and indeed measurements taken closer in time may be more similar than those taken at more different times. We also may take a single sample (e.g., biopsy) from a person but use it to report multiple outcomes; this is commonly done in microarray and fMRI experiments. If we use a microarray, the gene expression we see from some probe sets may be associated with the gene expression seen in other probe sets because the probe sets are either for the same gene or for genes that are associated with each other. Indeed, since many microarray chips routinely test for 10,000+ genes at a time, and some have multiple tests for the same gene on a chip, the importance of the correlations structures and multiple comparisons cannot be underestimated. Although correlation is occasionally discussed in the analysis of microarray data, it is a hot topic in fMRI data analyses. The voxels, defined registered areas on the fMRI image, have a correlation structure. Currently, some analysis methods ignore it, some methods impose a simple uniform structure that is not modified for different parts of the brain, and other methods attempt to fully model the correlation structure of the voxels.

Although the later methods may be the most accurate, they also require computing capabilities not commonly found at this point in time even for small numbers of study participants. Computing should catch up soon.

Another common problem that arises in microarray or other high-throughput experiments with large numbers of tests is the choice of error rates for sample size calculation. This calculation requires the consideration of the impact and consequences of multiple testing. Many researchers choose to use a significance level of 0.001 and power of 0.95 or higher in order to control the false discovery rate and select a candidate pool on which to follow up.[46] Especially in the high-throughput designs, a Bonferroni correction may make it impossible to have any statistically significant items in a study with a small sample size. Several methods that aim to limit the false discovery rate (FDR) may be employed, and new methods are frequently described in journals.

It also is important to remember that technical replicates help reduce laboratory variability, but independent biological replicates (i.e., samples from many different people) are important. The sample size calculations presented previously may be used with special care given to the choice of significance level and power, but in choosing which samples to collect and use, it is important to remember that the sample size estimates are for the number of independent samples. Microarrays and fMRI are two of several new areas with massive data sets for each individual sample that can hide the fact that often there are few independent samples. Data mining is common, but it too must take into account the correlation, multiple comparisons, and many other issues described previously. Likewise, we can perform sample size calculations in studies designed to check for clustering of genes. Currently, there are no hard and fast rules about the methods, except for the consensus that most methods are attempted with too few independent samples to uncover potentially complex biological structure. In the end, how many independent samples do we need to make a reasonable inference about a population? Consult your local statistician who specializes in the type of data in your study for details, new updates, and guidance. These are not just computing issues; in fact, with this much data for every specimen plenty of numbers can come roaring out of a computer, but we need to ensure they are the answers to the scientific questions of interest and that the experiment and analysis can be replicated.

5.6. Diagnostic Testing

In laboratory testing, we can construct a 2×2 table to relate the results of a diagnostic test (positive or

TABLE 15-2   HIV Test Results

Disease	Test Results		Total
	Positive	Negative	
Present	96	4	100
Absent	99	9801	9900

negative) to the biological truth (case or control status and presence or absence of a disease). Some key concepts are as follows: The *prevalence* is the proportion of subjects in a population with a particular disease. The *sensitivity* is the probability of testing positive given that a subject truly has a disease, whereas the *specificity* is the probability of testing negative given that the subject does not have the disease. These two concepts parallel the ideas of type I and type II error rates. Conversely, the *positive predictive value* (PPV) is the probability that a subject indeed has the disease given a positive test, whereas the *negative predictive value* (NPV) is the probability that a subject does not have the disease given a negative test.

To illustrate, suppose that 1% of a population is HIV-positive (prevalence = 1%), and suppose that a diagnostic test has 96% sensitivity and 99% specificity. Using these values for 10,000 subjects, we might observe the counts shown in Table 15-2).

In turn, the PPV is $96/(96 + 99) = 0.49$, whereas the NPV is $9801/(9801 + 4) \approx 1$. Thus, despite the high accuracy of the diagnostic test, a positive test is only 49% likely to correspond to an HIV-positive person, although a negative test almost certainly rules out HIV infection. This example shows that the interpretation of test results depends not only on the accuracy of the test but also on the prevalence of the disease in the population. For more information, see Pepe.[29]

6. CONCLUSION

Study design is part science and part art. Randomization methods and hypotheses need to match the question of interest and many choices are available. Sample sizes should be calculated to detect meaningful differences and for a number of different design parameters in order to account for the uncertainty in these parameters. Finally, it is wise to consult with a statistician at the earliest stages of planning a study to obtain help with study design, hypothesis generation, and appropriate sample size calculations. Timely collaboration with an eager statistician will help you and your studies succeed.

References

1. Zoon KC. The regulation of drugs and biological products by the Food and Drug Administration. In Gallin JI, Ognibene FP (ed.) *Principles and Practice of Clinical Research.* New York, Elsevier, 2007.
2. Moher D, Schulz KF, Altman DG, for the CONSORT Group. The CONSORT statement: Revised recommendations for improving the quality of reports of parallel-group randomised trials. *Lancet* 2001;357:1191–1194. Also published in *JAMA* 2001;285:1987–1991; and *Ann Intern Med* 2001;134:657–662.
3. Wichman A, Sandler AL. Institutional review boards. In Gallin JI, Ognibene FP (ed.) *Principles and Practice of Clinical Research.* New York, Elsevier, 2007.
4. McFarland HF, Frank JA, Albert PS, *et al.* Using gadolinium enhanced MRI lesions to monitor disease activity in multiple sclerosis. *Ann Neurol* 1992;32:758–766.
5. Albert PS, McFarland HF, Smith ME, Frank JA. Time series for counts from a relapsing remitting disease: Applications to modeling the disease course in multiple sclerosis. *Stat Med* 1994;13:453–466.
6. Stone LA, Frank JA, Albert PS, *et al.* The effect of interferon-beta on blood–brain barrier disruptions demonstrated by contrast-enhancing magnetic resonance imaging in relapsing remitting multiple sclerosis. *Ann Neurol* 1995;37:611–619.
7. Stone LA, Frank JA, Albert PS, *et al.* Characterization of MRI response to treatment with interferon beta 1b: Contrast enhancing MRI lesion frequency as a primary outcome measure. *Neurology* 1997;49:862–869.
8. Theodore WH, Albert P, Stertz B, *et al.* Felbamate monotherapy: Implications for antiepileptic drug development. *Epilepsia* 1995;36:1105–1110.
9. ISIS–4 Collaborative Group. ISIS–4: A randomized factorial trial assessing early oral captopril, oral mononitrate, and intravenous magnesium sulphate in 58,050 patients with suspected acute myocardial infarction. *Lancet* 1995;345:669–685.
10. Altman DG. *Practical Statistics for Medical Research.* Boca Raton, FL, Chapman & Hall, 1991.
11. Friedman LM, Furberg CD, DeMets DL. *Fundamentals of Clinical Trials,* 3rd ed. New York, Springer-Verlag, 1998.
12. Fleiss JL. *The Design and Analysis of Clinical Experiments.* New York, Wiley, 1999.
13. Agresti A. *Categorical Data Analysis,* 2nd ed. Hoboken, NJ, Wiley, 2002.
14. The Physicians' Health Study Group. Preliminary report: Findings from the aspirin component of the ongoing physicians' health study. *N Engl J Med* 1988;318:262–264.
15. Women's Health Initiative Study Group. Design of the Women's Health Initiative clinical trial and observational study. *Controlled Clin Trials* 1998;19:61–109.
16. Prentice RL, Pettinger M, Anderson GL. Statistical issues arising in the Women's Health Initiative. *Stat Med* 2005;61:899–910.
17. Manson JE, Hsia J, Johnson KC, *et al.,* Women's Health Initiative Investigators. Estrogen plus progestin and the risk of coronary heart disease. *N Engl J Med* 2003;349:523–534.
18. Korn EL. Nontoxicity endpoints in phase I trial designs for targeted non-cytotoxic agents. *J Natl Cancer Inst* 2004;96:977–978.
19. Parulekar WR, Eisenhauer EA. Phase I trial design for solid tumor studies of targeted non-cytotoxic agents: Theory and practice. *J Natl Cancer Inst* 2004;96:990–997.
20. Thall PF, Cook JD. Dose-finding based on efficacy-toxicity trade-offs. *Biometrics* 2004;60:684–693.

21. Hunsberger S, Rubinstein LV, Dancey J, Korn EL. Dose escalation trial designs based on a molecularly targeted endpoint. *Stat Med* 2005;24:2171–2181.

22. Thall PF, Simon R. Incorporating historical control data in planning phase II clinical trials. *Stat Med* 1990;9:215–228.

23. Simon R. Optimal two-stage designs for phase II clinical trials. *Control Clin Trials* 1989;10:1–10.

24. Simon R, Wittes RE, Ellenberg SS. Randomized phase II clinical trials. *Cancer Treat Rep* 1985;69:1375–1381.

25. Steinberg SM, Venzon DJ. Early selection in a randomized phase II clinical trial. *Stat Med* 2002;21:1711–1726.

26. Green S, Benedetti J, Crowley J. *Clinical Trials in Oncology*, 2nd ed. New York, Chapman & Hall, 2002.

27. Rosner B. *Fundamentals of Biostatistics*, 6th ed. Belmont, CA, Duxbury, 2005.

28. Piantadosi S. *Clinical Trials: A Methodologic Perspective*. New York, Wiley, 1997.

29. Pepe MS. *The Statistical Evaluation of Medical Tests for Classification and Prediction*. New York, Oxford, 2003.

30. Moore DS, Notz WI. *Statistics: Concepts and Controversies*, 5th ed. New York, Freeman, 2005.

31. Moore DS. *Introduction to the Practice of Statistics*, 5th ed. New York, Freeman, 2005.

32. Armitage P, Berry G, Matthews JNS. *Statistical Methods in Medical Research*. Oxford, Blackwell, 2001.

33. Draper NR, Smith H. *Applied Regression Analysis*, 3rd ed. New York, Wiley, 1998.

34. Clopper CJ, Pearson ES. The use of confidence or fiducial limits illustrated in the case of the binomial. *Biometrika* 1934;26: 404–413.

35. Agresti A, Coull BA. Approximate is better than "exact" for interval estimation of binomial proportions. *Am Stat* 1998;52: 119–126.

36. Borkowf CB. Constructing confidence intervals for binomial proportions with near nominal coverage by adding a single imaginary failure or success. *Stat Med* 2006;25:3679–3695.

37. van Belle G, Fisher LD, Heagerty PJ, Lumley TS. *Biostatistics: A Methodology for the Health Sciences*, 2nd ed. New York, Wiley, 2004.

38. Cohen J. *Statistical Power Analysis for the Behavioral Sciences*, 2nd ed. New York, Academic Press, 1990.

39. Guenther WC. Sample size formulas for normal T-tests. *Am Stat* 1981;35:243–244.

40. Magnesium in Coronaries (MAGIC) Trial Investigators. Early administration of intravenous magnesium to high-risk patients with acute myocardial infarction in the Magnesium in Coronaries (MAGIC) Trial: A randomized controlled trial. *Lancet* 2002;360: 1189–1196.

41. Lachin JM. Introduction to sample size determination and power analysis for clinical trials. *Controlled Clin Trials* 1981;2:93–113.

42. International Conference on Harmonisation. Guidance on statistical principles for clinical trials. International Conference on Harmonisation; E-9 Document. *Fed Reg* September 16, 1998;63:49583–49598. FR Doc 98-24754.

43. European Medicines Agency Committee for Proprietary Medicinal Products. Points to consider on adjustment for baseline covariates, CPMP/EWP/2863/99, London, May 22, 2003. Available at www.emea.eu.int/pdfs/human/ewp/286399en.pdf. [Accessed April 11, 2005]

44. van Belle G. *Statistical Rules of Thumb*. New York, Wiley, 2002.

45. Jovanovic BD, Levy PS. A look at the rule of three. *Am Stat* 1997;51:137–139.

46. Simon RM, Korn EL, McShane LM, Radmacher MD, Wright GW, Zhao Y. *Design and Analysis of DNA Microarray Investigations*. New York, Springer, 2004.

16

Design and Conduct of Observational Studies and Clinical Trials

JACK M. GURALNIK* AND TERI A. MANOLIO**

*Laboratory of Epidemiology, Demography, and Biometry, National Institute on Aging, National Institutes of Health, Bethesda, Maryland
**National Human Genome Research Institute, National Institutes of Health, Bethesda, Maryland

Epidemiologic study designs are rich and diverse, spanning studies involving single patients observed at the bedside to those conducted on a population-wide, national, or international basis. Epidemiologic studies may be purely observational, in which no true intervention occurs (other than the act of observation), or interventional, in which an educational or preventive effort, treatment, or diagnostic strategy is applied. It is important to recognize that even "unobtrusive" observation can still have a significant impact. Epidemiologic studies may also be controlled or uncontrolled, with controls most often being utilized in experimental studies testing one treatment against standard therapy or placebo. It is often useful to consider these studies as a hierarchy from simpler to more complex designs.

1. TYPES OF EPIDEMIOLOGIC STUDY DESIGNS

Table 16-1 gives an overview of the main types of epidemiologic studies. It divides them into descriptive studies, mainly providing information that characterizes an individual or population, and analytic studies, which are primarily aimed at answering questions about the relationships of study participant characteristics and disease outcomes. Although this division into descriptive and analytic studies is based on the main objectives of the studies and does help to organize the range of study designs, studies in both these categories can be used for both descriptive and ana-

lytic purposes. Unlike the other study designs, the ecological study assesses characteristics of populations, rather than individuals, and then compares correlations of these characteristics across populations. Observing the relationship between total cigarette consumption in different countries and lung cancer mortality rates in these countries is an example of an ecological study. At the level of the individual, the simplest study design is a description of a particular clinical phenomenon in a case report or case series. Slightly more complex is the cross-sectional survey, which provides estimates of disease prevalence in a defined group of subjects. Cross-sectional studies, while being the best way to describe populations, also serve a valuable role in studying relationships between the wide array of information often collected in these studies. Longitudinal cohort studies all begin with a baseline assessment, which is essentially a cross-sectional study, and this baseline is often used to study cross-sectional relationships while waiting for longitudinal data to become available.

Analytic studies listed in Table 16-1 are focused on answering questions about how specific characteristics are related to pathological outcomes. The two general types of observational approaches for this purpose are case–control and cohort studies. In the case–control study, persons with a particular disease or condition are compared to controls who do not have this condition to identify potential etiologic factors. The case–control study begins by characterizing disease status and then examining potential risk factors in those with and without the disease. This can be contrasted to the prospective cohort study, which begins by characterizing risk factor status and then ascertaining disease outcomes over time in those with and without a risk factor. Finally, clinical trials are designed to intervene on potentially modifiable risk factors to prevent or reduce the severity of disease outcomes.

Table 16-2 gives examples, using some landmark and some contemporary studies, of the major epidemiologic study designs. For the new epidemiologist, developing a favorite group of unambiguous examples of different study designs is quite advantageous. When evaluating published studies, it may sometimes be unclear exactly what study design is being utilized and having concrete examples at hand of the different study designs will help in understanding many newly encountered studies. The examples in Table 16-2 are discussed in more detail in the sections that focus on each design.

TABLE 16-1   An Overview of Study Designs

Descriptive studies
    Populations: Ecological (correlational) studies
    Individuals
        Case reports
        Case series
        Cross-sectional surveys
Analytic studies
    Observational studies
        Case–control studies
        Prospective cohort studies: concurrent (longitudinal) and
           nonconcurrent (historical)
    Intervention studies (clinical trials)

2. ECOLOGICAL (CORRELATIONAL) STUDIES

The ecological study utilizes data at the population level rather than the individual level. The example shown in Table 16-2 is of a study that used descriptive data on U.S. states' rates of coronary heart disease mortality and per capita cigarette sales and showed that across the states there was a significant correlation between these two rates.[1] Ecological studies are valuable because they can be done easily and quickly by using population data that has already been collected and seeking correlations between potential risk factors and various disease outcomes. The major disadvantage of this kind of study is that data for individuals

TABLE 16-2   Examples of Major Types of Epidemiologic Study Designs

Type of Study	Reference	Findings
Ecological study	Friedman GD. Cigarette smoking and geographic variation in coronary heart disease mortality in the United States. *J Chronic Dis* 1967;20:769–779	Coronary heart disease mortality rates in 44 states correlated with per capita cigarette sales in those states.
Case report and case series	Centers for Disease Control and Prevention. Pneumocystis pneumonia, Los Angeles. *MMWR* 1981;30:250–252	Initial report of five cases of *Pneumocystis* pneumonia in previously healthy, homosexual men; first report of AIDS epidemic.
Prevalence survey or cross-sectional study	Hedley AA, *et al.* Prevalence of overweight and obesity among U.S. children, adolescents, and adults, 1999–2002. *JAMA* 2004;291:2847–2850	Prevalence data on overweight and obesity using measured height and weight in National Health and Nutrition Examination Survey (NHANES).
Case-control or retrospective study	Herbst AL, *et al.* Adenocarcinoma of the vagina: Association of maternal stilbesterol therapy with tumor appearance in young women. *N Engl J Med* 1974;284:878–881	Case–control design was able to identify relationship of exposure to stilbestrol during mother's pregnancy with occurrence of rare tumor in female offspring many years later.
Nonconcurrent (historical) prospective cohort study	Plassman *et al.* Documented head injury in early adulthood and risk of Alzheimer's disease and other dementias. *Neurology* 2000;55:1158–1166.	Early life head trauma shown to be related to dementia in old age by using military medical records to identify World War II head trauma exposure group and nontrauma comparison group, with both groups traced and evaluated for dementia 50 years later.
Concurrent (longitudinal) prospective cohort study	Doll R, Hill AB. The mortality of doctors in relation to their smoking habits: A preliminary report. *Br Med J* 1954;228(1):1451–1455 Doll R, *et al.* Mortality in relation to smoking: 50 years of observations on male British doctors. *Br Med J* 2004;328:1519–1533	Prospective cohort study that showed early increase in risk of lung cancer and heart disease mortality in smokers and confirmed this over 50 years of follow-up.
Clinical trial	Steering Committee of the Physicians' Health Study Research Group. Final report on the aspirin component of the ongoing Physicians' Health Study. *N Engl J Med* 1989;321:129–135	Randomized, double-blind, placebo-controlled trial demonstrating that low-dose aspirin was associated with a significant reduction in risk of myocardial infarction.

are not available and there is no way to know that those with the risk factor are the ones who are actually getting the disease. This has been termed the *ecological fallacy*. Furthermore, potential confounding by other variables is difficult to assess in an ecological study. Populations in which a particular diagnostic test or preventive treatment is used more often may have lower rates of a disease, for example, but it may not be the test or treatment that accounts for this but, rather, the fact that a population with greater access to such care more often has other characteristics that lead to lower disease rates. Despite their limitations, ecological studies have been valuable as the first assessment of an association that was then studied using additional epidemiological designs.

3. CASE REPORTS AND CASE SERIES

3.1. Objectives and Design

The object of case reports and case series—the difference between them being that a case report describes a single case, whereas a case series presents several similar cases—is to make observations about patients with defined clinical characteristics. The design is a simple description of the clinical data, preferably from a very well-defined group of individuals, without reference to a comparison group. For example, the report of abdominal aortic aneurysm presenting as transient hemiplegia[2] is a case report, whereas the Centers for Disease Control and Prevention report cited in Table 16-2 of *Pneumocystis* pneumonia in previously healthy, homosexual men[3] is a case series. Observations in these reports should be comprehensive and detailed enough to permit recognition of similar cases by the reader. The report should include a clear definition of the phenomenon under study.

3.2. Observations and Analysis

The same definition should be applied equally to all patients in the series, and all observations should be made in as reliable and reproducible a method as possible. Findings are usually presented as needed to illustrate the phenomenon, such as frequency of a given "discrete" (i.e., present/absent) variable or mean or median of a continuous variable (e.g., age or blood pressure) in the study series. Important subgroups, such as those defined by sex or age, may need stratified data presentation. In a case series, analysis is limited to descriptive variables such as proportions or means with standard errors. Interpretations and conclusions should include a summary of the new phenomenon illustrated in the report, reference to previous, related observations, and suggestions of etiology or of further studies needed. An important question is whether the described series is representative of all patients with the disorder such that conclusions can be generalized. This is often difficult to determine in initial case reports and case series, and it may well call for other investigators to identify and describe similar cases.

3.3. Advantages and Disadvantages

Advantages of this design are that it is useful in forming hypotheses, planning natural history studies, and describing clinical experience. Of particular value is the use of the case report or case series, as in the *Pneumocystis* pneumonia example, to inform the medical community of the first cases of what could be an important emerging condition or disease. Very often, phenomena observed in clinical practice provide the first clues of more generalized etiologies or risks and provide valuable suggestions for hypothesis generation and further study. These studies are also easy and inexpensive to do in clinical settings. The disadvantages are primarily that selection of study patients may be biased, making generalization of results difficult; perhaps only the sickest or most typical (or most atypical) cases were included in the study. In addition, it may be unclear whether the confluence of findings was merely a chance happening or was truly characteristic of a new disease or syndrome. Case studies and case series provide important clues for further investigation in and of themselves, but if not reproduced they may merely represent interesting observations of which the astute clinician should be aware.

4. PREVALENCE SURVEYS OR CROSS-SECTIONAL STUDIES

4.1. Objectives and Design

The object of this design is to make observations about the presence of diseases, conditions, or health-related characteristics in a defined population at a specific point in time. These studies yield prevalence rates, defined as the number of persons with a disease or condition at a given time divided by the number in that population who are at risk for the condition at that time. For example, a general population survey that also assessed gynecologic conditions would only use the number of women in the survey as the denominator for estimates of the prevalence of these conditions. Prevalence surveys can also be used to characterize the disease and its spectrum of manifestations. Unlike incidence studies, which identify new cases of a condition over a set period of time, prevalence studies count all existing cases, whether they are of recent onset or long duration. The design involves (1) defining the population under study, (2) deriving a sample of that population, and (3) defining the characteristics being studied. The population under study could be, for example, "black adults over age 65 living in the United States in 2000" "workers in the beryllium industry between 1995 and 2004," or "public school children in Montgomery County in the school year 2000–2001." In clinical research, one might select a patient sample, such as "all treated hypertensives enrolled in Group Health Cooperative of Puget Sound between 1989 and 1996,"[4] although this obviously limits the generalizability of results. For very common diseases, such as hypertension or osteoarthritis, one can sample from a given age range in a defined geographic area, as was done in prominent epidemiologic studies in Framingham, Massachusetts, and Tecumseh, Michigan. One of the most extensive cross-sectional surveys in the United States is the National Health and Nutrition Examination Survey, with the example in Table 16-2 showing important prevalence estimates of overweight and obesity in the U.S. population.[5]

In general, it is neither feasible nor necessary to examine everyone in a given population at risk. Defined approaches for sampling[6] can be used to provide a random and reasonably representative sample of a population or population subgroup, conclusions from which can then be generalized to the base population from which they were drawn. Most important is to define the condition being studied for prevalence: What defines its presence or absence in a given study subject? For some conditions, it may be very difficult to determine if a condition is truly absent; atherosclerosis, for example, is so common and its manifestations at times so subtle that fairly extensive pathologic study (in an autopsy series)[7] or imaging (in living subjects)[8] may be required to ensure its absence. The definition of the condition or health characteristic under study in a prevalence survey should be standardized, reproducible, and feasible to apply on a large scale.

4.2. Observations and Data Analysis

Methods of data collection should be applied equally to all study participants. Although this sounds simple, it may not be so in practice; very elderly people may have impaired hearing or cognitive decline and may require a different approach to administration of study interviews, for example, than others. Such differences should be anticipated and every effort made to mini-

mize their potential impact on study results. At a minimum, the use of alternative methods of data collection in a given subject should be recorded and used as a variable in analysis.

Findings are presented as prevalence estimates (e.g., percent, cases per 100,000, or other similar proportion), with 95% confidence intervals calculated from the standard error of the estimate.* Frequency or mean levels of relevant factors may be compared in those with or without the prevalent condition, and data may be presented separately for important subgroups (e.g., those defined by age, sex, or coexisting conditions). Analysis is similar to that described for case–control studies (see later).

Conclusions in prevalence surveys are for the most part descriptive and hypothesis generating. In addition to descriptive findings, there may also be associative findings comparing prevalent cases and noncases on a variety of characteristics. For example, prevalent cases may be older or more often smokers or diabetics than noncases.

4.3. Advantages and Disadvantages

A major advantage of prevalence surveys is that when they are truly population based, they avoid many of the potential biases of case series, providing cases that are more representative of the general population. This is because case series by necessity involve people who have come to medical attention for one reason or another, or perhaps involve only the most severe cases or only those who have access to medical care. Population-based samples avoid these biases. In addition, although it is rarely inexpensive to do a prevalence survey unless it is a very small one, such studies are less expensive for common diseases than for rare ones because a smaller population sample will still provide reasonably stable prevalence estimates. Conditions with a prevalence of 1 in 1000 or 10,000 require very large samples and probably are not feasible for population-based cross-sectional studies.

Another advantage of prevalence surveys is that they are often of short duration. In addition, they can be addressed to specific populations of interest (e.g., workers in a given industry) and can examine a wide variety of exposures and outcomes simultaneously.

Disadvantages of this design include its unsuitability for rare diseases, as described previously, or for diseases of short duration. Since prevalence is proportional to incidence multiplied by duration, short-

---

*Standard error = $\sqrt{pq/n}$, where p is the prevalence, $q = 1 -$ prevalence, and n is the sample size.

duration diseases such as influenza may have a very high incidence but relatively low prevalence at any given time point. In addition, several types of bias may be operative, as described later for case–control studies, and high refusal rates may make accurate prevalence estimates impossible. Subjects who participate in these studies usually differ from those who do not, being more likely to be of higher socioeconomic status, better educated, and more health conscious or concerned than nonparticipants. Smokers and those practicing other high-risk health behaviors tend to have lower participation rates. If disease rates are substantially higher in nonparticipants than participants, then there will be an underestimation of the true prevalence of the disease. To gain insight into the potential bias resulting from nonparticipation, it is wise to attempt to characterize persons refusing to participate to the degree possible, if only through demographic characteristics available from the sampling frame. It is more important to track the numbers of total contacts and refusals to obtain an accurate estimate of participation rates. Epidemiologists tend to become uncomfortable with participation rates below approximately 80%, although rates exceeding this are difficult to achieve in population-based sampling.

Prevalence estimates are best derived from cross-sectional studies, but factors associated with a disease or condition can be assessed by both cross-sectional and case–control studies. However, if the main goal of a study is to assess factors associated with a disease or condition, it is important to recognize that this can be done with both cross-sectional and case–control designs. A disadvantage of cross-sectional studies is that they are more expensive and time-consuming in general than are case–control studies, particularly for rare diseases. If possible, it is often simpler to identify the cases (through hospitals, registries, etc.) and focus on recruiting them into the study rather than finding, for example, 300 cases in a population for a disease with a prevalence of 1 in 10,000, which would require the full participation of 3 million subjects. An additional disadvantage of cross-sectional studies, compared to longitudinal studies, is that the disease process may alter measures of related factors, such as blood pressure rising or (more commonly) falling immediately after a myocardial infarction. Finally, a cross-sectional study cannot address the temporal relationship between the measured factors and the development of disease for identification of potential causal factors. If a cross-sectional survey demonstrates an association between low cognitive function and temporal lobe size by cerebral magnetic resonance imaging, for example, one cannot determine from those data alone whether a small temporal lobe led to cognitive decline or the cognitive decline caused temporal lobe atrophy or, indeed, whether some third factor caused them both.

5. CASE–CONTROL STUDIES

5.1. Objectives and Design

Case–control studies are sometimes called *retrospective studies* because the approach is to identify persons with the disease of interest and then look backward in time to identify factors that may have caused it (Fig. 16-1). The object of a case–control study is to make observations about possible associations between the disease of interest and one or more hypothesized risk factors. The general strategy is to compare the frequency or level of potential risk factors between a representative group of diseased subjects, or cases, and a representative group of disease-free persons, or controls, derived from the same population. Although sometimes used for common diseases, case–control studies are best reserved for studying potential etiologies of rare diseases.

Unfortunately, looking "backward in time" can be difficult and prone to serious biases. If subjects are identified and studied in the present, without the availability of information collected previously, the researcher is forced to rely on subjects' memories, hospital records, or other nonstandard sources for information on past exposures. Many of the biases to which case–control studies are prone occur during this data collection step, as described later.

Three critical assumptions of case–control studies help to minimize the potential for bias. The first assumption is that cases are selected to be representative of all patients who develop the disease. This may be difficult when using a hospital series because patients treated at a tertiary referral center, for example, usually differ from those treated at smaller hospitals or those who do not seek care at all.

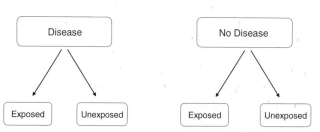

FIGURE 16-1 General strategy of case–control studies. Persons with or without disease are identified at the initiation of the study and information is collected looking backward in time to identify potential exposures.

The second assumption is that controls are representative of the general healthy population who do not develop the disease. A simple, but potentially problematic, approach is to select a random population sample and exclude the rare cases of disease it might include. Because cases are often selected from hospitalized patients, however, with the biases inherent in such patient samples, it may be wise to select controls from patients hospitalized for other conditions. In this way, the biases involved in people being hospitalized (they must have access to care, have survived the initial manifestations of disease, etc.) can be similarly operative in the controls, an example of *compensating bias*.[9] Care must be taken to ensure that the "other conditions" in the controls are not themselves a source of bias in the factors that may be associated with them.

The third assumption is that information is collected from cases and controls in the same way. This can be difficult, particularly if case status is known or obvious to the interviewer. Interviewers, particularly if they are aware of the study's hypothesis, may be more prone to seek exposure information from cases than from controls, so interviewers must be trained to ask questions and follow up positive or negative responses in the same way regardless of case status. Case–control studies that require invasive procedures for diagnosis, such as coronary angiography or tissue biopsy, obviously cannot expose controls to the risk of these procedures unless indicated clinically. Angiography-negative or biopsy-negative controls may solve that problem, but they have the disadvantage of not being representative of all persons without the disease. If the disease of interest is sufficiently rare, it might be safe to assume that a random sample of asymptomatic persons does not include any cases, or it contains so few that they are unlikely to influence the results. Ultimately, it is necessary to utilize only the information collected in identical manner from cases and controls in assessing potential etiologies. Information from diagnostic procedures limited to cases can be used for other investigations (e.g., descriptions of severity) but not for identification of etiologies.

Standard criteria should be used for selecting cases from a well-defined base population. If cases are to have angiography-defined coronary disease, for example, it is important to specify the general parameters of standard angiographic and radiographic technique; the reading methods, number of readers, and degree of agreement expected among them; the minimum degree of stenosis and number of vessels affected; and so on. Sources of cases can be case registries, admission records, pathology logs, laboratory logs, catheterization lists—preferably some common pathway that captures all potential cases. As in all other observational study designs, it is important to have as high a participation rate as possible to minimize biases resulting from nonresponse.

Definition of controls, and selection and recruitment of controls, is generally more difficult than for cases. The ideal control group probably does not exist, making potential biases in the controls one of the most common criticisms of case–control studies. Standard criteria should be used for selecting the controls from the same well-defined population as the cases and for ensuring to the degree possible that they are disease-free. Sources of controls include samples of the general population, such as neighborhood controls selected from the same census tract, telephone exchange, or zip code. Relatives or friends of cases should generally not be used as controls because they tend to be biased by their awareness of the disease, which may cause them to alter their behaviors or recollections. Genetic studies searching for linkage of a disease with measured genetic markers, however, will appropriately include family controls. Use of such controls is best limited to that purpose. Cost and accessibility should be considered in selection of controls because it is generally more difficult to motivate disease-free persons to join a study than those with disease who have a strong interest in determining its cause. Having more than one control per case may offer some improved statistical power, but little additional power is gained beyond having three or four controls per case.

One solution to the lack of a "perfect" control group is to utilize more than one type of control group. One control group might be selected from the same hospital as the cases; another control group might use neighborhood controls, in which each control is matched by neighborhood to a case. This approach is thought to be methodologically superior because the biases in one group may be minimized in the other and vice versa. Associations can be assessed in the two groups separately; often, very few differences are found, which strengthens the conclusions drawn.

Controls may be matched to cases for age, sex, or specific risk factors (e.g., smoking) if these are known to be related to disease and the intent is to identify additional potential etiologic factors. It is important to recognize, however, that once a factor is matched on, it cannot be examined in analysis because, by design, it will be the same in cases and controls. In addition, the difficulty in finding matching controls rapidly escalates with the number of factors matched upon. In general, unless one is certain that a given factor is related to disease etiology, it is probably better not to match on it so that it can be examined in analysis. If more than one control group is used, one group might be matched and another unmatched.

5.2. Observations and Data Analysis

In traditional retrospective case–control studies, data are collected in the present but look backward in time, either relying on recollections or records or assuming that current exposure measures are reflective of those present before the development of disease. This latter assumption is often not met because disease processes alter measures of risk factors (as described previously) that can make interpretation of case–control studies very challenging. As in other observational study designs, the factors to be observed and the conditions during the observation should be specified, using the same methods in cases and controls. The validity and reproducibility of measurement techniques should be established or assessed during the study.

Biases involved in case–control studies, and in many other epidemiologic study designs, have been reviewed in detail by Sackett.[10] The participant selection and recruitment stage is subject to several potential biases. Volunteer bias was discussed previously; persons who volunteer to participate in studies are generally different in important ways from those who do not. This is often referred to as the "healthy volunteer" effect. Prevalence/incidence bias is a particular kind of bias in which a late look at those who are exposed or affected remotely in the past will miss short-duration or fatal episodes. Myocardial infarction studies that include cases whose events were several years before entry to the study, for example, will not include those with very severe disease or those dying early from congestive heart failure or arrhythmias. Transient episodes, mild or silent cases, or cases in which evidence of the exposure disappears with disease onset (as may hypertension with the onset of congestive heart failure) can all contribute to bias in a case–control study. An estimate of the age of onset, or duration or severity of the case, can be helpful in this regard. Membership bias occurs because membership in a group may imply a degree of health that differs systematically from that of the general population. This is a particular problem with employed or migrant populations and is often referred to as the "healthy worker" or "healthy migrant" effect. It can be controlled by taking controls from the same worker or migrant population, but again, the degree to which the study findings may be generalized to the total population may be limited.

Other important biases that can occur in the data collection phase include diagnostic and exposure suspicion bias, recall bias, and family information bias. Diagnostic suspicion bias occurs when knowledge of a subject's prior exposure to a putative cause, such as

hormone replacement therapy, influences both the intensity and the outcome of the diagnostic process, such as screening for endometrial cancer. Exposure suspicion bias occurs when knowledge of a subject's disease status, such as the presence of mesothelioma, influences both the intensity and the outcome of a search for exposure to a putative cause, such as asbestos. It is closely related to recall bias, in which cases may intensively seek to remember any possible exposure that could have caused their illness, whereas controls do not think about these exposures at all. Family information bias occurs when the flow of information about exposures and illnesses within a family is stimulated by, or directed toward, a new case in its midst. This might involve, for example, a rare familial condition that is never mentioned until a family member begins to demonstrate some of the same symptoms.

Findings are presented in a "2 × 2" table with the exposure status in rows and the case–control status in columns (Table 16-3). Cell a represents the number of exposed cases, cell b the number of exposed controls, and the row total, $a + b$, all exposed subjects. Cell c is the number of unexposed cases, d the number of unexposed controls, and $c + d$ all unexposed subjects. Column totals $a + c$ and $b + d$ are the numbers of cases and controls, respectively. Comparison is made between the proportion of cases exposed, $a/(a + c)$, and the proportion of controls exposed, $b/(b + d)$. These two proportions can be compared using a chi-square test with one degree of freedom. Mean levels or distributions of continuous variables can also be compared between cases and controls using Student's t test for normally distributed variables or nonparametric tests for nonnormal variables.

Measures of association between exposures and case status include odds ratios and relative risks. Odds are related to probability (p); odds = $p/(1 - p)$. For example, if the probability of a horse winning a race is 50%, the odds of its winning are 1 to 1. If the probability is 25%, the odds are 1 to 3 for a win or 3 to 1 against a win. If the probability of a diseased person having been exposed is $a/(a + c)$, from Table 16-3, the odds of

TABLE 16-3 Presentation of Findings: The "2 × 2" Table

Exposure	Presence of Disease		Total
	Number with Disease	Number without Disease	
Present	a	b	$a + b$
Absent	c	d	$c + d$
Total	$a + c$	$b + d$	N

exposure are $[a/(a + c)]/[1 - \{a/(a + c)\}]$, which, multiplied by $(a + c)/(a + c)$, equals a/c. Similarly, the odds that a nondiseased person was exposed are b/d. Comparing the odds of exposure in a diseased person with the odds of exposure in a nondiseased person yields the odds ratio $(a/c)/(b/d)$ or ad/bc. The odds ratio is widely used in epidemiologic studies because it is the measure of association estimated in logistic regression methods.

A more familiar measure of association is the relative risk, which is risk in exposed persons $[a/(a + b)]$ divided by risk in the unexposed $[c/(c + d)]$. If the disease under study is rare, a is small compared to b, and c is small compared to d, so a and c contribute little to the denominators of $a/a + b$ and $c/c + d$. As a and c approach zero, $a + b$ approaches b, and $c + d$ approaches d. As the disease becomes increasingly rare, the relative risk approaches $[a/b]/[c/d]$ or ad/bc, the odds ratio. The odds ratio estimates the relative risk well if the disease is rare, but it is always further than the relative risk from unity; that is, it overestimates the magnitude both of harmful associations (relative risk > 1) and protective associations (relative risk < 1). A relative risk or odds ratio equal to 1 means that risk of exposure (or odds of exposure) is the same in those with or without disease; that is, there is no association between disease and exposure.

5.3. Advantages and Disadvantages

The major advantage of a case–control design is that it is likely the only practical way to study the etiology of rare diseases because rare diseases are difficult to study on a population basis. For example, Schlesselman[11] estimated that a cohort study of a condition occurring at a rate of 8 cases per 1000 would require observation of 3889 exposed and 3889 unexposed subjects to detect a potential twofold increase in risk. A case–control study, in contrast, would require only 188 cases and 188 controls. If the prevalence were lower, at 2 cases per 1000, cohorts of approximately 15,700 exposed and 15,700 unexposed subjects would be needed to detect a twofold increased risk, but a case–control study would still require only 188 cases and 188 controls. The tremendous value of a case–control control study in identifying a preventable risk factor in a newly emerging, but very rare, adenocarcinoma in young women is illustrated in the example given in Table 16-2.

A useful characteristic of these studies is that multiple etiologic factors can be studied simultaneously. If the key assumptions of the case–control study are met (cases are representative of all the cases, controls are representative of persons without the disease, and data are collected similarly in cases and controls), the associations and risk estimates are consistent with other types of studies. When case–control estimates are not consistent with those derived from other study designs, it is often because these assumptions have been violated.

Disadvantages of case–control studies are that they do not estimate incidence or prevalence. The denominator, or base population, from which the numerator, or cases, is drawn is often not known, so incidence and prevalence cannot be estimated. Relative risk is only indirectly measured by the odds ratio and may be biased if the disease is not rare. Selection, recall, and other biases may provide potentially spurious evidence of associations. Once associations are found in case–control studies, they must be examined for biologic plausibility in the laboratory and for consistency with estimates from other study designs before causality can be inferred. It is difficult in case–control studies to study exposures that are rare in the overall population unless the sample size is very large or the exposure turns out to be very common in the cases (as in the example of maternal stilbesterol exposure in Table 16-2). Finally, temporal relationships between exposure and disease can be difficult to document. For rare conditions, however, particularly those for which etiologic factors are being sought, case–control studies are the method of choice and in fact may be the only way of assessing potential risk factors.

A special type of case–control design avoids many of the potential pitfalls of classic case–control studies by selecting cases and controls from within a broader population sample established at some time before the onset of disease. This has become a particularly useful design in large-scale prospective studies with the development of effective collection and storage methods for biologic samples. Serum or plasma (or urine, DNA, etc.) can be collected and stored until a sufficient number of cases has accumulated to provide adequate study power. At that time, these baseline samples from the newly occurring cases can be thawed and measured, along with a comparison group of matched (or unmatched) controls, allowing a much more efficient approach to examining expensive or difficult-to-measure risk factors. This "nested case–control design" is used increasingly in large population studies[12,13] and avoids many of the biases involved in selection and data collection in cases and controls after the onset of disease. It has the disadvantage, however, that factors of interest must be able to be measured in stored samples, and that the condition must be common enough for a sufficient number of cases to develop within a reasonable time and in a cohort sample of reasonable size.

6. PROSPECTIVE OR LONGITUDINAL COHORT STUDIES

6.1. Objectives and Design

The object of prospective studies is to make observations about the association between a particular exposure or a risk factor and subsequent development of disease. They are "prospective" in that exposure or risk factor information is collected first and then disease outcomes accrue over time during a period of follow-up evaluation. Exposed and nonexposed groups are then compared for their rates of disease onset. Subjects can be selected for a particular exposure, such as uranium mining, along with a comparable group of nonexposed "controls," and both groups followed forward in time to determine numbers of disease events in exposed and unexposed subjects. More often, a representative sample of a particular geographic area is recruited and examined irrespective of status for a particular exposure, thus allowing many exposures to be studied simultaneously.

As shown in Table 16-1, prospective cohort studies can be divided into those that are concurrent (longitudinal) and those that are nonconcurrent (historical). In a *concurrent* prospective study, the population is defined at the time the study is initiated and followed into the future to determine disease incidence in relation to measured exposures. In a *nonconcurrent* prospective study, exposure information has been collected in a standardized way at some point in the past, and disease status is determined at the time the study is initiated. This is also known as a retrospective cohort study. As shown in the examples in Figure 16-2, a concurrent study would start in the present, with out-

comes ascertained in 2015. A nonconcurrent study, in contrast, might have started when exposure was measured in 1980, with outcomes ascertained in the present.

The nonconcurrent cohort approach can be used when access is available to good records of large population samples, such as military recruits, veterans, or airline pilots, and it is feasible to trace and obtain follow-up information on disease outcomes in these people. It has the advantage of providing long follow-up without waiting for time to pass to obtain disease outcomes. Table 16-2 shows the example of a study that used World War II medical records to identify persons with severe head trauma and nonhead trauma controls and then evaluated survivors 50 years later for dementia.[14] This type of study has the disadvantage of missing short-duration or fatal new cases (unless the past standardized data collection also included some comprehensive follow-up measures) that are typically missed in studies of prevalent cases, thereby introducing a prevalence–incidence bias. Such a bias might lead to identification of risk factors for disease that actually are risk factors for mild disease or better survival with the disease.

The concurrent prospective cohort design is by far the more common, generally because exposure information collected in the past is not sufficient to permit good risk definition in the present. Assumptions for this study design are that the exposed and nonexposed groups under study, or more typically the entire cohort under study, are representative of a well-defined general population. The absence of an exposure should also be well-defined. A traditional assumption of this design that is violated for a variety of risk factors is that exposure history is held constant over time; for example, smoking history as defined at baseline is assumed to be invariant over time. We know that this is unlikely to be true, however, because nonsmokers start smoking, current smokers stop, and past smokers sometimes resume. Such changes in exposure history may be dealt with by techniques of longitudinal data analysis that allow for varying exposure levels measured in the same individual over time.[15]

6.2. Observations and Data Analysis

In regard to observations, definitions of disease outcome should be well determined before the study's inception and held constant during the course of the study. This is often difficult to do in a very long-term study, however, because diagnostic approaches and techniques evolve over time. Criteria for myocardial infarction in 1948, for example, when the Framingham Heart Study began, were very different than they are

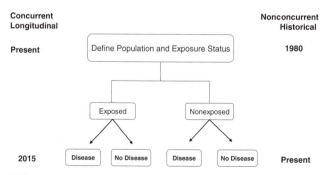

FIGURE 16-2 Concurrent and nonconcurrent prospective study design. In concurrent studies, people with or without exposures are identified at the initiation of the study and information is collected looking forward in time to identify disease outcomes. In nonconcurrent studies, exposure information has already been collected at some point in the past and participants are surveyed in the present to determine the presence or absence of disease.

now. The addition of cranial imaging has revolutionized the detection of stroke, substantially increasing its incidence and reducing its case fatality owing to increased detection of very mild cases. One approach to dealing with this is to continue using past measurement methods even though newer techniques are available. This may be scientifically unsound, however, if the new techniques are clearly an improvement and may not be possible if older techniques are replaced in clinical practice by subsequent advances in clinical practice. Even if a study chooses to retain older methodology in the examination protocols over which it has control, any follow-up information obtained from hospital records or outside sources of medical care will likely utilize the newer approaches. A solution for this problem is to use both old and new measures for a period to allow an assessment of their comparability and then perform an adjustment of the data over time (if possible) to account for the methodological change. Such comparisons and shifts occurred in methods for detection of myocardial infarction, for example, with the introduction of creatine kinase isoenzymes or, more recently, troponin measurement.[16] If an innovation in measurement is not clearly an improvement over standard methodology, however, one is better off keeping the approach constant; indeed, minor innovations may not be widely adopted and may eventually be abandoned by those who originally developed them.

Outcome variables vary in degree of confidence ("hardness") and reproducibility. Death is considered to be a "hard" outcome because it is an unambiguous state, compared to self-reported health status, which may be more variable. Specific cause of death may be more difficult to ascertain accurately because data are often not collected in a standard manner or are incomplete. Classification becomes increasingly difficult for outcomes without standard definitions or with many competing causes, such as angina or dysplasia.

As in all other study designs, standard criteria should be applied to exposed and nonexposed subjects to avoid bias in determining the outcomes. Definitions of disease should be reliable and reproducible so that, for example, different groups of investigators applying the same criteria should come to the same conclusions when reviewing the same case. Investigators in large-scale studies are well advised to perform reproducibility assessments of their outcome classification methods, particularly for difficult or critical end points, by sending the same case to more than one reviewer or repeating the classification in a blinded fashion some years after the initial classification.

Completeness of follow-up should be maintained at as high a rate as possible. Like initial nonrespon-

dents, persons who drop out of studies differ from those who remain and may introduce important biases. Prospective study findings can be presented in a 2×2 table as described previously, and they have the advantage of providing direct measures of relative risk since disease incidence is measured after exposures are assessed.

Data interpretation involves identification of associations with, or risk factors for, the disease outcomes under study and the larger groups to which the results may be generalized. In addition to statistical significance of a risk association with a particular exposure, the strength of the association (the magnitude of associated relative risk) and the prevalence of the exposure are important in determining the impact of the exposure on overall population risks of disease. The public health impact of a common risk factor that has a modest association with a disease may be larger than the impact of a rare risk factor that has a much stronger association with an outcome.

6.3. Advantages and Disadvantages

Prospective studies have the advantage that cases are incident, rather than prevalent, and may be more representative of all cases than are the prevalent (i.e., more long-standing) cases typically included in case–control studies. The design provides more information about natural history of the disease as well as direct estimates of incidence and relative risk. There are fewer potential sources of bias than in retrospective studies, although biases in participant selection, exposure assessment, and outcome ascertainment are hazards of either approach. In contrast to case–control studies, many disease outcomes can be studied with regard to their relationship to the exposure, but exposures must be identified and measured at the initiation of the study, before occurrence of the outcome. This provides the further advantage of firmly establishing the temporal relationships between exposure and disease. Whereas case–control studies may be the only way to study rare diseases, prospective cohort studies may be the only way to study a rare exposure, with persons with such an exposure gathered together at the beginning of the study.

A disadvantage of the concurrent cohort design is that study duration may be exceedingly long, making difficult the maintenance of consistent study methods and enthusiasm of staff and participants. In addition, follow-up of free-living populations may be expensive because people move or change their contact information and can be difficult to track. Large populations are often required, and rare diseases cannot be studied without extraordinarily large sample sizes.

7. CAUSAL INFERENCE IN OBSERVATIONAL STUDIES

Although observational studies are commonly conducted to identify potential risk factors for further investigation in interventional studies, definitive proof of causal nature (by demonstrating that risk is reduced when the risk factor is modified) may be difficult or impossible and may not be necessary for a number of reasons. Causal inferences can be drawn from solely observational data, however, as was done by Surgeon General Luther Terry in 1964 in implementing the very effective antismoking campaigns of that era.[17]

Epidemiologic evidence that supports drawing causal inferences is listed in Table 16-4. An association must first be statistically significant; that is, the less likely it is that chance is an explanation of the findings, the more likely the association is to be causal. Increasing strength of the association, as measured by relative risks or odds ratios in the 3 to 4 range, for example, rather than the 1.2 to 1.3 range, also increases the likelihood of a causal association. Dose–response relationships, in which higher levels of exposure are associated with higher levels of risk, provide very strong evidence for causal inferences. Smoking is a prime example: Lung cancer studies generally show that heavy smokers are at greater risk than lighter smokers, who in turn are at higher risk than nonsmokers.

The exposure obviously has to precede the development of the disease to be causal. Although several types of study designs, including case–control and cross-sectional surveys, often do not provide information on temporal sequence, prospective studies clearly do. Associations should be consistent internally within a data set—they should apply equally to men and women, to old and young, to employed and unemployed subjects, etc. Associations should also be replicated externally in other population samples. There should be biologic plausibility to the finding, with a reasonable (although possibly unproven) theory of a mechanism for the association. Also adding support to causation is experimental evidence, from laboratory animals or tissue preparations, that introduction of the exposure causes a phenomenon similar or related to the outcome under study.

Causal inferences are thus rarely drawn on the basis of a single study. They rely on a totality of evidence, from the laboratory, clinical realm, and population-based studies, for inferring a causal relationship. The key is that interventional studies are not always required to establish causality definitively. In some instances, the observational evidence may be so overwhelming that intervention studies would be unethical, particularly if the primary hypothesis is one of harm from the intervention, such as trials of the effects of smoking initiation on health.

8. CLINICAL TRIALS

The large majority of observational associations are not conclusive in themselves and require further investigation with interventional study designs such as clinical trials. Randomized large-scale clinical trials provide the strongest evidence both for the causal nature of a modifiable factor and for the effectiveness of modifying that factor on preventing disease outcomes. They should not be undertaken until there is a substantial body of knowledge suggesting that intervention may be effective but not so much evidence that conducting them would be considered unethical. This situation is sometimes referred to as "clinical equipoise." Deciding when to initiate a clinical trial can thus be extremely controversial, and success is often a matter of critical timing.

8.1. Objectives and Design

The object of a clinical trial is to determine whether a particular intervention, which can be a drug, a surgical procedure, a behavioral modification, or some combination of these, is associated with a change in the natural history of the disease, improved response over currently available therapy, or unacceptable side effects compared to other therapy. It is uncommon and arguably unethical to undertake a clinical trial simply to prove harm; if there is a strong body of evidence that one or more potential treatments is harmful, other avenues (e.g., case studies and postmarketing surveillance) should be undertaken to demonstrate that harm without placing new study subjects at risk. More often, there is evidence of both benefit and harm, and the key question is whether the risks outweigh the benefits. This was the case, for example, in the Cardiac Arrhythmia Suppression Trial (CAST), which was halted early due to increased fatality in the active treatment arm.[18]

TABLE 16-4   Evidence in Observational Studies That Supports Causation

1. Statistical significance
2. Strength of the association (odds ratio, relative risk)
3. Dose–response relationships
4. Temporal sequence of exposure and outcome
5. Consistency of the association (internal "validity")
6. Replication of results (external "validity")
7. Biologic plausibility
8. Experimental evidence

In randomized clinical trials, persons from a well-defined study population are assigned to treated or untreated groups at random and then observed for a specified period for the occurrence of well-defined end points. To undertake a clinical trial, there should be a substantial scientific foundation for effectiveness of two or more treatment strategies or of a new strategy compared to "usual care." None of the available strategies should be proven to be superior because withholding a proven treatment from randomized patients would be unethical.

More often, a treatment is known to be superior in one subgroup of patients and of questionable effectiveness in another subgroup. Long after treatment of high blood pressure was accepted as effective in reducing coronary heart disease and stroke in middle age, for example, this treatment was not routinely provided to older adults. At the time, elevated blood pressure in the elderly was thought to be a normal or compensatory phenomenon, and lowering of systolic hypertension in the elderly was viewed as potentially dangerous.[19] In that subgroup, it was thus ethical to randomize patients who, at the time, were not being treated at all (or were treated far less aggressively than younger patients) to receive active therapy or placebo, as was done in the Systolic Hypertension in the Elderly Program (SHEP).[20] This extraordinarily successful clinical trial demonstrated that, indeed, treatment of isolated systolic hypertension substantially reduced nearly every adverse cardiovascular end point examined.[21]

The scientific foundation for comparing two or more treatment strategies includes ecologic studies on populations, observational studies on individuals, and experimental data. Various epidemiologic study designs have been used in assessing the potential impact of dietary factors on diseases, and these findings were used, for example, in planning the Women's Health Initiative.[22] Diet and lifestyle tend to be relatively uniform within a culture, which makes it difficult to examine the impact of variations within a culture; across cultures, however, more striking comparisons may be possible and ecological studies, as noted previously, can point out differences that may be used as the basis of interventions. Experimental or laboratory data should also provide a strong foundation for the effectiveness of intervention, although such studies are often limited by the lack of suitable animal models in which to test a treatment.

8.2. Importance of Randomization

A key component of clinical trials is that they are nearly always randomized; that is, study subjects are assigned to treatment arms at random. Although studies that are not randomized are not by definition invalid, they are much more difficult to interpret and to keep free of bias. Investigators considering conducting nonrandomized trials should consult those with considerable experience in this challenging methodology to ensure that this approach is absolutely necessary and that such a study is designed to maximize its validity.

Randomization is a process for making a selection or treatment assignment in which there is associated, with every legitimate outcome in the assignment process, a known probability. Most commonly this means that every study subject has the same chance as every other study subject (within the same randomization scheme) of being assigned to a particular treatment.

The simplest and most common design is a two-armed trial, with half the subjects randomly assigned to receive one treatment and the other half to a second treatment or placebo. Every subject thus has an equal chance of being assigned to receive either treatment. More complex designs may involve the comparison of more than one "new" treatment to conventional therapy or placebo. In such cases, the size of the comparison group must be increased to provide similar levels of confidence for each of the multiple comparisons that will be made.

The primary reason for random assignment to treatment is to eliminate bias in treatment assignment. Such biases can readily find their way into nonrandomized treatment assignment by well-meaning investigators who may wait, for example, to enter a frail patient into a trial of a highly invasive therapy if they believe that invasive therapy is a likely assignment for that patient. Patients unable to tolerate any of the randomized assignments, of course, should not be entered in such trials; in a well-designed trial, such patients would be excluded on the basis of systematically applied inclusion and exclusion criteria.

Another important reason for random allocation is to ensure that any baseline differences in the treatment groups arise by chance alone. It is certainly possible for baseline differences to arise by chance (the occasional occurrence of unusual distributions is, after all, why people gamble) although its likelihood is diminished with large sample sizes. The University Group Diabetes Program (UGDP), for example, was hampered because the prevalence of cardiovascular disease (CVD) at baseline was higher in one treatment arm than in the others.[23] This group also had higher rates of CVD events during the follow-up period, but their higher baseline risk, rather than their treatment assignment, was suggested by critics as the reason for the observed differences. Although observed baseline differences in randomized groups can be adjusted for, the difficulty arises in baseline differences that are not observed or observable and that may influence the outcome of treatment. Such factors obviously cannot be adjusted for, and randomization must be relied on to ensure that any baseline differences in treatment arms arose by chance and were unrelated to treatment assignment.

Random allocation helps to ensure (although it does not guarantee) comparability of treatment groups not only on known confounders but also on unknown confounders. It also provides study groups with specific statistical properties regarding their baseline composition, which provides a statistical basis for certain tests of significance. Its main utility, however, is that it ensures that baseline characteristics are distributed by chance and are unrelated to treatment assignment.

Hallmarks of sound allocation schemes include (1) reproducibility of the order of allocation, (2) documentation of methods for generating and administering the allocation schedule, (3) features to prevent release of an individual treatment assignment until conditions for entry of that individual into the trial are fully satisfied, (4) masking ("blinding") of the assignment to everyone concerned until needed, (5) inability to predict future assignments from past assignments, and (6) procedures for monitoring departures from established protocols.[24]

Reproducibility of the order of allocation is particularly important in situations in which the integrity of the randomization process may be questioned, as occurred in the UGDP study mentioned previously. The ability of the investigators to reproduce their treatment allocation sequence for scrutiny by others might have allayed a number of concerns raised about that trial. Allocation schemes utilizing random number generators, for example, should specify the algorithm used and the "seed" provided to generate the initial random number. Methods for generating and administering the allocation schedule should be clearly defined and judiciously followed. Protocols should specify how investigators request a treatment assignment, from whom, and after fulfilling what conditions. Is the allocation mailed or transmitted by phone, fax, or modem? How long does the process take, and is it available 24 hours a day? Clear definition of the procedures for randomization, as well as training, certification, and monitoring of study personnel in carrying out these procedures, is essential for ensuring a sound allocation scheme and a bias-free trial.

Sound allocation schemes include methods to prevent release of a treatment assignment until essential conditions for entry into the trial are satisfied. This is critical because once study subjects are randomized into a clinical trial, they must remain in that trial for analysis purposes (even if they must be withdrawn from active treatment) without exception. Exclusion of patients after randomization invalidates the randomization procedure and threatens the validity of the entire trial. Although it is not uncommon for exclusions from data analyses after randomization to be reported in clinical trial results published in high-quality journals, such exclusions should raise serious concerns about the study's conduct, analysis, and interpretation.

An essential condition for entry of a patient into a trial is that the patient be eligible to receive any of the study treatments that he or she might be allocated. Occasionally, conditions arise after randomization that prevent patients from receiving the treatment to which they were allocated, but they still must be included in the analysis by their treatment assignment. This approach is referred to as analysis by "intention to treat" even if treatment was not received and is a hallmark of well-designed clinical trials. Because including patients in analysis who did not actually receive the treatment by necessity weakens the ability to detect differences between randomly assigned groups, it is important to minimize such problems by determining eligibility before a treatment assignment is released. If the trial is not completely "masked" ("blinded"), as occurs in trials with obvious interventions such as surgical therapy or lifestyle modifications, assignment should remain masked to everyone concerned until it is needed. An interesting illustration of this is a possibly apocryphal story of an interventional study in which randomization was to an extensive diagnostic procedure or no procedure. House officers were asked to retrieve randomization envelopes in order from a box, but the envelopes were not identical—those containing an allocation to the diagnostic intervention included many forms that made the envelopes much thicker than the control allocations, which included only a card saying "control." When house officers came across these envelopes in the midst of a busy day or night, there was a tendency for them to avoid "randomizing" a patient if the next envelope was a thick one, or even to take the envelope behind it. More secure ways of conveying treatment allocations at the time of randomization of a patient are certainly possible, and every effort should be made to use them.

Sound allocation schemes also ensure that future assignments cannot be predicted from past assignments. Treatments, especially in an unmasked trial, should not be allocated in an alternating or otherwise defined sequence because each treatment assignment can then be predicted from past ones. To ensure that consecutive runs of numerous patients allocated to one treatment do not arise by chance (which can introduce some important temporal biases into a trial), some allocation schemes are designed in balanced "blocks" of four, six, or eight patients, half of whom receive each treatment. In such cases, particularly when treatment assignment is unmasked, the blocks must also be randomly assigned so that a series of, for example, two assignments to treatment "A" followed by one to treatment "B" cannot then be used to determine that the next treatment assignment will necessarily be to B. An effective approach would be to have some blocks of six, some of four, and some of eight patients so that, on average, allocation was equal, but no single treatment assignment could be predicted from past ones.

Proper implementation of established procedures for treatment allocation should be actively monitored. Frequency of allocation to treatment arm, by investigator, by study center, and by the study as a whole, should be routinely monitored. Evidence of baseline differences in allocated groups, or of frequent withdrawals from active treatment once treatment is assigned, should trigger review of randomization and data collection procedures to ensure that baseline data are completely collected and all eligibility criteria are definitively determined before an assignment is made.

There are few acceptable alternatives to randomization, although many "pseudo-random" schemes do appear in the clinical literature. Nonrandom systematic schemes, such as assignment based on odd versus even numbered admission days or sequentially as patients are entered, are to be avoided because they allow prediction of treatment assignment before eligibility is determined. Determination of treatment assignment before enrollment, as discussed previously, may subtly bias recruitment decisions and approaches in individual patients. Approaches based on odd or even Social Security number or hospital numbers are to be avoided for the same reason.

Common misconceptions about randomization include the perception that a "haphazard" or nonrandom procedure is the same as a truly random one. Another is that randomization ensures comparable study groups—although this is often true, it is not guaranteed, particularly with small study groups. A third is that differences in the baseline composition of the study groups are evidence of a breakdown in the randomization process, but again, this is not necessarily the case. Ability to reproduce the randomization scheme is key in demonstrating that baseline differences arose by chance. A fourth misperception is that it is possible to test for "randomness," but in reality all that can be done is to estimate probabilities that a distribution arose by chance. Finally, many investigators and reviewers believe that a study that does not involve random allocation is invalid. As discussed previously, this is not necessarily so, but such studies are much more difficult to perform and require expert attention to minimizing bias and ensuring their validity.

9. CONDUCT OF EPIDEMIOLOGIC STUDIES: THE STUDY PROTOCOL

9.1. Importance of the Study Protocol

A well-written study protocol is probably the key ingredient in conducting a good epidemiologic study or clinical trial. It undoubtedly is also a key ingredient in doing good laboratory work, but having full documentation of all procedures prior to beginning work that in many cases is exploratory may be impractical in a laboratory. In contrast, in clinical and epidemiologic studies there are often multiple investigators and clinical sites, and lack of standardization of procedures prior to study initiation is a recipe for disaster. Therefore, a protocol, which describes the key elements of the study, and a manual of operations (or manual of procedures), which contains full and detailed documentation of all operational procedures, assessment protocols, and data management procedures, are critical in ensuring a high-quality study.

The study protocol provides a "road map" for the performance of the study so that everyone involved in it, as well as those who will interpret the findings after it is completed, will understand what is being or has been done. In addition, it forces the investigator to anticipate problems before a study begins, when changes in the design can be implemented without impact on the validity of data already collected. It also facilitates communication with potential collaborators, reviewers, or funding agencies. Abbreviated protocols of a few pages can be helpful in this regard and are useful in drafting the consent documents in which the investigators communicate to their most important partners—the study participants—exactly what will be done. A well-written study protocol also assists in manuscript preparation.

Protocols generally have five key components: (1) a background, or rationale; (2) specific objectives, or three-to-five key aims of the study; (3) a concise statement of the design; (4) a summary of the methods, both for conducting the study and for analyzing the data; and (5) a section on the responsibility of investigators, not only for mishaps with patients but also for authorship and presentation (Table 16-5). The first two are relatively self-explanatory and are not unique to

TABLE 16-5   Components of Study Protocol

1. Background and rationale
2. Specific objectives (three-to-five aims of study)
3. Concise statement of design
4. Methods and analysis
     Definition of patient population; inclusion and exclusion
       criteria
     Definition of outcomes; primary vs. secondary outcomes
     Description of treatment
     Recruitment approach
     Data collection
     Informed consent
     Data analysis: primary outcome, associations to be studied,
       techniques to be used
5. Policy on oral or written presentation of results, responsibilities
     of investigators

epidemiologic studies. As in all scientific work, a clear statement of the primary study hypothesis or research question is essential; its logical and precise formulation is probably the single most important step in developing a successful study design.

9.2. Statement of Design

The design statement should describe concisely what the study will do, including a definition of the study participants, the exposure or intervention (if any), and the outcomes. Examples of concise statements of design are "an observational study of decline in pulmonary function in persons living in heavily industrialized areas compared to those living in non-industrialized areas" or "a randomized trial of regional versus general anesthesia on rates of postoperative pneumonia in patients undergoing peripheral vascular grafting."

9.3. Study Sample: Inclusion and Exclusion Criteria

The methods section of a protocol should include a definition of the study population, which should be as specific as possible. Care should be taken to avoid making the definition too restrictive, by requiring difficult to assess inclusion criteria or by having many frequently occurring exclusion criteria. Examples of concise descriptions of the study population are "all patients undergoing radical prostatectomy at the Hammersmith Hospital in 2001" or "a random sample of adults aged 65 and older living in Hagerstown, Maryland." Definitions narrowed to the level of, for example, angiographically defined coronary disease or chronic obstructive pulmonary disease with no other chronic disease limit the potential study population substan-

tially, making recruitment of subjects and generalization of the results difficult. In a case–control study, it is important to describe carefully both the cases and the controls. Following the concise description of the study population should be a list of inclusion and exclusion criteria allowing the investigators to recruit the sample described.

Inclusion criteria should be as specific as necessary to isolate the condition under study, again without unnecessarily constricting the available study population. Participants must have at least one inclusion criterion, but some studies require two or more criteria (age 60 or older and diabetic) or a minimum number of criteria from a larger possible list (e.g., "two of the following: age over 60, diabetes, hypertension, and electrocardiographic abnormalities"). Inclusion criteria generally include the disease or condition under study plus some demographic criteria, such as age, sex, area of residence, and recent hospitalization.

Exclusion criteria are primarily for patient safety and secondarily for avoiding potential confounding factors or missing data. Patients must not have any exclusion criterion; if patients may still be accepted into the study in the presence of one of the stated criteria, it is not an exclusion criterion. A common error early in study development is listing unnecessarily broad exclusion criteria; one should be very selective in identifying exclusions. Some conditions may not be definite exclusions but may raise concerns for the investigators as to whether, for example, lower doses of interventional agents might be needed or more frequent follow-up evaluations for side effects might be indicated. Some of these may be issues that can be dealt with by modifying the study protocol to ensure participant safety without excluding these patients outright, thus enhancing the generalizability of the study. It is also wise to allow some room for clinical judgment (within specific bounds of patient safety and scientific necessity), particularly in a clinical trial in which randomization will minimize biases owing to judgment of individual practitioners. Specification of exclusion criteria is much less of a concern in observational studies owing to the reduced potential impact on participant safety, although the need for clear specification and concerns regarding ability of participants to undergo study procedures obviously pertain.

Exclusion criteria generally involve conditions making the study either difficult or impossible to conduct, such as participants in whom one treatment or another is inappropriate or unethical. The Coronary Artery Surgery Study (CASS), a classic study of medical versus surgical therapy conducted in the late 1970s, excluded patients with left main coronary disease because such patients had been previously proven to benefit from surgery. Randomizing them to receive medical rather than surgical therapy would thus have been unethical. Although inclusion and exclusion criteria are defined at the outset of a study, and with luck remain constant throughout, at times results of other studies definitively answer a question in a subgroup, and that information must then be utilized for the benefit of the study subjects. In the CASS example, had the trial begun before the results on left main coronary artery disease were available and patients with this condition had been randomized to receive medical treatment, they would have been offered surgical treatment but retained in the medical arm for analysis as necessitated by intention to treat analysis. Avoiding such situations is obviously desirable in planning a trial. Information about ongoing trials whose outcome could affect the study in progress should be sought before beginning a study.

Other exclusions are often for logistic reasons, such as excluding protected groups (prisoners and the mentally impaired) or those unable to communicate with the study staff because of language barriers or cognitive impairment. Subjects hospitalized emergently or outside of the working hours of the study staff might need to be excluded, or they might not, but special provisions may be needed in the study protocol to accommodate them. Circumstances making determination of the outcome difficult or impossible should also be considered for exclusion. Studies of electrocardiographic evidence of ischemia often will exclude patients with left bundle branch block or other repolarization abnormalities because these conditions can complicate the measurement of ischemia. Unfortunately, such exclusion criteria can specifically affect a particular segment of the population desired for study.

A study of the effect of an exercise intervention on patients with hypertension, for example, listed as an exclusion criterion the presence of certain nonspecific electrocardiogram abnormalities. Such abnormalities were more common in African American hypertensives than in Caucasians, and application of that criterion was leading to exclusion of a large number of African American participants who were very relevant to the study hypotheses. On further examination, it did not appear that this exclusion criterion was critical as long as patients were carefully monitored during the study and it was removed for both African Americans and Caucasians, leading to greatly facilitated recruitment and generalizability of results without compromising patient safety. As another example, a study of lipid lowering and anticoagulation after coronary artery bypass grafting had specific entry criteria for high-density lipoprotein and low-density lipoprotein

cholesterol that were much less common in women than in men, leading to exclusion of a large number of women who probably could and should have been included. It is thus important to consider the impact of exclusion criteria developed for one group of patients when applied to other subgroups, particularly subgroups of great scientific importance to the study. It is best to make such determinations before initiation of recruitment rather than part way through a study, although review of frequency of inclusion and exclusion criteria in a study encountering problems in recruitment may help to identify criteria that are problematic and might bear revisiting.

9.4. Mistakes Concerning the Study Sample

Common mistakes concerning the study sample, other than unnecessary exclusion criteria or needlessly restrictive inclusion criteria, include making plans for a study without any reliable data on patient availability. Recruitment estimates should not be based on impressions or recollections; if possible, availability should be estimated from the same sampling frame or source as will be used to recruit subjects into the study (e.g., admission logs and catheterization schedules). Other problems include unrealistically optimistic timetables for recruitment or, more commonly, no recruitment goals at all, with plans merely stated to recruit as many as possible in as short a time as possible. It is critical to estimate the number of patients who must be recruited per week or per month to meet overall study goals and then to set a timetable for meeting those goals, allowing an initial period of slow startup. Substantially falling behind on recruitment goals, especially early in a study, should prompt a reevaluation of recruitment strategies as well as inclusion and exclusion criteria. In a multisite study, site-specific recruitment should be followed, and problems in sites that are not meeting recruitment goals addressed, with improvement in these sites potentially coming from applying strategies that are working well for the high recruiting sites.

Another problem is revising sample size calculations to make them consistent with recruitment realities, by increasing the estimated effect size or outcome rate or compromising study power. Event rates and effect sizes should be monitored throughout the recruitment periods by an unblinded monitoring panel to ensure that initial assumptions are being met, but these may be as likely to increase the size of the study sample as reduce it if initial assumptions were overly optimistic.

9.5. Definition of Outcome

Outcome definitions should be as specific and clear as possible. In clinical trials, it is common to define one primary outcome on which sample sizes and recruitment strategies are based and several secondary outcomes that may be of interest but do not represent the primary study question. Although focusing a clinical trial on a single outcome may seem inefficient, definition of a single overriding goal ensures that conflicting results for multiple end points do not muddle the interpretation of an intervention's effectiveness. For an intervention with multiple competing outcomes, such as coronary heart disease and breast cancer after hormone replacement therapy, the primary end point may need to be death from any cause or some other combination to account for multiple effects. Otherwise, one runs the risk of having a significant benefit in outcome A cause premature termination of a trial in which outcome B remains to be definitively tested. Definition of the outcome thus requires careful consideration of competing risks and of potential adverse effects. A trial is then generally sized for that outcome and stopped based only on that outcome, unless some unexpected and severe adverse effect necessitates a premature termination.

Secondary outcomes may be of considerable interest and should be defined, but they are not used in estimating sample size or, unless an effect is very great, in deciding when to end a study. The primary outcome in the Systolic Hypertension in the Elderly Program,[21] for example, was fatal and nonfatal stroke, but there was also great interest in determining whether treatment of isolated systolic hypertension had an impact on coronary heart disease morbidity and mortality, congestive heart failure, or total mortality. These were all defined as secondary outcomes and listed *a priori* to avoid the pitfalls of multiple testing and potentially spurious associations.

Outcome definitions may come from standard clinical definitions; textbooks may be a good source but often they are not specific enough to be useful in a research study. Criteria may be needed for a "definite" outcome, versus one that is "probable" or "possible." Consensus conferences or recognized expert bodies can also be useful in defining an outcome, as in the Walter Reed panel that defined the currently used classification system for AIDS. Definitions can also be drawn from a previous widely recognized study, such as the Framingham definition of congestive heart failure or the SHEP definition of stroke. Changes in diagnostic strategies over time may also need to be taken into account, such as the impact of troponin measurement in defining myocardial infarction and the use of magnetic resonance imaging in detecting stroke, as mentioned previously.

9.6. Definition of Treatment

The intervention or treatment should be specified as clearly as possible without unnecessarily constricting patient management. To the degree that investigators and reviewers can agree that a given treatment regimen is reasonable and appropriate for most patients, it should be specified, as well as the bounds within which good clinical judgment should prevail. Application of the intervention is usually very clearly specified, including schedules and criteria for initiating and increasing therapy, as needed, or decreasing it in the face of adverse effects. Concurrent medications, procedures, or other interventions that are permitted or disallowed during the course of an ongoing clinical trial (e.g., medications that would interact with a study's active treatment) should also be specified in the treatment protocol. Criteria for withdrawal from active treatment or other deviations from treatment protocols should also be specified. Criteria and schedules for drug withdrawal and reintroduction in patients possibly suffering an adverse reaction should be specified. Situations necessitating permanent discontinuation of study drug should also be listed, recalling that patients in whom study drug has been withdrawn must still be included in intention to treat analyses.

9.7. Masking

Masking, sometimes referred to as "blinding," of treatment assignment is utilized to diminish bias among patients and investigators in assessing the effects of an intervention. Masked protocols should clearly specify who is to be masked, why, how, and to what. If masking the treatment assignment is necessary and feasible, it is generally advisable for it to be designed so that as few people as possible know the treatment assignment. If complete masking is not possible, one should at least try to mask patients and those ascertaining the outcome of treatment. Often, this is facilitated by having outcomes ascertained by a blinded subgroup of investigators other than those directly involved in recruiting and managing the interventions in study subjects. If subjects know their treatment assignment, they are told not to reveal it to the blinded assessors.

The effectiveness of masking should be assessed for the patient in single-masked studies and for both patient and investigator in double-masked studies. Patients can be asked what drug they believe they are (or were) receiving and the basis for their beliefs. The effectiveness of the "mask" can be determined by estimating whether they are correct more often than would be expected by chance. Because of the strong nature of the placebo effect, the majority of patients in placebo-controlled trials often believe they are receiving active drug. Although there may be nothing that can be done about patients identifying their treatment, an assessment of the effectiveness of masking can be useful in interpreting study findings.

Criteria for unmasking should also be specified, as well as the specific persons to whom unmasked information will be provided if needed. The need for unmasking often arises in clinical care situations, when a patient is hospitalized acutely or in need of surgical intervention. Physicians are trained to learn everything they can about their patients, particularly what medicines they are taking, and admitting a sick patient on an unknown "study drug" can cause considerable anxiety among care providers. Often, however, the actual treatment assignment will have no impact on treatment plans or will affect them so minimally that treatment can be tailored to accommodate any of the study drug possibilities. Working with a physician to determine the course of action if the patient were assigned to drug X in doses Y or Z, versus drug Q in doses R or S, often demonstrates to them that unmasking really is not necessary. In the interests of patient care, of course, if a physician insists on knowing a treatment assignment, the unblinded assignment should be provided. Efforts should be made, however, to limit the dissemination of that information, particularly to the patient and to the study investigative team. Having the treatment assignment provided by someone other than those involved in study monitoring and outcome assessment is strongly advisable; coordinating center personnel can often fulfill this role well. Contact information on masked study medication bottles should list this central contact if possible rather than one of the investigators directly involved in follow-up of the particular patient.

Observational studies do not involve a randomized treatment assignment, but there is a risk of bias in ascertainment of the outcome if risk factor status is known. Something as apparently straightforward as reading an echocardiogram or cerebral imaging study can be significantly confounded by participant age, in which a finding that would be clearly abnormal in a 45-year-old is viewed clinically as "normal for age" in an 85-year-old. "Aging changes" often are neither normal nor inevitable with age, and research assessments (as opposed to clinical assessments) are much better made in a standardized format without regard to participant age or other characteristics. The best way to ensure this is to mask all extraneous information to those ascertaining the outcome so that the only information provided to them is the minimum essential to providing a standardized assessment.

9.8. Data Collection

Methods for data collection should be as specific as possible, in sufficient detail to allow another investigator to enter into the study or to reproduce it at any time. Details should include the data to be collected and how they are collected, a timetable for follow-up evaluation, specifics of laboratory methods, and so forth, and all these details should be documented in a manual of operations or manual of procedures. For example, the Multiple Risk Factor Intervention Trial investigators agreed that glucose levels were to be monitored but did not specify in the protocol whether this should be plasma glucose or serum glucose. Some centers used one measure and some another, and major differences among centers were noted, which led to the discovery of the lack of standardization. Although this was not a serious problem and was eventually corrected, one would prefer to define these issues at the outset and ensure that all centers are using the same approach. Standardization of laboratory procedures across sites can also be facilitated through the use of a central laboratory.

9.9. Recruitment

Recruitment is one of the most difficult aspects of a clinical study; despite the challenges of defining the question, designing the protocol, obtaining funding, and analyzing the data, the success of recruitment makes or breaks most clinical trials. Several "facts of life" need to be kept in mind.[24] Early estimates of patient availability are almost uniformly unrealistically high. The likelihood of achieving a prestated recruitment goal is small and takes a major effort. Patients presumed eligible for the study during the planning phase can be expected to mysteriously disappear as soon as a study starts. Recruitment can be expected to be more difficult, to cost more, and almost always to take more time than anticipated.

Preparatory steps in recruitment include collecting reliable data to estimate availability of patients. As discussed previously, matching of cases and controls requires a much larger population of potential controls than are expected to be needed to select from that control population the persons who meet all the matching criteria. There should be a general recruitment approach and an outline of steps in the process. Particularly important is identifying some kind of common contact point or "bottleneck" through which eligible persons pass. In hospital studies, this can be the preadmission area, the catheterization laboratory, or some other common point. The approach for identifying potential eligible persons and following them through every step in the recruitment process should be clearly laid out. Contacts necessary for recruitment should be identified, such as the admissions clerical staff, primary care physician, or dietitian.

One of the more common mistakes in recruitment is competing with private physicians for patients. Physicians should not believe that study investigators are going to take away their patient, provide treatment without their knowledge (other than the study intervention), or otherwise interfere with the patient's relationship with the primary physician. This may be changing somewhat in the era of managed care, but likely not for the better. Physicians are now more pressed for time than ever and anything they perceive as likely to complicate their management or increase the time needed with a patient will not be viewed favorably. Physicians can be important "gatekeepers" for participation of their patients, and most patients will consult their physicians if they are considering participating in a long-term study. Critical groundwork should be done with local physicians and medical societies to ensure their support of a proposed study and to make the protocol as unobtrusive to them as possible. Because primary physicians are often relied on to provide follow-up information, involving them in the protocol development process and actively soliciting their interest and participation are important steps in conducting a successful study.

Another common mistake in recruitment is providing basic care rather than referring patients back to their primary care physician. It is very important that study personnel not be viewed by participants as providing primary care, not only because personnel in most studies are not equipped to do so but also because patients deserve to receive care from physicians concerned only with their welfare, not by investigators who also have concerns related to ensuring the success of their study. Although many investigators are able to balance these (at times) competing needs successfully, it is quite difficult to do and is probably best avoided, particularly for junior investigators. However, investigators do have the obligation to report abnormal findings to participants and their physicians (with the participants' consents, of course) and to refer participants back to their physician for follow-up evaluation and treatment. Defining what is "abnormal" in the

context of a research study is often very difficult, especially when assessments are being done that are experimental or not traditionally used clinically. Most protocols include sections on "alert values and referrals" to ensure that specific abnormalities detected in the course of a research study are followed up in the best interests of the patients involved.

Failure to maintain adequate contact with referring physicians is a major error and happens more commonly than one might expect. Courtesy reports of examination findings and contacts (again, with participants' written consents for release of such information to their physicians) go a long way in maintaining the interest of both patient and physician in continued or future participation in research studies.

9.10. Data Analysis

The analysis section of a protocol should define the primary outcome, the key associations to be studied, important confounding factors (also known as *covariates*) to be included in analysis, and the analytic methods to be used. Confounders are factors that are distributed differently in the exposed and unexposed groups and may confuse the relationship between exposure and outcome; they may make an association appear to be present where none exists, mask an association that truly does exist, or change the apparent magnitude or direction of a relationship. To be a confounder, a factor must be related to both the exposure and the outcome. This is a strong reason for assigning treatment at random. Randomization removes any associations between potential confounders and treatment assignment, other than those caused by chance. Data analysis plans should also consider what variables will be used in performing and publishing research on the study. It is very common to collect too much information from participants that ultimately never is used. If one cannot identify a distinct testable hypothesis that will use specific data, it probably is not worth collecting the information. The data analysis section of the protocol should include a brief overview of the statistical methods to be used in data analysis and statistical packages planned for use should be identified.

Protocols should always include policies on oral or written presentation of results and responsibilities of investigators so that everyone involved knows how to go about proposing an article, identifying coauthors, and obtaining clearance for journal submission. Although this may sound very bureaucratic, few problems arise more commonly or cause more ill will than disputes about authorship. Defining policies at the outset, and even mapping out key articles and agreeing on an equitable distribution of first- and co-authorships, is the best approach to avoiding bitter disputes later in a study.

9.11. Protocol Modifications in the Course of a Study

Study protocols should be revised as needed. Drafts, "final" versions, and revisions should be dated on every page, with replacement pages provided for important updates.

Some of the problems that can arise during the course of a study that might necessitate protocol revision were noted previously, such as identification of a subgroup that clearly benefits from one of the randomized study treatments. In addition, measurements can "drift" over time owing to changes in personnel or in laboratory or reading methods, and quality control procedures should be in place to address this. Changes can occur in standards of care in the community, leading to the gradual adoption or abandonment of one of the study treatments. "Drop-ins" to active treatment occurred frequently in SHEP, for example, as the treatment of isolated systolic hypertension became more accepted by the medical community. By the end of that study, approximately half the randomized patients were receiving "open-label" or known treatment to lower their blood pressure while still receiving study drug, and still the study was able to show a result. Problems such as these do arise and not all can be anticipated. It is required to have a monitoring panel, whether it is a group of unblinded investigators for a small clinical trial or a formal outside monitoring group, known as a data safety monitoring board, for a larger trial. This panel is critical for ensuring that the study is adhering to ethical guidelines, making suggestions for protocol changes, and recommending stopping the study if necessary.

9.12. Data Management

Every study participant should have his or her own study record that should be stored in a locked area when not in use. Participant confidentiality is a critical issue and a growing one in the current era of informatics and large linked databases. Each participant should have a unique study number for use as an identifier; participants' names should not be used as identifiers for any study materials, although names are needed for making periodic contacts and obtaining follow-up information. Names should not be in the database or on coding forms. If data are collected at multiple points (various clinic stations or in hospitals or laboratories),

separate forms should be developed, and a system for tracking completion of forms should be implemented.

9.13. Subgroup Analysis

Subgroup analysis is an important aspect of design and analysis of clinical trials and one that can lead to misinterpretation of data. Subgroup analyses are often performed when no overall effect is found for a trial. They can also be used to search for high-risk or unusual groups with a marked treatment effect. It may be possible to identify some subgroups that respond very well to beta blockers, for example, or others that respond very poorly to bronchodilators. Although identification of such subgroups can be useful, one must be careful of "data dredging," or examining many subgroups until a significant effect is found. Such investigations can lead to identification of spurious associations, which is why subgroup analyses should be identified *a priori*, as described previously. Data dredging may become a concern in drug trials whose results have a potentially large impact on the financial interests of the drug manufacturer. An intriguing analysis was performed in CAST in which persons born under a particular astrological sign had a significant, demonstrable benefit of treatment despite an overall detrimental effect of treatment in the study as a whole. Observing many subgroups will often lead to a significant association by chance alone, and such spurious associations are best avoided by defining subgroup analyses at the outset of a trial and ensuring that they have a strong biologic plausibility. Subgrouping variables should be limited to baseline characteristics to ensure that they were not affected by the study treatment. More stringent significance testing ($p < 0.01$ or 0.005 rather than $p < 0.05$) should generally be applied to subgroup analysis, especially if the number of hypotheses tested is large. Findings in *a posteriori* or "data-driven" subgrouping variables should be supported in some other way before they are published, either through replication in other data sets or evidence of compatible findings in published studies, biologic plausibility, or experimental findings. Methods and procedures for conducting subgroup analyses should be reported, and conclusions should be drawn very cautiously.

References

1. Friedman GD. Cigarette smoking and geographic variation in coronary heart disease mortality in the United States. *J Chronic Dis* 1967;20:769–779.
2. Joo JB, Cummings AJ. Acute thoracoabdominal aortic dissection presenting as painless, transient paralysis of the lower extremities: A case report. *Emerg Med* 2000;19:333–337.

3. Centers for Disease Control and Prevention. Pneumocystis pneumonia, Los Angeles. *MMWR* 1981;30:250–252.

4. Klungel OH, *et al.* Control of blood pressure and risk of stroke among pharmacologically treated hypertensive patients. *Stroke* 2000;31:420–424.

5. Hedley AA, *et al.* Prevalence of overweight and obesity among U.S. children, adolescents, and adults, 1999–2002. *JAMA* 2004;291:2847–2850.

6. Lilienfeld AM, Lilienfeld DE. *Foundations of Epidemiology*, 3rd ed. New York, Oxford University Press, 1980.

7. Strong JP, *et al.* Prevalence and extent of atherosclerosis in adolescents and young adults: Implications for prevention from the Pathobiological Determinants of Atherosclerosis in Youth Study. *JAMA* 1999;281:727–735.

8. Newman AB, *et al.* Coronary artery calcification in older adults with minimal clinical or subclinical cardiovascular disease. *J Am Geriatr Soc* 2000;48:256–263.

9. Schlesselman JJ. *Case–Control Studies: Design, Conduct, and Analysis*. New York, Oxford University Press, 1982.

10. Sackett DL. Bias in analytic research. *J Chronic Dis* 1979;2:51–63.

11. Schlesselman JJ. *Case–Control Studies: Design, Conduct, and Analysis*, pp. 17–19. New York, Oxford University Press, 1982.

12. Ridker PM, Hennekens CH, Miletich JP. G20210A mutation in prothrombin gene and risk of myocardial infarction, stroke, and venous thrombosis in a large cohort of U.S. men. *Circulation* 1999;99:999–1004.

13. Roest M, *et al.* Heterozygosity for a hereditary hemochromatosis gene is associated with cardiovascular death in women. *Circulation* 1999;100:268–273.

14. Plassman BL, *et al.* Documented head injury in early adulthood and risk of Alzheimer's disease and other dementias. *Neurology* 2000;55:1158–1166.

15. Zeger SL, Liang KY, Albert PS. Models for longitudinal data: A generalized estimating equation approach. *Biometrics* 1988;44:1049–1060.

16. Laurion JP. Troponin I: An update on clinical utility and method standardization. *Ann Clin Lab Sci* 2000;30:412–421.

17. U.S. Department of Health, Education, and Welfare. *Smoking and Health: Report of the Advisory Committee to the Surgeon General*. Washington, DC, Public Health Service, 1964.

18. Echt DS, *et al.* Mortality and morbidity in patients receiving encainide, flecainide, or placebo. The Cardiac Arrhythmia Suppression Trial. *N Engl J Med* 1991;324:781–788.

19. Kannel WB. Clinical misconceptions dispelled by epidemiological research. *Circulation* 1995;92:3350–3360.

20. The Systolic Hypertension in the Elderly Program (SHEP) Cooperative Research Group. Rationale and design of a randomized clinical trial on prevention of stroke in isolated systolic hypertension. *J Clin Epidemiol* 1988;41:1197–1208.

21. The Systolic Hypertension in the Elderly Program (SHEP) Cooperative Research Group. Prevention of stroke by anti-hypertensive drug treatment in older persons with isolated systolic hypertension. Final results of the Systolic Hypertension in the Elderly Program (SHEP). *JAMA* 1991;265:3255–3264.

22. Prentice RL, Sheppard L. Dietary fat and cancer: Consistency of the epidemiologic data, and disease prevention that may follow from a practical reduction in fat consumption. *Cancer Causes Control* 1990;l:81–97.

23. Prout TE, *et al.* The UGDP controversy. Clinical trials versus clinical impressions. *Diabetes* 1972;21:1035–1040.

24. Meinert CL. *Clinical Trials: Design, Conduct and Analysis*. New York, Oxford University Press, 1986.

17

Small Clinical Trials

MITCHELL B. MAX

Pain and Neurosensory Mechanism Program, National Institute of Dental and Craniofacial Research, National Institutes of Health, Bethesda, Maryland

1. INTRODUCTION

During my years as a student and resident in internal medicine and neurology in the 1970s, bench-oriented mechanistic research held all the glamour. Clinical trials appeared to be a rather dull final step in the medical discovery process. Whereas new basic research methodologies were emerging every month, the methods of clinical trials seemed to be centuries old, perhaps dating to Francis Bacon's writings about controlled experiments or to British naval physicians' trials of scurvy cures. I assumed the secrets of clinical trial methods were all recorded in large dusty tomes written by statisticians, and that I had about the same chance of saying something fresh about clinical trial methods as I did about Exodus.

After 20 years of carrying out clinical trials in the treatment of chronic pain, along with colleagues in basic neuroscience, my view of the state of clinical trial methodology has changed. The main point that I hope to convey in this chapter is that there are great opportunities for aspiring clinical researchers in each disease area to become innovators in clinical trial methods. Clinical trials, which were rarely performed before World War II, are still a relatively new tool of medicine. The major opportunity for innovation is that the standard methods laid down in the 1950s predate the dissection of virtually every human disease into mechanistic subgroups and are quite inefficient in telling us which patient should get which treatment. Those who can begin to solve these problems will have a fascinating time and be in great demand.

In addition to this attempt to entice a few readers into full-time careers as clinical trialists, the chapter will explore several other ideas that have been consistently useful to myself and my research fellows in designing trials, including the difference between explanatory and pragmatic orientation in clinical trial design, approaches to studying groups of patients with heterogeneous disease mechanisms, the importance of placebo responses, and the concept of "assay sensitivity" in clinical trials. I will make no attempt to cover all of the major technical issues of clinical trial design. The latter task requires a small book of its own. Such books exist and are essential reading for any aspiring clinical trials specialist.[1] I also will assume a background in elementary biostatistics. Finally, my hands-on experience has been limited to clinical trials of pain treatments, so many of my examples will be drawn from that field. However, if the reader tests my claims against the clinical trial literature in his or her disease of interest, I predict that he or she will find the challenges and opportunities to be similar.

2. WHERE WE ARE IN THE SHORT HISTORY OF CONTROLLED CLINICAL TRIALS

The following historical exercise may be a morale booster for any clinical research fellow, as it was for me a decade ago while I was writing a review of analgesic trials methods.[2] In the stacks of any medical library, select a clinical research journal in your disease of interest and trace back the controlled trials through the 1960s, 1950s, and late 1940s. It will be easy to find trials in the later decades, and as you precess to the 1950s, you will encounter review articles spreading the new gospel of the controlled trial. As you travel further back in time to 1948 or 1949, controlled trials vanish. Instead, there are just a series of open-label observations or statements of opinion about therapies. Histories of controlled clinical trials[3,4] point out that before the British Medical Research Council's 1948 study of streptomycin in tuberculosis, controlled trials such as Lind's study of limes in scurvy or Louis' study of bloodletting in pneumonia were the exception. Medical therapeutics was dominated by the opinion of authorities, professors who were reluctant to bend their views of diagnosis or treatment to those of their rivals, as would be required in multicenter trials.

As the last step in this library exercise, read in detail some of the reviews of clinical trial methods from the 1950s.[5,6] Compare the methods described in them to current clinical trial practice. I found it striking that our current clinical trial methods are well described in these 50-year-old reviews.

This exercise was heartening to me in several respects. First, I realized that modern clinical trial methods had just been made up approximately 50 years ago by thoughtful but not unapproachably brilliant clinicians. One can still talk with some of them at scientific meetings, and they will encourage you to improve on their methods. The second realization is that these methods were based on the assumption that one is treating a rather uniform disease such as the infectious processes that were the targets of many of the early randomized trials. The underlying statistical methods came largely from studies of gambling and agriculture, where dice are dice and wheat is wheat. In contrast, our current conversations with basic scientists focus on the many mechanisms that produce similar disease phenotypes, and we seek to develop superselective treatments aimed at subduing one disease mechanism without side effects. As I will elaborate later in the chapter, the standard parallel group clinical trial is rather clumsy and inefficient in detecting treatment responses in an otherwise unidentifiable small subset of patients with a particular disease mechanism.

These historical considerations, and my experience doing clinical trials, made it clear to me more than a decade ago that a second generation of clinical trial methods incorporating considerations of mechanism is long overdue. The opportunities for clinical innovation have grown since then because of the declining numbers of new clinical investigators and the explosion in mechanistic hypotheses emerging from basic science laboratories.

3. EXPLANATORY VERSUS PRAGMATIC ORIENTATION IN CLINICAL TRIALS: IMPLICATIONS FOR STUDY DESIGN

Schwartz and Lellouch,[7] who characterized two different purposes of clinical trials that they called "explanatory" and "pragmatic," articulated one of the most useful distinctions for the design of clinical trials of all types. An explanatory approach seeks to elucidate a biological principle. The study population is considered to be a model from which one may learn principles of pharmacology or physiology—principles that are likely to shed light on a variety of clinical problems. A pragmatic approach, in contrast, focuses on the question, "What is the better treatment in the particular clinical circumstances of the patients in the study?"

As an illustration of how these approaches to design differ (Table 17-1), consider a hypothetical analgesic that animal studies had shown to be effective in models of visceral pain:

Main question: The explanatory researcher is interested in the question, "How is visceral pain processed in the human central nervous system?" The pragmatic researcher might be asking, "In everyday practice, what is likely to be the best treatment for pain caused by malignant tumors involving the abdominal viscera?"

Patient choice: The explanatory researcher might select only a small subset of cancer patients in whom there was unequivocal radiological proof of hollow viscus involvement and no other lesions that might be causing pain, whereas the pragmatic researcher might open the study to patients with abdominal pain in the presence of lesions of the hollow viscus, retroperitoneum, and/or spine, where it was not entirely clear how much of the apparent visceral pain was referred from another site.

Treatment selectivity: An explanatory approach would use a treatment with a specific receptor target or perhaps a localized injection into a specific nervous system site—for example, an intraspinal injection of a mu or delta opioid analgesic agonist drug. A pragmatic approach seeks the clinical favorite, even if the treatment hits many receptors at many sites—for example, a combination of an anti-inflammatory medication with methadone, an opioid that also blocks N-methyl-D-aspartate (NMDA) glutamate receptors.

Dose: The explanatory investigator tries to maximize the therapeutic response by selecting a high dose and monitoring patients frequently. In many cases,

TABLE 17-1 "Explanatory" versus "Pragmatic" Orientations of Clinical Trials: Effect on Design Choices in Hypothetical Visceral Cancer Pain Trial

| Design Issue | Orientation of Clinical Trial | |
	Explanatory	Pragmatic
Main question	How is visceral pain processed in the CNS?	What is the best treatment in clinical practice?
Patient choice	Selective:	Inclusive:
	Diagnostic imaging shows only visceral lesions.	Visceral lesions most prominent of lesions likely to be causing pain.
Treatments	Pharmacologically specific: e.g., muß vs. delta opioid receptor agonists, given spinally	Clinical favorites, including combinations or "dirty" drugs: e.g., oral methadone + NSAID
Dose	High; often fixed	Titrate as in clinic
Treatment supervision	Optimal	As in clinical practice
Controls	Placebo	Other active medications
Analysis	Completers	Intent-to-treat

CNS, central nervous system; NSAID, nonsteroidal anti-inflammatory drug.

patients are individually titrated to the maximum dose tolerated. In contrast, the pragmatic investigator might choose an intermediate dose and provide the looser supervision common in clinical practice. Neither of these approaches gives very good information about the optimal dose to use in practice. For this, one needs a prospective dose–response study, in which one randomly assigns patients among a variety of doses. In studies in which patients are individually titrated to effect, one can analyze outcomes at a variety of dose levels as one proceeds, but this method tends to err on the high side; for example, investigators overestimated the optimum dose for several antihypertensives by a factor of 10.[8]

Treatment supervision: In an explanatory study, this will be rigorous in order to make possible the intensive and precise treatment that will optimally test the hypothesis. In a pragmatic study, supervision between clinic visits will mimic that in everyday practice—that is, minimal in most cases.

Control groups: An explanatory approach will usually mandate a placebo because even small amounts of pain relief over the placebo response may provide information about the mechanisms of visceral pain transmission and relief. A pragmatic approach, in contrast, generally compares the new treatment to the best treatment in clinical use. Placebo comparisons may still be desirable in such studies, particularly when there is no significant difference between the study drug and standard control, but detection of a small therapeutic effect is of less interest.

Data analysis: In an explanatory trial, a few patients who discontinue the study medication after the first dose because of unpleasant side effects would provide no data about the biological effects of repeated dosing and are therefore excluded from the main analysis; all patients should be analyzed in a secondary analysis, however, because some reviewers might be interested in this result. In a pragmatically oriented trial, the primary analysis should be an "intent-to-treat" analysis, including either all patients who were randomized or all patients who received at least one dose, because treatment failures due to side effects will weigh into the clinician's choice of treatment.

The dichotomous explanatory/pragmatic schema is an oversimplification, of course. The investigator usually wishes to address both theoretical and practical concerns. This distinction may, however, offer a useful perspective for making design choices in complex cases.

4. ISSUES IN SMALL CLINICAL TRIALS THAT EXAMINE BOTH DISEASE MECHANISM AND TREATMENT EFFICACY

Consider the challenges usually facing the investigator designing a small (e.g., 100 patients or fewer) single-center clinical trial. Typically, this would be one of the first studies of a new treatment's effects in a disease condition, often termed a phase II trial. The orientation is almost certainly explanatory—the main purpose is to gather insights about pathophysiology and infer principles of treatment.

The investigator's main challenge is that many different mechanisms can generate the dysfunction found within most diagnostic categories, whether one is dealing with cancer, heart failure, arthritis, depression, epilepsy, or chronic pain. A major thrust of clinical research is to identify disease mechanisms in individual patients, whether by genotyping of the patient or the affected organ, performing functional imaging, or performing electrophysiological or biochemical tests. If one can use these tests to fill the trial with patients with the same disease mechanisms, this meets the explanatory ideal, but this is rarely the case. More commonly, tests to distinguish mechanisms are works in progress, and one hopes to use drug responses to shed additional light on these possible mechanistic distinctions. Many new treatments are targeted to one particular mechanism, which may exist in only a modest proportion of the enrolled patients. Even if the treatment benefits certain patients without a hint of toxicity, a study of a mixed group may lack power to statistically detect an effect because the success in responders is averaged with the lack of effect in those with other disease mechanisms.

4.1. Correlate Intensive with Simple Assessments of Disease Mechanisms

Given that we are firing our magic bullet into a barrel of mixed mechanisms, how can we proceed in a way most likely to make sense of the results? First, one should apply as much of the latest technology as one can afford for discerning mechanisms to every patient in the trial. However, one should also bear in mind that later investigators trying to replicate this result in a larger group or clinicians applying the results in practice will not be able to do a positron emission tomography scan, for example, on every patient. Therefore, one should also include some easily measured variables that may correlate with disease mechanism. In chronic pain studies, for example, one might prospectively assess the response of pain components with different evoking stimuli, qualities, or temporal profile.[9] Many of these assessments require no more than paper and pencil.

4.2. Maximize Treatment Effect and Minimize Variance

The other principles of optimizing study design will become clear if we study the standard formula for clinical trial sample size:

$$N = [\sigma^2 f(\alpha,\beta)]/\Delta^2,$$

where N is the number of patients in each treatment group, σ is the standard deviation of the primary outcome measure, $f(\alpha,\beta)$ is a function of the alpha and beta error one is willing to accept, and Δ is the differ-

ence between treatment effects that one wishes to be able to detect.

Alpha is conventionally chosen as 5% and beta as 10 or 20%. N has practical upper limits. I have found that when I require a fellow to enroll more than 60 patients in a clinical trial, we run out of money several times before study completion and my fellow rarely speaks to me again. The equation shows that there are only two moves one can make to keep the N within manageable limits. The first is to increase the expected treatment effect. The other move is to decrease σ^2, the experimental variance.

4.2.1. Increasing the Treatment Effect

Note in the previous formula that sample size increases exponentially as one seeks to increase discriminating power. That is, to halve the size of a treatment difference one can detect, one must quadruple the N. As described in the discussion of explanatory trials, the investigator may choose to maximize the dose or intensity of treatment or optimize the choice of patients with susceptible mechanisms. The investigator should bear in mind this steep N vs. $1/\Delta$ relationship when choosing the main question for the study. The easiest effect to detect reliably is that between an effective drug and placebo. If one compares two effective drugs, the difference is most commonly one-fourth to one-half of the drug–placebo difference, therefore requiring 4–16 times the sample size. A popular question raised in brainstorming sessions of pain scientists is whether groups of chronic pain patients with differing clinical features (e.g., definite neuropathic pain vs. others) differ in response to a given drug, or whether one class of analgesic is better than another in a given group. Not surprisingly, these questions are almost impossible to answer in single-center studies without the use of crossover designs, pharmacokinetic tailoring, or other special maneuvers because the differences in effects between two proven treatments are usually small.[10] On the other hand, contemplation of the sample size formula may encourage single-center researchers to do drug combination studies because combining treatments of different mechanisms may increase the treatment effect beyond that of the individual drugs.[11,12]

4.2.2. Decreasing the Variance

The Duchess of Windsor once said that one could never be too thin or too rich. In the same vein, one can never remove too much of the variance in an explanatory trial. One can use these efficiencies to reduce the sample size or to detect smaller treatment differences between patient subgroups that will aid inferences about mechanisms. There are several approaches to decreasing clinical trial variance.

4.2.2.1. Decrease Pharmacokinetic Variability

If one gives a group of patients a fixed dose of a drug, variations in distribution and metabolism will cause plasma concentrations to vary severalfold. Because of additional variations in the link between plasma concentration, concentration at the site of action, and physiological effect, the variation in effect will be even wider. Many patients will have insufficient drug to benefit, and the resulting smaller mean effect and greater variability will deliver a double blow to the power of the study.

Some clinical pharmacologists[13] have advocated concentration-controlled clinical trials. One may either calculate the patient's pharmacokinetic parameters after a single dose of the drug and use these values to design repeated dose regimens or measure plasma concentrations during repeated dosing and use those to adjust the dose. These designs are particularly attractive for drugs that may have delayed but life-threatening toxicities, such as many cancer chemotherapies,[14] immunosuppressants,[15] and anti-infective agents.[16]

Concentration-controlled designs are ideal for studies of drug combinations. An example is a study in which Coda et al.[17] showed that the monoamine releasing drug fenfluramine enhances morphine analgesia. Figure 17-1 shows that after determining subjects' morphine kinetics after a single-dose infusion, the investigators programmed a computer-controlled pump to precisely maintain each of three morphine concentrations for 45 minutes while the subjects reported on the intensity of a standard set of painful electrical stimuli. Subjects took part in 4 days of testing to complete a "2 × 2" factorial design. On 2 of the days they received morphine and on the other 2 saline, and in each of these pairs they were randomly assigned to receive either fenfluramine or placebo (Fig. 17-1, bottom). With a sample size of only 10, Coda et al. were able to show that the addition of fenfluramine doubled the effect of a given dose of morphine.

Such computer-controlled infusions also would be useful during functional imaging studies that aim to explore the effect of a drug on physiological function. These experiments often require multiple types of control observations over time to explore, for example, physiological function with rest and several types of stimulation, with and without drug. In such studies, investigators usually give a bolus of intravenous drug, which may give differing tissue concentrations during the subsequent observations.[18] Rapid attainment and

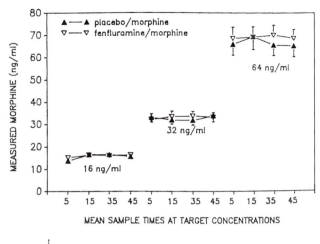

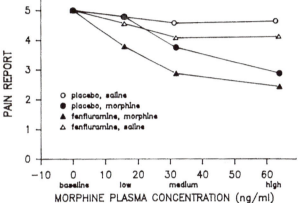

FIGURE 17-1 Reducing clinical trial variability related to pharmacokinetics.[17] (*Top*) Accuracy of tailored morphine infusions, designed from individuals' previously measured kinetic parameters after a morphine bolus. Mean plasma morphine concentrations in 10 subjects are plotted against time. Every 45 minutes, the pump was reset to double the plasma morphine concentration. Vertical bars indicate standard errors. (*Bottom*) Analgesic effects of placebo fenfluramine, morphine, and morphine plus fenfluramine on ratings of electrical tooth pulp stimuli in normal subjects. Subjects participated in four testing sessions and were randomly assigned to one of the treatments in each. During each session, subjects rated the experimental pain stimuli before morphine (or saline) infusion and again at morphine (or saline) infusion rates targeted to produce morphine plasma concentrations of 16, 32, or 64 ng/ml. The addition of fenfluramine to morphine approximately doubled analgesic potency relative to morphine alone.

maintenance of a steady-state plasma drug concentration would considerably decrease variability and improve the controlled comparisons.

Some pharmacologists have pointed out limitations of the concentration-controlled trial in repeated-dose clinical trials. Determination of drug concentrations and dose adjustment may be cumbersome for outpatients. Because of variations in the link between drug concentration and effect among patients, some patients may still have little response. Ebling and Levy[19] suggested that one can get more information by varying the concentration during each patient's treatment until one attains two predetermined levels of effect. Alternatively, if toxicities are reversible and immediately apparent, one can dispense with blood concentration measurement and optimize response by using the simple "sledgehammer principle" discussed in the section on explanatory studies—increase the dose to the highest level the patient can tolerate. Sheiner and Steimer[20] reviewed additional ways in which pharmacokinetic considerations can improve clinical trials.

4.2.2.2. Decrease the Variability in Measurement of the Primary Outcome Variable

Decreasing the variability in measuring the primary outcome is often a powerful and inexpensive way to stimulate the pace of therapeutic advance in an entire field, yet this issue sometimes escapes scrutiny. I use measurement of chronic pain as an example. In a MEDLINE search of clinical trials of treatments for osteoarthritis and rheumatoid arthritis published between 1991 and 1993, I found that 21/23 trials defined the pain outcome as single subjective pain rating at the end of the treatment period. This is likely to be a very inefficient method for assessing the effects of a treatment for arthritis because most kinds of pain fluctuate considerably over time (Fig. 17-2).[21] Measurement of pain at a single time point is an inefficient estimator of the average level of the symptom. Jensen and McFarland[22] studied 200 patients with mixed chronic pain syndromes and found that because of day-to-day fluctuations in pain, a single rating of pain correlated only modestly with a "gold standard" for the week's pain—an average of over 100 hourly ratings. This correlation coefficient of 0.74 improved to 0.96–0.98 if twice daily ratings for 1 week were used to estimate the true average. In a clinical trial, a similar improvement in estimation of an actual treatment effect would permit a reduction in the sample size by approximately half.

If the major clinical trial outcomes in your disease of interest are associated with substantial variation in measurement, I urge you to examine the literature to determine if methods for minimizing variance in clinical trials have been rigorously examined. In the pain field, it was remarkable that the work of hundreds of research psychologists had rarely addressed the practical question of "Which scale allows the maximum power or minimum sample size in a clinical trial?" Bellamy et al.[23,24] illustrated how multicenter clinical trials groups can systematically approach the comparison of outcome measures.

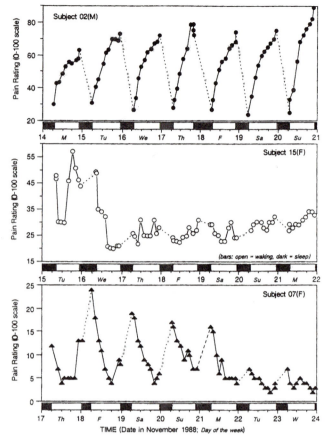

FIGURE 17-2 Variation of pain over time in three patients with osteoarthritis of the knee.[21] Note the considerable variation within and between days. If the outcome of a clinical trial were pain measured at a single arbitrarily chosen time point, the variation in pain might statistically overwhelm a worthwhile treatment effect. Data from other chronic pain populations[22] have suggested that taking an average of 7–14 pain measurements over 1 week might prevent most of this loss of power.

4.2.2.3. Dissect Out Covariates with Significant Effects on the Primary Outcome

In the sample size formula, only unexplained sources of variation contribute to the variance, σ^2. If one studies a disorder well enough to know the predictors of outcome, one can remove those components from this unexplained error term. For example, Jung et al.[25] reported that rash duration, age, sex, the presence of a prodrome, and the severity of pain and of the acute rash explain 23% of the variance in the occurrence of postherpetic neuralgia. Since the sample size formula shows N to be proportional to the variance, this additional knowledge would reduce the size of a study aimed at preventing postherpetic neuralgia by approximately 23%. Covariates such as surgical trauma and prior opioid exposure have been reported to improve the sensitivity of analgesic clinical trials.[26]

Assessment of genetic polymorphisms that affect individuals' treatment response offer promise in explaining part of the outcome variance in many disease areas.[27,28]

4.2.2.4. Use Crossover Designs Wherever Possible

In a parallel group (also termed "completely randomized") design, each patient receives a single treatment. In a crossover design, each patient receives some (incomplete block) or all (complete block) of the treatments being studied.

Crossover Designs. In diseases with outcomes that revert to baseline after treatment is stopped, there are several obvious advantages to crossover designs. Clinical trials that use subjective outcomes often require large sample sizes because detection of a drug effect must compete with many other causes of variation in outcome: the nature of the lesion causing the patient's disease, his or her psychological makeup, interaction with the study personnel, etc. Much of this between-patient variation can be eliminated by using a crossover design, in which treatment comparisons are largely or entirely within the same patient.[29–32] Because of this reduction in variance, and because each patient is used several times, crossover studies often have greater statistical power than parallel group designs that include 5 or 10 times the number of patients.[29] This is an important practical advantage, particularly when studies are performed in a single center. Despite the previous warnings about the difficulty of detecting differences between two active drugs, Raja et al.[33] were able to show that opioids were superior to tricyclic antidepressants and placebo in postherpetic neuralgia by entering 74 patients into a three-period crossover study. A parallel group study of similar N would have failed to show the treatment difference. Crossover designs are particularly suited to episodic conditions such as dysmenorrhea or migraine.[34]

Such advantages notwithstanding, there may be problems with the use of crossover designs. First, if treatment-induced changes in the major outcome are not quickly reversed when treatment is withdrawn, crossover designs are inappropriate. Examples include successful treatment of major depression, relief of pain by permanent nerve block, or remission of cancer or infection. Next, change in underlying disease over time may introduce great variability into patient responses, thereby undermining the major potential advantage of the crossover design. This necessitates that the total duration of the crossover study be short enough to ensure that such within-patient variation will be less than the variation already existing between the patients enrolled. Because of the added length of

crossover studies, changes in the underlying disease as well as logistical factors and voluntary withdrawals usually result in a higher dropout rate than in parallel group studies. Although the greater power of the crossover approach may compensate for a higher dropout rate, reviewers may doubt the general applicability of the results of a study completed by a minority of the patients entered. Experience with one or two crossover studies in the population of interest will predict whether a crossover design will improve efficiency and suggest the optimal length.

In the past two decades, the major concern with crossover studies has been the possibility of bias produced by unequal "carryover effects." Carryover effects are changes in the efficacy of treatments resulting from treatments given in earlier periods; they may be mediated by persistence of drug or metabolites, changes in brain or peripheral tissues caused by the treatment, or behavioral or psychological factors. Statisticians have most energetically attacked the two-treatment, two-period design (2×2; Fig. 17-3). Critics claim that results may be difficult to interpret whenever the treatment effect differs for the two periods. In this event, one cannot distinguish with any certainty whether this is due to

a carryover effect (persistence of a pharmacological or psychological effect of the first treatment into the second period);

a "treatment × period interaction" (the passage of time affects the relative efficacy of the treatments; e.g., by the second period, patients who initially received placebo might be too discouraged to respond to any subsequent treatment); or

a difference between the groups of patients assigned the two different orders of treatment.

Because of these concerns, regulatory agencies have been particularly reluctant to rely on data from such designs.

Fortunately, these statistical difficulties are largely limited to the 2×2 case (and Senn[32] argues that these

Standard 2 x 2	Alternative 1	Alternative 2	Alternative 3
A–B	A–B	A–B–B	A–B–B
B–A	B–A	B–A–A	B–A–A
	A–A		A–B–A
	B–B		B–A–B

FIGURE 17-3 Examples of crossover designs used to compare two treatments, A and B. Many statisticians have criticized the two-period, two-treatment design (*left*) for insensitivity in detecting carryover effects. (*Right*) These three designs are examples of alternative designs that are better able to distinguish treatment from carryover effects.

difficulties have been exaggerated). If the investigator adds several other treatment sequences (Fig. 17-3, alternative 1) or a third treatment period (Fig. 17-3, alternatives 2 and 3), unbiased estimates of treatment effects are possible even in the presence of various types of carryover effects,[30,31] although the statistical analysis becomes quite complicated for some designs. For studies involving three or more treatments, there are a variety of designs that allow these effects to be distinguished.

My current view is that the relative brevity, simplicity, and superior power of the 2×2 design makes it attractive for single-center studies in situations in which previous experience suggests that there is no significant carryover effect. After trying some of the alternative designs for two treatment studies, I have returned to using the 2×2 design. If one is doing studies for regulatory review, one may wish to seek expert advice about the regulators' current statistical thinking. Statisticians who were trained during the decades when crossover studies were scorned may argue for the easier parallel group design, but given the great historical success of crossover designs in analgesic studies, investigators should know the issues and be prepared to dig in their heels.

Enriched Enrollment Designs. A variant of the crossover design, the "enriched enrollment" design, may be useful in studying treatments to which only a minority of patients respond.[35] If the results are not statistically significant in a conventional clinical trial, one cannot retrospectively point at the responders and claim that the treatment accounted for their relief. One can, however, enter responders into a second prospective comparison or, for extra power, a series of comparisons between treatment and placebo (Fig. 17-4). If the results of the second trial considered alone are statistically significant, this suggests that the patients' initial response was not just due to chance. Although statistically defensible, enriched enrollment designs are open to the criticism that prior exposure to the treatment may defeat the double-blind procedure (particularly with treatments that have distinctive side effects) and sometimes result in spurious positive results. Leber and Davis[36] argue that unblinding effects and other biases accounted for much of the treatment effect that led to Food and Drug Administration (FDA) approval of tacrine in Alzheimer's disease. Another caveat is that positive results from an enriched population of drug responders can no longer be generalized to the entire patient population—they just suggest that a subpopulation of responders exists. However, these concerns may be overstated, argues the FDA's Robert Temple,[37] who believes that enriched enroll-

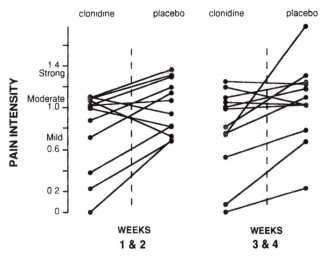

FIGURE 17-4 Enriched enrollment design in a study of the pain-relieving effects of transdermal clonidine in diabetic neuropathy.[35] In an initial crossover trial of 40 patients, the mean difference between drug and placebo treatments was not significant. Twelve of those 40 patients who got more relief with drug than placebo were treated with two subsequent pairs of drug and placebo treatment, each drug being given for 1 week. In the graph, each line represents a patient. Although the treatment order within each treatment pair was randomly assigned, the results are presented in a uniform order for clarity. The results confirm the responsiveness of most of the 12 patients to clonidine ($p = 0.01$) and suggest that a subset of patients have a distinct set of pain mechanisms responsive to adrenergic agonists.

ment studies may play a valuable role in drug development:

> The first task for some agents is to find any group in which the drug can be shown to work. Exactly whom it works in and how to select the patients it is most likely to work in are important, but are refinements that follow demonstration that it works at all.

Other types of multiple crossover studies may be used to get the most information in an initial study of a treatment.[37] For example, in a two-stage design, Sang et al.[38] first entered patients with diabetic neuropathy pain into a placebo-controlled crossover study of dextromethorphan vs. placebo in which drug doses were individually maximized. Apparent responders were offered enrollment in a prospective randomized double-blind dose–response study in which they received either placebo or 25, 50, or 100% of the previous maximum dose. This stratagem yielded both evidence of drug efficacy and prospectively derived data about the dose–response curve for dextromethorphan analgesia.

n-of-1 or Single Case Designs. "n-of-1" or "single case" designs are another variation of crossover studies. Researchers interested in rare disorders or those able to make many mechanistic distinctions may be interested in studying the response of single patients. However, a single crossover of a drug vs. a placebo in one patient has very little power to distinguish a real effect from chance variation. One can increase the power by randomizing the patient to multiple rounds of each treatment. A common design is to give three-to-five pairs of drug vs. placebo and randomize the order within each pair. Figure 17-5 shows an example of a trial of amitriptyline in a patient with fibromyalgia, in which the patient had a higher (defined as more favorable) symptom score on active drug than on placebo in three successive trials.[39]

Recent reports on n-of-1 studies differ with regard to their use and analysis. Some authors have emphasized their usefulness to primary clinicians for guiding practice in individual patients.[40,41] For example, some pediatric psychiatrists have advocated multiple placebo-controlled crossovers of a stimulant before committing a child to long-term treatment for attention deficit disorder.[42] Senn[43] has argued that it may be a disservice to the patient to base treatment only on the individual patient's data because random variation may give a false-positive or false-negative result. He and others have suggested statistical analyses that consider the group response along with the single patient response in making individual treatment decisions.

For published reports of individual patients, there is no doubt that a response replicated three-to-five times under randomized, double-blinded, placebo-controlled conditions is superior to the usual case

report claiming that a patient got better after treatment with drug X without a control treatment or replication. However, most published trials one encounters when searching for "n-of-1 studies" report pooled analyses from a small group of patients given repeated crossovers. This type of analysis has several advantages over statistical analyses of single patient responses. First, one does not need to apply a Bonferroni or other statistical correction for the multiple patients. Second, with multiple crossovers in a group trial, one can distinguish several components of the overall variance—variance due to treatment vs. control, between-patient variation in response to the same treatment ("patient-by-treatment interaction term"), and within-patient variation in response to the same treatment. The latter term is the component used as the random error term in computing statistical significance. Carving the first two components out of the random error adds power not available in the single crossover group trial, in which between-patient variation in response to the same treatment is lumped into the random error.[43]

Parallel Group Designs. Parallel study designs are preferable when there are strong concerns about carryover effects or when the natural history of the disorder makes progression changes likely during the period required for a crossover study. Between-patient variability is the major problem posed by parallel group designs, and several approaches have been suggested to mitigate its impact.[44] For example, baseline pain scores may be subtracted from the treatment scores to yield pain intensity difference scores, or they may be treated as a covariate. This often eliminates a large part of the variance, thereby increasing the power of treatment comparisons.

The investigator should also make an effort to balance the treatment groups for variables that predict response, whenever these predictors are known or suspected. If one wishes to examine response in specific subgroups, assignments must also be balanced appropriately. Groups can be balanced using stratification or various techniques of adaptive randomization.[1,45] In studies with sample sizes typical of single-center trials, 20–40 patients per group, these methods can significantly increase the power of a study if the prognostic variables are well chosen and the statistical methods take the balancing method into account.[44] With sample size more than 50 per group, the randomization process alone is likely to balance out most variables.[46] If stratification is not feasible, post hoc covariate analyses or other statistical techniques may be an acceptable substitute if the variables in question are distributed fairly evenly among the treatment groups.

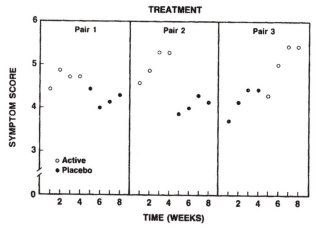

FIGURE 17-5 Results of an n-of-1 randomized controlled trial comparing three pairs of amitriptyline and placebo treatment in a patient with fibromyalgia. Symptom score is plotted against time. High scores correspond to less pain-related impairment. This patient had a statistically significant improvement in symptoms with amitriptyline ($p = 0.03$, paired t test).[39]

5. PLACEBO RESPONSES IN CLINICAL TRIALS

Placebo, which means "I shall please" in Latin, is a term applied either to a remedy that does not affect the "specific mechanisms" of the disease in question or to the favorable response that the treatment often elicits. Scientists and philosophers have wrestled with this concept for generations, resulting in many learned volumes.[47,48] I focus on the implications of placebo responses for clinical trials.

5.1. What Diseases and Symptoms Respond to Placebos?

Spiro[48] critically reviewed the clinical literature and concluded that placebos can affect subjective ratings of symptoms and function, as well as some physiological measurements that depend on smooth muscle function, including blood pressure, airway resistance, and gastrointestinal motility. Clinicians' ratings of physical findings and performance often respond to placebos. Individual studies have shown frequent placebo responses in Parkinson's disease,[49,50] Alzheimer's disease,[51,52] and schizophrenia[53] that were larger than clinicians would expect to see from spontaneous visit-to-visit fluctuations. In contrast, there is no rigorous evidence that macroscopic structural lesions of organs such as malignant tumors or arterial stenoses respond to placebos.

5.2. The Placebo Response Is the Friend of the Clinician and the Enemy of the Disease Mechanism-Oriented Investigator

As discussed previously, the most important goal of the explanatory clinical investigator is to maximize the ratio of the specific treatment effect to the experimental variation. Large placebo responses oppose this goal in two respects. First, the "specific treatment effect" is inferred to be the difference between improvement shown by patients on the treatment and those on a placebo. That is, in most patients who respond to a specific treatment, part of the response is a placebo response. In cases in which the placebo effect is large, a "ceiling effect" may limit the amount of incremental difference that can be seen with a specific treatment. Second, placebo responses, and the nature of the interaction between placebo and specific treatment responses, may vary greatly among individuals with different backgrounds, cognitive styles, etc. Therefore, as the mean size of the placebo response increases, the

experimental variance may increase, with corresponding loss of power.

The recent psychopharmacology literature offers a revealing debate about placebo responses, because in recent years large placebo effects have caused many trials of novel antidepressants and anxiolytics to fail. Some experts warn investigators to avoid psychotherapeutic intervention and to keep warm contact with the patient to the minimum needed to ensure patient compliance.[54] In my first year at the National Institutes of Health, I was shocked when my psychologist colleagues told me to avoid "being too helpful" to my clinical trial patients while they were in a study. One really cannot escape from this irony. Although our research group has not tried to suppress our natural affinity for the patients, we postpone many of our non-study-related therapeutic efforts until after patients complete the drug trial. To counteract the desire of the patient to please us with a positive report, we emphasize that we are unsure of the value of the experimental treatment and need to know the brutal truth if it does not work.

Sullivan[55] explored the paradox that when clinical investigators dismiss the placebo response as a nuisance to be contained, they impoverish scientific conceptions of healing. An alternative view is that a better understanding of placebo responses will reveal "specific mechanisms" of the healing interaction. This interesting research agenda will not have simple answers. Initial reports suggesting that placebo analgesic responses after surgery can be reduced to endorphin secretion[56] have been refuted by the finding that placebo analgesia is not reduced in magnitude by pretreating patients with large doses of naloxone.[57] Placebo responses undoubtedly involve brain centers for language, sensation, mood, movement, and anticipation of the future—that is, most of the brain and every bodily system under its control.

5.3. Placebo Response vs. "Regression to the Mean"

Figure 17-6 summarizes data from two large placebo-controlled dose–response studies of irbesartan, an antihypertensive.[58] During the first 2 weeks of treatment, diastolic pressure dropped by a mean of 4 mmHg in patients treated with placebo capsules and 5–10 mm in patients treated with 25–300 mg irbesartan. Was the 4-mm drop a "placebo response"? A plausible alternative explanation was that this improvement reflects the phenomenon of "regression to the mean." In chronic disorders with fluctuating symptoms and signs, patients are more likely to volunteer for studies and qualify for entry when their disease is, by chance,

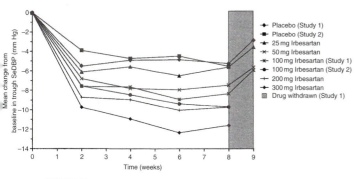

FIGURE 17-6 Placebo response or regression to the mean? Diastolic blood pressure is plotted against time in 889 patients randomized to treatment with placebo or various doses of irbesartan, an investigational antihypertensive drug.[58] At 2 weeks, diastolic blood pressure decreased in all treatment groups. From these data alone, it is not clear whether the drop in the placebo group was due to a placebo response or to regression to the mean—that is, random change in patients who were enrolled at a time when their blood pressure was higher than usual. The rebound of blood pressure after withdrawal of placebo (*shaded bar*) suggests that these patients' blood pressure is sensitive to placebo influences.

in a worse period. Conversely, after study entry, there will be a tendency for them to improve just by random variation. In the studies illustrated in Figure 17-6, an increase in blood pressure when patients were taken off placebo at week 8 suggests blood pressure was responsive to the patients' expectations of a drug effect during treatment. Another way to distinguish placebo response from regression to the mean is to include a "no treatment group" as well as a placebo group. One may infer that improvement in the no treatment group is regression to the mean, and the additional improvement in the placebo group is the placebo response.

Hrobjartsson and Gotzche[59] used this strategy to measure placebo responses in 156 published clinical trials that included both a placebo group and a no treatment group. They concluded that most or all of what is commonly considered placebo response is really regression to the mean, except perhaps in studies of pain, anxiety, and other outcomes reported by the subject.

5.4. Can One Identify and Exclude "Placebo Responders" from Clinical Trials?

Investigators in many fields have tried to identify characteristic "placebo responders" and exclude them from trials, with mixed conclusions. In analgesic studies carried out in the early 1950s, several leading research teams concluded that they were unable to sort out such a subgroup;[60] given repeated single doses of placebo interspersed with doses of opioids, more than

80% of patients with surgical or cancer pain reported analgesia from at least one dose of placebo. In other disease areas, however, the quest to identify placebo responders has continued in the form of single blind placebo "run-in" periods preceding randomization.

Quitkin *et al.*[61] and Nierenberg *et al.*[62] performed careful analyses of several clinical trial cohorts of depressed patients and replicated a finding of distinct patterns of response to tricyclic antidepressants. Mood improvements in the first 2 weeks that then fluctuate and eventually relapse are common in both drug-treated and placebo-treated patients and are inferred to be placebo responses, in contrast to steady improvements with onset after 2 weeks, which are virtually limited to the drug groups. These investigators have argued for using a short placebo run-in period to exclude patients with a marked placebo response and to stratify and statistically correct the outcomes of patients with lesser degrees of improvement during the run-in.[63]

Other psychiatric investigators consider placebo run-ins unhelpful.[54,64,65] They object that this maneuver wastes time, is deceptive in intent, and does not work—clinicians emit subliminal cues that the placebo run-in offers no real treatment, which dampen patients' response, whereas a much larger placebo effect occurs at the time of the real randomization. Montgomery[65] and Schweizer and Rickels[54] propose the alternative of a longer baseline observation period to exclude patients with mild or rapidly cycling mood disorders. In a review of methods in irritable bowel syndrome trials, Hawkey[66] points out another liability of placebo run-in periods in spontaneously fluctuating disorders. By excluding patients whose symptoms have decreased by chance during the run-in period, one tends to be left with patients whose symptoms have worsened by chance, increasing the improvement that will occur in the real study because of regression to the mean. After weighing all of these arguments, however, the reader should keep in mind that there are no data from any disease area directly comparing the statistical efficiency of trials with and without single blind placebo run-in periods. This would be a worthwhile enterprise in any disease area. An ideal design, which I could not find represented in published studies, would continue to treat and follow patients who appeared to remit during the run-in period.

Other investigators have suggested that because placebo responses are less durable than specific therapeutic responses, lengthening trial duration might increase the treatment–placebo difference.[67] Spiller[67] suggests that placebo response drops off after 12 weeks in irritable bowel syndrome, whereas Quitkin *et al.*[61] observe that in antidepressant trials, even a 6-week trial period is long enough for many placebo respond-ers to relapse. However, lengthening a study increases the cost and the number of dropouts. Moreover, some placebo responses are durable. A variety of major sur-gical procedures that later proved to be useless, includ-ing gastric freezing for duodenal ulcers and actual or sham internal mammary artery ligation for angina pectoris, were initially reported to improve or elimi-nate the pain of 60–100% of patients for 1 year after surgery.[68]

5.5. "Unblinding" and Placebo Effects

All agree that patient and clinician expectations contribute to the placebo effect. Many studies have shown that subjects who notice side effects after taking a pill will report more improvement than those who feel no side effects. To minimize such bias, one must strive to maximize the effectiveness of blinding proce-dures and determine if patients can guess their study assignment by the appearance, taste, or side effects of the treatments.[69] In studies of drugs that have unmis-takable side effects, some investigators use "active pla-cebos" that mimic the side effects of the analgesic (Fig. 17-7).[70,71] It is not clear whether one needs to exactly match the magnitude of the side effects of the two

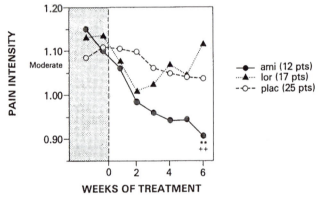

FIGURE 17-7 Side effects may bias patients toward spurious reports of efficacy. This figure shows pain intensity vs. time in a parallel group comparison of amitriptyline (ami), lorazepam (lor), and lactose placebo (plac) in patients with postherpetic neuralgia.[71] Both active drugs but not lactose placebo produced moderate to severe sedation in all patients. Amitriptyline was superior to the inert placebo, reaching statistical significance in week 6. Patients taking lorazepam initially reported pain reduction, during the time in which sedation was most pronounced, but this effect dissipated after the first few weeks. It is possible that patients who noted sedation thought they were on a strong analgesic and that this belief biased them toward reporting pain relief. To improve blinding and reduce this potential bias, we have subsequently used small doses of lorazepam, benztropine, and other drugs to mimic side effects of experimental medications.

treatments to eliminate this bias. A single-dose com-parison of several different drugs in postherpetic neu-ralgia suggested that most of the side effect-induced placebo response occurs with the detection of the first mild symptom.[72] Assigning control groups for behav-ioral interventions is more challenging than for drug studies, and this was reviewed by Whitehead.[73]

The magnitude of the "active placebo" effect is cur-rently under heated debate in psychiatry. Critics of the antidepressant literature suggest that the apparent benefit of antidepressant drugs is largely or even com-pletely a result of the bias inherent in comparing drugs with prominent side effects to inert placebos.[74] Quitkin et al.[75] have ably defended the efficacy of antidepres-sants, but after the exchange of arguments, it is still not clear from the data presented exactly how much side effects bias the main outcomes.

Potential drawbacks of active placebos include the possibility that the active drug included in the placebo may worsen the underlying symptom, thereby con-tributing to a false-positive result; improve the symptom, making it more difficult for the experimen-tal drug to show an effect; or cause adverse reactions. If one has evidence that the active placebo does not affect the target symptom and chooses the lowest dose to produce some symptoms similar to the experimen-tal drug, I think that the resulting protection against accepting a truly ineffective drug outweighs the risks.

The effectiveness of double-blinding may be checked by administering a brief questionnaire to the patient and study nurse (or another member of the research staff who has frequent contact with the patient and may have enough knowledge of the study to guess the treatment assignment). One should ask the subject to guess the identity of the treatment from the list of pos-sibilities and to give the reason for the guess.[69] Impor-tant reasons are side effects and therapeutic effects. It is not clear, however, that patients or researchers can accurately identify the reason for their guess.

The results of such a questionnaire are clearest in the case in which a drug produces immediate side effects but a delayed therapeutic effect, as is seen with antidepressant drugs and preventive medications of any kind. In that case, the questionnaire should be administered at an early time point, and identification of the treatments more frequently than chance sug-gests incomplete blinding of the study. A correct iden-tification of the study drug based on therapeutic effect, however, can occur in a perfectly blinded study of an effective drug. Whatever the treatment, more patients with clinical improvement than without will tend to guess that an active treatment was given. One can factor out most of this bias if one stratifies the analysis to compare the frequency of correct responses within each level of therapeutic benefit reported by the patient.[76]

Other arguments that a positive result was not due to unblinding include (1) showing that the frequency of side effects or the number of patients who stopped drug escalation because of side effects were similar in drug and comparator groups and (2) showing that the positive therapeutic result persists in groups matched for the occurrence of side effects.[77] Despite universal agreement that high-quality clinical trial reports should assess the quality of blinding, few current reports do so.[78]

5.6. Placebos, Positive Controls, and the Concept of "Assay Sensitivity"

To minimize the risks of false positives and negatives, researchers and drug regulators have developed a distinct logic regarding the choice of controls and the interpretation of clinical trials. This framework, illustrated in Figure 17-8, is now routinely applied in determining the validity of single-dose analgesic trials,[2,79] and it is sometimes applied by regulatory officials to other conditions.[80,81]

Although the simplest of the classic designs consists of two treatments—the test medication and a placebo—many trials also include a standard "positive control," previously shown effective in that condition. To demonstrate the value of these controls, consider a hypothetical analgesic study comparing the putative analgesic drug X to a morphine positive control and a placebo (Fig. 17-8a). Using summed pain relief scores over the 4-hour study period as the measure of analgesia, drug X tended to be slightly but not statistically significantly more effective than morphine, and both drug X and morphine were statistically superior to placebo. The conclusions are straightforward: Drug X

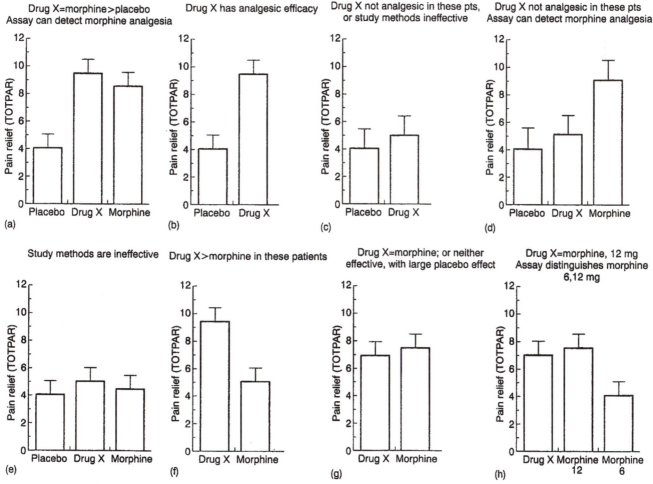

FIGURE 17-8 Placebo and "positive controls" in the interpretation of clinical trials, using the example of an acute analgesic trial (see text). The symbol > denotes "statistically significantly greater than," and = denotes "not significantly different from." TOTPAR, "total of the pain relief" scores over a given period.[2]

is an effective analgesic, and the study methods were sufficiently sensitive to distinguish morphine from placebo.

The omission of a positive control does not fatally flaw the study if drug X is superior to placebo (Fig. 17-8b), although one cannot be certain about the strength of the effect. The positive control serves as a yardstick against which to compare the magnitude of the analgesia produced by drug X. Should drug X fail to produce more analgesia than the placebo, however, the omission of the positive control would render the study uninterpretable (Fig. 17-8c). One cannot reliably conclude that drug X is ineffective in this condition. Perhaps the drug is truly analgesic in patients with this condition, but the study methods were too insensitive to observe this effect. This could happen because patients were too stressed by the clinical setting to respond to medication, the pain questionnaires were insensitive, the procedures of the nurse–observer were variable or confusing, or merely because of random variation. If a morphine positive control were included and shown superior to both placebo and drug X (Fig. 17-8d), this would validate the study methodology and indicate that drug X was not analgesic in this population. Alternatively, if as commonly occurs in real-world clinical trials, morphine produced no more analgesia than drug X and the placebo (Fig. 17-8e), one could conclude that the study methods were inadequate to show the effects of even a strong analgesic.

What are the consequences of omitting the placebo and comparing drug X only to a standard analgesic? As in the previous case, this omission is less damaging when the assay shows a difference between the two treatments. The data in Figure 17-8f suggest that drug X is an effective analgesic in this population, although the proportion of analgesia attributable to the placebo effect cannot be determined for either drug X or morphine. If the responses to drug X and standard analgesic were similar, however (Fig. 17-8g), interpretation would be troublesome. The data might reflect either that drug X and morphine were both effective analgesics or that neither were effective and there was a large placebo effect.

If the use of a placebo group is difficult, an alternative approach is to use a second dose level of the standard treatment. Figure 17-8h shows that morphine 12 mg surpassed morphine 6 mg, demonstrating the sensitivity of the study methods and implying that the effects of both drug X and morphine 12 mg were not merely placebo effects.

In addition to doses of a test drug, a standard treatment, and a placebo, many clinical trials include additional treatment groups or controls that are chosen to further elucidate the major research question.

For example, one might add additional dose levels of the standard treatment, both to serve as a comparative yardstick and to verify that the study methods can separate high from moderate doses of a standard.

Whatever one's disease area of interest, one may wish to test the soundness of proposed research designs by graphing the possible outcomes of the trial as in Figure 17-8. If the conclusion given a particular outcome is ambiguous, consider additional treatment groups that would distinguish among the alternative explanations. The addition of treatment or control groups is costly, however. One must either recruit more patients or reduce the size of each treatment group, lessening the statistical power of the comparisons. In many cases, particularly where negative results will not be of great interest, researchers may choose to omit controls whose main value is to clarify the interpretation of the negative result.

5.7. Placebo Treatment in Extended Studies

In brief studies of symptomatic treatments, placebos are often ethically justified because patients understand that they can terminate the study and take additional medication at any time.[82] In actual practice, many patients experience some placebo relief, and most tolerate the study for the one or two hours needed to evaluate the response to a single dose of drug or placebo.

Chronic studies are a different matter, however. Patients will not tolerate unrelieved severe symptoms for days at a time if effective treatment (e.g., opioids for cancer pain) exists. In studies of the treatment of structural disease, it is obvious that one cannot ethically give a placebo alone if that could cause permanent harm. Therefore, in these situations, the only feasible way to conduct placebo-controlled studies may be to give both placebo and active treatment groups as an add-on treatment, in which all patients are already on optimal doses of a standard treatment. This is the usual design for the development of new antiepileptic drugs.[81] In addition, there are specific approaches to the incorporation of placebo and positive controls in repeated dose analgesic studies.[79]

Although the illustrations in Figure 17-8 used an analgesic trial as an example, the principles are relevant to the general issue of the ethics of using placebos in clinical trials. Rothman and Michels[83] have argued that it is never appropriate to use a placebo when a known effective treatment exists—new treatments should be compared to the standard treatment. Although this may be true in cases in which withholding the known treatment poses major risks of irreversible harm (e.g., studies of treatments of cancer or serious infections), many clinical scientists have pointed out that Rothman and Michels' argument might impair the early development of many treatments, when proof of principle for a weak treatment is needed to continue efforts to improve the treatment.[80,84–86] Moreover, a finding that a new treatment is equivalent to a standard treatment, in the absence of a placebo group, leaves open the possibility that neither was effective in that particular trial and that natural history or placebo effects explained the results. As discussed previously, such a study may produce spurious evidence for the new drug's efficacy and lead to widespread use of an ineffective medication. Miller and Shorr[87] have articulated a middle ground between Rothman and Michels and their critics.

6. CONCLUSION

I hope that I have shown the reader opportunities and tools for improving clinical trial methods. The reader will undoubtedly be able to find others. When I was a relative beginner, obsessed with finding the techniques that would make for the "killer experiment," a veteran scientist advised me, "Just take an important question, hold it before you for years, and you will eventually find gold." An important question that we all face is how to develop individualized treatment regimens aimed at distinct mechanisms of disease. The investigator who learns to increase the power of therapeutic trials to illuminate responses in mechanistically defined subsets of patients will not only improve clinical treatment but also help ensure that human phenomena and concerns will steer the extraordinary engine of basic biomedical research.

References

1. Friedman LM, Furberg CD, DeMets DL. *Fundamentals of Clinical Trials*, 3rd ed. Littleton, MA, PSG Publishing, 1996.
2. Max MB, Laska EM. Single-dose analgesic comparisons. In Max MB, Portenoy RK, Laska EM (eds.) *The Design of Analgesic Clinical Trials*, pp. 55–95. New York, Raven Press, 1991.
3. Lilienfeld AM. Ceteris paribus: The evolution of the clinical trial. *Bull History Med* 1982;56:1–18.
4. Marks HM. Notes from the underground: The social organization of therapeutic research. In Maulitz RC, Long DE (eds.) *Grand Rounds: 100 Years of Internal Medicine*, pp. 297–336. Philadelphia, University of Pennsylvania Press, 1988.
5. Modell W, Houde RW. Factors influencing clinical evaluation of drugs; with special reference to the double-blind technique. *JAMA* 1958;167:2190–2198.
6. Beecher HK. *Measurement of Subjective Responses: Quantitative Effects of Drugs*. New York, Oxford University Press, 1959.

7. Schwartz D, Lellouch J. Explanatory and pragmatic attitudes in therapeutic trials. *J Chronic Dis* 1967;20:637–648.
8. Temple R. Dose–response and registration of new drugs. In Lasagna L, Erill S, Naranjo CA (eds.) *Dose–Response Relationships in Clinical Pharmacology*, pp. 145–170. Amsterdam, Elsevier, 1989.
9. Woolf CJ, Decosterd I. Implications of recent advances in the understanding of pain pathophysiology for the assessment of pain in patients. *Pain* 1999;Suppl. 6:S141–S147.
10. Rasmussen PV, Sindrup SH, Jensen TS, Bach FW. Therapeutic outcome in neuropathic pain: Relationship to evidence of nervous system lesion. *Eur J Neurol* 2004;11(8):545–553.
11. Gilron I, Orr E, Tu D, Peter O'Neill J, Zamora JE, Bell AC. A placebo-controlled randomized clinical trial of perioperative administration of gabapentin, rofecoxib and their combination for spontaneous and movement-evoked pain after abdominal hysterectomy. *Pain* 2005;113:191–200.
12. Gilron I, Bailey JM, Tu D, Holden RR, Weaver DF, Houlden RL. A placebo-controlled randomized clinical trial of morphine, gabapentin and their combination for chronic pain in diabetic neuropathy and postherpetic neuralgia. *N Engl J Med* 2005;352:1324–1334.
13. Peck CC. Concentration-controlled versus concentration defined clinical trials—A reply [Letter]. *Clin Pharmacol Ther* 1993;53:385–387.
14. Johnston A, Holt DW. Concentration-controlled trials: What does the future hold? *Clin Pharmacokinet* 1995;28:93–99.
15. Russ GR, Campbell S, Chadban S, *et al.* Australian Rapamune-Tacrolimus Study Group. Reduced and standard target concentration tacrolimus with sirolimus in renal allograft recipients. *Transplant Proc* 2003;35(3 Suppl.):115S–117S.
16. Fletcher CV, Anderson PL, Kakuda TN, *et al.* Concentration-controlled compared with conventional antiretroviral therapy for HIV infection. *AIDS* 2002;16:551–560.
17. Coda BA, Hill HF, Schaffer RL, Luger TJ, Jacobson RC, Chapman CR. Enhancement of morphine analgesia by fenfluramine in subjects receiving tailored opioid infusions. *Pain* 1993;52:85–89.
18. Adler LJ, Gyulai FE, Diehl DJ, Mintun MA, Winter PM, Firestone LL. Regional brain activity changes associated with fentanyl analgesia elucidated by positron emission tomography. *Anesth Analg* 1997;84:120–126.
19. Ebling WF, Levy G. Population pharmacodynamics: Strategies for concentration- and effect-controlled clinical trials. *Ann Pharmacother* 1996;30:12–19.
20. Sheiner LB, Steimer J-L. Pharmacokinetic/pharmacodynamic modeling in drug development. *Annu Rev Pharmacol Toxicol* 2000;40:67–95.
21. Bellamy N, Sothern RB, Campbell J. Rhythmic variations in pain perception in osteoarthritis of the knee. *J Rheumatol* 1990;17:364–372.
22. Jensen MP, McFarland CA. Increasing the reliability and validity of pain intensity measurement in chronic pain patients. *Pain* 1993;55:195–203.
23. Bellamy N. Pain measurement. In Bellamy N (ed.) *Musculoskeletal Clinical Metrology*, pp. 65–76. Dordrecht, The Netherlands, Kluwer, 1993.
24. Bellamy N, Campbell J, Syrotuik J. Comparative study of self-rating pain scales in rheumatoid arthritis patients. *Curr Med Res Opin* 1999;15:121–127.
25. Jung BF, Johnson RW, Griffin DRJ, Dworkin RH. Risk factors for postherpetic neuralgia in patients with herpes zoster. *Neurology* 2004;62:1545–1551.

26. Max MB, Portenoy RK, Laska EM (eds.). *The Design of Analgesic Clinical Trials. Advances in Pain Research and Therapy*, vol. 18. New York, Raven Press, 1991.

27. Askmalm MS, Carstensen J, Nordenskjold B, *et al*. Mutation and accumulation of p53 related to results of adjuvant therapy of postmenopausal breast cancer patients. *Acta Oncol* 2004; 43:235–244.

28. Lee DK, Currie GP, Hall IP, Lima JJ, Lipworth BJ. The arginine-16 beta2-adrenoceptor polymorphism predisposes to bronchoprotective subsensitivity in patients treated with formoterol and salmeterol. *Br J Clin Pharmacol* 2004;57:68–75.

29. Louis TA, Lavori PW, Bailar JC, Polansky M. Crossover and self-controlled designs in clinical research. *N Engl J Med* 1984;310:24–31.

30. Jones B, Kenward MG. *Design and Analysis of Cross-Over Trials*, 2nd ed. London, Chapman & Hall, 2005.

31. Ratkowsky DA, Evans MA, Alldredge JR. *Cross-Over Experiments: Design, Analysis, and Application*. New York, Dekker, 1993.

32. Senn S. *Cross-Over Trials in Clinical Research*, 2nd ed. Chichester, UK, Wiley, 2002.

33. Raja SN, Haythornthwaite JA, Pappagallo M, *et al*. A placebo-controlled trial comparing the analgesic and cognitive effects of opioids and tricyclic antidepressants in postherpetic neuralgia. *Neurology* 2002;59:1015–1021.

34. Lipton RB, Bigal ME, Stewart WF. Clinical trials of acute treatments for migraine including multiple attack studies of pain, disability, and health-related quality of life. *Neurology* 2005;65(Suppl.):50–58.

35. Byas-Smith MG, Max MB, Muir J, Kingman A. Transdermal clonidine compared to placebo in painful diabetic neuropathy using a two-stage "enriched" enrollment trial design. *Pain* 1995;60:267–274.

36. Leber PD, Davis CS. Threats to the validity of clinical trials employing enrichment strategies for sample selection. *Contr Clin Trials* 1998;19:178–187.

37. Temple RJ. Special study designs: Early escape, enrichment, studies in non-responders. *Commun Statist Theory Methods* 1994;23:499–531.

38. Sang CN, Booher S, Gilron I, Parada S, Max MB. A randomized, placebo-controlled trial of dextromethorphan and memantine in painful diabetic neuropathy and postherpetic neuralgia. *Anesthesiology* 2002;96:1053–1061.

39. Guyatt GH, Heyting A, Jaeschke R, Keller J, Adachi JD, Roberts RS. N of 1 randomized trials for investigating new drugs. *Contr Clin Trials* 1990;11:88–100.

40. Guyatt GH, Keller JL, Rosenbloom D, Adachi JD, Newhouse MT. The n-of-1 randomized controlled trial: Clinical usefulness. Our three-year experience. *Ann Intern Med* 1990;112:292–299.

41. Pope JE, Prashker M, Anderson J. The efficacy and cost effectiveness of N of 1 studies with diclofenac compared to standard treatment with nonsteroidal anti-inflammatory drugs in osteoarthritis. *J Rheumatol* 2004;31:140–149.

42. Kent MA, Camfield CS, Camfield PR. Double-blind methylphenidate trials: Practical, useful, and highly endorsed by families. *Arch Pediatr Adolesc Med* 1999;153:1292–1296.

43. Senn S. *Statistical Issues in Drug Development*. Chichester, UK, Wiley, 1997.

44. Lavori PW, Louis TA, Bailar JC, Polansky M. Designs for experiments—Parallel comparisons of treatment. *N Engl J Med* 1983;309:1291–1298.

45. Therneau TM. How many stratification factors are "too many" to use in a randomization plan? *Contr Clin Trials* 1993;14:98–108.

46. Meinert CL. *Clinical Trials: Design, Conduct, and Analysis*. New York, Oxford University Press, 1986.

47. White L, Tursky B, Schwartz GE (eds.). *Placebo: Theory, Research, and Mechanism*. New York, Guildford, 1985.

48. Spiro HM. *Doctors, Patients, and Placebos*. New Haven, CT, Yale University Press, 1986.

49. Shetty N, Friedman JH, Kieburtz K, Marshall FJ, Oakes D, Parkinson Study Group. The placebo response in Parkinson's disease. *Clin Neuropharmacol* 1999;22:207–212.

50. McRae C, Cherin E, Yamazaki TG, *et al*. Effects of perceived treatment on quality of life and medical outcomes in a double-blind placebo surgery trial. *Arch Gen Psychiatry* 2004;61:412–420.

51. Spencer CM, Noble S. Rivastigmine. A review of its use in Alzheimer's disease. *Drugs Aging* 1998;13:391–411.

52. Kawas CH, Clark CM, Farlow MR, *et al*. Clinical trials in Alzheimer disease: Debate on the use of placebo controls. *Alzheimer Dis Assoc Disord* 1999;13:124–129.

53. Montgomery SA. Alternatives to placebo-controlled trials in psychiatry: ECNP consensus meeting. *Eur Neuropsychopharmacol* 1999;9:265–269.

54. Schweizer E, Rickels K. Placebo response in generalized anxiety: Its effect on the outcome of clinical trials. *J Clin Psychiatry* 1997;58(Suppl. 11):30–38.

55. Sullivan MD. Placebo responses and epistemic control in orthodox medicine. *J Med Philos* 1993;18:213–231.

56. Levine JD, Gordon NC, Fields HL. The mechanism of placebo analgesia. *Lancet* 1978;23:654–657.

57. Gracely RH, Dubner R, Wolskee PJ, Deeter WR. Placebo and naloxone can alter post-surgical pain by separate mechanisms. *Nature* 1983;306:264–265.

58. Pool JL, Guthrie RM, Littlejohn TW, *et al*. Dose-related antihypertensive effects of irbesartan in patients with mild-to-moderate hypertension. *Am J Hypertension* 1998;11:462–470.

59. Hrobjartsson A, Gotzsche PC. Is the placebo powerless? Update of a systematic review with 52 new randomized trials comparing placebo with no treatment. *J Intern Med* 2004;256:91–100.

60. Houde RW, Wallenstein SL, Beaver WT. Clinical measurement of pain. In de Stevens G (ed.) *Analgetics*, pp. 75–122. New York, Academic Press, 1965.

61. Quitkin FM, Stewart JW, McGrath PJ, *et al*. Further evidence that a placebo response to antidepressants can be identified. *Am J Psychiatry* 1993;150:566–570.

62. Nierenberg AA, Quitkin FM, Kremer C, Keller MB, Thase ME. Placebo-controlled continuation treatment with mirtazapine: Acute pattern of response predicts relapse. *Neuropsychopharmacology* 2004;29:1012–1018.

63. Quitkin FM, McGrath PJ, Stewart JW, *et al*. Placebo run-in period in studies of depressive disorders: Clinical, heuristic and research implications. *Br J Psychiatry* 1998;173:242–248.

64. Trivedi M, Rush J. Does a placebo run-in or a placebo treatment cell affect the efficacy of antidepressant medications? *Neuropsychopharmacology* 1994;11:33–43.

65. Montgomery SA. The failure of placebo-controlled studies: ECNP consensus meeting. *Eur Neuropsychopharmacol* 1999;9:271–276.

66. Hawkey CJ. Irritable bowel syndrome clinical trial design: Future needs. *Am J Med* 1999;107(5A):98S–102S.

67. Spiller RC. Problems and challenges in the design of irritable bowel syndrome clinical trials: Experience from published trials. *Am J Med* 1999;107(5A):91S–97S.

68. Turner JA, Deyo RA, Loeser JD, Von Korff M, Fordyce WE. The importance of placebo effects in pain treatment and research. *JAMA* 1994;271:1609–1614.

69. Moscucci M, Byrne L, Weintraub M, Cox C. Blinding, unblinding, and the placebo effect: An analysis of patients' guesses of treatment assignment in a double-blind trial. *Clin Pharmacol Ther* 1987;41:259–265.

70. Greenberg RP, Fisher S. Seeing through the double-masked design: A commentary. *Contr Clin Trials* 1994;15:244–246.

71. Max MB, Schafer SC, Culnane M, Smoller B, Dubner R, Gracely RH. Amitriptyline, but not lorazepam, relieves post-herpetic neuralgia. *Neurology* 1988;38:1427–1432.

72. Max MB, Schafer SC, Culnane M, Dubner R, Gracely RH. Association of pain relief with drug side-effects in post-herpetic neuralgia: A single-dose study of clonidine, codeine, ibuprofen, and placebo. *Clin Pharmacol Ther* 1988;43:363–371.

73. Whitehead WE. Control groups appropriate for behavioral interventions. *Gastroenterology* 2004;126:S159–S163.

74. Kirsch I, Sapirstein G. Listening to Prozac but hearing placebo: A meta-analysis of antidepressant medication. *Prevention Treatment* 1998;1. [On line journal: http://journals.apa.org/prevention/volume1/pre0010002a.html]

75. Quitkin FM, Rabkin JG, Davis J, Davis JM, Klein DF. Validity of clinical trials of antidepressants. *Am J Psychiatry* 2000;157:327–337.

76. Shlay JC, Chaloner K, Max MB, *et al*. A randomized placebo-controlled trial of a standardized acupuncture regimen and amitriptyline for pain caused by HIV-related peripheral neuropathy. *JAMA* 1998;280:1590–1595.

77. McArthur JC, Yiannoutsos C, Simpson DM, *et al*. A phase II trial of nerve growth factor for sensory neuropathy associated with HIV infection. *Neurology* 2000;54:1080–1088.

78. Fergusson D, Glass KC, Waring D, Shapiro S. Turning a blind eye: The success of blinding reported in a random sample of randomised, placebo controlled trials. *Br Med J* 2004;328(7437):432.

79. Max MB. The design of clinical trials of treatments for pain. In Max MB, Lynn J (eds.) *Interactive Textbook of Clinical Symptom Research.* Bethesda, MD, National Institute of Dental and Craniofacial Research, 2000. [On line textbook: http://painconsortium.nih.gov/symptomresearch/chapter_1/index.htm]

80. Temple RJ. When are clinical trials of a given agent vs. placebo no longer appropriate or feasible? *Contr Clin Trials* 1997;18:613–620.

81. Leber PD. Hazards of inference: The active control investigation. *Epilepsia* 1989;30(Suppl. 1):S57–S63.

82. Levine RJ. The need to revise the Declaration of Helsinki. *N Engl J Med* 1999;341:531–534. [Comments in *N Engl J Med* 1999;341:1851–1853]

83. Rothman KJ, Michels KB. The continuing unethical use of placebo controls. *N Engl J Med* 1994;331:394–398.

84. Charney DS, Nemeroff CB, Lewis L, *et al*. National depressive and manic-depressive association consensus statement on the use of placebo in clinical trials of mood disorders. *Arch Gen Psychiatry* 2002;59:262–270.

85. Loder E, Goldstein R, Biondi D. Placebo effects in oral triptan trials: The scientific and ethical rationale for continued use of placebo controls. *Cephalalgia* 2005;25:124–131.

86. Fleischhacker WW, Czobor P, Hummer M, Kemmler G, Kohnen R, Volavka J. Placebo or active control trials of antipsychotic drugs? *Arch Gen Psychiatry* 2003;60:458–464.

87. Miller FG, Shorr AF. Unnecessary use of placebo controls: The case of asthma clinical trials. *Arch Intern Med* 2002;162:1673–1677.

18

Large Clinical Trials and Registries— Clinical Research Institutes

ROBERT M. CALIFF

Duke Clinical Research Institute, Durham, North Carolina

1. INTRODUCTION

Medical practice has entered the era of "evidence-based medicine," characterized by an increasing societal belief that clinical practice should be based on scientific information in addition to intuition, mechanistic reasoning, and opinion. As our society has increasingly recognized that unfettered use of technology will lead to limitless increases in cost, the only rational way to allocate resources is to understand whether competing therapeutic approaches provide clinical benefit and, if so, the cost required to achieve that benefit. Simultaneous with the realization that expansion of medical finances is not limitless, the huge societal investment in biotechnology is beginning to pay off in the form of many potential new approaches to treating disease. Therefore, with current methodology, the need for evidence is increasing faster than the resources are being made available to perform the studies.

2. HISTORY

The first randomization was performed by Fisher in 1926 in an agricultural study.[1] In developing analysis of variance, he recognized that experimental observations must be independent and not confounded to allow full acceptance of the statistical methodology. He therefore randomized different plots to different approaches to the application of fertilizer. Amberson has been credited with the first randomization of patients in a 1931 trial of tuberculosis therapy in 24 patients, using a coin toss to make treatment assignments.[2] The British Medical Research Council trial of streptomycin in the treatment of tuberculosis marked the modern era of clinical trials in 1948.[3] This trial established principles for the use of randomization in large numbers of patients, and it set guidelines for administration of the experimental therapy and objective evaluation of outcomes.

In the past decade, computers have enabled rapid accumulation of data from thousands of patients in studies conducted throughout the world. Peto, Yusuf, Sleight, and Collins developed the concept of the large simple trial in the First International Study of Infarct Survival (ISIS-I),[4] beginning with the concept that only by randomizing 10,000 patients could the beneficial effects of beta blockers be understood. The development of client server architecture provided a mechanism for aggregating large amounts of data and distributing the data quickly to multiple users. The most recent advances in the development of the World Wide Web provide an opportunity to share information instantaneously in multiple locations throughout the world. Finally, the recognition that measurement of adherence to proven diagnostic and treatment approaches improves patient outcome has led to the broad adoption of electronic health records and computerized provider order entry, as well as organized approaches to implementation of evidence-based medicine.

3. PHASES OF EVALUATION OF THERAPIES

Evaluating therapies and interpreting the results as they are presented requires an understanding of the goals of the investigation; these goals can be conveniently categorized using the nomenclature used by the Food and Drug Administration (FDA) to characterize the phase of investigation in clinical trials (Table 18-1). Although all trials should heed the lessons from large, pragmatic trials concerning adequate sample size and avoidance of unnecessary complexity, the specific issues of large, pragmatic trials do not become important until potential therapies are subjected to phase III or IV trials. The first two phases are focused on initial evaluation for evidence of frank toxicity, obvious clinical complications, and physiological support for the intended mechanism of action of the therapy. In these phases, attention to detail is critical and should take priority over simplicity (although detail for no good purpose is a waste of resources, regardless of the phase of the trial).

The third phase, commonly referred to as the "pivotal" phase, evaluates the therapy in the relevant clinical context, with the goal of determining whether the therapy should be used in clinical practice. For phase III, the relevant end points include measures that can be recognized by patients as important: survival time, major clinical events, quality of life, and cost. A well-designed clinical trial with a positive effect on clinical outcomes justifies serious consideration for a change in clinical practice and certainly provides grounds for regulatory approval for sales and marketing.

After a therapy is approved by regulatory authorities and in use, phase IV begins. Traditionally, phase IV has been viewed as the monitoring of the use of a therapy in clinical practice, with a responsibility of developing more effective protocols for the use of that therapy, based on inference from observations and reporting of adverse events. In addition, phase IV is used to develop new indications for drugs already approved for a different use. The importance of this phase has evolved from the recognition that many circumstances experienced in clinical practice will not have been encountered in randomized trials completed at the time of regulatory approval. Examples of phase IV studies include the evaluation of new dosing regimens, as in several ongoing comparisons of low-dose versus high-dose angiotensin-converting enzyme inhibition in patients with heart failure, and the prospective registries of use of therapies such as the National Registry of Myocardial Infarction, Can Rapid Risk Stratification of Unstable Angina Patients Suppress Adverse Outcomes with Early Implementation of the ACC/AHA Guidelines, and the Society of Thoracic Surgeons Database.[5-7] As the array of effective therapies has increased, phase IV is viewed as a time to compare one effective marketed therapy against another. In some cases, this need arises because of changing doses or expanding indications for a therapy; in other cases, the phase III trials did not provide the relevant comparisons for a particular therapeutic comparison.

4. CRITICAL GENERAL CONCEPTS

With rare exceptions, the purpose of a phase III or phase IV clinical trial, registry, or outcome study is to estimate what is likely to happen to the next patient if one treatment strategy or the other is chosen.

TABLE 18-1   Phases of Evaluation of New Therapies

Phase	Features	Purpose
I	First administration of a new therapy to patients	Exploratory clinical research to determine if further investigation is appropriate
II	Early trials of new therapy in patients	To acquire information on dose–response relationship, estimate incidence of adverse reactions, and provide additional insight into pathophysiology of disease and potential impact of new therapy
III	Large-scale comparative trial of new therapy versus standard of practice	Definitive evaluation of new therapy to determine if it should replace current standard of practice; randomized controlled trials required by regulatory agencies for registration of new therapeutic modalities
IV	Monitoring of use of therapy in clinical practice	Postmarketing surveillance to gather additional information on impact of new therapy on treatment of disease, rate of use of new therapy, and more robust estimate of incidence of adverse reactions established from registries

Adapted from Antman EM, Califf RM. Clinical trials and meta-analysis. In Smith TW (ed.) *Cardiovascular Therapeutics*, p. 679. Philadelphia, Saunders, 1996.

TABLE 18-2   Questions to Ask When Reading and Interpreting the Results of a Clinical Trial

Are the results of the study valid?
  Primary guides
    Was the assignment of patients to treatment randomized?
    Were all patients who entered the study properly accounted for at its conclusion?
    Was follow-up complete?
    Were patients analyzed in the groups to which they were randomized?
  Secondary guides
    Were patients, their clinicians, and study personnel blinded to treatment?
    Were the groups similar at the start of the trial?
    Aside from the experimental intervention, were the groups treated equally?
What were the results?
  How large was the treatment effect?
  How precise was the treatment effect (confidence intervals)?
Will the results help me in caring for my patients?
  Does my patient fulfill the enrollment criteria for the trial? If not, how close is the patient to the enrollment criteria?
  Does my patient fit the features of a subgroup in the trial report? If so, are the results of the subgroup analysis in the trial valid?
  Were all the clinically important outcomes considered?
  Are the likely treatment benefits worth the potential harm and costs?

To assess the degree to which the proposed study enhances the ability to understand what will happen to the next patient, the investigator must be aware of an array of methodological and clinical issues. Although this task requires substantial expertise and experience, the issues can be considered in a broad framework. The simplest but most essential concepts for understanding the relevance of a clinical study to practice are validity and generalizability. Table 18-2 illustrates an approach to these issues, developed by the McMaster group, to be used when reading the literature.

4.1. Validity

The most fundamental question about a clinical trial is whether the result is valid. Are the results of the trial internally consistent? Would the same result be obtained if the trial were repeated? Was the trial design adequate, including blinding, end point assessment, and statistical analyses? Of course, the most compelling evidence of validity in science is replication. If the results of a trial or study remain the same when the study is repeated, especially in a different clinical environment by different investigators, the results are likely to be valid.

4.2. Generalizability

Given a valid clinical trial result, it is equally important to determine whether the findings are generalizable. Unless the findings can be replicated and applied in multiple practice settings, little has been gained by the trial with regard to informing clinical practice. Since it is impossible to replicate every clinical study in practice, it is especially important to understand the inclusion and exclusion criteria for patients entered into the study and to have an explicit understanding of additional therapies that the patients received. For example, studies done on "ideal" patients without comorbid conditions or on young patients without severe illness can be misleading when the results are applied to clinical practice since the rate of poor outcomes, complications, and potential drug interactions could be much higher in an older population with more comorbidities. Of increasing concern in this regard are children and the very elderly.[8,9] In both age groups, the findings of clinical trials are unlikely to be easily extrapolated to effective clinical practice, especially with regard to dosing.

5. EXPRESSING CLINICAL TRIAL RESULTS

The manner in which the results of clinical research are reported can profoundly influence the perception of practitioners evaluating the information to decide which therapies to use. A clinical trial will produce a different degree of enthusiasm about the therapy tested when the results are presented in the most favorable light. To guard against this problem, investigators should report clinical outcome trials in terms of both relative and absolute risk reductions, including confidence intervals for point estimates. Even when exact results are provided in addition to the risk reduction so that the practitioner could reconstruct the results in different ways, the primary method of presentation has a major effect on perception.[6] Multiple studies have demonstrated that physicians are much more likely to recommend a therapy when the results are presented as a relative risk reduction rather than as an absolute difference in outcomes.[10,11] This appears to happen because the relative risk reductions result in larger numbers, even though they are reporting exactly the same clinical phenomenon. This sobering problem points out one of the most important features of large, pragmatic trials; because they try to answer questions that will directly change patient care, the audience for the results will often far exceed the local community of experts and often will include generalist physicians, lay people, and the press. Planning is critical in order to handle these issues appropriately.

One important metric for reporting the results of pragmatic clinical trials is the number of poor outcomes prevented by the more effective treatment per 100 or 1000 patients treated. This measure, the number needed to treat (NNT), translates results for specific populations studied into public health terms by quantifying how many patients would need to be treated to create a specific health benefit. The absolute difference can be used to assess quantitative interactions—that is, significant differences in the number of patients needed to treat to achieve a degree of benefit with a therapy as a function of the type of patient treated. An example is the use of thrombolytic therapy: The Fibrinolytic Therapy Trialists' (FTT) collaboration demonstrated that 37 lives are saved per 1000 patients treated when thrombolytic therapy is used in patients with anterior

ST segment elevation, whereas only 8 lives are saved per 1000 patients with inferior ST segment elevation (Fig. 18-1).[12] The direction of the treatment effect is the same, but the magnitude of the effect is different.

Two other important aspects of the NNT calculation that should be considered are the duration of treatment needed to achieve the benefit and the number needed to harm (NNH). Although it is intuitively less impressive to save a life per 100 patients treated over 5 years versus saving a life per 100 patients treated in 1 week, this issue is often forgotten. The NNH can be simply calculated, just as the NNT is calculated.

This approach becomes more complex with end points that are not discrete, such as exercise time. One approach to expressing trial results when the end point is a continuous measurement is to define the minimal clinically important difference (the smallest difference that would lead practitioners to change their practices)

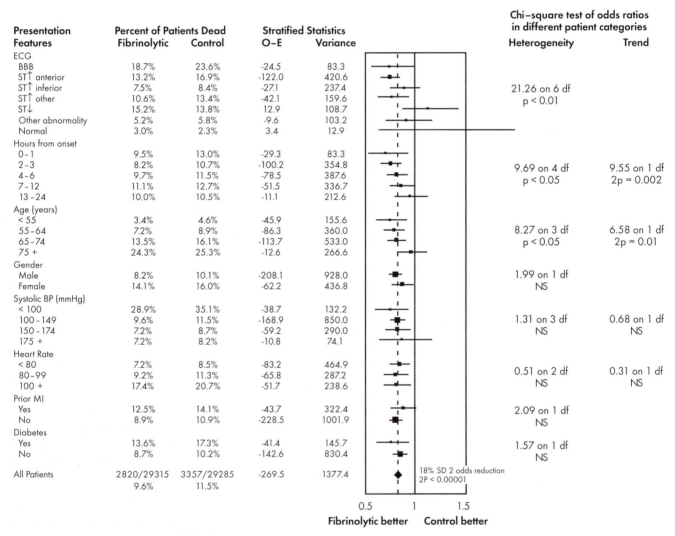

FIGURE 18-1 Proportional effects of fibrinolytic therapy on mortality.

Randomized Controlled Trials
Summary Measures of Treatment Effect

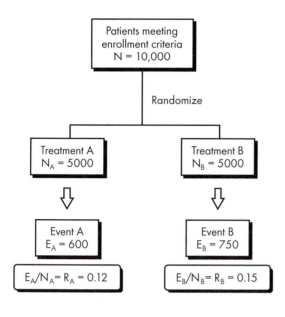

	Event	No Event	
A	E_A = 600	4400	5000
B	E_B = 750	4250	5000
	1350	8650	10,000

Statistical Tests of Rx Effect

1. χ^2 = 19.268 $\varnothing$ p < 0.001
2. Fischer Exact Test: p < 0.001
3. Comparison of Proportions: z = 4.360 $\varnothing$ p < 0.001

Statements Describing Rx Effect

1. Relative Risk R_A/R_B = 0.80
2. Relative Risk Reduction = (1 − Relative Risk) = 0.20
3. Odds Ratio = $\dfrac{R_A/(1 - R_A)}{R_B/(1 - R_B)}$ = 0.77
4. Absolute Risk Difference = ($R_B - R_A$) = RD = 0.03
5. Numbers Needed to Treat = (1/Abs. Risk Diff.) = 33

FIGURE 18-2 Summary measures of treatment effect.

and to express the results in terms of the NNT to achieve that minimal clinically important difference. Another problem with NNT and NNH occurs when the trial on which the calculation is based is not a generalizable trial enrolling patients who are likely to be treated in practice. Indeed, when relevant patients (e.g., elderly patients or those with renal dysfunction) have been excluded, these simple calculations can be misleading.

The relative benefit of therapy, on the other hand, is the best measure of the treatment effect in biological terms. This concept is defined as the proportional reduction in risk resulting from the more effective treatment, and it is generally expressed in terms of an odds ratio or relative risk reduction. The relative treatment effect can be used to assess qualitative interactions, which represent statistically significant differences in the direction of the treatment effect as a function of the type of patient treated. In the FTT analysis, the treatment effect in patients without ST segment elevation is heterogeneous compared with that of patients with ST segment elevation.[12] Figure 18-2 gives the calculations for commonly used measures of treatment effect.

A particularly useful display of data is the odds ratio plot (Fig. 18-3). Both absolute and relative differ-

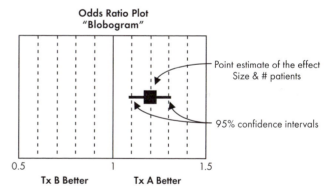

FIGURE 18-3 Odds ratio plot.

ences in outcome should be expressed in terms of point estimates and confidence intervals. This type of display gives the reader a balanced perspective since both the relative and the absolute differences are important, as well as the confidence in the estimate. Without confidence intervals, the reader has difficulty ascertaining the precision of the estimate of the treatment effect. The goals of a large, pragmatic trial include (1) the enrollment of a broad array of patients so that the effect of treatment in different types of patients can be assessed and (2) the enrollment of enough patients with enough events to make the confidence intervals narrow and definitive. Using an odds ratio or risk ratio plot, the investigator can quickly create a visual image that defines the evidence for homogeneity or heterogeneity of the treatment effect as a function of baseline characteristics.

6. CONCEPTS UNDERLYING TRIAL DESIGN

As experience with multiple clinical trials accumulates, some general concepts seem worth emphasizing. These generalities do not always pertain, but they serve as useful guides to the design or interpretation of trials. Failure to consider these general principles often leads to a faulty design and failure of the project.

6.1. Treatment Effects Are Modest

The most common mistake in designing clinical trials is overestimation of the expected treatment effect. Most individuals heavily involved in therapeutic development cannot resist the temptation of assuming that the pathway being targeted is the most important contributor to patient outcome. Unfortunately, relative reductions in adverse clinical outcomes exceeding 25% are extremely uncommon.

When treatments affecting outcome have been assessed, small trials typically greatly overestimate the effect observed in subsequent larger trials. The reasons for this observation are not entirely clear. One important factor is a publication bias against studies reporting negative findings.[13] Of the many small studies performed, the positive ones tend to be published. A second factor could be analogous to regression to the mean in observational studies: When a variety of small trials are done, only those with a substantial treatment effect are likely to be continued into larger trials. Of course, in most cases the small trials have so much uncertainty in estimating the treatment effect that the true effect of many of the promising therapies is overestimated, whereas the effect of some of the therapies showing little promise based on point estimates from small studies is underestimated. Thus, when larger studies are completed, giving a more reliable estimate of treatment effect, the estimate of benefit tends to regress toward average.

The Global Utilization of Streptokinase and rt-PA for Occluded Coronary Arteries (GUSTO-I) trial used an extensive process to devise the expected sample size.[14] An expected effect was calculated using all previously published data on the relationship between coronary perfusion on an angiogram and mortality in patients with ST segment-elevation myocardial infarction. A panel of experts was then assembled both in Europe and in the United States to determine the mathematical calculations and the differences that would be needed to create a clinically meaningful benefit. In the end, both approaches yielded a value of a 14% relative difference (1 life saved per 100 patients treated) or a 14% reduction in relative risk of death, whichever was smaller. The trial was then sized to detect these differences, and a difference of 15% on a relative basis and 1% on an absolute basis was observed when the trial was completed.

The implications of this principle are that sample sizes need to increase by a significant (perhaps logarithmic) amount, and a registry of all clinical trials is needed so that all evidence generated from human clinical trials will be available to the public. This issue of a clinical trials registry has been a topic of great public interest.[15] The National Library of Medicine appears likely to be the repository for this registry, which presumably will be required for all clinical trials, regardless of funding sources.

6.2. Qualitative Interactions Are Uncommon

A reversal of treatment effect as a function of baseline characteristics is unusual. Many training programs have taught clinicians that many therapies are effective only in very select subsets of the population, yet there are few examples demonstrating such a targeted effect (the emerging field of pharmacogenomics may change this principle, as underlying gene defects in common diseases may be specifically altered by highly specific therapies). This principle has important implications for the amount of data collection in well-designed clinical trials. There is a tendency to collect voluminous data on the chance that the treatment may be effective only in a small group of patients; this rarely happens, however, and even if it did, the chances of detecting such an interaction are quite low. The main study is typically powered to detect a clinically meaningful effect, thereby leaving little power to detect the same effect in a smaller sample. Of course, when there is a compelling reason to look for a difference (e.g., response to therapy as a function of a known biological modifier of the disease response), it should be done. A useful exercise is to fix the amount of data that can be collected, thereby forcing the experts to defend their proposed ancillary data collection relative to the proposals of others. In a large trial, adding a single data item can add hundreds of thousands of dollars to the study budget.

6.3. Quantitative Interactions Are Common

When therapies are beneficial for specific patients with a given diagnosis, they are generally beneficial to most patients with that diagnosis. However, therapies commonly provide a differential absolute benefit as a function of the severity of the patient's illness. Given the same relative treatment effect, the number of lives saved or events prevented will be greater when the therapy is applied to patients with a greater underlying risk. Examples of this concept include the greater benefit of angiotensin-converting enzyme inhibitors in patients with markedly diminished left ventricular function, the larger benefit of thrombolytic therapy in patients with anterior infarction, and the greater benefit of bypass surgery in older patients compared to younger patients. Most often, these sorts of measures are the same ones that would be gleaned to characterize the population in clinical terms, so the extra cost of data ascertainment and recording is small.

This same principle also seems to hold for harm. Elderly patients patients with multiple comorbidities, and patients with renal dysfunction often have the highest risk of adverse drug effects. If these patients are excluded from clinical trials, the true risks will not be known when the treatment enters practice, and accurate assessment of risk through current methods of postmarketing assessment will be difficult, if not impossible.[16]

6.4. Unintended Biological Targets Are Common

Therapies are appropriately developed by finding an alterable pathophysiological pathway or target and by exploiting that concept using a model that does not involve the intact human. Despite all good intentions, proposed therapies frequently either work via a different mechanism than the one for which the therapy was devised, or affect an entirely different system. An example is thrombolytic therapy for myocardial infarction, which was developed using coronary thrombosis models; unfortunately, this therapy also affects the intracranial vessels. Inotropic therapies for heart failure were developed using measures of cardiac function, but many of these agents, which clearly improve cardiac function acutely, also cause an increase in mortality, perhaps due to a detrimental effect on the neurohormonal system. Several new agents for treatment of diabetes mellitus were developed to alter pathways of glucose uptake, but unanticipated effects on liver cells have been encountered. The effect of the phenter-

mine and fenfluramine combination on cardiac valves was unexpected. Major problems with myonecrosis led to the withdrawal of cerivastatin from the market,[17] and an extensive public debate resulted from the withdrawal of several COX-2 inhibitors after billions of dollars in sales.[18] These examples point to the societal need to evaluate therapies in broad populations of patients before making them available to the public, rather than relying on surrogate end points in small numbers of patients.

6.5. Interactions among Therapies Are Not Predictable

Many common diseases can be treated with multiple therapies with some degree of benefit. Yet, clinical trials seldom evaluate more than one treatment simultaneously. Evidence indicates that this may be an error. When abciximab was developed, its pharmacodynamic and pharmacokinetic interactions with heparin were easily characterized. However, the interaction of the two drugs with regard to clinical effect was simply not known. A series of sequential clinical trials demonstrated that when full-dose abciximab was combined with a lower than normal dose of heparin, the bleeding rate in the setting of percutaneous intervention dropped to the same level as full-dose heparin alone, and the efficacy unexpectedly improved compared to full-dose abciximab and standard-dose heparin. This result was simply not predictable from the known biology and pharmacology of these agents.

The ongoing controversy about aspirin and angiotensin-converting enzyme inhibitors (ACEIs) raises this issue. Both therapies are beneficial in patients with cardiovascular disease, but evidence from physiological studies suggests that the prostaglandin effects of aspirin may nullify some of the vascular effects of ACEIs. Retrospective evaluations of clinical trials have been equivocal. There may be many other interactions that could be discovered through carefully designed factorial trials.

6.6. Long-Term Effects May Be Unpredictable

The concept that the short-term and longer term effects of therapy may differ is easiest to grasp when evaluating surgical therapy. Patients routinely take a risk of operative mortality and morbidity in order to achieve longer term gain. This principle also holds for some acute medical therapies. Fibrinolytic therapy actually increases the risk of death in the first 24 hours while exerting a mortality benefit from that point forward. The recent controversy over hormone replacement therapy points out that a treatment could be detrimental for more than 1 year, and then a benefit could accrue.

7. GENERAL DESIGN CONSIDERATIONS

When reading the results of a clinical study or designing a study, the purpose of the investigation is critical to placing the outcome in context. Those who design the investigation have the responsibility of constructing the project and presenting the results in a manner reflecting the intent of the study. In a small phase II study, an improvement in a surrogate pathophysiological outcome is exciting and could easily lead the investigator to overstate the clinical implications of the finding. Similarly, megatrials with little data collection seldom give useful information about disease mechanisms unless carefully planned substudies are performed. The structural characteristics of trials can be characterized as a function of the attributes discussed in the following sections.

7.1. Pragmatic versus Explanatory

Most clinical trials are designed to demonstrate a physiological principle as part of a chain of causality of a particular disease. Such trials, termed explanatory trials, need only be large enough to prove or disprove the hypothesis being tested. Major problems have arisen because of the tendency of those doing explanatory trials to generalize the findings into recommendations about clinical therapeutics.

Trials designed to answer questions about which therapies should be used are called pragmatic trials. These trials should have clinical outcomes as the primary end point, so that when the trial is complete, the result will inform the practitioner and the public about whether using the therapy in the manner tested will result in better clinical outcomes than the alternative approaches. These trials generally require much larger sample sizes to arrive at a valid result and a more heterogeneous population to be generalizable to populations treated in practice.

The decision about whether to perform an explanatory trial or a pragmatic trial has major implications for the design of the study. When the study is published, the reader must also take into account the intent of the investigators since the implications for practice or knowledge will vary considerably depending on the type of study. The organization, goals, and structure of the large pragmatic trial may be understood best by comparing the approach that might be used in an explanatory trial with the approach used in a large pragmatic trial.[19] These same principles are important in designing disease registries.

7.2. Entry Criteria

In an explanatory trial, the entry criteria should be carefully controlled so that the particular measurement of interest will not be confounded. For example, a trial designed to determine whether a treatment for heart failure improves cardiac output should study patients who are stable enough for elective hemodynamic monitoring. Similarly, in a trial of depression, patients who are likely to return and who can provide the data needed for depression inventories are sought. In contrast, in a pragmatic trial, the general goal is to include patients who represent the population seen in clinical practice and whom the organizers of the study believe can make a plausible case for benefit in outcome. From this perspective, the number of entry and exclusion criteria should be minimized since the rate of enrollment will be inversely proportional to the number of criteria. In this broadening of entry criteria, particular effort is made to include patients with severe disease and comorbidities since they will likely be treated in practice.

Thus, an explanatory trial focuses on very specific criteria to elucidate a biological principle, whereas a large pragmatic trial should employ inclusion criteria that mimic what would happen if the treatment were to be employed in practice.

7.3. Data Collection Form

The data collection form provides the information on which the results of the trial are built; if an item is not included on the data collection form, it will obviously not be available at the end of the trial. On the other hand, the likelihood of collecting accurate information is inversely proportional to the amount of data collected. In an explanatory trial, patient enrollment is generally not an issue since small sample sizes are needed. However, in a pragmatic trial there is almost always an imperative to enroll patients as quickly as possible. Thus, a fundamental concept in pragmatic trials is to keep the data collection form as brief as possible.

The ultimate example of this concept is the ISIS approach of collecting only enough data to fill a single page in a clinical trial.[4] This approach has allowed the enrollment of tens of thousands of patients in mortality trials with no reimbursement to the health care providers enrolling patients. From this work have come some of the most important findings in cardiovascular disease (beta blockers reduce mortality in acute myocardial infarction, aspirin reduces mortality in acute myocardial infarction, and fibrinolytic therapy is broadly beneficial in acute myocardial infarction). Regardless of the length of the data collection form, it is critical to include only information that will be useful in analyzing the trial outcome.

7.4. Ancillary Therapy

Decisions about the use of nonstudy therapies in a clinical trial are critical to its validity and generalizability. Including therapies that will interact in a negative way with the experimental agent could ruin the chance to detect a clinically important treatment advance. Especially in a physiological experiment, interfering with the primary question by using another therapy would be a serious problem.

Alternatively, in a pragmatic trial the goal is to evaluate the therapy in the context in which it will be used. Since clinical practice is not managed by a prespecified algorithm, and many confounding situations can arise, evaluation of the experimental therapy in the context of such an approach is likely to give an unrealistic approximation of the likely impact of the therapy in clinical practice. For this reason, unless a specific detrimental interaction is known, pragmatic trials avoid prescribing particular ancillary therapeutic regimens. One exception is the encouragement (but not the requirement) to follow clinical practice guidelines if they exist for the disease being addressed by the trial.

7.5. Multiple Randomization

Until recently, enrolling a patient in multiple simultaneous clinical trials was considered to be ethically questionable. The origin of this ethical concern is unclear, but it seems to have arisen from a general impression that clinical research exposes patients to risks they would not experience in clinical practice, implying greater detriment from more clinical research and violation of the principles of beneficence and justice if a few subjects took this risk for the benefit of the broader population. More recently, the concept has been proposed that when the best treatment is not known, randomization is desirable (the uncertainty principle). Stimulated by the apparent need to develop multiple therapies simultaneously in HIV-AIDS treatment, the concept of multiple randomization has been reconsidered. Furthermore, access to clinical trials is increasingly recognized as a benefit rather than a burden, partially because of improved patient care in general in clinical trials.

Factorial trial designs represent a specific approach to multiple randomizations with advantages from statistical and clinical perspectives. Because most patients are now treated with multiple therapies, the factorial design represents a clear method to determine whether therapies add to each other, work synergistically, or nullify the effects of one or both therapies being tested. As long as a significant interaction does not exist between the two therapies being tested, both can be tested in a factorial design with a sample size similar to the size needed to test one therapy.

7.6. Pick the Winner

There is no other effective way to develop therapies than measuring intermediate physiologic end points in the early phases of study. Based on favorable physiologic responses, some therapies are brought forward for more extensive studies, and others are eschewed. However, other approaches to winnowing the possible doses or intensities of therapy must be developed after initial physiological evaluation since these physiological end points are unreliable. One such approach is the "pick the winner" approach. In this design (Fig. 18-4), several doses or intensities of the therapy are devised, and at regular intervals during the trial an independent data and safety monitoring committee evaluates clinical outcomes with the goal of dropping arms of the study according to prespecified criteria.

8. LEGAL AND ETHICAL ISSUES

8.1. Medical Justification

Each of the proposed treatments in the trial must be within the realm of currently acceptable medical prac-

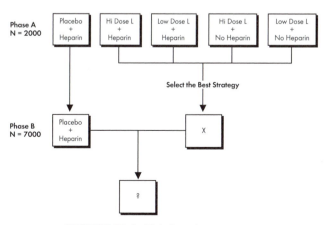

FIGURE 18-4 Pick-the-winner strategy.

tice for the specific medical condition of the patient. Difficulties with consideration of medical justification typically arise in two areas: Studies are generally initiated because there is reason to believe that one therapeutic approach is better than another, and many currently accepted therapies have never been subjected to the type of scrutiny being applied to new therapies. These factors create a dilemma for the practitioner, who may be uncomfortable with protocols that require a change in standard practice. An approach to this dilemma, called the "uncertainty principle," has been proposed. In this scheme, the clinician should be comfortable offering entry into a study at any time he or she is substantially uncertain about whether one of the treatments is in fact better. The patient, of course, is given the opportunity to review the situation and make a decision, but for most patients the physician's recommendation will be a critical factor in deciding whether to participate in a study.

8.2. Groups of Patients versus Individuals

The ethical balance typically depends on the good of larger numbers of patients versus the good of individuals involved in the trial. Examples are accumulating in which a therapy appeared to be better based on preliminary results or small studies and then was shown to be inferior based on adequately sized studies.[20] These experiences have led some authorities to argue that clinical practice should not change until a highly statistically significant difference in outcome is demonstrated.[21] Indeed, the standard for acceptance of a drug for labeling by the cardiorenal group at the FDA is two adequate and well-controlled trials, each independently reaching statistical significance. If the alpha for each trial is 0.05, an alpha value of 0.0025 (0.05×0.05) would be needed for both to be positive. The counterargument is that the physician advising the individual patient should let the patient know which treatment is most likely to lead to the best outcome. In fact, Bayesian calculations could be used to provide running estimates of the likelihood that one treatment is better. In the typical general construct of large pragmatic trials, however, this approach is not taken: Applying the ethical principles enumerated previously, an effort is made to accrue enough negative outcomes in a trial that a definitive result is achieved with a high degree of statistical significance and narrow confidence intervals.

An increasing area of confusion is the distinction between clinical investigation and measures taken to improve the quality of care as an administrative matter. The argument has been made that the former requires individual patient informed consent, whereas the latter

falls under the purview of the process of medical care and does not require individual consent.

Several special situations must be considered in studies in an emergency medical situation, which often allows insufficient time for explaining the research project in exacting detail and for obtaining informed consent. In treating acute stroke or myocardial infarction, the time to administration of therapy is a critical determinant of outcome, and time spent considering participation in a protocol could increase the risk of death. Accordingly, the use of an abbreviated consent to participate followed by a more detailed explanation later during the hospitalization has been sanctioned. Collins and Peto have made a compelling case that the slow, cumbersome informed consent form used in the United States in ISIS-2 actually resulted in the unnecessary deaths of a large number of patients with acute myocardial infarction.[22]

An even more complex situation occurs in research concerning treatment of cardiac or respiratory arrest. Clinical investigation in this field almost came to a halt because of the impossibility of obtaining informed consent. After considerable national debate, such research is now being done only after careful consideration by the community of providers and citizens about the potential merits of the proposed research.

A situation at least as complex exists for patients with psychiatric disturbances. Considerable discussion continues about the appropriate circumstances in which to obtain consent and to continue the patient in the trial as his or her clinical state changes.

8.3. Blinding

Blinding is essential in most explanatory trials since the opportunity for bias is substantial. In most pragmatic trials, blinding is also greatly preferred to reduce bias in the assessment of outcome. Single blinding refers to blinding of the patient (but not the investigator) to the therapy being given. Double blinding refers to blinding of both the patient and the investigator, whereas triple blinding refers to a double-blinded study in which the committee monitoring the trial is also blinded to which group is receiving which treatment. Despite the rarity of deceit in clinical research, examples of incorrect results due to bias in trials without blinding[23] and with single-blind studies reinforce the value of blinding.[24]

However, when blinding would prevent a true test of a treatment strategy, other methods must be used to ensure objectivity. The clearest example is a trial of surgical versus medical therapy; in this situation, the patient and the primary physician cannot remain blinded. A similar situation exists when the administration of one therapy is markedly different than the other. In some cases, a "double-dummy" technique (in which the comparative therapies each have a placebo) can be used, but often this approach leads to too much complexity.

Given the large number of effective therapies, an increasing problem will be the lack of availability of placebo. Manufacturing a placebo that cannot be distinguished from the active therapy and that cannot affect outcome is a complex and expensive effort. Often when a new therapy is compared with an old therapy, or two available therapies are compared, one of the commercial parties will not cooperate since the manufacturer of the established therapy has nothing to gain by participating in a comparative trial with a new therapy. Since a placebo needs to mimic the active therapy enough that the blind cannot be broken, the successful performance of a placebo-controlled trial depends on the cooperation and participation of the manufacturers of both therapies.

In other circumstances, blinding is simply not possible, particularly when the intervention is behavioral or surgical. Interestingly, in some circumstances sham surgical incisions have been used successfully to ensure that high-cost, high-risk surgical procedures were being evaluated with maximum objectivity.

8.4. End Point Adjudication

One of the complexities of large pragmatic trials arises when blinding is not feasible or desirable. In order to minimize the chance for bias, every effort must be made to ensure that randomization is proper (not revealed so that investigators can direct patients to particular therapies)[25] and that end points are obtained fairly. End point ascertainment methods include blinded observers at the sites and clinical events adjudication committees that can review objective data in a blinded manner independent of the site judgment.

Since most important end points (other than death) require a judgment, unbiased assessment of end points is essential in trials without treatment blinding. This point has been made vividly in trials of cardiovascular devices. In the initial Coronary Angioplasty versus Excisional Atherectomy Trial comparing directional coronary atherectomy with balloon angioplasty, the majority of myocardial infarctions were not noted on the case report form, despite electrocardiographic and enzymatic evidence of these events.[26] Even in a blinded trial, the end points of myocardial infarction, recurrent ischemia, and new or recurrent heart failure are recorded subjectively enough that independent judgment is thought to be helpful in most cases.[27]

8.5. Intensity of Intervention

When a therapeutic intervention is tested, one must always consider whether its intensity is appropriate. This issue is quite obvious in the dosing of drugs. In recent trials of direct thrombin inhibitors, a twofold error in the dosing of hirudin resulted in a significant increase in the risk of intracranial hemorrhage.[28–30] Perhaps even more important, when the target range for activated partial thromboplastin time (aPTT) was modestly increased for heparin therapy, the actual observed aPTT increased by only 8 seconds, but the intracranial hemorrhage rate with heparin increased to an unacceptable range. Correction of the dosages of hirudin and heparin brought clinical outcomes for both agents to the acceptable range.[31,32]

This same issue also exists in behavioral or policy interventions. A trial using prognostic and normative information to assist in end-of-life decision making, the Study to Understand Prognoses and Preferences for Outcomes and Risks of Treatments (SUPPORT), failed to change behavior, perhaps because the strength of the intervention was not adequate to truly affect the practitioners.[33] The major strategic question is how to design appropriate explanatory studies to define the most likely effective strength of the intervention prior to embarking on a large pragmatic trial.

8.6. Surrogate End Points

The quest to circumvent the need for large sample sizes in clinical trials continues to fuel interest in surrogate markers. The hope has been that small studies could be used to develop pathophysiological constructs to determine the strength of the intervention for definitive evaluation or to replace the need for a definitive intervention trial altogether. Unfortunately, this approach has led to a number of therapeutic misadventures (Table 18-3). Antiarrhythmic drugs were developed based on their ability to reduce ventricular arrhythmias on ambulatory monitoring. When the Cardiac Arrhythmia Suppression Trial was terminated prematurely because of a higher mortality with therapies that had been shown to reduce ventricular arrhythmias on monitoring, it became clear that this surrogate marker was inappropriate.[34] Similarly, studies developing dosing for heart failure therapies have used improvement in cardiac output as a surrogate marker. A succession of inotropic (milrinone and ibopamine) and vasodilator (flosequinan and prostacyclin) compounds have been shown to improve hemodynamics in the short term, but the results of long-term therapeutic trials have been disastrous.[35] Recently, concern has been raised about blood pressure lowering drugs; two compounds that are equally effec-

TABLE 18-3   Speculation on Reasons for Failure of Surrogate End Points

Disease and Intervention	End Points Surrogate	End Points Clinical	Reason for Failure[a] A	B	C	D
Cardiologic disorder						
Arrhythmia						
Encainide; flecainide	Ventricular arrhythmias	Survival		+		++
Quinidine; lidocaine	Atrial fibrillation	Survival		+		++
Congestive heart failure						
Milrinone; flosequinan	Cardiac output; ejection fraction	Survival		+		++
Elevated lipid levels						
Fibrates; hormones; diet; lovastatin	Cholesterol levels	Survival		+		++
Elevated blood pressure						
Calcium channel blockers	Blood pressure	Myocardial infarction; survival		+		++
Cancer						
Prevention						
Finasteride	Prostate biopsy	Symptoms; survival	++[b]			
Advanced disease						
Fluorouracil plus leucovorin	Tumor shrinkage	Survival		+		++
Other diseases						
HIV infection or AIDS						
Antiretroviral agents	CD4 levels; viral load	AIDS events; survival		+	+	+
Osteoporosis						
Sodium fluoride	Bone mineral density	Bone fractures	+			+
Chronic granulomatous disease						
Interferon-γ	Bacterial killing; superoxide production	Serious infection			++	

[a]A, surrogate end point not in causal pathway of the disease process; B, of several causal pathways of the disease, the intervention only affects the pathway mediated through the surrogate; C, the surrogate is not in the pathway of the intervention's effect or is insensitive to its effect; D, the intervention has mechanisms of action that are independent of the disease process.

[b]In settings in which only latent disease is prevented.

AIDS, acquired immunodeficiency syndrome; HIV, human immunodeficiency virus; +, likely or plausible; ++, very likely.

Adapted from Fleming TR, DeMets DL. Surrogate end points in clinical trials: Are we being misled? *Ann Intern Med* 1996;125:607.

tive in lowering blood pressure may have very different effects on mortality and other major clinical outcomes.[36]

These lessons about surrogate end points have important implications for clinicians and for clinical trial design. Titrating therapy to a physiologic end point may or may not be the correct approach to improving the outcome of the patient. In the administration of oral beta-blocking agents to patients with heart failure, hemodynamics commonly deteriorate before they improve. Thus, physiologic surrogates as therapeutic targets in individual patients should be validated in populations before they are accepted as standard practice.

8.7. Conflict of Interest

The concept of an investigator completely free of bias is a theoretical ideal that is not achievable. The degree of bias or conflict of interest can be considered in a graded fashion. Investigators should not have a direct financial interest in an industry sponsor of a clinical trial. Paid consultancies are also considered to be beyond the scope of acceptable relationship with industry. Compensation for work done on a clinical research project should be reasonable for the work performed, and it should be handled through an explicit contract. Perhaps the most universal and common conflict in clinical investigation is the bias of the investigator because of a belief in a particular concept. Blinding greatly reduces this risk, but the vast majority of clinical studies cannot be blinded. Failure to keep an open mind about the basic results of the investigation can cause the researcher to miss critical discoveries.

Several documents have explicitly laid out the guidelines for governing conflict of interest (Table 18-4). In addition, attention has focused on the responsibility of those who write editorials to be free of any conflict of interest.[37]

TABLE 18-4 Conflict-of-Interest Guidelines in Cardiovascular Clinical Trials and
Medical Organizations

	Stock, Equity, Interest	Consultancy	Honoraria, Educational Program Payments	Travel Expenses	Financial Time Window
Multicenter cardiovascular trials					
Post-CABG	No	No	Not addressed	Not addressed	Until date of publication
BARI	No	No	Not addressed	Not addressed	Not addressed
TIMI phases III and IV	No	No	Not addressed	Not addressed	1 year after presentation
GUSTO	No	No	No	No	1 year after publication
Medical organizations					
American Medical Association	No	Disclosure	Disclosure	Not addressed	
NIH/ADAMHA (rejected)[a]	No	No	No	Not addressed	
American College of Cardiology (ACC)	Disclosure to ACC if >$10,000	Disclosure to ACC if >$10,000	Disclosure to ACC if >$10,000	Not addressed	
Harvard Medical School	No	No	Not addressed	Not addressed	
British Cardiac Society	Disclosure with publication	Disclosure with publication	Disclosure with publication	Not addressed	
American Federation for Clinical Research	No	Disclosure with lectures	Disclosure with lectures	Not addressed	
American Heart Association (AHA)	Disclosure if invited speaker, committee	Disclosure if invited speaker, committee	Not permitted when representing AHA	Not addressed	

[a]Proposed guidelines September 1989.

BARI, Bypass Angioplasty Revascularization Investigation; CABG, coronary artery bypass graft; GUSTO, Global Utilization of Streptokinase and rt-PA for Occluded Coronary Arteries; NIH/ADAMHA, National Institutes of Health/Alcohol, Drug Abuse, Mental Health Administration; No, not permitted; TIMI, Thrombolysis in Myocardial Infarction.

Adapted from Topol EJ, *et al.* Patient safety and conflict of interest in clinical trials. *J Am Coll Cardiol* 1992;19:1123–1128.

8.8. Special Issues with Device Trials

Trials of medical devices raise special issues that deserve careful consideration. In comparisons of devices with other devices or medical therapy, the orientation of the clinician implanting the devices is often complicated by the fact that the technical skill of the clinician is an integral component of the success or failure of the therapy. Therefore, failure of therapy can be interpreted as a failure of the physician as well as the device. Obviously, in most device trials blinding of therapy is also impossible. For these reasons, particular focus on methodology is required in the assessment of clinical outcomes in device trials. Ideally, clinical outcomes should be assessed by a blinded review mechanism, and studies should be designed by groups including investigators who do not have a particular interest in the device-related outcomes but who have expertise in the disease-specific outcomes or clinical research methodology.

9. HYPOTHESIS FORMULATION

9.1. Primary Hypothesis

Every clinical study should have a primary hypothesis. The goal of the study design is to develop a hypothesis that allows the most important question from the viewpoint of the investigators to be answered without ambiguity. This issue is obvious in clinical trials, but in observational studies the appropriate approach to the problem is much less clear. Often, the investigator is tempted to "dredge" the data; no method of tracking multiple analyses exists to develop considerations related to multiple hypothesis testing.

9.2. Secondary and Tertiary Hypotheses

The data collection form provides an information infrastructure for organizing questions to be answered by the trial. A number of secondary hypotheses will be of interest to investigators, including analyses of the relationship between patient characteristics and treatment effect.

In addition to answering questions about the therapy being evaluated, the study can address questions concerning other aspects of the diagnosis, treatment, or outcomes of the disease. Constructing pathophysiological substudies embedded in larger clinical outcome studies has been especially rewarding. The GUSTO-I trial convincingly demonstrated the relationship between coronary perfusion, left ventricular function, and mortality in a systematic substudy.[38]

Finally, many interesting issues about medical practice can be addressed through ancillary studies of clinical trials. Comparisons of outcomes in Canada and the United States[39,40] and regional variations in the United States[41] have provided insight into medical practice.

9.3. Intention to Treat

One of the most important concepts in the interpretation of clinical trials is that of intention to treat. Exclusion of patients who were randomized into a trial leads to bias that cannot be quantified; therefore, the results of the trial cannot be interpreted with confidence.

The purpose of randomization is to ensure the random distribution of any factors, both known and unknown, that might affect the outcomes of the patients randomly allocated to one treatment or the other. Any post-randomization deletion of patients weakens the assurance that the randomized groups are at equal risk before treatment. Nevertheless, there are several common situations in which it may be reasonable to drop patients from analysis.

In blinded trials, when patients are randomized but do not receive the treatment, it is reasonable to have a study plan that would drop these patients from the primary analysis. The plan can call for substitution of additional patients to fulfill the planned sample size. When this happens, extensive analyses must be done to ensure that there was no bias in determining who was not treated. In unblinded trials, dropping patients who do not receive the treatment is treacherous and should not be allowed. Similarly, withdrawing patients from analysis after treatment has started cannot be permitted in trials designed to determine whether a therapy should be used in practice since the opportunity to "drop out without being counted" does not exist when a therapy is given in practice.

10. PUBLICATION BIAS

Clinical trials with negative findings are much less likely to be published than those with positive results. Approximately 85% of studies published in medical journals report positive results.[42] In a sobering analysis, Simes[43] found that a review of published literature showed combination chemotherapy for advanced ovarian cancer to be beneficial, whereas a review of published and unpublished trials together showed that the therapy had no significant effect. Dickerson and colleagues[44] found substantial evidence of negative reporting bias in a review of clinical trials protocols submitted to Oxford University and Johns Hopkins University. In particular, industry-sponsored research with negative results was unlikely to be published.

Awareness of this publication bias should lead to several specific actions by researchers and practitioners. First, researchers must strive to make clinical research results available to the scientific community through publication, regardless of whether those results are consistent with preconceived notions. Second, the reader must be cautious in interpreting studies showing positive results out of the context of independent confirmation; certainty that unpublished negative results from another investigation do not exist cannot be ensured in most cases. Finally, researchers working on systematic overviews must use all available means to search for both published and unpublished findings; since positive results are much more likely to be published, combining only published findings can lead to the wrong conclusion.

11. STATISTICAL CONSIDERATIONS

11.1. Type I Error and Multiple Comparisons

The hypothesis testing in a clinical study may be thought of as setting up a "straw man" that the effects of the two treatments being compared are identical. The goal of statistical testing is to determine whether this straw man hypothesis should be accepted or rejected based on probabilities. The type I error (alpha) is the probability of rejecting the null hypothesis when it is correct. Since clinicians have been trained in a simple, dichotomous mode of thinking, as if the p value was the only measure of probability, the type I error is generally designated at an alpha level of 0.05. However, if the same question is asked repeatedly, or if multiple subgroups within a trial are evaluated, the likelihood of finding a "nominal" p value of less than

0.05 increases substantially.[45] When evaluating the meaning of a p value, the clinician should be aware of the number of tests of significance performed and the importance placed on the p value by the investigator as a function of multiple comparisons.

11.2. Type II Error and Sample Size

The type II error (beta) is the probability of inappropriately accepting the null hypothesis (no difference in treatment effect) when a true difference in outcome exists. The power of a study (1-beta) is the probability of rejecting the null hypothesis appropriately. This probability is critically dependent on (1) the difference in outcomes observed between the treatments and (2) the number of primary end points. A common error in thinking about statistical power is to assume that the number of patients determines the power; rather, it is the number of end points.

The precision with which the primary end point can be measured also affects the power of the study; end points that can be measured precisely require fewer patients. An example is the use of sestamibi estimated myocardial infarct size. Measuring the area at risk before reperfusion and then measuring final infarct size can dramatically reduce the variance of the end point measure by providing an estimate of salvage rather than simply infarct size.[46] As is often the case, however, the more precise measure is more difficult to obtain, leading to great difficulty in finding sites that can perform the study; in many cases, the time required to complete the study is as important as the number of patients needed. This same argument is one of the primary motivators in the detailed quality control measures typically employed when instruments are developed and administered in trials of behavioral therapy or psychiatry.

For studies using physiological end points, using the continuous measure generally will increase the power to detect a difference. In restenosis trials, the number of patients needed to detect a reduction in diameter stenosis below 50% is greater than the number of patients needed to detect a difference in the mean or median diameter stenosis or minimal luminal diameter.

A review of the *New England Journal of Medicine* in 1978 determined that 67 of 71 negative studies had made a significant (more than 10% chance of missing a 25% treatment effect) type II error, and that 50 of the 71 trials had more than a 10% chance of missing a 50% treatment effect.[47] Unfortunately, the situation has not improved sufficiently since that time. The most common reasons for failing to complete studies with adequate power include inadequate funding for the project and loss of enthusiasm by the investigators.

It is highly desirable to have a power of at least 80% when conducting a clinical trial; 90% power is preferable. Discarding a good idea or a good therapy because of a study that had little chance of detecting a true difference is obviously an unfortunate circumstance. One of the most difficult concepts to grasp is that a study with little power to detect a true difference not only has little chance of demonstrating a significant difference in favor of the better treatment but also the direction of the observed treatment effect is highly unpredictable because of random variation with small samples. There is an overwhelming tendency to assume that if the observed effect is in the wrong direction in a small study, the therapy is not promising, whereas if the observed effect is in the expected direction but the p value is insignificant, the reason for the insignificant p value is an inadequate sample size. We can avoid these problems by designing and conducting adequately sized clinical trials.

Observational comparisons are at least as likely as randomized trials to include too few patients. However, observational studies rarely include power calculations. The same type of calculations commonly used in randomized trials can be used to place an observed effect with a p value > 0.05 in perspective. By discussing the minimal clinically important difference and providing the reader with an estimate of the probability of finding such a difference if it existed, the author of an observational study can place the study in much sharper perspective.

11.3. Equivalence

The concept of equivalence will become increasingly important in today's cost-conscious environment. Where an effective therapy already exists, the substitution of a less expensive (but clinically equivalent) therapy is attractive. In these positive control studies, substantial effort is required to define equivalence. Sample size estimates require the designation of a difference below which the therapies would be considered equivalent and above which one therapy would be considered superior to the other. Sample sizes are often larger than the requirements to demonstrate if one therapy is clearly superior to the other.

A common example of an issue that will increasingly arise concerns the substitution of a less expensive treatment for another that is already known to be effective. Table 18-5 gives the sample sizes for ensuring that a new treatment does not increase the risk of an event by 1, 2, or 3% for a disease with a 10% event rate with an already effective treatment.

TABLE 18-5   Equivalence Same Size Estimates

- Standard treatment mortality estimated at 10%
- A new, less expensive therapy is developed
- How many patients will it take to prove "equivalence" (alpha = .05; beta = .10)?

Old Treatment No Worse Than	No. of Patients
1% increase in mortality	32,582
2% increase in mortality	8,575
3% increase in mortality	3,998

Clinicians must be wary of studies that are designed with a substantial type II error resulting from an inadequate number of end points, with the result that the two treatments are thought to be equivalent because the *p* value is greater than 0.05. This approach could lead to a gradual loss of effectiveness of therapy for cardiovascular conditions. If we were willing to accept that a therapy for acute myocardial infarction with 1% higher mortality in an absolute sense was "equivalent," and we examined four new less expensive therapies that met those criteria, we could cause a significant erosion of the progress in reducing acute myocardial infarction mortality.

Several major cardiovascular trials have been based on the concept of equivalence. The Bypass Angioplasty Revascularization Investigation (BARI) was predicated on the hypothesis that percutaneous intervention would not increase the 5-year mortality rate beyond 2%.[48] The International Joint Efficacy Comparison of Thrombolytics study examined the hypothesis that reteplase retained at least half the benefit of streptokinase in reducing mortality compared with conservative therapy.[49] Several thrombolytic trials, such as the Assessment of the Safety and Efficacy of a New Thrombolytic (ASSENT-II) study, investigating novel mutant plasminogen activators tested the hypothesis that the new agent is equivalent to accelerated alteplase in terms of mortality effect.[50]

As more positive control trials are being done, a greater appreciation is also being developed for the concept that an equivalence trial need not be as large as previously believed if the new treatment is indeed slightly better than the old treatment. With only a modest trend toward benefit, the sample size required to rule out a clinically important negative effect can be quite small.

11.4. Sample Size Calculations

The critical step in a sample size calculation, whether for a trial to determine a difference or to test for equiv-

alence, is the estimate of the minimally important clinical difference (MID). By reviewing the proposed therapy in comparison with the currently available therapy, the investigators should endeavor to determine the smallest difference in the primary end point that would change clinical practice. Practical considerations may not allow a sample size large enough to evaluate the MID, but the number should be known. In some cases, the disease may be too rare to enroll enough patients, whereas in other cases the treatment may be too expensive or the sponsor may not have enough money. Once the MID and the financial status of the trial are established, the sample size can be determined easily from a variety of published computer algorithms or tables. It is useful for investigators to produce plots or tables to enable them to see the effects of small variations in event rates or treatment effects on the needed sample size. In the GUSTO-I trial,[14] the sample size was set after a series of international meetings determined that saving an additional 1 life per 100 patients treated with a new thrombolytic regimen would be a clinically meaningful advance. With this knowledge, and a range of possible underlying mortality rates in the control group, a table was produced demonstrating that a 1% absolute reduction (difference of 1 life per 100 treated) or a 15% relative reduction could be detected with 90% certainty by including 10,000 patients per arm.

12. META-ANALYSIS AND SYSTEMATIC OVERVIEWS

Regardless of the goal of performing adequately sized clinical trials, clinicians are often faced with therapeutic dilemmas in which there is not enough evidence to be certain of the best treatment. The basic principle of combining medical data from multiple sources seems intuitively appealing since this approach results in greater statistical power. However, the trade-off is the assumption that the studies being combined are similar enough that the combined result will be valid. Inevitably, this assumption rests on expert opinion.

Table 18-6 provides an approach to reading meta-analyses. The most common problems with meta-analyses are combining studies with different designs or outcomes and failing to find unpublished negative studies. There is no question about the critical importance of a full literature search, as well as involvement of experts in the field of interest to ensure that all relevant information is included. Statistical methods have been developed to help in the assessment of systematic

TABLE 18-6   How to Read and Interpret a Meta-Analysis

Are the results of the study valid?
  Primary guides
    Does the overview address a focused clinical question?
    Are the criteria used to select articles for inclusion
      appropriate?
  Secondary guides
    Is it unlikely that important, relevant studies were missed?
    Is the validity of the included studies appraised?
    Are the assessments of studies reproducible?
    Are the results similar from study to study?
What are the results?
  What are the overall results of the review?
  How precise are the results?
Will the results help me in caring for my patients?
  Can the results be applied to my patient?
  Are all clinically important outcomes considered?
  Are the benefits worth the risks and costs?

publication bias.[51] Another complex issue involves the assessment of the quality of individual studies within a systematic overview. Statistical methods have been proposed for differential weighting as a function of quality,[52] but these have not been adopted on a wide scale.

The methodology of the statistical evaluation of pooled information has recently been a source of tremendous interest. The fixed effects model assumes that the trials being evaluated are homogeneous with regard to estimate of the outcome; given the uncertainties expressed previously, the assumption of homogeneity seems unlikely. Accordingly, a random effects model has been developed that considers not only the variation within trials but also the random error between trials.[53]

An interesting approach to meta-analyses, termed cumulative meta-analysis, has been developed.[54] With this approach, as data become available from new trials, they are combined with findings of previous trials with the calculation of a cumulative test of significance. In theory, this approach should allow the medical community to determine the point at which the new therapy should be adopted into practice. Another variation on the theme of meta-analysis is meta-regression, a method allowing the evaluation of covariate effects within multiple trials to explain heterogeneity in observed results.

The apparent lack of congruence between the results of meta-analyses of small trials and subsequent results of large trials has been a source of substantial confusion. Meta-analyses of small trials found that both magnesium therapy and nitrates provided a substantial (>25%) reduction in the mortality of patients with

myocardial infarction.[55] The large ISIS-4 trial found no significant effect on mortality of either treatment.[56] Although many causes have been posited for these discrepancies, a definitive explanation does not exist. The major message seems to be that large numbers of patients are needed to be certain of the effect of a therapy. Guidelines for reading meta-analyses are given in Table 18-6.

13. UNDERSTANDING COVARIATES AND SUBGROUPS

Because of the insatiable curiosity of clinicians and patients about whether different responses to treatment may be seen in different types of patients, an analysis of trial results as a function of baseline characteristics is inevitable. Traditionally, this analysis has been performed using a subgroup analysis, in which the treatment effect is estimated as a function of baseline characteristics taken one at a time (e.g., age, sex, or weight). This approach has been called a "false-positive result machine" but might just as well be referred to as a "false-negative result machine." The false positives are generated because of the problem of multiple comparisons; by chance alone, a significant difference will be apparent in at least 1 in 20 subgroups even if there is absolutely no treatment effect. In 1980, Lee et al.[45] randomly split a population of 1073 into two hypothetical treatment groups (the treatments were actually identical) and found a difference in survival in a subgroup of patients, with a p value of <0.05.[45]

At the same time, given the large number of patients needed to demonstrate an important treatment effect, dividing the population into subgroups markedly reduces the power to detect differences when they are real. Consider a treatment that reduces mortality 15% in a population equally divided between men and women, with a p value for the treatment effect of 0.03. If the treatment effect is identical for men and women, the approximate p value will be 0.06 within each subgroup since each group is half as large. It would obviously be foolish to conclude that the treatment was effective in the overall population but not in men or women.

A more appropriate and conservative method would be to develop a statistical model predicting outcome with regard to the primary end point for the trial and then evaluate the effect of the treatment as an effect of each covariate after adjusting for the effects of the general prognostic model. This type of analysis, known as a treatment by covariate interaction analysis, assumes that the treatment effect is homogeneous in subgroups examined unless a definitive difference is observed.

An example of this approach occurred in the Prospective Randomized Amlodipine Survival Evaluation (PRAISE) trial,[57] which observed a reduction in mortality with amlodipine in patients with idiopathic dilated cardiomyopathy but not in patients with ischemic cardiomyopathy. This case was particularly interesting because this subgroup was prespecified to the extent that the randomization was stratified. However, the reason for the stratification was that the trial designers expected that amlodipine would be ineffective in patients without cardiovascular disease; the opposite occurred. Responsibly, the trial organization mounted a confirmatory second trial. In the completed follow-up trial (PRAISE-2) the special benefit in the idiopathic dilated cardiomyopathy group was not replicated.

In the BARI trial,[48] a post hoc analysis showed a significant benefit of bypass surgery in patients with treated diabetes mellitus but not in other patients. This analysis had not been specified before the trial started enrollment, nor had the randomization been stratified. However, the data and safety monitoring committee had asked for an analysis of this issue based on concerns raised in an acute revascularization trial.

The test for interaction has been said to have limited power so that a borderline significant result should attract clinical interest, although there is inadequate experience with this test to be comfortable about interpretation.

14. THERAPEUTIC TRUISMS

A review of recent clinical trials points out that many commonly held beliefs about clinical practice need to be challenged. If these assumptions are shown to be less solid than previously believed, a substantial change in the pace of clinical investigation will be needed.

Frequently, medical trainees have been taught that variations in practice patterns are inconsequential. The common observation that different practitioners treat the same problem in different ways has been tolerated because of the general belief that these differences did not matter. Clinical trials have demonstrated, however, that small changes in practice patterns for epidemic diseases can have a sizable impact. An example is the extreme variation in recommendations regarding the preferred aPTT for patients treated with unfractionated heparin anticoagulation. Based on the pathophysiological surrogate of arterial patency,[58,59] the GUSTO investigators adjusted the recommended aPTT upward in the transition from GUSTO-I to GUSTO-IIa. The average 8-second increase in aPTT resulted in a doubling of the rate of intracranial hemorrhage in patients treated with thrombolytic therapy and heparin.[28] When the heparin dose was reduced in GUSTO-IIb, the intracranial hemorrhage rate reproduced that observed in GUSTO-I.[31] Clinical trials have demonstrated that small changes in practice patterns for epidemic diseases can have a sizable impact.[60]

Another ingrained belief of medical training is that observation of the patient will provide evidence for changing treatment. Although no one would dispute the importance of following symptoms, many acute therapies have effects that cannot be judged in a short time, and many therapies for chronic illness prevent adverse outcomes in patients with very few symptoms. For example, in treating acute congestive heart failure, inotropic agents improve cardiac output early after initiation of therapy but lead to a higher risk of death. Beta blockers cause symptomatic deterioration acutely but appear to improve long-term outcome. Mibefradil was effective in reducing angina and improving exercise tolerance, but it also caused sudden death in an alarming proportion of patients, leading to its removal from the market.

Similarly, the standard method of determining the dose of a drug has been to measure physiological end points. In a sense, this technique represents a surrogate end point approach. No field has more impressively demonstrated the futility of this approach than that involving the treatment of heart failure. A variety of vasodilator and inotropic therapies have been shown to improve hemodynamics in the acute phase but subsequently were shown to increase mortality. The experience with heparin and warfarin has taught us that large numbers of patients are required to understand the relationship between the dose of a drug and clinical outcome.

Finally, the maxim "do no harm" has been a fundamental tenet of medical practice. However, most biologically potent therapies cause harm in some patients while helping others. The recent emphasis on the neurological complications of bypass surgery provides ample demonstration that a therapy that saves lives can also lead to complications in individuals.[61] Intracranial hemorrhage resulting from thrombolytic therapy exemplifies a therapy that is beneficial for populations but has devastating effects on some individuals. Similarly, beta blockade causes early deterioration in many patients with heart failure, but the longer term survival benefits are documented in multiple clinical trials. The patients who are harmed can be detected easily, but those patients whose lives are saved cannot be detected.

15. STUDY ORGANIZATION

Whether the investigator is contemplating a large or small trial, the general principles of organization of the study should be the same (Fig. 18-5). A balance of interest and power must be created to ensure that after the trial is designed, the experiment can be performed without bias and the interpretation will be generalizable.

15.1. Executive Functions

15.1.1. The Steering Committee

In a large trial, the steering committee is a critical component of the study organization. This group designs, executes, and disseminates the study. A diverse steering committee, providing multiple points of view representing biology, biostatistics, and clinical medicine, is more likely to organize a trial that will withstand external scrutiny. This same principle holds for small trials; an individual investigator, by organizing a committee of peers, can avoid egocentric thinking about a clinical trial.

The principal investigator plays a key role in the function of the trial as a whole, and a healthy interaction with the steering committee can provide a stimulating exchange of ideas on how best to conduct a trial. The principal trial statistician is also crucial in making final decisions about study design and data analysis. An executive committee can be useful, providing a small group to make real-time critical decisions for the trial organization. This committee should typically include the sponsor, principal investigator, statistician, and key representatives from the steering committee and the data coordinating center.

15.1.2. The Data and Safety Monitoring Committee

The data and safety monitoring committee (DSMC) is constructed to oversee the safety of the trial from the point of view of the patients being enrolled. The DSMC should include clinical experts, biostatisticians, and, sometimes, medical ethicists; these individuals should have no financial interest, emotional attachment, or other investment in the therapies being studied. Committee members have access to otherwise confidential data during the course of the trial, allowing decisions to be made on the basis of information that, if made available to investigators, could compromise their objectivity. The DSMC also carries an increasingly scrutinized ethical obligation to review the management of the trial in the broadest sense, in conjunction with each institutional review board, to ensure that patients are treated according to ethical principles.

The role of the DSMC has become a topic of significant global interest. Little has been published about the function of these groups, yet they hold considerable

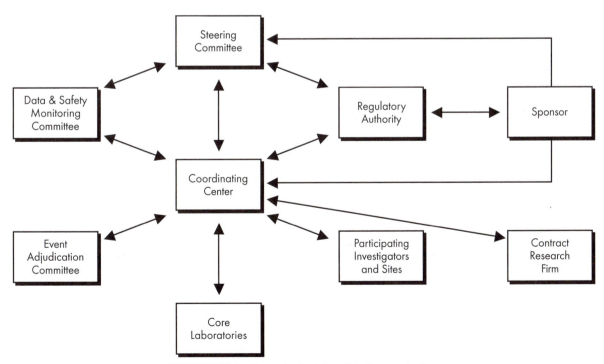

FIGURE 18-5 General principles of study organization.

power over the functioning of clinical trials. The National Cancer Institute (NCI) has published guidelines for DSMCs in NCI-funded trials.

15.1.3. The Institutional Review Board

The institutional review board (IRB) continues to play a critical role in the conduct of all types of clinical research. Approval by the IRB is generally required for any type of research, even if the research is not funded by an external source. The IRB should consist of physicians with expertise in clinical trials as well as representatives with expertise in medical ethics and representatives of society in the community in which the research is being conducted. As with the DSMC, the IRB function has come under scrutiny, especially from government agencies charged with ensuring the protection of human subjects.

Several types of studies are typically exempted from the IRB process, including studies of public behavior, research on educational practices, and studies of existing data in which the research data cannot be linked to individual subjects. Surveys and interviews may also be exempted when the subjects are not identified and the data have a very low likelihood of leading to a lawsuit, financial loss, or reduced employability of the subject.

15.1.4. Regulatory Authorities

Government regulatory authorities have played a major role in the conduct of clinical research. Requirements by the FDA and other national health authorities provide the rules by which industry-sponsored clinical trials are conducted. In general, regulatory requirements include interpretation of fundamental guidelines to ensure adherence to human rights and ethical standards. The FDA and equivalent international authorities are charged with ensuring that drugs and devices that are marketed are safe and effective (a charge with broad leeway for interpretation). Importantly, in the United States there is no mandate to assess comparative effectiveness or cost-effectiveness.

15.1.5. Industry or Government Sponsors

Having provided funding for the study, the sponsor of a clinical trial understandably prefers to be heavily involved in the conduct of the study. Worldwide, the majority of clinical investigation is now done either directly by the pharmaceutical or medical device industry or indirectly by for-profit clinical research organizations. This approach seems reasonable and desirable for explanatory trials, but pragmatic trials, if not overseen by an independent steering committee,

run a greater risk of bias because the sponsor of a study has a large financial stake in the success of the therapy being tested. Even in the case of government sponsorship, trials are frequently performed as a result of political agendas, with much to be gained or lost for individuals within the scientific community depending on the result. All of these issues speak to the advantage of a diverse steering committee to manage the general functioning of a large pragmatic clinical trial.

15.2. Coordinating Functions

The coordinating functions of large pragmatic trials may be viewed as a whole as in Fig. 18-6. The fundamental functions are intellectual and scientific leadership, site management, and data management. These core functions are supported by a number of administrative functions, including information technology, finance, human resources, contracts management, pharmacy and supplies distribution, and randomization services. Given the magnitude of large trials, each project is dependent on the administrative and leadership skills of a project manager and a principal investigator. A major weakness in any one of these functions can lead to a failure of the entire effort, whereas excellence in all components creates a fulfilling and exciting experience.

15.2.1. Intellectual Leadership

The roles of the principal investigator and chief statistician are critical to the success of the trial organization. Not only must these leaders provide conceptual

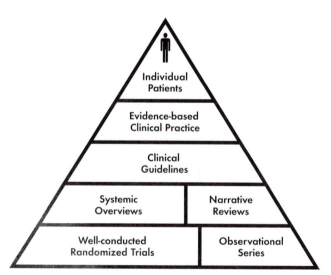

FIGURE 18-6 Pyramid describing the functions and interrelationships involved in coordinating large pragmatic trials.

expertise but also their knowledge of successful approaches to operational concepts in the real world can be the difference between a successful trial and a failure. In large trials, a small change in protocol or addition of one more visit or testing procedure can add huge amounts to the cost. The larger the trial, the greater the economy of scale in materials, supplies, and organization. For example, a simple protocol amendment can take months and cost hundreds of thousands of dollars (and even more in terms of the delay) to successfully go through multiple national regulatory authorities and hundreds of IRBs. Often, the intellectual leaders of a trial are not in touch with the practical implications of their decisions.

15.2.2. Data Coordinating Center

The data coordinating center (DCC) is responsible for coordinating the collection and cleaning of data for the clinical trial. In this role, the DCC must comply with an increasing number of regulations governing both the quality of the data and its confidentiality. Furthermore, the DCC should produce routine reports that allow the trial organization and the DSMC to oversee the conduct of the trial and ensure that the question the human subjects volunteered to answer is being addressed properly. The DCC must be able to harness data from multiple formats, including standard paper data records, remote data entry, and information from third-party computer sources.

15.2.3. Site Management Organization

In large trials, managing the conduct of the sites is a major effort. It requires training and communications programs and also regulatory affairs to ensure compliance with federal and nonfederal guidelines. In large trials, international enrollment is typically needed, and the organization must be able to provide in-service education and study monitoring in multiple languages while also complying with regulations from multiple national authorities.

Given the imperative to initiate and complete trials efficiently, site management groups are increasingly concerned with maintaining good relations with clinical sites that perform well in clinical trials. These relationships are often fostered by ongoing educational programs aimed at increasing the quality of participation at the sites and rewarding the personnel by supplementing their knowledge of conducting and interpreting clinical trials. In addition, metrics are being implemented to measure functions such as recruitment rates, protocol deviations, data quality,

and personnel turnover. Sites that perform well are selected for future trials to increase efficiency.

15.3. Supporting Functions

15.3.1. Information Technology

Large trials are increasingly dependent on a successful information platform. A competitive coordinating center is dependent on first-rate information technology expertise to maintain communication, often on a global basis.

15.3.2. Finance

Even in relatively simple, low-paying trials, an effective financial system is critical to success. Study budgets are typically divided, with approximately half of the funds going to the sites performing the study and half going to coordinating efforts, with this money frequently split among multiple contractors and subcontractors. Since payments to the sites typically depend on documented activities at the sites, the flow of cash needs to be carefully regulated to avoid either overpayment or underpayment. Furthermore, the coordinating staff needs to be carefully monitored to ensure that study funds are appropriately allocated to get the work done without overspending.

15.3.3. Human Resources

The staff required to conduct large pragmatic trials comprises a diverse group of employees with different needs. Information technology expertise in particular is difficult to acquire and maintain in this very competitive environment. The second most difficult group of employees to find and retain is qualified project leaders. The knowledge base required and the skills needed are extraordinary.

15.3.4. Contracts Management

For better or worse, our global society is increasingly directed by legal contracts. In a typical large clinical trial, a huge number of contracts must be in place, and an entourage of lawyers is busily looking out for the interests of each entity. The sponsor typically will contract with entities to coordinate portions of the trial. The number of coordination contracts depends on whether the primary coordination is done internally within an industry or government sponsor, or contracted out to one or more contract research organizations. Each participating site then has a contract with the sponsor, the coordinating organization, or both.

15.3.5. Pharmacy and Supplies

The production and distribution of study materials, including those required for inservice, and actual supplies, such as investigational drugs and devices, require considerable expertise. The knowledge required ranges from practical skills such as knowing how to package materials for maximum understanding by the sites to expertise in "just-in-time" distribution across international boundaries and working knowledge of the mountains of regulations regarding good clinical practice and good manufacturing practice for clinical trials.

15.3.6. Randomization Services

A fundamental principle of large pragmatic trials is that proper randomization will balance for baseline risk, including both known and unknown risk factors, to allow for an unbiased comparison of treatments. In large multicenter trials, this issue takes on tremendous complexity. Because sealed envelopes are notoriously prone to tampering in large, geographically distributed trials, central randomization has been viewed as superior. This can be accomplished by either telephone randomization or, increasingly, an interactive voice randomization service (IVRS). IVRS has the advantage of providing instantaneous access to global networks of investigators and automatic recording of patient characteristics at the time of randomization.

15.3.7. Project Management

Within the context of the sponsoring organization with its ongoing priorities, the coordinating entities with their ongoing priorities, and the sites with their ongoing priorities, someone must ensure that the individual project is completed on time and on budget. This responsibility is typically shared by the principal investigator, the project manager, and a sponsor representative (a project officer for government grants and contracts and a scientific or business manager for industry trials). This task should ideally fall to people with skills in organizational management, finance, regulatory affairs, medical affairs, leadership, and personnel management. Individuals with the skills to carry out these responsibilities are difficult to find. Interestingly, few educational programs are in place to train these people despite the huge shortage of qualified individuals.

16. INTEGRATION INTO PRACTICE

Because the goal of clinical investigation is to improve the care of patients, integrating the findings of a clinical investigation into practice must be undertaken carefully. The old method of each practitioner reading the literature and making individual decisions is inadequate. Recognition of this deficit has led to a variety of efforts to synthesize empirical information into practice guidelines (Fig. 18-7). These guidelines may be considered as different paths to the top of the mountain, with several different routes acceptable as long as the difficulty and likelihood of success is known. In addition, large efforts such as the Cochrane collaboration[62] are attempting to make available systematic overviews of clinical trials in most major therapeutic areas.

This effort has been integrated into a "cycle of quality" construct in which disease registries form the basis for capturing continuous information about the quality of care for populations.[63] Within these populations, clinical trials with adequate size and performed in relevant study cohorts can lead to definitive clinical practice guidelines. These guidelines can then form the basis for performance measures that are used to capture the quality of care delivered. Ultimately, gaps in clinical outcomes in this system can be used to define the need for new technologies and behavioral approaches. Increasingly, the linkage of interoperable electronic health records, professional–society-driven quality efforts, and patient/payer-driven interest in improving outcomes is leading to a system in which clinical trials are embedded within disease registries so that the total population can be understood and the implementation of findings into practice can be measured.[64]

17. CONTROVERSIES AND PERSONAL PERSPECTIVE

17.1. Governmental Regulation

During the past several years, as the United States' health care system has evolved into a business model, controversy has arisen over the role of government regulation of clinical research. Regulation comes in two forms: (1) regulatory authorities charged with ensuring the safety and efficacy of products sold for medical purposes and (2) ethical regulation of investigation done for the public good. Although the usual business approach is to assume that the marketplace will sort out beneficial therapies from those that do not work, history belies this belief in terms of medical therapies. Determining medical benefit is complex, and observation made in ignorance of the principles of controlled clinical trials is inadequate; without requirements for controlled trials, many detrimental therapies

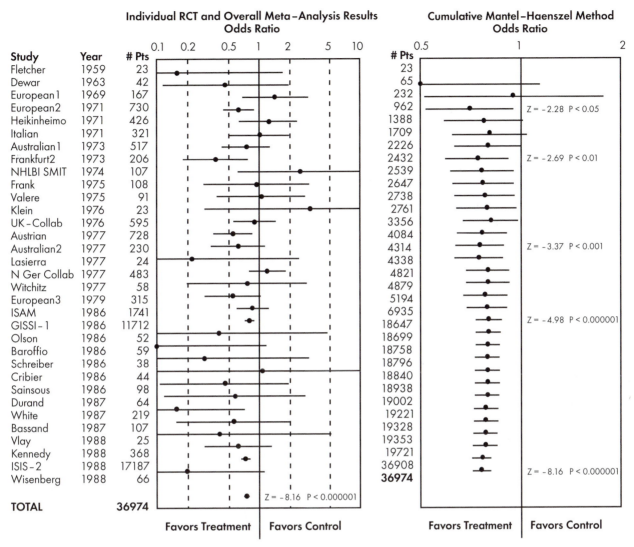

FIGURE 18-7 The mountain of evidence.

would likely still be in use. In areas in which outcome studies have not been required, the potential public risk is unknown and may be substantial. The initial formation of the FDA occurred because of the sale of pharmaceutical products that were lethally contaminated, and every new regulation has resulted from a similar tragedy. The mandate of the FDA extends only to the point of determining that a therapy is safe and effective compared with no therapy, however, and does not include the determination of which therapy is more effective when two effective therapies are available. Additionally, the mandate with regard to pharmaceuticals is more stringent, with a general requirement for substantial evidence of clinical benefit from randomized controlled trials, whereas device regulations allow less definitive evidence of clinical benefit. The avalanche of new food additives and

behavioral therapies has been exempted from the scientific demonstration of efficacy. Once a pharmaceutical or device product is on the market, the FDA does rigorously control the advertising and promotion that can be done with that product, but the FDA can deal only with errors of commission rather than errors of omission. Thus, a therapy that lowers blood pressure and has been shown in adequate controlled trials to lower mortality and stroke rates can advertise that benefit. A comparative therapy that lowers blood pressure but has not been shown to reduce mortality and stroke can be advertised as being effective in lowering blood pressure without mentioning the absence of clinical outcome trials.

My belief is that the major onus is on the medical profession to ensure that adequate clinical trials are done to demonstrate which therapies are most effective. The industry that develops new therapies is obligated only to meet the regulations for being marketed. Managed care organizations are focused on cost rather than outcome, and in many cases incentives exist to impede new therapies from clinical use when they increase expense. Only with the active involvement of health care providers will the necessary studies be done to allow the best information to be available to make choices among active therapies.

17.2. Composite and Surrogate End Points

The number of important clinical questions far exceeds our ability to address them. For this reason, there is a great temptation to perform clinical trials using composite end points (a combination of several clinical outcomes in one end point) and surrogate end points. Composite end points have the advantage of creating an outcome scale that allows for categorical or continuous measures that add power to the study. However, the less important end points can drive the difference, when small differences in a more important end point such as mortality could not possibly be detected. Similarly, surrogate end points are hampered by constant uncertainty about whether the surrogate will work, particularly since the difference between therapies might relate to an outcome dictated by an unexpected effect of the therapy. For these reasons, surrogates and composites should be regarded as intermediate approaches to determining which therapies need to be definitively tested. One exception is when a minor modification is made to a drug or device and a consensus exists that the modification will not fundamentally alter the effect of the treatment. Practitioners and investigators need to eschew the temptation to substitute surrogates for adequate measures of health outcomes in sufficiently large populations. Instead, the clinical community should focus on more efficient methods of aggregating large amounts of information at a lower cost.

17.3. Randomized Trials versus Observational Studies

Randomized trials and observational studies represent different approaches to answering questions. Traditionally, the randomized trial has been hampered by enrollment of selected patients who are perhaps not representative of clinical practice and by concern about human experimentation. Given the current organization of medicine, the number of therapeutic questions will unavoidably far exceed the number of randomized trials that can be performed. Having been involved

in numerous randomized trials and observational treatment comparisons, I believe that much more effort is needed to expand the ability to perform randomized trials. Until the technology revolution allowed the rapid accumulation of information from multiple sites and the introduction of the large pragmatic trial methodology, it seemed that observational studies were needed to fill in the huge gaps in clinical therapeutics.[65] With current capabilities, the major role of observational studies should be to generate hypotheses, to fill in information about small variations in practice, and to deal with the extrapolations necessary to inform decisions about chronic disease management. Because most patients live for years with chronic diseases, questions such as what type of lipid lowering agent to use, how to lower blood sugar in diabetes, how to treat depression in the teenage years, and whether to use antibody-based therapies chronically for arthritis will have long-term implications requiring extrapolation beyond the time frame that can reasonably be measured. The technology half-life is now so short that long-term randomized trials run the risk of being historical artifacts.

17.4. Sharing of Information

Given the for-profit nature of the medical products and pharmaceutical industry and the increasingly financial orientation of the health care delivery system, a risk exists that important medical information will not be available. The financial incentive in the medical products industry has stimulated tremendous creativity and should be preserved. A potential conflict exists, however, between the professional ethic of the health care provider to share information to improve the delivery of health care in broader terms and the need of financially driven delivery systems to maintain a competitive edge. Ironically, this potential conflict emerges at a time when the ability to rapidly share data and interpretation of data has improved in an exponential fashion. Patients expect that when medical research is done, the information will be shared; one could argue that information sharing to advance health in general is a critical part of the ethical contract between provider and patient when informed consent is obtained.

Recent U.S. government legislation has mandated a national registry for clinical trials involving "serious and life-threatening diseases." This mandate has the potential for making all clinical trials public enough that results will be made available to avoid damaging reflections on the sponsor, even if the results are unfavorable to a particular product or belief.

18. THE FUTURE

Because of the continuing explosion of knowledge about human biology and society's inability to increase the financial outlay for the unfettered use of biology and technology, the future of clinical investigation is bright. Without quantitative information, there is no rational method to make decisions about what will be supported in medical practice and what will be eschewed.

During the next several years, practitioners will make increasing use of electronic health records that will generate computerized databases to capture information at the point of care. Early efforts in this area, focusing on procedure reports to meet mandates from payers and quality reviewers, will be replaced by systems aimed at capturing information about the entire course of the patient's encounter with the health care system. Although the impetus for this approach will come from those who pay for medical care, practitioners will find the information to be useful for justifying rational medical care and improving the overall efficiency with which care is delivered. Multimedia presentations will allow the clinician to view medical records and imaging studies simultaneously in the clinic or in the hospital. In order to efficiently exchange information, the nomenclature of diagnosis, treatments, and outcomes will progressively become standardized.

The computerized management of information will be part of an inevitable coalescence of practitioners into integrated health systems. In order to efficiently care for populations of patients at a reasonable cost, practitioners will work in large, geographically linked, economically interdependent groups. This integration of health systems will propel outcomes research into a new era in which strategies of care can be tested over time and refined in continuous learning processes for health care providers.

Although integrated health care systems will provide the structure for medical practice, global communications will provide mechanisms to quickly answer questions about diagnosis, prevention, prognosis, and treatment of common and uncommon diseases. The ability to aggregate information about thousands of patients in multiple health systems will change the critical issues facing clinical researchers. Increasingly, attention will be diverted from efforts to obtain data, and much effort will be required to develop efficient means of analyzing and interpreting the types of information that will be available.

Ultimately, leading practitioners will band together in global networks oriented toward treating illnesses of common interest. When a specific question requiring randomization is identified, the studies will be much simpler because the randomization can simply be added to the computerized database and information that currently requires construction of a clinical trials database will be immediately accessible without additional work. Information systems will be designed to provide continuous feedback of information to clinicians, supporting rational decisions about therapy. In essence, a continuous series of observational studies will be in progress, assessing outcomes as a function of diagnostic processes and therapeutic strategies.

References

1. Fisher RA, Mackenzie WA. Studies of crop variation: II. The manurial response of different potato varieties. *J Agric Sci* 1923;13:315.
2. Lilienfeld AM. Ceteris paribus: The evolution of the clinical trial. *Bull Hist Med* 1982;56:1–18.
3. Medical Research Council. Streptomycin treatment of pulmonary tuberculosis. *Br Med J* 1948;2:769–782.
4. ISIS-1 (First International Study of Infarct Survival) Collaborative Group. Randomised trial of intravenous atenolol among 16,027 cases of suspected acute myocardial infarction: ISIS-1. *Lancet* 1986;2:57–66.
5. Rogers WJ, Bowlby LJ, Chandra NC, *et al.* Treatment of myocardial infarction in the United States (1990 to 1993). Observations from the National Registry of Myocardial Infarction. *Circulation* 1994;90:2103–2114.
6. Forrow L, Taylor WC, Arnold RM. Absolutely relative: How research results are summarized can affect treatment decisions. *Am J Med* 1992;92:121–124.
7. Society of Thoracic Surgeons Database. Available at www.sts.org/sections/stsnationaldatabase/. Accessed November 30, 2005.
8. Roberts R, Rodriguez W, Murphy D, Crescenzi T. Pediatric drug labeling: Improving the safety and efficacy of pediatric therapies. *JAMA* 2003;290:905–911.
9. Alexander KP, Peterson ED. Evidence-based care for all patients. *Am J Med* 2003;114:333–335.
10. Bobbio M, Demichelis B, Giustetto G. Completeness of reporting trial results: Effect on physicians' willingness to prescribe. *Lancet* 1994;343:1209–1211.
11. Naylor CD, Chen E, Strauss B. Measured enthusiasm: Does the method of reporting trial results alter perceptions of therapeutic effectiveness? *Ann Intern Med* 1992;117:916–921.
12. Fibrinolytic Therapy Trialists' (FTT) Collaborative Group. Indications for fibrinolytic therapy in suspected acute myocardial infarction: Collaborative overview of early mortality and major morbidity results from all randomised trials of more than 1000 patients. *Lancet* 1994;343:311–322.
13. Olson CM, Rennie D, Cook D, *et al.* Publication bias in editorial decision making. *JAMA* 2002;287:2825–2828.
14. The GUSTO Investigators. An international randomized trial comparing four thrombolytic strategies for acute myocardial infarction. *N Engl J Med* 1993;329:673–682.
15. DeAngelis C, Drazen JM, Frizelle FA, *et al.* Clinical trial registration: A statement from the International Committee of Medical Journal Editors. *N Engl J Med* 2004;351:1250–1251.

16. Gross R, Strom BL. Toward improved adverse event/suspected adverse drug reaction reporting. *Pharmacoepidemiol Drug Saf* 2003;12:89–91.

17. Davidson MH. Rosuvastatin safety: Lessons from the FDA review and post-approval surveillance. *Expert Opin Drug Saf* 2004;3:547–557.

18. Topol EJ. Arthritis medicines and cardiovascular events—"house of coxibs." *JAMA* 2005;293:366–368.

19. Tunis SR, Stryer DB, Clancy CM. Practical clinical trials: Increasing the value of clinical research for decision making in clinical and health policy. *JAMA* 2003;290:1624–1632.

20. Lo B, Fiegal D, Cummins S, Hulley SB. Addressing ethical issues. In Hulley SB, Cummings SR (eds.) *Designing Clinical Research*, pp. 151–157. Baltimore, Williams & Wilkins, 1988.

21. Peto R, Collins R, Gray R. Large-scale randomized evidence: Large, simple trials and overview of trials. *J Clin Epidemiol* 1995;48:23–40.

22. Collins R, Peto R. Introducing new treatments for cancer: Practical, ethical and legal problems. In Williams CJ (ed.) *The Ethics of Clinical Trials*. New York, Wiley, 1992.

23. Karlowski TR, Chalmers TC, Frenkel LD, *et al.* Ascorbic acid for the common cold: A prophylactic and therapeutic trial. *JAMA* 1975;231:1038–1042.

24. Henkin RI, Schechter PJ, Friedewald WT, *et al.* A double-blind study of the effects of zinc sulfate on taste and smell dysfunction. *Am J Med Sci* 1976;272:285–299.

25. Farr BM, Gwaltney JM Jr. The problems of taste in placebo matching: An evaluation of zinc gluconate for the common cold. *J Chronic Dis* 1987;40:875–879.

26. Harrington RA, Lincoff AM, Califf RM, *et al.* for the CAVEAT Investigators. Characteristics and consequences of myocardial infarction after percutaneous coronary intervention: Insights from the Coronary Angioplasty versus Excisional Atherectomy Trial (CAVEAT). *J Am Coll Cardiol* 1995;25:1693–1699.

27. Mahaffey KW, Granger CB, Tardiff BE, *et al.* for the GUSTO-IIb Investigators. End point adjudication by a clinical events committee can impact the statistical outcome of a clinical trial: Results from GUSTO-IIb [Abstract]. *J Am Coll Cardiol* 1997;29(Suppl. A):410A.

28. GUSTO IIa Investigators. Randomized trial of intravenous heparin versus recombinant hirudin for acute coronary syndromes. *Circulation* 1994;90:1631–1637.

29. Antman EM. Hirudin in acute myocardial infarction. Safety report from the Thrombolysis and Thrombin Inhibition in Myocardial Infarction (TIMI) 9A Trial. *Circulation* 1994;90:1624–1630.

30. Molhoek GP, Laarman GJ, Lok DJ, *et al.* Angiographic dose-finding study with r-hirudin for the improvement of thrombolytic therapy with streptokinase (HIT-SK). *Eur Heart J* 1995;16(Suppl. D):33–37.

31. The GUSTO IIb Investigators. A comparison of recombinant hirudin with heparin for the treatment of acute coronary syndromes. *N Engl J Med* 1996;335:775–782.

32. Antman EM. Hirudin in acute myocardial infarction. Thrombolysis and Thrombin Inhibition in Myocardial Infarction (TIMI) 9B trial. *Circulation* 1996;94:863–865.

33. Connors AF, Jr., Speroff T, Dawson NV, *et al.* The effectiveness of right heart catheterization in the initial care of critically ill patients. *JAMA* 1996;276:889–897.

34. Pratt CM, Moye L. The Cardiac Arrhythmia Suppression Trial: Implications for anti-arrhythmic drug development. *J Clin Pharmacol* 1990;30:967–974.

35. Packer M. Calcium channel blockers in chronic heart failure: The risks of "physiologically rational" therapy [Editorial]. *Circulation* 1990;82:2254–2257.

36. Furberg CD, Wright JT, Davis BR, *et al.* Major cardiovascular events in hypertensive patients randomized to doxazosin vs chlorthalidone—The Antihypertensive and Lipid-Lowering Treatment to Prevent Heart Attack Trial (ALLHAT). *JAMA* 2000;283:1967–1975.

37. Angell M, Kassirer JP. Editorials and conflicts of interest [Editorial]. *N Engl J Med* 1996;335:1055–1056.

38. GUSTO Angiographic Investigators. The effects of tissue plasminogen activator, streptokinase, or both on coronary artery patency, ventricular function, and survival after acute myocardial infarction. *N Engl J Med* 1993;329:1615–1622.

39. Rouleau JL, Moye LA, Pfeffer MA, *et al.* for the SAVE Investigators. A comparison of management patters after acute myocardial infarction in the United States. *N Engl J Med* 1993;328:779–784.

40. Mark DB, Naylor CD, Hlatky MA, *et al.* Use of medical resources and quality of life after acute myocardial infarction in Canada and the United States. *N Engl J Med* 1994;331:1130–1135.

41. Pilote L, Califf RM, Sapp S, *et al.* Regional variation across the United States in the treatment of acute myocardial infarction. *N Engl J Med* 1995;333:565–572.

42. Dickersin K, Min YI. Publication bias: The problem that won't go away. *Ann N Y Acad Sci* 1993;703:135–146.

43. Simes RJ. Publication bias: The case for an international registry of clinical trials. *J Clin Oncol* 1986;4:1529–1541.

44. Dickersin K, Chan S, Chalmers TC, *et al.* Publication bias and clinical trials. *Controlled Clin Trials* 1987;8:343–353.

45. Lee KL, McNeer JF, Starmer CF, Harris PJ, Rosati RA. Clinical judgment and statistics: Lessons from a simulated randomized trial in coronary artery disease. *Circulation* 1980;61:508–515.

46. Gibbons RJ, Christian TF, Hopfenspirger M, Hodge DO, Bailey KR. Myocardium at risk and infarct size after thrombolytic therapy for acute myocardial infarction: Implications for the design of randomized trials of acute intervention. *J Am Coll Cardiol* 1994;24:616–623.

47. Freiman JA, Chalmers TC, Smith H Jr, Kuebler RR. The importance of beta, the type II error and sample size in the design and interpretation of the randomized control trial: Survey of 71 negative trials. *N Engl J Med* 1978;299:690–694.

48. BARI Investigators. Comparison of coronary bypass surgery with angioplasty in patients with multivessel disease. *N Engl J Med* 1996;335:217–225.

49. The International Joint Efficacy Comparison of Thrombolytics (INJECT) Investigators. Randomised, double-blind comparison of reteplase double-bolus administration with streptokinase in acute myocardial infarction: Trial to investigate equivalence. *Lancet* 1995;346:329–336.

50. Van de Werf F, Adgey J, Ardissino D, *et al.* Single-bolus tenecteplase compared with front-loaded alteplase in acute myocardial infarction: The ASSENT-2 double-blind randomised trial. *Lancet* 1999;354:716–722.

51. Begg C, Berlin J. Publication bias: A problem in interpreting medical data. *J R Stat Soc A* 1988;151:419–445.

52. Detsky A, Naylor C, O'Rourke K, *et al.* Incorporating variations in the quality of individual randomized trials into meta-analysis. *J Clin Epidemiol* 1992;45:255–265.

53. Berkey C, Hoaglin D, Mosteller F, *et al.* A random-effects regression model for meta-analysis. *Stat Med* 1995;14:395–411.

54. Lau J, Antman EM, Jimenez-Silva J, Kupelnick B, Mosteller F, Chalmers TC. Cumulative meta-analysis of therapeutic trials for myocardial infarction. *N Engl J Med* 1992;327:248–254.

55. Antman E. Randomized trial of magnesium for acute myocardial infarction: Big numbers do not tell the whole story. *Am J Cardiol* 1995;75:391–393.

56. ISIS-4 (Fourth International Study of Infarct Survival) Collaborative Group. ISIS-4: A randomised factorial trial assessing early oral captopril, oral mononitrate, and intravenous magnesium sulphate in 48,050 patients with suspected acute myocardial infarction. *Lancet* 1995;345:669–685.

57. Packer M, O'Connor CM, Ghali JK, *et al.* for the PRAISE Study Group. Effect of amlodipine on morbidity and mortality in severe chronic heart failure. *N Engl J Med* 1996;335: 1107–1114.

58. Hsia J, Kleiman NS, Aguirre FV, Chaitman BR, Roberts R, Ross AM. Heparin-induced prolongation of partial thromboplastin time after thrombolysis: Relation to coronary artery patency. *J Am Coll Cardiol* 1992;20:31–35.

59. ISIS-2 (Second International Study of Infarct Survival) Collaborative Group. Randomised trial of intravenous streptokinase, oral aspirin, both, or neither among 17,187 cases of suspected acute myocardial infarction: ISIS-2. *Lancet* 1988;2:349–360.

60. Arnout J, Simoons M, de Bono D, Rapold HJ, Collen D, Verstraete M. Correlation between level of heparinization and patency of the infarct-related coronary artery after treatment of acute myocardial infarction with alteplase (rt-PA). *J Am Coll Cardiol* 1992;20:513–519.

61. Roach GW, Kanchuger M, Mangano CM, *et al.* Adverse cerebral outcomes after coronary bypass surgery. *N Engl J Med* 1996;335:1857–1863.

62. Sackett DL, Oxman A, eds. *The Cochrane Collaboration Handbook.* Oxford: The Cochrane Collaboration, 1995.

63. Califf RM, Peterson ED, Gibbons RJ, *et al.* Integrating quality into the cycle of therapeutic development. *J Am Coll Cardiol* 2002;40:1895–1901.

64. Welke KF, Ferguson TB Jr, Coombs LP, *et al.* Validity of the Society of Thoracic Surgeons National Adult Cardiac Surgery Database. *Ann Thorac Surg* 2004;77:1137–1139.

65. Califf RM, Pryor DB, Greenfield JC Jr. Beyond randomized clinical trials: Applying clinical experience in the treatment of patients with coronary artery disease. *Circulation* 1986;74: 1191–1194.

19

Using Secondary Data in Statistical Analysis

BRADLEY D. FREEMAN,* STEVEN BANKS,† AND CHARLES NATANSON†

*Washington University School of Medicine, St. Louis, Missouri
†Critical Care Medicine Department, National Institutes of Health Clinical Center, Bethesda, Maryland

In 1976, Glass coined the term *meta-analysis* to describe "the statistical analysis of a large collection of results from individual literature for the purpose of integrating their respective findings."[1] More generally, meta-analysis refers to any systematic statistical method for combining data from independent clinical studies for two basic purposes. The first is to determine if similar treatment effects exist for a therapy among independent clinical studies and, if so, to estimate a net effect for this therapy. In this way, meta-analysis overcomes the limitation of interpreting a number of small, underpowered clinical trials. Alternatively, if treatment effects differ substantially for a therapy among independent clinical studies, the second purpose of a meta-analysis is to examine factors that may explain these differing effects. The statistical techniques commonly used in meta-analysis originated in the 1930s when Fisher and others, working in agricultural science, developed methods for extracting and analyzing data derived from a large number of individual experiments.[2–5] The recent popularity of this technique follows from the large increase in the number of clinical trials published in the past several decades.[6] Many of these studies, despite addressing nearly identical questions, often reached inconsistent conclusions. Techniques of secondary data analysis have been developed to resolve these inconsistencies, to more accurately quantify the effectiveness of the therapies evaluated, and to generate new hypotheses for further clinical testing.[7]

The purpose of this chapter is to briefly review the general concepts and potential limitations of meta-analysis. Furthermore, we illustrate how a meta-analysis is performed by applying this technique to studies examining the use of anti-inflammatory therapies in sepsis. Although these methods may be used for both observational and experimental studies, we restrict our discussion to the focus of this book, the analysis of clinical trials.

1. TECHNIQUES OF META-ANALYSIS

The steps involved in performing a meta-analysis are formulating the question; identifying pertinent studies; assessing the characteristics and quality of these studies for inclusion or exclusion; and extracting, analyzing, and reporting the data.

1.1. Formulating the Question

As with clinical research in general, the first step in a meta-analysis is to formulate the question for study and determine that it can be approached by this technique. The validity and importance of the meta-analysis are contingent on this first step. A poorly conceived research hypothesis will usually lead to a meta-analysis of dubious value.

1.2. Identifying Studies for Meta-Analysis

Analogous to subject selection in patient-oriented research, the protocols for study inclusion in a meta-analysis should be prospectively formulated, systematic, and explicit. Identification of published studies usually begins with a search of personal reference files and electronic and online databases such as MEDLINE, Current Contents, Best Evidence, Cochrane, and HealthSTAR. The title and abstract of studies identified by these methods are perused to exclude any that lack relevance. The full texts of the remaining articles are retrieved and thoroughly studied. Likewise, the reference lists of these articles are reviewed to identify potentially pertinent publications not retrieved by these techniques.

A problem inherent in any method of article retrieval is *publication bias*. Publication bias refers to the phenomenon that studies published in peer-refereed journals are much more likely to report statistically significant results than are studies that report either a nonsignificant or a null conclusion.[8] Consequently, published studies may not be representative of all studies that have been conducted addressing a specific clinical problem, may overrepresent those studies showing a "positive" result, and may partly underlie the discrepancies between some meta-analyses and large randomized trials.[9] The potential for publication bias may be limited by attempting to identify and include trials that are not published in peer-refereed journals, such as those presented at industry-sponsored or academic forums.[8]

1.3. Defining Eligibility Criteria and Data Abstraction

Meta-analyses of experimental studies ideally include randomized controlled trials that are similar in nature and in which the diagnosis, outcome, patient characteristics, and treatment groups are unambiguously defined. Frequently, however, studies satisfying these exacting standards are either nonexistent or limited in number, forcing investigators to be less discriminating in the studies selected for analysis. In one extreme, all available studies, regardless of size, design, or quality, could be included in the analysis. Although this approach may result in an analysis that is broadly representative of the studies addressing a specific question, the inclusion of trials that are poorly designed or conducted may compromise accuracy. Alternatively, exclusion of studies for methodologic reasons may increase the statistical validity of the analysis but may limit the ability to generalize the findings. In practice, an investigator must *a priori* establish inclusion and exclusion criteria, specifically addressing such features as trial design, size, treatment protocols, patient characteristics, outcomes, and follow-up evaluation.[10] Furthermore, some authors advocate minimizing the effects of variable quality by the use of weighting techniques.[11] Once a study has been selected for inclusion, data should be extracted, preferably onto structured forms that have been pretested to ensure interobserver reliability and quality of the information.

1.4. Data Analysis

To compare studies, a common measure of treatment effect must be determined. The odds ratio (with accompanying confidence interval) is the metric most frequently used to represent net treatment effect when the outcome is one of two possible states (e.g., survival or death). In addition to providing a point estimate of therapeutic effectiveness, comparison of the magnitude and direction (e.g., positive or negative) of odds ratios among studies provides an indication of statistical homogeneity or heterogeneity. Studies that have odds ratios that are similar in both magnitude and direction are considered statistically homogenous.[7,12] However, statistical homogeneity and clinical homogeneity are not synonymous. Thus, the use of an odds ratio to summarize the effect of treatment for a group of statistically homogeneous studies should be further justified by determining if certain systematic differences exist between trials. That is, in addition to reporting an odds ratio to describe an overall effect of treatment, investigators should also report whether a significant statistical interaction exists between the intervention under study and relevant trial characteristics (e.g., it must be determined that the odds ratio is an accurate estimate of treatment effect independent of factors such as study design, number of patients enrolled, and comorbid illness).[13]

The treatment effects associated with studies that are statistically heterogeneous are dissimilar in their direction and/or magnitude.[7,12] Combining statistically heterogeneous studies for the purposes of calculating a single estimate of treatment effect, such as an odds ratio for survival, produces a result of questionable validity. However, as in the case of analysis of statistically homogeneous studies, it is often informative to examine if a significant interaction exists between study variables (e.g., size, design, or severity of illness) and treatment effect. This may provide insight as to how the collection of studies being analyzed may be described by multiple point estimates or odds ratios. Furthermore, examination of factors that may have produced interactions with the treatment under investigation may not only provide insight into why heterogeneity between the studies exists but also suggest additional hypotheses for further testing.

1.5. Complete Enrollment of Studies

In designing a clinical study, the number of patients who must be enrolled to address the question of interest with the desired degree of statistical certainty and power is established *a priori*. That is, after enrollment is completed and the study is analyzed, it is not acceptable to simply enroll additional patients and re-analyze the data. Likewise, once a set of studies is identified, the data are abstracted, and a meta-analysis is performed, it is not appropriate to alter *a priori* established inclusion criteria. However, if a sufficient number of studies addressing the same question are subsequently published following the completion of a meta-analysis, an argument can be made to include these studies in a separate analysis to determine if the previous conclusions are supported.[10]

2. META-ANALYSIS OF CLINICAL TRIALS OF ANTI-INFLAMMATORY AGENTS IN SEPSIS

As one example of the use of meta-analysis to address questions raised by a large body of research, we illustrate the application of this technique to clinical trials examining the use of anti-inflammatory agents in patients with sepsis.[14–25]

2.1. Background: The Role of Inflammation in Mediating Sepsis

One of the most intensely studied areas in the field of sepsis in recent decades has been the role of inflammation in mediating this syndrome. Interest in this area arose from extensive work in animal models in which the administration of mediators of the inflammatory response were found to produce hemodynamic and physiologic derangements indistinguishable from those occurring in severe sepsis.[26–29] Subsequently, it was shown that agents with anti-inflammatory activity diminished inflammation and improved outcome in these same animal models.[30–45] These studies culminated in the hypothesis that sepsis resulted from an excessive inflammatory reaction to infection.[14,46–51]

2.2. Formulating the Question

The hypothesis that sepsis is the result of an exaggerated inflammatory response has been tested in a large number of clinical trials in which mediator-specific anti-inflammatory agents have been administered to patients with this syndrome. In contrast to the findings in animal models, however, these agents were largely found to lack benefit or actually showed harm clinically.[25,46–48] The questions we sought to answer by meta-analysis were twofold. First, by pooling these various studies, would we increase statistical power sufficiently to show benefit, harm, or lack of effect with these agents, thereby confirming or refuting a role for inflammation in sepsis? Second, would the use of these techniques allow us to identify factors related to patients or trial design that would influence the effectiveness of these agents?

2.3. Identifying Studies for Meta-Analysis

We queried two databases (Embase and MEDLINE) to identify prospective studies of nonglucocorticoid agents with a purported anti-inflammatory activity in patients with sepsis.[47,48] To limit the effects of publication bias, we also reviewed the proceedings of scientific and industry-sponsored meetings devoted to sepsis held during the corresponding time frame.

2.4. Defining the Eligibility Criteria for the Meta-Analysis and Abstracting the Data

We included data from any prospective study that included a group that could be considered as a control population. We abstracted the following information from each study: inclusion and exclusion criteria, study design, treatment protocols, and survival rates. We identified 21 studies involving patients with sepsis and septic shock.[50,52–72] We excluded 3 of these studies because they lacked either survival data or a control group.[52–54] In the 18 remaining clinical studies, septic patients were treated with six different classes of anti-inflammatory agents: bradykinin antagonists, platelet-activating factor antagonists, monoclonal antibodies directed against tumor necrosis factor-α (TNF), prostaglandin antagonists, soluble TNF receptors, and interleukin-1 receptor antagonists.[50,55–75]

2.5. Analyzing the Data: Determining Homogeneity

In performing our analysis, it was necessary for us to determine whether the studies we had identified were sufficiently homogeneous so as to allow for estimation of a common net therapeutic effect. To do this, we first examined the effects of dose within each of the clinical trials in which multiple doses of an agent were administered. If no significant dose-dependent effects were found, we pooled these groups for analysis. Conversely, if dose-dependent treatment effects were identified within a study, we compared these treatment groups to all other treated patients in the analysis. If this comparison showed that no difference existed, then the treatment groups were combined. Alternatively, if a treatment difference was detected, the groups were analyzed separately.

Eleven of the 18 trials we identified examined multiple doses of anti-inflammatory agents.[55,59–64,67,69–71] In 9 of these 11 studies, no dose-dependent effects were found.[55,59–63,69–71] In 1 clinical trial, two doses of the study agent, anti-TNF monoclonal antibody, had effects on survival that differed significantly from each other but not from the other agents included in the analysis.[64] Thus, the data from patients receiving anti-TNF monoclonal antibody were pooled, regardless of dose. In contrast, a trial of a high-molecular-weight soluble TNF receptor showed that dose escalation was associated with a mortality rate that differed significantly from all other treated patients in the analysis (51 vs. 36%, respectively; $p < 0.05$).[48,67] Thus, the data from the patients receiving high doses of this agent were analyzed both separately and in conjunction with the data from the other patients in the analysis. We next determined that the studies conducted for each of the six classes of agents were sufficiently alike that they could be pooled for analysis. Likewise, when we compared the effects of the six classes of agents with each other, we found that they too could be combined for analysis. This allowed us to determine that the use of anti-inflammatory agents in patients with sepsis, both including and excluding patients receiving high doses of soluble TNF receptor, was associated with a small but statistically significant reduction in mortality (odds ratio, 1.12; 95% confidence interval, 1.01–1.25; $p = 0.04$). This finding is consistent with the hypothesis that the sepsis syndrome results, in part, from excessive inflammation.

2.6. Examining Interactions between Study Characteristics and Treatment Effects

The use of a single estimate of treatment effect in a meta-analysis should be further justified by determining that significant interactions do not exist between study characteristics and the therapy under evaluation.[13] Our data set allowed us to explore three factors that potentially influenced the effectiveness of anti-inflammatory therapies in patients with sepsis: trial size, study design, and control group mortality rate. We found that large studies—those enrolling more than 500 patients—consistently showed a small trend toward benefit with these agents, similar to the findings of our meta-analysis.[76] In contrast, small studies—those enrolling 250 patients or less—had variable results and were as likely to show benefit with anti-inflammatory treatment as harm (Fig. 19-1).[76] This effect of study size is consistent with the hypothesis that anti-inflammatory agents produce a marginal beneficial effect, and that only large studies are sufficiently powered to consistently demonstrate this benefit and small studies are underpowered and more susceptible to sampling error. We also examined the effect of study design (e.g., double blind vs. single blind) in our analysis. In theory, double-blind studies should have less potential for bias than single-blind studies. We found that whereas single-blind studies tended to be smaller and more variable in outcome, both double-blind and single-blind studies suggested the same small beneficial effect with these agents.[76] Thus, study design did not appear to alter the overall treatment effect in our analysis. Finally, we examined whether there was an interaction between the mortality rate in patients serving as controls and treatment effect in our analysis. We reasoned that variation in severity of illness, as reflected in variability in control group mortality rate, may influence the effectiveness of anti-inflammatory therapy. We found that overall, the mortality rate in control groups was constant, approximately 36%, for all the agents we examined.[48,76] The consistency of the control mortality rate indicated that these studies enrolled patients with comparable severity of illness and risk of death. Thus, control group mortality rates did not influence the effectiveness of anti-inflammatory therapy in these studies. In summary, through examination of factors that potentially interacted with treatment effect, we were both better able to understand the data in our analysis and confirm the validity of using a single point estimate to summarize this data.

3. CONCLUSIONS

Meta-analysis has evolved as a technique useful for summarizing a large number of clinical trials and for resolving discrepancies raised by these trials. There are many similarities between randomized controlled trials and meta-analyses. Both randomized trials and meta-analyses are designed to answer a scientifically valid question. Likewise, both techniques require that the patients or studies included, the data collected, and the analysis performed be prospectively planned, and that the results obtained be analyzed for factors that may potentially interact with treatment effect. Finally, both techniques deal with populations, not with single individuals. Thus, clinicians must use discretion when applying the conclusions derived from both these techniques to the individual patient.

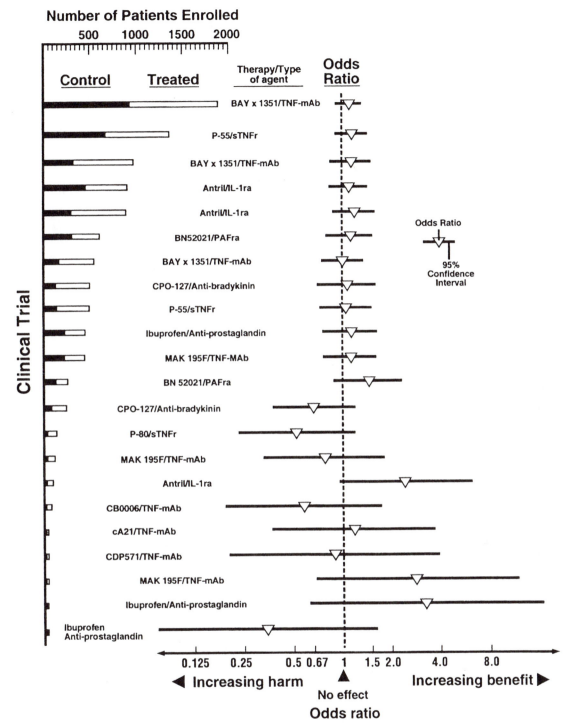

FIGURE 19-1 Survival odds ratios and 95% confidence intervals for clinical trials of anti-inflammatory agents in patients with sepsis. These trials are ranked in order by the total number of patients enrolled. The larger clinical trials (enrollment > 250 patients) showed comparable small beneficial effects. The smaller trials enrolling < 250 patients were equally likely to show beneficial or harmful trends. The results of these smaller trials are not inconsistent with those of the larger trials, as can be seen by the overlap of the 95% confidence intervals. Rather, the estimate of treatment effect in these smaller trials is less accurate because of sampling error. Modified from Natanson et al.,[76] with permission.

References

1. Glass GV. Primary, secondary, and meta-analysis research. *Education Res* 1976;5:3–8.
2. Fisher RA. *Statistical Methods for Research Workers*. London, Oliver and Boyd, 1932.
3. Tippet LHC. *The Methods of Statistics*. London, Williams & Northgate, 1931.
4. Pearson ES. The probability integral transformation for testing goodness of fit and combining independent tests of significance. *Biometrika* 1938;30:134–148.
5. Cochran WG. Problems arising in the analysis of a series of similar experiments. *J R Stat Soc Suppl. B* 1937;4:102–118.
6. Petitti DB. Introduction. In *Meta-Analysis, Decision Analysis, and Cost-Effectiveness Analysis*, 2nd ed., pp. 1–12. New York, Oxford University Press, 2000.
7. Lau J, Joannidis J, Schmid CH. Summing up evidence: One answer is not always enough. *Lancet* 1998;351:123–127.
8. Petitti DB. Information retrieval. In *Meta-Analysis, Decision Analysis, and Cost-Effectiveness Analysis*, 2nd ed., pp. 43–67. New York, Oxford University Press, 2000.
9. LeLorier J, Gregoire G, Benhaddad A, LaPierre J. Discrepancies between meta-analyses and subsequent large randomized, controlled trials. *N Engl J Med* 1997;337:536–542.
10. Pogue J, Yusuf S. Overcoming the limitations of current meta-analysis of randomized controlled trials. *Lancet* 1998;351:47–52.
11. Petitti DB. Advanced issues in meta-analysis. In *Meta-Analysis, Decision Analysis, and Cost-Effectiveness Analysis*, 2nd ed., pp. 75–93. New York, Oxford University Press, 2000.
12. Petitti DB. Statistical methods in meta-analysis. In *Meta-Analysis, Decision Analysis, and Cost-Effectiveness Analysis*, 2nd ed., pp. 94–118. New York, Oxford University Press, 2000.
13. Brand R, Kragt H. Importance of trends in the interpretation of an overall odds ratio in the meta-analysis of clinical trials. *Stat Med* 1992;ll:2077–2082.
14. Natanson C, *et al.* Selected treatment strategies for septic shock based on proposed mechanisms of pathogenesis. *Ann Intern Med* 1994;120:771–783.
15. Ziegler EJ, *et al.* Treatment of gram-negative bacteremia and shock with human antiserum to a mutant *Escherichia coli*. *N Engl J Med* 1982;307:1225–1230.
16. McCutchan JA, Wolf JL, Ziegler EL, Braude AL. Ineffectiveness of single-dose human antiserum to core glycolipid (*Escherichia coli* J5) for prophylaxis of bacteremic, gram-negative infection in patients with prolonged neutropenia. *J Suisse Med* 1983;113(Suppl.):40–55.
17. Baumgarner JD, *et al.* Prevention of gram-negative shock and death in surgical patients by antibody to endotoxin core glycolipid. *Lancet* 1985;ll:59–63.
18. The J5 Study Group. Treatment of severe infectious purpura in children with human plasma from donors immunized with *Escherichia coli* J5: A prospective double-blind study. *J Infect Dis* 1992;165:695–701.
19. Calandra T, *et al.* and the Swiss–Dutch J5 Immunoglobulin Study Group. Treatment of gram-negative septic shock with human IgG antibody to *Escherichia coli* J5: A prospective, double-blind, randomized trial. *J Infect Dis* 1988;158:312–319.
20. The Intravenous Immunoglobulin Collaborative Study Group. Prophylactic intravenous administration of standard immune globulin as compared with core-lipopolysaccharide immune globulin in patients at high risk of postsurgical infections. *N Engl J Med* 1992;327:234–240.
21. Greenman RL, *et al.* A controlled clinical trial of E5 murine monoclonal IgM antibody to endotoxin in the treatment of gram-negative sepsis. *JAMA* 1991;266:1097–1102.
22. Bone RC, *et al.* A second large controlled clinical study of E5, a monoclonal antibody to endotoxin: Results of a prospective, multicenter, randomized, controlled trial. *Crit Care Med* 1995;23:994–1006.
23. Ziegler EJ, *et al.* Treatment of gram-negative bacteremia and septic shock with HA-1A human monoclonal antibody against endotoxin. *N Engl J Med* 1991;324:429–436.
24. McClosky RV, *et al.* and the CHESS Trial Study Group. Treatment of septic shock with human monoclonal antibody HA-1A. *Ann Intern Med* 1994;121:l–5.
25. Freeman B, Natanson C. Anti-inflammatory therapies in sepsis and septic shock. *Exp Opin Invest Drugs* 2000;9:1651–1663.
26. Weil MH, Maclean CD, Fisseher MB, Spink WW. Studies of the circulatory changes in the dog produced by endotoxin from gram-negative microorganisms. *J Clin Invest* 1956;35:1191–1198.
27. Natanson C, *et al.* Endotoxin and tumor necrosis factor challenges in dogs simulate the cardiovascular profile of human septic shock. *J Exp Med* 1989;169:823–832.
28. Waage A, Espevik T. Interleukin-1 potentiates the lethal effects of tumor necrosis factor/cachectin in mice. *J Exp Med* 1988;167:1987–1992.
29. Okusawa S, *et al.* Interleukin-1 induces a shock like state in rabbits: Synergism with tumor necrosis factor and the effect of cyclooxygenase inhibition. *J Clin Invest* 1988;81:1162–1172.
30. Ohlsson K, *et al.* Interleukin-1 receptor antagonist reduces mortality from endotoxin shock. *Nature* 1990;348:550–552.
31. Wakabayashi G, *et al.* A specific receptor antagonist for interleukin-1 prevents *Escherichia coli*-induced shock in rabbits. *FASEB J* 1991;5:338–343.
32. Fisher E, *et al.* Interleukin-1 receptor blockade improves survival and hemodynamic performance in *Escherichia coli* septic shock, but fails to alter host responses to sublethal endotoxemia. *J Clin Invest* 1992;89:1551–1557.
33. Beutler B, Milsark IW, Cerami AC. Passive immunization against cachetin/tumor necrosis factor protects mice from lethal effect of endotoxin. *Science* 1985;229:869–871.
34. Suitters AJ, *et al.* Differential effect of isotype on efficacy of anti-tumor necrosis factor-α chimeric antibodies in experimental septic shock. *J Exp Med* 1994;179:849–856.
35. Bagby GJ, *et al.* Divergent efficacy of antibody to tumor necrosis factor-α in intravascular and peritonitis models. *J Infect Dis* 1991;163:83–88.
36. Mathison JC, Wolfson E, Ulevitch RJ. Participation of tumor necrosis factor in the mediation of gram negative bacterial lipopolysaccharide-induced injury in rabbits. *J Clin Invest* 1988;81:1925–1937.
37. Fiedler VB, *et al.* Monoclonal antibody to tumor necrosis factor-α prevents lethal endotoxin sepsis in adult rhesus monkeys. *J Lab Clin Med* 1992;120:574–588.
38. Emerson TE, *et al.* Efficacy of monoclonal antibody against tumor necrosis factor-α in an endotoxemic baboon model. *Circ Shock* 1992;38:75–84.
39. Eskandari MK, *et al.* Anti-tumor necrosis factor antibody therapy fails to prevent lethality after cecal ligation and puncture or endotoxemia. *J Immunol* 1992;9:2724–2730.
40. Silva AT, Bayston KF, Cohen J. Prophylactic and therapeutic effects of a monoclonal antibody to tumor necrosis factor-α in experimental gram-negative shock. *J Infect Dis* 1990;162:421–427.
41. Jesmok G, *et al.* Efficacy of monoclonal antibody against human recombinant tumor necrosis factor in *E. coli*-challenged swine. *Am J Pathol* 1992;141:1197–1207.

42. Hinshaw LB, *et al.* Survival of primates in LDjoo septic shock following therapy with antibody to tumor necrosis factor (TNF-α). *Circ Shock* 1990;30:279–292.

43. Tracey KJ, *et al.* Anti-cachectin/TNF monoclonal antibodies prevent septic shock airing lethal bacteremia. *Nature* 1987;330:662–664.

44. Hinshaw LB, *et al.* Lethal *Staphylococcus* awrews-induced shock in primates: Prevention of death with anti-TNF antibody. *J Trauma* 1992;33:568–573.

45. Opal SM, *et al.* Efficacy of a monoclonal antibody directed against tumor necrosis factor in protecting neutropenic rats from lethal infection with *Pseudomonas aeruginosa. J Infect Dis* 1990;161:1148–1152.

46. Freeman BD, Natanson C. Clinical trials in sepsis and septic shock. *Curr Opin Crit Care* 1995;l:349–357.

47. Freeman BD, Eichacker PQ, Natanson C. The role of inflammation in sepsis and septic shock: A meta-analysis of both clinical and pre-clinical trials of anti-inflammatory therapies. In Gallin J, Snyderman R (eds.) *Inflammation: Basic Principles and Clinical Correlates.* 3rd ed., pp. 965–976. Philadelphia, Lippincott Williams & Wilkins, 1999.

48. Zeni F, Freeman BD, Natanson C. Anti-inflammatory therapies to treat sepsis and septic shock: A reassessment. *Crit Care Med* 1997;25:1095–1100.

49. Quezado ZMN, Banks SM, Natanson C. New strategies for combating sepsis: The magic bullets missed the mark . . . but the search continues. *Trends Biotechnol* 1995;13:56–63.

50. Sevransky JE, Natanson C. Published clinical trials in sepsis: An update. *Sepsis* 1999;3:11–19.

51. Depietro MR, Natanson C. Sepsis trials: What have we learned? *Clin Pulmon Med* 1999;6:367–377.

52. Saravolatz LD, Wherry JC, Spooner C. Clinical safety, tolerability, and pharmacokinetics of murine monoclonal antibody to human tumor necrosis factor-α. *J Infect Dis* 1997;169:214–217.

53. Exley AR, Cohen J, Boorman W. Monoclonal antibody to TNF-α in severe septic shock. *Lancet* 1990;335:1275–1276.

54. Boillot A, Capellier G, Racadot E. Pilot clinical trial of an anti-TNF-α monoclonal antibody for the treatment of septic shock. *Clin Intensive Care* 1995;6:52–56.

55. Rodell TC, Foster C. Sepsis data show negative trend in second phase II sepsis trial. Press Release: Cortech, 7000 North Broadway, Denver, CO 80821, July 18, 1995.

56. Fein AM, *et al.* Treatment of severe systemic inflammatory response syndrome and sepsis with a novel bradykinin antagonist, Deltbant (CP-0127). *JAMA* 1997;277:482–487.

57. Dhainaut JFA, *et al.* Platelet-activating factor receptor antagonist BN 52021 in the treatment of severe sepsis: A randomized, double-blind, placebo-controlled, multicenter clinical trial. *Crit Care Med* 1994;22:1720–1728.

58. Dhainaut JF, *et al.* Confirmatory platelet-activating factor receptor antagonist trial in patients with severe gram-negative bacterial sepsis: A phase II, randomized, double-blind, placebo-controlled, multicenter trial. *Crit Care Med* 1998;26:1963–1971.

59. Kay CA. Can better measures of cytokine responses be obtained to guide cytokine inhibition. Knoll AG, Ludwigshafen, Germany, 1997. Presentation and handout. Cambridge Health Institutes' Designing Better Drugs and Clinical Trials for Sepsis/SIRS: Reducing Mortality to Patients and Suppliers. Washington, DC, February 20–21, 1996.

60. Reinhart K, *et al.* Assessment of the safety and efficacy of the monoclonal anti-tumor necrosis factor antibody fragment, MAK 195F, in patients with sepsis and septic shock: A multicenter, randomized, placebo-controlled, dose-ranging study. *Crit Care Med* 1996;24:733–742.

61. Fisher CJ, *et al.* Influence of an anti-tumor necrosis factor monoclonal antibody on cytokine levels in patients with sepsis. *Crit Care Med* 1993;21:318–327.

62. Dhainaut JFA, *et al.* CPD571, a humanized antibody to human tumor necrosis factor-α: Safety, pharmacokinetics, immune response, and influence of the antibody on cytokine concentrations in patients with septic shock. *Crit Care Med* 1995;23:1461–1469.

63. Abraham E, *et al.* Efficacy and safety of monoclonal antibody to human tumor necrosis factor-α in patients with sepsis syndrome. *JAMA* 1995;273:934–941.

64. Cohen J, Carlet J. INTERSEPT: An international, multicenter, placebo-controlled trial of monoclonal antibody to human tumor necrosis factor-α in patients with sepsis. *Crit Care Med* 1996;26:1431–1440.

65. Abraham E, Anzueto A, Guiterrez G. Double-blind randomized controlled trial of monoclonal antibody to human tumour necrosis factor in treatment of septic shock. *Lancet* 1998;351:929–933.

66. Clark MA, Plank LD, Connolly AB. Effect of a chimeric antibody to tumor necrosis factor on cytokine and physiologic responses in patients with severe sepsis—A randomized clinical trial. *Crit Care Med* 1998;26:1650–1659.

67. Fisher CJ, *et al.* Treatment of septic shock with the tumor necrosis factor receptor: Fc fusion protein. *N Engl J Med* 1996;334:1697–1702.

68. Abraham E. Cytokine modifiers: pipe dream or reality? *Chest* 1998;113:224S–227S.

69. Abraham E, *et al.* p55 tumor necrosis factor receptor fusion protein in the treatment of patients with severe sepsis and "septic shock." *JAMA* 1997;277:1531–1538.

70. Fisher CJ, *et al.* Initial evaluation of human recombinant interleukin-1 receptor antagonist in the treatment of sepsis syndrome: A randomized, open-label, placebo-controlled multicenter trial. *Crit Care Med* 1994;22:1–21.

71. Fisher CJ, *et al.* Recombinant human interleukin-1 receptor antagonist in the treatment of patients with sepsis syndrome. *JAMA* 1994;271:1836–1843.

72. Opal SM, Fisher CJ, Dhainaut JF. Confirmatory interleukin-1 receptor antagonist trial in severe sepsis: A phase II, randomized, double-blind, placebo-controlled, multicenter trial. *Crit Care Med* 1997;25:1115–1124.

73. Bernard GR, *et al.* The effects of ibuprofen on the physiology and survival of patients with sepsis. *N Engl J Med* 1997;336:912–918.

74. Bernard GR, Reines FID, Halushka PV. Prostacydin and thromboxane A2 formation is increased in human sepsis syndrome. *Am Rev Respir Dis* 1991;144:1095–1101.

75. Haupt MT, and the Ibuprofen Study Group. Effect of ibuprofen in patients with severe sepsis: A randomized, double-blind, multicenter study. *Crit Care Med* 1991;19:1339–1347.

76. Natanson C, Esposito C, Banks S. The siren's song of confirmatory sepsis trials: Selection bias and sampling errors. *Crit Care Med* 1998;26:1927–1931.

20

An Introduction to Survival Analysis

LAURA LEE JOHNSON* AND JOANNA H. SHIH**

*Office of Clinical and Regulatory Affairs, National Center for Complementary and Alternative Medicine, National Institutes of Health, Bethesda, Maryland

**Biometric Research Branch, Division of Cancer Treatment and Diagnosis, National Cancer Institute, National Institutes of Health, Bethesda, Maryland

The goal of this chapter is to introduce some commonly used statistical methods for the analysis of survival time data in medical research. Survival data consist of two pieces of information for each subject: the time under observation and the ultimate outcome at the end of that time. The analysis of survival time data is complicated by the fact that the follow-up length is often different for each participant, and the event of interest, such as myocardial infarction, is often not observed in all the subjects by the end of the study. For those participants in whom the event is not observed, what is known is that their survival times are longer than their time spent in the study, but their exact survival times are unknown. More advanced statistical methodologies than those presented in the previous chapters are needed to analyze such data.

This chapter describes features of survival time data, defines the true or underlying survival function, and introduces the product-limit estimator for the survival function. Next, it presents several approaches for comparing two survival curves, a summary of stratified analysis and Cox's regression analysis. The chapter concludes with a few remarks on more advanced topics of potential interest.

1. FEATURES OF SURVIVAL DATA

In survival analysis, the main interest focuses on the time taken for some dichotomous event to occur. Although the term *survival* is used, the event of interest is not limited to death or failure. It can be any dichotomous event, such as nonfatal myocardial infarction

(MI), adverse events, computer crashes, bursting of a balloon filling with air—any definable event. Survival time is defined as the time from some fixed starting point (time origin) to the onset of the event. In animal studies, often the starting time point is the same for all subjects. In contrast, in controlled clinical trials, the starting point is the time that the participants enter the study, which may vary for each participant. In epidemiology, the time origin may be birth or time of first exposure.

There are two key features of survival data. First, the length of follow-up varies among participants. For example, for a study with a fixed ending date, participants entering the study late would have shorter follow-up time than those entering the study early. Second, the event of interest is almost never observed in all subjects at the end of study. The survival time is called *censored* if the event is not observed at the end of the study to indicate the period of observation was cut off before the event occurred. This type of censoring is the most common and is referred to as *right censoring*. There are various reasons for censoring to occur. One common reason is that the study ends before the event occurs. Such censoring is called *administrative censoring*. Other reasons for censoring include patient withdrawal from the study and loss of contact. Censoring for reasons unrelated to the outcome of the study is called *independent censoring*. In all the methods presented in this chapter, the assumption of independent censoring is required.

The diagram in Figure 20-1 and Table 20-1 are commonly used to illustrate the features of survival data. Patient accrual occurs in the first 6 months of the study.

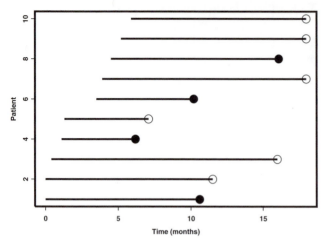

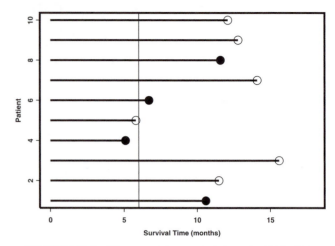

FIGURE 20-1 Diagram of patient accrual and follow-up from the data from Table 20-1. *Solid circles*, uncensored observation; *open circles*, censored observation.

FIGURE 20-2 Diagram of the survival times for Table 20-1.

TABLE 20-1   Data from First Hypothetical Example

Patient No.	Time at Entry (Months)	Time at Death or Censoring (Months)	Dead (D) or Censored (C)	Survival Time (Months)
1	0.0	10.6	D	10.6
2	0.0	11.5	C	11.5
3	0.4	16.0	C	15.6
4	1.1	6.2	D	5.1
5	1.3	7.1	C	5.8
6	3.5	10.2	D	6.7
7	3.9	18.0	C	14.1
8	4.5	16.1	D	11.6
9	5.2	18.0	C	12.8
10	5.9	18.0	C	12.1

After that, participants are monitored for a minimum of 12 months. The total length of the study is 18 months. This is an example of a study in which the total possible follow-up time will vary, in this case between 12 and 18 months, among study participants based on when a participant entered the study. The earliest accrued patients are being observed for the longest time. Figure 20-1 illustrates the *staggered entry* of participants into the study during the 6-month accrual period. Many survival studies have this pattern of participant accrual. This assumes those who enter the study at any given time are a random sample of those in the population still at risk at that time. This assumption is important in choosing how to estimate the hazard function discussed later in the chapter. As in any study, the sample used needs to be as similar to the population of interest as possible. In the case of survival analysis, we may refer to this as homogeneity

of time and participant. This is further discussed in Section 3.2.1.

Looking at Figure 20-1 more closely, we see the first participant was recruited at the beginning of the study (time 0) and had an event at approximately month 10. The second participant was also recruited at the beginning of the study and censored at approximately month 11. The survival time for each participant is obtained by subtracting the time of entry from either the time of the event or the time of last follow-up. Figure 20-2 rearranges Figure 20-1, moving the lines so all the survival times start from time 0. Figure 20-2 illustrates the survival time of each participant and provides a simpler picture than the first diagram in comparing the survival times among the participants. This would not necessarily be a better way to look at the data if time homogeneity, i.e., having similar participants enter the study at all time points, did not apply. For additional reading, similar examples can be found in introductory textbooks.[1-4]

2. SURVIVAL FUNCTION

The survival function, denoted by $S(t)$, is the probability of an individual surviving at least until time t, where $0 \leq S(t) \leq 1$. If the survival function is known from theory or empirical observation, then we can use it to understand the survival experience of a population at various time points. For example, if S represents the survival experience of post-MI patients, then we can understand the probability of surviving 3 years post-MI. However, knowledge of the survival experience is usually limited to a sample of individuals rather than the whole population, meaning we do not observe the survival function except with a complete census.

We use our sample to estimate the survival function and make inferences from this estimate to the population of interest. Note that in statistics we use the phrase "survival experience" to denote a statistical function of the time to an event, not as a phrase to describe life or its quality at the end of life.

2.1. Kaplan–Meier Product-Limit Estimator

The standard estimator of the survival function proposed by Kaplan and Meier[5] is called the *product-limit estimator*. This estimator is obtained by taking the product of a sequence of conditional probabilities. The Kaplan–Meier curve can also be referred to as a life table or actuarial analysis. Although the names are used interchangeably in some fields, these three items differ based on how precisely times are recorded in the data set. For our purposes, we will discuss only the Kaplan–Meier. Before formally defining the estimator, consider a simple example in Table 20-2.

2.1.1. Calculation and Formula for an Estimate

In Table 20-2 at time 5, there are 20 individuals at risk, of whom 2 have the event. Thus, the estimate of the probability of surviving beyond time 5 is equal to 1 − (number having the event at time 5/number at risk at time 5), which is $1 - 2/20 = 0.90$. At time 6, 18 individuals are at risk but no one has an event. Therefore, among those at risk at time 6, the probability of surviving beyond time 6 is 1. To be at risk at time 6, these individuals must have been at risk at time 5. Thus, their probability of survival beyond time 6 is equal to the product of the probability of surviving time 5 and the conditional probability of surviving time 6 given having survived time 5, which is $0.90 \times 1 = 0.90$. Three individuals were censored between time 6 and time 10, and thus they are not at risk at time 10. That leaves 15 individuals at risk at time 10: 20 − 2 who had an event before time 10, minus 3 more who were censored before time 10. The estimate of the probability of surviving beyond time 10, conditional on having survived up to that point, is equal to 1 − (number having the

event at time 10/number at risk at time 10), or $(1 - 1/15)$. The probability of survival beyond time 10 is equal to the product of the probability of surviving time 9 and the conditional probability of surviving time 10 given having survived time 9, written as $0.90 \times (1 - 1/15) = 0.84$ in Table 20-2. The later survival probabilities were obtained the same way as described previously. The product-limit estimate from this example is plotted in Figure 20-3. It is a step function in which steps occur at the observed survival times. Notice that the survival curve remains flat at the censored times. Many statistical graphing packages place a tick mark on the curve to indicate that a censoring has occurred.

Formally, the product-limit estimator is defined as

$$\hat{S}(t) = \prod_{i:t_i \leq t} (1 - f_i/r_i), \qquad (20.1)$$

where the t_i's are the ordered observed survival times or censoring times from the sample, f_i is the number of events at a time t_i, and r_i is the number of individuals at risk at t_i.

2.1.2. Calculation of Variance

Since $\hat{S}$ is an estimator, it varies from sample to sample. Such variability is measured by the variance

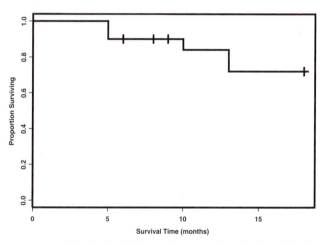

FIGURE 20-3 Product-limit survival curve from the data in Table 20-2.

TABLE 20-2 Construction of a Product-Limit Estimator

Time, t_i	No. at Risk, r_i	No. of Events, f_i	Product-Limit Estimator	Cumulative No. Censored, $m_c(t_i)$	Cumulative No. of Events, $m_e(t_i)$
0	20	0	1.00	0	0
5	20	2	$1 - 2/20 = 0.90$	0	2
6	18	0	$(1 - 0/18) \times 0.90 = 0.90$	0	2
10	15	1	$(1 - 1/15) \times 0.90 = 0.84$	3	3
13	14	2	$(1 - 2/14) \times 0.84 = 0.72$	3	5

of $\hat{S}$ that, when the sample size is large, can be approximated using the Greenwood estimator

$$\text{Var}\left(\hat{S}(t)\right) \approx \hat{S}(t)^2 \sum_{i:t_i \le t} \frac{f_i}{r_i(r_i - f_i)}. \qquad (20.2)$$

With the product-limit estimate and its variance, one can make inferences about the survival probability at various time points. For example, we can construct a 95% confidence interval for $S(13)$ from the preceding example. The variance of $\hat{S}(13)$ is approximately equal to 0.0115. Thus, the 95% confidence interval at $t = 13$ is $(0.72 - 1.96 \times \sqrt{0.0115}, 0.72 + 1.96 \times \sqrt{0.0115}) = (0.51, 0.93)$.

The product-limit estimator of the data in Table 20-1 is plotted in Figure 20-4. There are six tick marks on the curve because six survival times are censored. The probability of surviving beyond 6 months is 0.9 with variance 0.009.

The Greenwood variance estimator tends to underestimate the variance in the tails of the survival distribution. Other variance estimators, including those by Borkowf,[6] attempt to alleviate this problem. The interested reader may consult a statistician for more information.

2.2. Comparing Two Survival Functions

We consider hypothesis tests for comparing the survival functions from two independent samples. These samples need to be defined based on events that happen before the follow-up period. There are two methods of comparison. The first is to compare the two survival functions at a prespecified time point. For example, in a clinical trial patients are randomized to either treatment A or treatment B. One might be

interested in comparing the survival experience between the two groups at one year postrandomization. Alternatively, instead of comparing the survival at a fixed time point, one may want to compare the overall survival experience. In this case, the comparison is made across the entire range of the survival times. We first consider the fixed time point comparison. Then we introduce the log-rank test statistic to compare the overall survival.

2.2.1. Comparing Two Survival Functions at a Given Time Point

Formally, the hypothesis to be tested is formulated as

$$H_0 : S_1(t) = S_2(t), \qquad (20.3)$$

where t is a prespecified time point. To test this hypothesis, we calculate the product-limit estimate at time t for each sample, $\hat{S}_1(t)$, $\hat{S}_2(t)$, and form the test statistic

$$Z = \frac{\hat{S}_1(t) - \hat{S}_2(t)}{\text{SE}(\hat{S}_1(t) - \hat{S}_2(t))'} \qquad (20.4)$$

where $\text{SE}\left(\hat{S}_1(t) - \hat{S}_2(t)\right) = \sqrt{\text{var}\left(\hat{S}_1(t)\right) + \text{var}\left(\hat{S}_2(t)\right)}$. Z varies from sample to sample, and thus it has a sampling distribution. When the two samples are independent and the sample size is large, Z has approximately a standard normal distribution under the null hypothesis. If the Z value is in the upper or lower $100 \times \alpha/2\%$ (for a two-sided test) or $100 \times \alpha\%$ (for a one-sided test) of the standard normal distribution, we reject the null hypothesis and conclude that the two survival probabilities at time t are not equal. Otherwise, we fail to reject the null hypothesis, and hence there is insufficient evidence to conclude that the two survival probabilities are different. In addition, one can calculate the p-value[7] for the test statistic. If the p-value is smaller than α, then we conclude that the two survival probabilities are different.

2.2.2. Comparing Two Survival Functions Using the Whole Curve: Log-rank Test

The comparison presented in the previous section is not completely satisfactory for the following reasons, given by Pocock.[8] First, the time point chosen for comparison is usually, although not always, arbitrary. Second, one may tend to choose the time point post hoc where the largest difference occurs and exaggerates the survival difference. If someone chose to measure differences at 3.69 years, it would look suspicious, but at 1, 2, 4, 6 years, it would look and actually might be perfectly reasonable. Third, one is not making

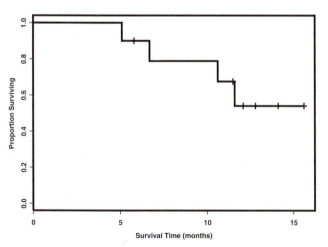

FIGURE 20-4 Product-limit survival curve from the data in Table 20-1.

full use of the precise survival times for each patient. The log-rank test[9] is the most common method for comparing the overall survival experience between two groups. The hypothesis to be tested is $H_0 : S_1(\cdot) = S_2(\cdot)$ where the dot means the whole range of the survival times on the curve. Thus, rather than an arbitrary fixed time point, we will test whether the two independent samples come from the same population regarding the overall survival experience. The log-rank test also does not make use of the precise times of event but, rather, uses the ranks of the times.

The log-rank test essentially compares the observed number of events with the number to be expected if the two groups have the same underlying true survival curves. Specifically, we arrange the distinct survival times from the two groups in an ascending order, excluding censored survival times. Let $\{t_1, t_2, \ldots, t_k\}$ denote these ordered survival times. At each time $t_j, j = 1, \ldots, K$, we construct a 2×2 table:

	No. Dead	No. Surviving	Total
Group 1	a_j	b_j	$a_j + b_j$
Group 2	c_j	d_j	$c_j + d_j$
Total	$a_j + c_j$	$b_j + d_j$	n_j

If the null hypothesis is true, then the expected number of deaths in group 1 at time j, denoted by $E(a_j)$, is equal to

$$E(a_j) = \text{total \# of events at } t_j \times \frac{\text{total \# at risk in group 1}}{\text{total \# at risk}}$$

$$= (a_j + c_j)(a_j + b_j)/n_j,$$

$$(20.5)$$

and the variance of a_j is equal to

$$\text{Var}(a_j) = \frac{(a_j + b_j)(a_j + c_j)(b_j + d_j)(c_j + d_j)}{(n_j - 1)n_j^2} \quad (20.6)$$

We form K using such 2×2 tables and calculate the test statistic using the results from these tables:

$$Z = \frac{\sum_{j=1}^{K} a_j - \text{E}(a_j)}{\sqrt{\sum_{j=1}^{K} \text{Var}(a_j)}}. \quad (20.7)$$

Z has approximately the standard normal distribution under H_0. If Z is in the upper or lower $100 \times \alpha/2\%$ ($100 \times \alpha\%$ for one-sided test) of the standard normal distribution, then we reject H_0 and conclude that the two samples are not from the same population. In addition, we can calculate the p value. If the p value is smaller than α, then the null hypothesis is rejected. Otherwise, there is insufficient evidence to conclude that the two survival functions are different. Variations to the log-rank test exist for multiple groups and differentially

weighting events at different parts of the time scale. Different variations used on the same data set may lead to different answers. Additionally, log-rank methods should be cautiously interpreted if the survival curves cross.

2.2.3. Example 1: Chronic Active Hepatitis Study

Kirk et al.[10] randomized 44 patients with chronic active hepatitis to either prednisolone or an untreated control group. Pocock analyzed these data in detail. Their survival times are listed in Table 20-3. The product-limit estimates are plotted in Figure 20-5. It shows that the patients treated with prednisolone have a better survival experience overall. We consider two-tailed hypothesis tests with $\alpha = 5\%$. We first compare the survival proportions at 5 years (60 months), where the product-limit estimate is 0.41 in the control group and 0.82 in the prednisolone group. The standard error of the difference of these two estimates is 0.1322. Using Eq. (20.4), the test statistic Z equals $(0.41 - 0.82)/0.1322 = -3.08$. Because Z is lower than -1.96, the 2.5th percentile of the standard normal distribution, we reject the null hypothesis and conclude that patients receiving prednisolone have better survival probability at 5 years than untreated patients.

TABLE 20-3 Survival Data for 44 Patients with Chronic Active Hepatitis

Survival times (months) in the control group
  2, 3, 4, 7, 10, 22, 28, 29, 32, 37, 40, 41, 54, 61, 63, 71, 127+, 140+, 146+, 158+, 167+, 182+
Survival times (months) in the prednisolone group
  2, 6, 12, 54, 56+, 68, 89, 96, 96, 125+, 128+, 131+, 140+, 141+, 143, 145+, 146, 148+, 162+, 169, 173+, 181+

Censored survival times are indicated by +.

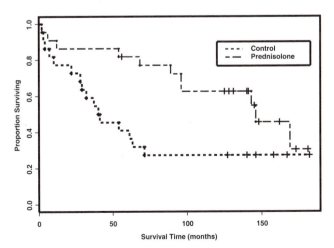

FIGURE 20-5 Product-limit survival curves from the chronic active hepatitis study.

Next, we compare the overall survival experience between the two groups using the log-rank test. To illustrate how the log-rank test works, consider one survival time at 10 months. At this time point, there are 18 patients at risk in the control group, of whom 1 dies, and there are 20 at risk in the prednisolone group, of whom none dies. The expected number of events in the control group, under the null hypothesis, is 0.47 with variance 0.25. We need to repeat this calculation for every observed death time and use these results to form the Z statistic. The Z statistic equals 2.16, which is greater than 1.96, the 97.5th percentile of the standard normal distribution. Thus, we conclude that patients treated with prednisolone have better overall survival than those untreated.

2.2.4. Stratified Log-Rank Test

In a clinical study, it is sensible to collect information on the participant's personal characteristics (e.g., age, sex, and ethnicity) and history of disease and treatment. Some of these factors may be related to the participant's survival experience. For example, young people usually have a better survival experience than old people. If the prognostic factors are related to survival and they are balanced between the two treatment groups, then the log-rank test presented in the previous section is appropriate. Otherwise, the log-rank test may be biased. That is, it is likely that the two populations have the same survival experience, but the log-rank test declares they are different. *Confounding* is a change in the relationship between the outcome and treatment due to imbalance in another predictor variable. The difference in survival observed in the two groups actually comes from the difference in the prognostic factors rather than the treatment. To alleviate the bias, one solution is to compare the survival difference between the two treatment groups within each level of the prognostic factors. For example, if the survival experience is dramatically different for patients with a history of MI compared to patients without a history of MI, then we may want to compare the survival difference among patients with MI and without MI separately. By doing the analysis within each level of the prognostic factors, any observed difference in survival can be attributed to the treatment effect.

The stratified log-rank test is the log-rank test that accounts for the difference in the prognostic factors between the two groups. Specifically, we divide the data according to the levels of the significant prognostic factors and form a stratum for each level. At each level, we arrange the survival times in ascending order and calculate the observed number of events, expected number of events, and variance at each survival time

as we would do in the regular log-rank test. Let a_{ij} denote the number of events in group 1 in the ith stratum at time t_{ij}. Then the stratified log-rank test has the following form:

$$Z = \frac{\sum_{i=1}^{m} \sum_{j=1}^{K} a_{ij} - E(a_{ij})}{\sqrt{\sum_i \sum_j Var(a_{ij})}}, \qquad (20.8)$$

where m is number of strata, K_i is number of survival times in the ith stratum, $E(X)$ is the mean of X, and $Var(X)$ is the variance of X. Under the null hypothesis of no difference in survival between the two groups, and when the sample size in each stratum is large, Z has approximately the standard normal distribution.

2.2.5. A Few Cautious Notes

The introductory books[1-4] mentioned previously have more information on each of these items; however, we briefly discuss three common mistakes people make when first using survival analysis in hopes of helping the reader avoid them.

Notice we never mentioned *mean* survival. We cannot sensibly look at mean survival time when some survival times are censored. The same is true when looking at the proportion who survived a given amount of time, such as one year, when not all participants were followed for one year. We might be able to read the *median* survival derived from the Kaplan–Meier if the sample's curve drops below 0.5.

Additionally, when comparing two survival curves, caution is in order. The log-rank test will use all of the data and is appropriate in many cases. Visual inspection of curves is not appropriate. What may appear to be a large difference between two curves at later time points may be an illusion. If confidence intervals are drawn, it will become clear; due to smaller amounts of data on the right-hand side of the curve, the confidence intervals will widen. As discussed in Chapter 15,[7] a large difference in point estimates does not automatically mean a statistically significant difference.

We re-emphasize that survival analysis comparison groups need to be defined on factors known before treatment. Investigators may wish to compare responders to nonresponders. Such analyses, creating comparison groups based on factors known only after the treatment has begun, may give misleading results. It is strongly recommended that expert advice be obtained before pursuing such analyses.

2.3. Proportional Hazards Model

The stratified log-rank test is a useful method for comparing the survival between two treatment groups while accounting for the effects of prognostic factors. However, it has some limitations. First, quantitative prognostic factors must be categorized to form strata. Second, if there are many prognostic factors, each with several levels, the number of strata can quickly become large with few patients in each stratum. This results in loss of power in the stratified test. Finally, whether we use the unstratified or stratified log-rank test, it is primarily a significance test and does not estimate the magnitude of the difference. For these reasons, we need a method that allows for both categorical and continuous prognostic factors that also is able to provide the estimate of the treatment difference. The proportional hazards model proposed by Cox[11] aims at achieving these goals. Specifically, it uses regression methods to model the shape of the hazard function with one empirical part that depends on time and a second exponential part that depends on the other covariates.

2.3.1. Calculation and Formulas

Consider a patient at risk of an event after being followed for time t. The hazard at time t represents the instantaneous probability that the patient will have an event before any subsequent time $t + \delta$, $\delta > 0$. Mathematically, the hazard function, $h(t)$, is the derivative of $\log\{S(t)\}$, so the hazard function for a patient at risk at time t is defined as

$$h(t) =$$

$$\lim_{\delta \to 0} \frac{\text{probability of having an event before time } t + \delta}{\delta}.$$

(20.9)

The hazard can be interpreted as an instantaneous event rate. A somewhat more intuitive, simpler interpretation is to remember the following: higher hazards—worse survival; lower hazards—better survival. The proportional hazard has the representation

$$h(t) = h_0(t) \exp(\beta_1 x_1 + \ldots + \beta_p x_p), \quad (20.10)$$

where $h_0(t)$ is called the baseline hazard; β_k, $k = 1, \ldots, p$ are regression coefficients; and x_k, $k = 1, \ldots, p$ are prognostic factors. If there is no prognostic factor present in the hazard function, then $h(t)$ is the same as the baseline hazard. A proportional hazard means that the change in a prognostic factor results in a proportional change of the hazard on a log scale. To demonstrate this, consider only one prognostic factor, x with the associated coefficient β. The log of the hazard at time t for $x = a$ is $\log h_0(t) + \beta \times a$, and when $x = b$ is $\log h_0(t) + \beta \times b$. The difference of the hazards on the log

scale is $(a - b) \times \beta$, which does not change with t. The difference of the hazards is proportional to the change in x, in this case $a - b$, and the "proportion" is equal to the regression coefficient β. In this type of modeling, a function of the regression coefficient β, $\exp(\beta)$, is referred to as the *relative risk*. Commonly, x represents the treatment indicator and is given a value of 1 if a patient is assigned to the investigational treatment, and x is given a value of 0 if the patient is assigned to the control treatment. Then β is used to measure the magnitude of the treatment difference because $\exp(\beta) = h_0(t) \exp(\beta \times 1)/h_0(t) \exp(\beta \times 0)$ represents the hazard ratio in the two treatment groups. If $\beta = 0$, then the hazard ratio is 1, and thus the two groups have the same survival experience.

The regression coefficients are usually unknown and need to be estimated from the data. It is complex to estimate these regression coefficients. However, statistical software to estimate the hazard function is commonly available. When the sample size is large, the estimate of each regression coefficient approximately follows a normal distribution. In addition, we can test the effect of prognostic factors. Let β and $\mathrm{SE}(\beta)$ denote the estimate of β and standard error. The Z statistic

$$Z = \frac{\hat{\beta}}{\mathrm{SE}(\hat{\beta})} \quad (20.11)$$

has approximately the standard normal distribution under the null hypothesis $H_0 : \beta = 0$—that is, where the prognostic factor has no effect.

2.3.2. Example 2: Study of Melanoma Patients

The University of Oklahoma Health Sciences Center performed a nonrandomized study of 30 melanoma patients comparing the immunotherapies Bacillus Calmette-Guerin (BCG) and *Corynebacterium parvum* for their abilities to prolong remission and survival times for melanoma patients.[4] These patients receiving either BCG or *C. parvum* had tumor resection before the treatment began. One primary goal of this study was to assess the effects of prognostic factors on the disease-free survival (i.e., survival without relapse). Prognostic factors collected included age, sex, and disease stage. The goal was to compare these two regimens for their abilities to prolong remission and survival times. Here, the survival time is the minimum time to death and time to relapse. The last follow-up date was April 15, 1977. The estimation results are presented in Table 20-4. Neither the treatment nor any of the prognostic factors are significant when comparing the absolute Z values to the upper

TABLE 20-4   Estimation Results from the Study of
Melanoma Patients

	$(\hat{\beta})$	SE $(\hat{\beta})$	Z
Age	0.01	0.01	0.79
Sex	0.33	0.57	0.58
Disease stage	1.41	1.07	1.32
Treatment	0.32	0.54	0.61

2.5 percentiles of the standard normal distribution (= 1.96). Thus, we conclude that from this sample we fail to declare significant difference in survival for patients of different ages, sex, disease stage, and treatment received.

3. SPECIAL CONSIDERATIONS

The following are topics this chapter cannot cover in detail but that are frequently raised in the literature and in collaborating and consulting with clinicians. The interested reader may consult a statistician or read a survival textbook for more information on topics of interest.

3.1. Changes over Time

3.1.1. Time-Varying Coefficients or Time-Dependent Hazard Ratios

In Cox regression, the concept of proportional hazards is important. It means that the relative risk of an event, or β in the regression model [Eq. (20.10)], is constant over time. If we do not have proportional hazards, then the regression coefficient β should be modeled over time and referred to as a *time-varying coefficient*. For long-term clinical and cohort studies, it may be important to check if treatment group hazard ratios vary with time. There are ways to analyze data in this manner and good upfront study design is important.

3.1.2. Time-Dependent Covariates

Covariates may change their values over time. Such variables are referred to as *time-dependent covariates*. Time-dependent covariates may be used in Cox models, but with extreme caution. One important point is that when using time-dependent covariates, the standard Cox model typically cannot be used to predict the survival curve over time. For more information, see Fisher and Lin.[12]

3.2. Dependent or Informative Censoring

When a participant's condition influences the censoring time, dependent censoring occurs. Dependent or informative censoring is not independent of the event of interest; for example, participants dropping out of a weight loss study because they have stopped exercising. If they dropped out because they moved out of the area, that would be considered independent or noninformative censoring. Depending on exactly what is occurring, there are several methods that can be employed. Two common problems that frequently occur in medical studies are changes in inclusion/exclusion criteria during study accrual and a participant no longer being followed for a study's event of interest because another event has occurred.

3.2.1. Changes in Inclusion/Exclusion Criteria and Nonindependent Censoring

Assume Table 20-1 provides data for a study that for the first 4 months had study entry criteria of ages 20–50 years. After slow recruitment, the age criteria were expanded to allow ages 20–70 years. However, recruitment continued for only 2 more months. This meant older subjects only were allowed to enroll during months 5 and 6, so older subjects could not have 18 months of follow-up because they were not allowed to enter at the beginning of the study accrual period. As a result, the Kaplan–Meier estimate will be biased because censoring is not independent. At large t, the risk sets will not include older participants because they were not recruited early enough and therefore will be censored early. The hazard function will be biased too small for larger t, thus $\hat{S}$, the estimated survival, will be larger than the true population survival function at large t. Regarding the other survival model, Cox proportional hazards, we can avoid the bias. If age is a covariate in the Cox model, then the Cox model will not be biased. The requirement for independent censoring in the Cox model is that censoring must be independent given the covariates. With age in the model, the Cox model is conditioned on age and the coefficient estimates from the Cox model will not be biased. In short, we always need to ask what population we are making reference to and how well the subjects in the study correspond to that population.

3.2.2. Competing Risks

An individual under study may experience one of several different events of interest. If the occurrence of one event precludes the occurrence of another event, then these events are *competing risks*. For example, if the event of interest is cancer diagnosis, then death from MI is a competing risk. The two events may not be independent; in fact, they may be strongly associated. Competing risks may be handled by redefining the hazard function for the proportional hazards model. In our cancer study, we need to censor the individual at the time of MI death. That person did not have the event we were interested in, but he or she is no longer at risk for the event of interest. A practical interpretation of the hazard is $h(t)$ = "the risk of cancer among those at risk of cancer at time t." This excluded the MI deaths and anyone else no longer at risk of cancer at time t. However, it should be noted that the Kaplan–Meier survival estimate does not have a sensible interpretation for competing risks.

3.2.3. Left and Interval Censoring

An adolescent smoking study that begins to follow students in the 6th grade may find some students have already begun smoking. Unless reliable information can be obtained about the date the current smokers began to smoke, the age the 6th grade students started smoking is *left censored*. For these students the event times (age when smoking started) are unknown, although it is known that the starting ages are younger than the students' ages when the study began. Left censoring usually can be avoided by careful study design. *Interval censoring* occurs when an event is known to have occurred between two time points, but the exact date is unknown. If smoking is surveyed yearly from 6th grade forward, the exact smoking start date may be unknown but it would be known that a student was not smoking at the beginning of 7th grade but was smoking by the beginning of the 8th grade. Interval censoring is common in screening studies. Both left and interval censoring need careful attention during study design and data analyses.

3.3. Recurrent Events Analysis

Many medical outcomes are recurrent. Examples of recurrent events include tumor recurrences, lung exacerbations in cystic fibrosis patients, and hospital admissions. Modeling events such as these is called *recurrent events analysis*. The inference goals of the recurrent events analysis may vary depending on the applications. For example, one might be interested in studying the treatment effect on the entire recurrent events process or the dependency of the recurrent event rates on the previous event history. Various statistical methods have been developed for the analysis of recurrent event time data, some of which are described by Kalbfleish and Prentice.[13]

3.4. Sample Size

Sample size calculations for survival analyses are substantially different than explained in Chapter 15.[7] The idea is that the power of a study depends not on the original sample size but on the number of events and the amount of subject-time observed. An event may never happen for some subjects, so the sample size is based on the number of events needed. Then, working backwards, we estimate the number of subjects needed. There are a variety of formulas that may be used, and these are outlined in survival analysis textbooks and a few papers.[14–17] Information that will be needed to compute sample size is similar to that needed for the standard two-sample difference of means example described in Chapter 15.[7] To determine sample size, we need to know power, α, one-tailed or two-tailed test, plus other study design information.

4. CONCLUSION

Survival analysis makes inference about event rates as a function of time. The two primary methods to estimate the true underlying survival curve are Kaplan–Meier and Cox regression. Kaplan–Meier is simple and supports stratification factors but cannot evaluate covariates. The Cox model does provide a framework for making inferences about covariates and requires proportional hazards, although it is quite flexible when used and interpreted correctly. Independent censoring, either directly in the Kaplan–Meier or given covariates in the Cox model, is a requirement for consistent unbiased estimates. Survival analysis can handle right censoring, staggered entry, recurrent events, competing risks, and much more as long as we have available representative risk sets at each time point to allow us to model and estimate event rates. Statistical methods for survival analysis remain an active area of research and collaboration among statisticians and their colleagues, and the reader will benefit from joining in this process.

References

1. Friedman LM, Furberg CD, DeMets DL. *Fundamentals of Clinical Trials.* New York, Springer-Verlag, 1998.
2. Altman DG. *Practical Statistics for Medical Research.* New York, Chapman & Hall, 1991.
3. Hosmer DW, Lemeshow S. *Applied Survival Analysis: Regression Modeling of Time to Event Data.* New York, Wiley, 1999.
4. Lee ET. *Statistical Methods for Survival Data Analysis.* New York, Wiley–Interscience, 1992.
5. Kaplan EL, Meier P. Nonparametric estimation from incomplete observations. *J Am Stat Assoc* 1958;53:457–481.

6. Borkowf CB. A simple hybrid variance estimator for the Kaplan–Meier survival function. *Stat Med* 2005;24:827–851.

7. Johnson LL, Borkowf CB, Albert PS. An introduction to biostatistics: Randomization, hypothesis testing, and sample size. In Gallin JI, Ognibene FP (eds.) *Principles and Practice of Clinical Research*. New York, Elsevier, 2007.

8. Pocock SJ. *Clinical Trials: A Practical Approach*. New York, Wiley, 1983.

9. Mantel N, Haenszel W. Statistical aspects of the analysis of data from retrospective studies of disease. *J Natl Cancer Inst* 1959;22:719–748.

10. Kirk AP, Jain S, Pocock S, Thomas HC, Sherlock S. Late results of the Royal Free Hospital prospective controlled trial of prednisolone therapy in hepatitis B surface antigen negative chronic active hepatitis. *Gut* 1980;21:78–83.

11. Cox DR. Regression models and life tables. *J R Stat Soc Ser B* 1972;34:187–220.

12. Fisher LD, Lin DL. Time-dependent covariates in the Cox proportional-hazards regression model. *Annu Rev Public Health* 1999;20:145–157.

13. Kalbfleish JD, Prentice RL. *The Statistical Analysis of Failure Time Data*. New York, Wiley, 2002.

14. Freedman LS. Tables of the number of patients required in clinical trials using the logrank test. *Stat Med* 1982;1:121–129.

15. Schoenfeld DA. Sample size formula for the proportional hazards regression model. *Biometrics* 1983;39:499–503.

16. Schoenfeld DA, Richter JR. Nomograms for calculating the number of patients needed for a clinical trial with survival as an endpoint. *Biometrics* 1982;38:163–170.

17. Simon RM, Korn EL, McShane LM, *et al. Design and Analysis of DNA Microarray Investigations*. New York, Springer, 2004.

21

Measures of Function and Health-Related Quality of Life

LYNN H. GERBER

Center for Study of Chronic Illness and Disability, College of Health and Human Services, George Mason University, Fairfax, Virginia

1. INTRODUCTION TO FUNCTION AND QUALITY OF LIFE

There are many approaches to assessing the health of a person. Most include measurements in several domains. The term *domain* is used to describe the physical, psychological, and social aspects of an individual's activities. For example, measures of physical impairments, which might include anatomical or physiological abnormalities, are thought to contribute to overall function and quality of life. In addition, measures of limitations of function, such as gait velocity or prehension, are done because they provide an accurate and quantitative picture of what an individual is able to perform. Measures of disability are also acknowledged as being useful because they place the abnormalities within the context of an individual's daily routines.

Recently, there has been much interest in evaluating what constitutes meaningful and valued daily activity to an individual. This interest has been driven by two converging processes in U.S. health care: (1) a strong consumer movement, in part fueled by easy access to information, which has spearheaded rising expectations for personal health, and (2) a need for a more systematic and comprehensive approach to health status measurements.

In 1948, the World Health Organization (WHO) prompted a major departure from the disease-driven orientation previously adopted to define the concept of health and to assess outcomes. WHO stated that "physical, mental, and social well-being and not merely absence of disease" defines health.[1] The definition helped set the conceptual framework for what constitutes treatment goals, thereby acknowledging the importance of using multidimensional outcome measures that would include domains of physical, mental, and social health and measures of function and disability. The relationships among these domains provide many investigational opportunities. These may be causally connected or associated through complex societal and economic or other relationships not necessarily disease dependent. Although it had been generally accepted that impairment drives disability, which in turn results in poorer quality of life, these relationships are neither linear nor unidirectional. Muscle atrophy may result from disuse as well as from a neurological impairment. If the atrophy is a result of disuse, reversibility is more likely. Treatment aimed at reducing disability is likely to reverse the impairment and result in increased muscle mass and strength.

Instruments to measure these phenomena have improved significantly during the past decade. They have become more quantitative and more reliable. Most have been standardized, have been tested for content validity and inter- and intrarater reliability, and have been studied for application to specific diseases. They have been expanded to include measures of personally held beliefs and values in order to understand the impact of disease on behaviors such as coping, participation, and motivation. This has increased the variety of information obtained and raised confidence in data collected from quality of life (QOL) measures, both self-report and observer-administered instruments. WHO devised a new classification scheme that helps define the domains to be assessed and from which data can be collected in a consistent and reliable way.[2] This has been updated[3] and altered to reflect scientific advances in physiology and genetics as well as the importance of the individual with respect to his or her unique needs within the context of the individual's own environment. The newer approach, derived from the International Classification of Functioning Disability and Health (ICF), incorporates some aspects of the theoretical framework developed by Nagi,[4] who created an expectation that health care must go beyond managing morbidity and reducing mortality. The health care system is committed to helping reduce the burden of disease, but it has become increasingly aware of patient priorities, which include the desire to be independent, maintain valued activity, and have a sense of well-being in all aspects of daily life. The ICF incorporates many of these features into its classification scheme and is being increasingly used worldwide. It has four domains of measurement: (1) body functions, which include cardiovascular, hematological, digestive, neuromusculoskeletal, and voice/speech; (2) body structures, which include nervous system, gastrointestinal, systems pertaining to movement, and genitourinary; (3) activities and participation, which include learning and applying knowledge, communication, mobility, and self-care; and (4) environment, products, and technology, which include natural environment, human-made changes, and support.

In the United States, the most frequently used outcome measure, which is not a QOL measure, assesses functional level and burden of care following discharge from the hospital. The Functional Impact Measure (FIM)[5] has seven levels of assessment and 18 items designed to measure patient disability and rehabilitation functional outcome.

This chapter discusses approaches and commonly used measures to assess health-related QOL (HRQOL) and functional activity.

2. DEFINITION OF QUALITY OF LIFE

The model of well-being espoused by WHO assumed a broad view of health that helped establish the need for devising a patient-centered basis for assessment. Early discussions suggested that we should be measuring well-being, a state dependent on physical and functional status and the degree of family support, social activity, and friendships; personal achievements and philosophy; and financial adequacy and work achievements. The Centers for Disease Control

and Prevention defined QOL as the perception of physical and mental health over time (www.cdc.gov/brfss).

In the health care literature, QOL is linked with function and/or health status, frequently referred to as HRQOL. Functional status is the degree to which an individual can perform chosen roles without limitation in three key domains: physical, social, and psychocognitive function. These roles are seen within the context of the unique needs of the person. For example, an individual with rheumatoid arthritis might have evidence of 10 swollen, hot joints with pain on palpation. This describes the impairment. The individual is unable to walk 50 feet because of pain, limited joint motion, and fatigue. Her inability to walk is the disability. She is unable to get to work and cannot hold a job. Her unemployment status is her societal limitation or handicap. Function reflects elements of disability and handicap in this system but does not include measures of personal values, goals, or expectations.

Typically, clinical examinations measure physical and biological phenomena, mood, and mental status. The correlations between these measures and function are not always high, and they omit questions pertaining to one's status with respect to cultural and value systems and goals. Since they assess different domains, all of which are thought to be important, clinical trials should include measures of all. In fact, the Food and Drug Administration (FDA) requires functional measures to be included in clinical trials to demonstrate efficacy. Functional measures and HRQOL indicators are measures of different but complementary phenomena, and a substantial body of data supports the view that physical findings and disease severity do not always correlate with patient self-report about QOL.[6]

What establishes an instrument as a QOL or HRQOL instrument is the component of measuring patient satisfaction. Instruments have been devised that rate specific activities based on their value to an individual and also assess the impact of these activities on the individual's feelings of satisfaction and competence. Controversy exists about whether HRQOL is more about life than health and whether health care should even include the domains that often highly influence QOL, such as social, financial, and societal dimensions. As treatment choices become broader and individuals participate more in decision making, an ever increasing number of studies include HRQOL measurements as outcomes, and investigators suggest they provide meaningful information that informs clinical practice and is helpful in decision making.

Table 21-1 contrasts the differences between health status measures and HRQOL.

TABLE 21-1   Characteristics of Measures of Health-Related Instruments

Quality of life
    Subjective
    Composite, multiple domains
    Self-report
    Well-being model
    Generic
    Sociomedical
    Produces a profile or indicator
    Value or satisfaction measures
Health status
    Objective or subjective
    Single item or composite
    Observer measured or self-report
    Sickness or well-being model
    Disease specific or generic
    Diagnostic or prognostic use
    Produces a score or health index
    May have value or satisfaction measures

TABLE 21-2   Requirements for All Measures

Practical: Administration time not burdensome, low risk to patient. Complexity of measures and scoring easily interpretable.
Validity: Measures what is intended to measure.
    Content validity: Comprehensiveness of sampling of questions.
    Criterion validity: Its congruence with "gold standard."
    Construct validity: In quality of life and health status measures, there is no gold standard. Therefore, correlations and factor analyses are used to determine how well the items accord in measuring common items. Construct validity is determined based on an accumulation of such correlations, usually from several studies.
Reliable: Measures are consistent, having little measurement error. Able to distinguish between patients.
    Internal consistency: The demonstration that similar questions are shown to correlate highly on repeated measure. How well the questions measure the same theme.
    Reproducible: Test–retest is stable (Cronbach's is a measure of this).
Responsive: Measure is sensitive to change, has good evaluative properties. Assessment at the low and high ends of satisfaction and function is adequate for the population studied.

3. HEALTH-RELATED QUALITY OF LIFE MEASURES

3.1. Uses of Quality of Life Measures

Not all studies or clinical trials need to have all domains measured. For example, a drug trial assessing the appropriate dosage or toxicity of a new antibiotic (phase I trial) or one that is seeking pilot information about the feasibility of using a new device might need an instrument that is focused only on a physiologic measure. Sometimes, investigators are interested in disease severity or its extent rather than its impact. *Health status* is a term that refers to the degree to which an individual has a disease or symptom. Health status measurements, not HRQOL measures, quantify the severity or degree of illness.

Use of HRQOL measures in clinical trials is established,[7] and these are being used for a variety of purposes. They have been used to (1) assess health needs of individuals or groups; (2) assess outcomes of treatments, health promotion, and disease prevention programs; (3) assess cost-effectiveness; (4) develop health policy; (5) monitor general health of groups of patients; and (6) influence an individual's choice of a specific treatment plan.[8]

These instruments have recently risen in favor and have been used more frequently during the past decade because of several important factors. A substantial number of instruments have been standardized, validated, and have good psychometric properties whose theoretical framework has been tested and found to be methodologically sound (Table 21-2).

In addition, patients are demanding more personalized care. They want their individualized functional needs addressed. They want "customer" satisfaction. There is an increasingly wide selection of treatment options for patients that requires education about risks and benefits that may impact on function and QOL, as well as morbidity and mortality. The FDA requires investigators to measure treatment impact on function, not only on disease activity, in trials of investigational drugs.[9]

3.2. Structure of Quality of Life Measures

Many schemes have been proposed that purport to identify the necessary and sufficient domains for HRQOL indices. Because HRQOL cannot, strictly speaking, be observed, it must be inferred from behaviors. To do this, these measures usually use self-reports that relate symptoms (e.g., pain and fatigue) with aspects of physical, mental, and social function (disability) and often demographic, vocational, and environmental impact on life (handicap). This helps us devise a set of characteristics of symptoms or a profile unique to individuals. HRQOL ultimately assigns value to this profile, helping answer the question, "Have we added life to years?"[10]

There are several examples of the domains such measures include. One is the five "Ds" of death, disease, disability, discomfort, and dissatisfaction.[11] Another includes genetics, anatomy/physiology, physical function, mental function, and health potential.[12] Two additional measure disease, physical well-being, psychological well-being, social well-being, and general HRQOL;[13] and clinical status, physical functioning/well-being, mental functioning/well-being, social/role functioning/well-being, and general health perceptions and satisfaction.[14] One construct that is particularly useful identifies five core domains: opportunity, health perceptions, functional status, impairment, and death/duration of life.[15] Several taxonomies have been developed that present an organizational strategy for considering how these multidimensional QOL and functional measures may be classified.[16–18]

The methodologies and suitability of HRQOL measures have been and remain carefully scrutinized. In constructing these measures, several issues must be considered. First, which items should be included in the instrument? Often, this decision is made based on what the outcome of the study might look like or how the assessment will be used. For example, some instruments designed to assess effectiveness of care will have to be able to distinguish among differences in subjects at a moment in time or be able to measure change over time (even in those without significant disease or illness). They may have to predict future outcome resulting from a new versus an established treatment.[19]

The second issue pertains to the methodologic or measurement characteristics of the instrument. Some use single indicators such as mortality. This is frequently considered a "hard"[7] end point, very accurately recorded, leaving little doubt about verification. Using mortality as the "gold standard" for health outcomes ignores the fact that we have changed our health outcome expectations. The outcomes sought are not exclusively survival, because our population wishes to be disease free, disability free, and have a good QOL. As such, because mortality does not measure what many believe is the desired outcome of health care; it is not a sensitive indicator of health. Composite measures are needed to detect health changes as well as to generate policy models. The measurement characteristics can be presented as an index. Examples include Arthritis Impact Measurement Scales (AIMS)[20] and the Functional Life Index (FLI).[21] The data can also be summarized into a profile that describes each domain or area of measure separately. The Sickness Impact Profile (SIP) is an example of this type of instrument.[22]

The third consideration is to decide how to "weight" the responses. When a measure has several components, relative value needs to be assigned to each. Weighting can be assigned by investigators through a Delphi process or consensus. It can also be assigned based on frequency of responses—those items more commonly selected get higher scores. It can be weighted based on preference of the respondents.

The SIP[22] is an excellent and frequently used HRQOL profile and has become a gold standard for health-related measures. The purpose of this index is to describe and quantify the impact that disease has on a person's behavior. Its application is to measure the outcomes of care, which could be applied to health care program planning as well as patient management. The SIP was created through an interview process of those healthy and sick to obtain descriptors of behavioral change that was associated with sickness. Sickness connotes a person's experience of illness and was learned from the effect the change in status from well to sick had on daily routines and feelings. Categories of activities were established, and statements describing behavior were listed under each category. Twelve categories were selected (sleep/rest, eating, work, home management, recreation/leisure, ambulation, mobility, self-care, social interaction, alertness, emotional behavior, and communication). Statements within categories included "I laugh or cry suddenly," "I walk shorter distances or stop to rest often," and "I am not working." Respondents checked only those statements that described them. One hundred evaluators reviewed the questions and participated in intervals designed to assess what they thought the relative impact or importance would be of each item under consideration. Each category was scored separately. There were correlations between SIP scores and criterion variables in a variety of diseases and other health indices ($R > 0.5$ in almost all). Reliability (test, retest, and internal consistency) was very high ($R > 0.8$).

3.3. Criteria for Selection of HRQOL Measures

Many instruments currently in use meet the criteria described previously for reliability, validity, and ease of use. In making the proper selection for a particular situation, a number of additional considerations should be included. Some HRQOL measures have been designed with specific populations of patients in mind and have not been applied to a broader range of patients/subjects. This is true for both specific diagnostic and age groups and is relevant for level of disability. Most indices are developed using a disease severity, not a functional limitation orientation.

Ease of scoring is a factor for clinical investigators. The clarity and number of questions must be considered for many populations, especially those with cognitive impairment or nonnative English speakers.

The purpose for which the data are being collected should influence selection. HRQOL measures used as clinical outcome measures to assess treatment efficacy must include domains that the investigator believes will likely be affected by successful treatment. They should not be too general. These indices are not specific enough to be used as measures of drug toxicity (phase I), may enhance information from phase II trials, but are most frequently used for phase III/IV trials.

Some HRQOL measures are designed to determine the needs of given populations, whereas others are designed to monitor patients over time or assess the impact of a new health care program on a population of individuals [e.g., AIMS and the European Organization for Research and Treatment of Cancer (EORTC), which tend to be more disease-specific than generic]. It is best to review how a particular instrument has been used and then make the best fit for the needs of a specific trial.

The method of administration should be carefully considered. Many of the instruments are self-administered. This is certainly efficient; however, misunderstanding questions and failure to complete all questions may result from this type of administration. A telephone survey or an in-person interview may better ensure the completeness of the data, but they require more resources including training interviewers; hence, they are more costly to perform. In both situations, the interviewer may unwittingly influence the answer by interpreting the question in a unique way. The caregiver or personal representative of the patient/subject may be the source of information. This is an efficient means for obtaining information; however, there are often perceptual differences between the patient and the caregiver, and correlations between the same test administered to the patient or caregiver are not always strong.

3.4. Specific Quality of Life Instruments

There are 11 instruments, selected from more than 100, that illustrate the kinds of measurements that can be used to measure HRQOL in the clinical research setting. These were selected because they have good psychometric properties, have been used commonly, are generally easy to administer, and have a variety of applications. A summary of selected properties is provided in Table 21-3. These are generic (disease-neutral) measures.

Other measures of note—AIMS,[20] FLI,[21] the Quality of Life Index (QOLI),[23] and EORTC[24]—were designed to assess patients with a specific disease. AIMS measures domains of function for those with arthritis, and the others do the same for cancer patients. Arthritis and cancer diagnoses share a number of features; they are common, chronic, and complex and are associated with varying and fluctuating degrees of disability.

These disease-oriented measures were designed to assess the influence that treatment has on patient outcomes in multiple domains relevant to either the cancer or the arthritis patient. QOLI has also been used for program assessment. AIMS and QOLI have been used extensively in clinical trials. The domains selected by the instrument designers were influenced by existing instruments and how they did or did not meet the anticipated needs for health assessment. Each of these measures has a significant proportion of questions at the performance or functional level and fewer measures of symptoms, feelings, or perceptions.

The Functional Status Questionnaire,[25] Dartmouth Cooperative Measure,[26] Duke–UNC Health Profile,[27,28]

TABLE 21-3 General HRQOL Instruments

Name	Administration	No. of Items	Population	Objective
15D	S	15	16 years	HRQOL only
Assessment of Quality of Life (AQOL)	S, P	15	Adult	Include economic data
Child Health Questionnaire (CHQ)	S, P	Parents, 50; Youth, 87	5–17 years	HRQOL in children
Duke Health Profile	S	17	Adult	HRQOL and health status
Functional Status Questionnaire (FSQ)	S	34	Adult	Principal, psychosocial, and role function in ambulation
Health Assessment Questionnaire (HAQ)[a]	S, P	20 2 visual analog scales	Adult	Ambulation, ADL
McMaster Health Index	S, P	59	Adult	HRQOL and social/emotional status
Nottingham Health Profile	S	45	Adult	Patients' perceived emotional, social, and physical health
Ferrans & Powers QLI	S	66	Adult	QOL in terms of satisfaction
SF-36	S	36	> 14 years	Measures health concepts
Youth QOL	S	13	11–18 years	QOL

[a]Also has a child-based questionnaire (CHAQ).
ADL, activities of daily living; P, phone interviews; S, self-administered.

and McMaster Index[29] were originally developed for a primary care or outpatient, ambulatory population. Treatment outcome was of interest, but the instruments had to be able to measure the outcomes resulting from the delivery of health services and provide good quantitation of health rather than disease. Each of these measures included domains measuring social, physical, and mental and general health. They also included more measures of perceived health, self-esteem, thoughts about future health and personal goals, and overall QOL.

The SIP, Nottingham Health Profile,[30,31] and SF-36[32,33] were designed to assess a wide variety of patient populations, be relatively disease neutral, and span a broad age range of patients receiving care in many different settings. Of the three, only SIP has no measures of satisfaction or feelings. Its construct was designed to assess the patients' perceptions of loss of functions, and in a sense it is entirely structured around perception of how the current status differs from a previous one. It has no value qualifiers about this loss or change in functional status. It is scored by comparison with a varied sample. The other two, however, include several domains that address general perceptions and emotional reactions to illness or change in functional status. They can be used to assess change over time.

The Older American Resources and Services Functional Assessment[34] and Comprehensive Assessment and Referral Evaluation[35] are among the more complex and difficult to administer. They were designed to evaluate the elderly and are heavily weighted toward asking questions about services needed among individuals as well as groups of elderly patients. These measures are administered by an observer or an informant, take 45 minutes to 1 hour to administer, and have 144 and 369 questions, respectively.

Two additional instruments are also recommended for your consideration. They are designed for general populations and commonly used: the Quality of Well Being Scale[36] and the 15D.[37]

Additional information about many of these instruments is available at www.rar.duhs.duke.edu. Researchers can obtain permission to use certain instruments (e.g., SF-36) at www.qmetric.com/products/assessments/license.

4. OTHER INSTRUMENTS TO CONSIDER

This chapter discusses the development and generation of HRQOL indices and profiles. There are a significant number of valid, reliable, and sensitive instruments that measure health-related events from a

less value-driven or less comprehensive approach with respect to domains. Sometimes, these are more appropriate for application to a clinical trial and may be more specific to the unique needs of a particular patient population or have greater sensitivity. These include activity of daily living scales, usually focused on self-care and mobility. The PULSES[38] or Barthel Index[39] are examples, as is the Health Assessment Questionnaire designed for arthritis patients.[40] Some instruments measure only one domain, such as social health,[41] psychological well-being,[42] or life satisfaction.[43] Many measure a particular symptom, such as depression,[44,45] or a particular function, such as mental status.[46] There are a host of visual analog scales to measure fatigue, pain, and global health. An excellent review of these instruments is available.[47]

5. IMPORTANCE OF QUALITY OF LIFE MEASURES FOR HEALTH CARE

The universal application of the controlled clinical trial has established a standard by which the efficacy of therapeutic interventions is judged. These trials are essential for determining whether a treatment is effective in improving health. We are also required to assess the treatment's cost, whether this is the most effective way to use resources, and what value there is to individuals, their families, and society to use such treatments. The WHO defines healthiness as "a state of complete physical, mental, and social well-being, not merely the absence of disease." There has been much discussion about the need to factor quality and not solely quantity of life in important decisions. Second, decisions about striving to cure chronic, complex, serious illnesses should be weighed against how this would impact on function and HRQOL. Answers to these questions, which are now factored into decisions made by regulatory agencies, the pharmaceutical industry, legislators, and individual patients, require instruments designed to measure how treatment impacts what is valued by patients, and the instrument must provide relevant, reliable, valid, and sensitive data. Significant effort by NIH has been directed at improving this methodology (www.nihpromis.org).

There remain some unanswered questions and some deficiencies in the state of HRQOL development, its methodologies, and its applications.[48,49] The following are some of the issues in need of resolution. Many of the HRQOL assessments use multiple and different domains. Should these be restricted? Should the methods of quantification and valuation be standardized? Should all domains be totaled or should (can) each component be separately analyzed? Who should assess—the patient, caregiver, health care professional, family, or all? Should HRQOL measures be required for all trials? Should they be incorporated into clinical practice? How do cultural differences influence HRQOL assessment? How do we translate information obtained into improved clinical practice, assessment of risk, cost, and health care policy?

The use of HRQOL measures has become more prevalent during the past decade. All indications suggest this trend will continue, and health care practice as well as policy will be influenced by data that attempt to address questions of values as they relate to health.

References

1. World Health Organization. *World Health Organization Constitution. Basic Documents.* Geneva, World Health Organization, 1948.

2. World Health Organization. *International Classification of Functioning, Disabilities and Health*, 1C1DH-2. Geneva, World Health Organization, 1997.

3. World Health Organization. *The International Classification of Impairments, Disabilities and Health.* Available at http://who.int/icidh/index.html.

4. Nagi SZ. Congruency in medical and self-assessment of disability. *Ind Med* 1969;8:27–36.

5. Granger CV, Hamilton BB, Keith RA, Zielezny M, Sherwin FS. Advances in functional assessment for medical rehabilitation. *Topics Geriatr Rehabil* 1986;1:59–74.

6. Corinsky KE, *et al.* Health status vs. quality of life in older patients. Does the distinction matter? *Am J Med* 1999;106:435–440.

7. Sanders C, Egger M, Donavan J, *et al.* Reporting on quality of life in randomized controlled trials: A bibliographic study. *Br Med J* 1998;317:1191–1194.

8. Ware JE, *et al.* Choosing measures of health status for individuals in general populations. *Am J Public Health* 1981;71:620–625.

9. Shoemaker D, *et al.* A regulatory perspective. In Spilker B (ed.) *Quality of Life Assessments in Clinical Trials*, pp. 193–207. New York, Raven Press, 1990.

10. Patrick DL, Erickson P. What constitutes quality of life? Concepts and dimensions. *Clin Nutr* 1988;7:53–63.

11. Sanazaro PJ, Williamson JW. End results of patient care: A provisional classification based on reports by internists. *Med Care* 1986;6:123–130.

12. Bergner M. Measurement of health status. *Med Care* 1985;23:696–704.

13. Roberge R, Berthelob JM. The Health Utility Index: Measuring health differences in Ontario by socioeconomic status. *Health Rep* 1995;7:25–32.

14. Stewart AL. The medical outcomes study framework of health indicators. In Stewart AL, Ware JE (eds.) *Measuring Functioning and Well Being.* Durham, NC, Duke University Press, 1992.

15. Patrick DL, Erickson P. Assessing health-related quality of life for clinical decision making. In Walker S, Rosser R (eds.) *Quality of Life Assessment: Key Issues in the 1990s.* Dordrecht, The Netherlands, Kluwer, 1993.

16. Guyatt GH, Veldhuyzen Van Zanten SO, Feeny DH, Patrick DL. Measuring quality of life in clinical trials: A taxonomy and review. *Can Med Assoc J* 1998;140:1441–1448.

17. Guyatt GH, Feeny DH, Patrick DL. Measuring health related quality of life. *Ann Int Med* 1993;118:622–629.

18. Farquhar M. Definition of quality of life: A taxonomy. *J Adv Nurs* 1995;22:502–508.

19. Kirschner B, Guyatt G. A methodological framework for assessing health indices. *J Chronic Dis* 1985;38:27–36.

20. Meenan RF, Gertman PM, Mason JH. Measuring health status in arthritis: The Arthritis Impact Measurement Scale. *Arthritis Rheum* 1980;23:146–152.

21. Schipper H, *et al.* Measuring the quality of life in cancer patients: The Functional Life Index—Cancer: Development and validation. *J Clin Oncol* 1984;2:472–483.

22. Bergner M, *et al.* The Sickness Impact Profile: Development and final revision of a health status measure. *Med Care* 1981;19:787–805.

23. Spitzer WO, *et al.* Measuring the quality of life of cancer patients: A concise QL index for use by physicians. *J Chronic Dis* 1981;34:585–597.

24. Sprangers MAG, *et al.* The European Organization for Research and Treatment of Cancer approach to quality of life assessment: Guidelines for developing questionnaire modules. *Qual Life Res* 1993;2:287–295.

25. Jette AM, *et al.* The Functional Status Questionnaire: Reliability and validity when used in primary care. *J Gen Intern Med* 1986;1:143–149.

26. Nelson EC, *et al.* Assessment of function in routine clinical practice: Description of the COOP chart method and preliminary findings. *J Chronic Dis* 1987;40(Suppl. l):555–635.

27. Parkerson GR, Jr, *et al.* The Duke–UNC Health Profile: An adult health status instrument for primary care. *Med Care* 1981;19:806–828.

28. Parkerson GR, Broadhead WE, Tse CK. The Duke Health Profile: A 17-item measure of health and dysfunction. *Med Care* 1990;28:1056–1072.

29. Chambers LW, Sackett DL, Goldsmith CH. Development and application of an index of social function. *Health Sew Res* 1976;11:430–441.

30. Hunt SM, *et al.* A quantitative approach to perceived health status: A validation study. *J Epidemiol Community Health* 1980;34:281–286.

31. Hunt SM, *et al.* The Nottingham Health Profile: Subjective health status and medical consultations. *Soc Sci Med* 1981;15:221–229.

32. Ware JE, Sherbourne CD. The MOS 36-item Short Form Health Survey, SF-36. Conceptual framework and item selection. *Med Care* 1992;30:473–483.

33. Stewart AL, Hays RD, Ware JE. The MOS Short Form General Health Survey. *Med Care* 1988;26:724–735.

34. Fillenbaum GG, Smyer MA. The development, validity and reliability of the OARS Multidimensional Functional Assessment Questionnaire. *J Gerontol* 1981;36:428–434.

35. Gurland B, *et al.* The Comprehensive Assessment and Referral Evaluation (CARE)—Rationale, development and reliability. *Int J Aging Hum Dev* 1977;8:9–42.

36. Fanshel S, Bush JW. A health status index and its application to health services outcomes. *Operations Res* 1970;18:1021–1065.

37. Sintonen H. An approach to measuring and valuing health states. *Soc Sci Med* 1981;15C:55–65.

38. Granger CV, Albrecht GL, Hamilton BB. Outcome of comprehensive medical rehabilitation: Measurement by

PULSES Profile and Barthel Index. *Arch Phys Med Rehabil* 1979;60:145–154.

39. Mahoney FI, Wood OH, Barthel DW. Rehabilitation of chronically ill patients: The influence of complications on the final goal. *South Med J* 1958;51:605–609.

40. Fries JF, *et al.* Measurement of patient outcomes in arthritis. *Arthritis Rheum* 1980;23:137–145.

41. Katz MM, Lyerly SB. Methods for measuring adjustment and social behavior in the community. *Psychol Rev* 1963;13:503–535.

42. Goldberg DP, Hillier VT. A scaled version of the General Health Questionnaire. *Psychol Med* 1979;9:139–145.

43. Neugarten BL, Havighurst RJ, Tobin SS. The measurement of life satisfaction. *J Gerontol* 1961;16:134–143.

44. Beck AT, *et al.* An inventory for measuring depression. *Arch Gen Psychiatry* 1961;4:561–571.

45. Radloff LS. The CES-D Scale: A self-report depression scale for research in the general population. *Appl Psychol Measure* 1977;1:385–401.

46. Folstein MF, Folstein SE, McHugh PR. "Mini-Mental State" a practical method for grading the cognitive state of patients for the clinician. *J Psychiatr Res* 1975;12:189–198.

47. McDowell I, Newell C. *Measuring Health: A Guide to Rating Scales and Questionnaires*, 2nd ed. Oxford, Oxford University Press, 1996.

48. Leplege A, Hunt S. The problem of quality of life in medicine. *JAMA* 1997;278:47–50.

49. Muldoon MF, Barger SD, Flory JD, *et al.* What are quality of life measurements measuring? *Br Med J* 1998;316:542–545.

22

Overview of Technology Development

BRUCE GOLDSTEIN

Office of Technology Transfer, National Institutes of Health, Rockville, Maryland

1. INTRODUCTION

The changes during the last 20 years in the dynamics of scientific progress generally, and in the biomedical arena in particular, have been as dramatic as the changes wrought upon a landscape by a river altering course, flooding some regions and carving others. Inexorably, the ground that had been solid crumbles, and new shores emerge. For those who have established the foundations of their research careers in the realm of pure academia, the new landscape lacks many of the familiar landmarks and paths. Although many people find such changes disturbing, confusing, or simply aggravating, the most successful researchers will have to learn to navigate the new terrain.

As is discussed in more detail in Chapter 23, one of the major forces precipitating the changes in the manner of scientific development occurred in the law of patents. First, in 1980, the Supreme Court ruled that life-forms created through recombinant-DNA technology could be protected by patents. Second, in 1982, Congress created a special appeals court, the Federal Circuit Court of Appeals, to hear specific kinds of cases, including patent law. This court has clarified much of patent law and made enforcing patents far more practical than it had been. Third, and most relevant to this chapter, Congress passed a series of laws in the early 1980s (with important, subsequent amendments) that enabled the transfer of some of the government's rights to inventions to nongovernment parties.[1] The combination of these events dramatically accelerated the development of the scientific field now called biotechnology, and it started the legal field that today

is broadly called "technology transfer," among other things.[2]

People are largely unaware of all the various tools used to accomplish the transfer of technology. Ask people who have heard about technology transfer, and many will reply that it involves lawyers arranging for large corporations to license government-owned patents (a topic covered in another chapter). Ask them how technology transfer impacts their research, and they are likely to say "not at all." The river is still carving new territories, however, and sooner rather than later, most of the pure researchers will be forced to navigate the new terrain. Research agreements, inventions, patent licenses, material transfers, confidentiality, software, copyrights, trademarks, and many other, perhaps even more unfamiliar things loom—and pitfalls, deep enough to swallow a career or two, exist. To add another layer of confusion, the perspectives of for-profit industries, nonprofit/academic groups, and government about technology development are significantly different from each other.

In this chapter, to identify the new landmarks and map the terrain, a purely fictional scenario is described, relating a series of hypothetical events. Then, using the scenario as a backdrop, some of the various tools are examined in turn, with a focus on why, when, and how each is used appropriately. The causes of the more common snags will also be discussed so that those problems caused by divergent perspectives may be avoided. Hopefully, at the conclusion, the features of the new landscape will appear as opportunities—ways to enhance and enable research—rather than as obstacles.

2. SCENARIO: DISASTERS WAITING TO HAPPEN

Meet Gillian Niher, M.D., Ph.D. She has developed a stellar reputation as an up-and-coming neuronal researcher. Her focus has been on therapies for neural injuries, primarily peripheral nerves. From a teaching position at Smallville Medical School, she found a tenure-track position at the National Institutes of Health (NIH) in a lab with facilities in the NIH Clinical Center. Unfortunately, she was stuck for ideas for her next blockbuster study; although generally interested in a variety of cutting-edge technologies, she had not yet settled on one. Then, her very close college friend, Alan Prophet, Ph.D., came to Bethesda on a business trip and visited her. Over lunch, Alan told Gillian about his gene-therapy research at Tate State University (a private institution in Maryland that does not rely on grants from NIH to support its bioscience research, but several projects are funded by industry).

Alan mentioned that Tate State sponsored "spin-off" companies for professors who invent new bioscience products. He also mentioned that he was named as a co-inventor on a recently issued patent on the genetic sequence of a newly discovered neuronal growth factor. With support from Tate State, Alan and his colleagues created a small company called Neurion to develop this gene. They had found some support from a group of venture capitalists, who received a large share of corporate control in exchange for financing. The company had already succeeded using the gene in several *in vitro* models. They also had recently done some toxicity and efficacy tests in injured rats and rabbits, but the results were not yet public. Alan invited Gillian to visit Neurion's facilities, and Gillian excitedly agreed.

Two weeks later, she went to Neurion's small facilities near the Tate State campus. When she arrived, Alan told her that before he could give her a tour, she would have to sign a form the lawyers drafted to make sure trade secrets stayed secret, and Gillian agreed to comply. Then, Alan showed her preliminary data that demonstrated the growth factor was surprisingly effective in stimulating neuron regrowth, either when the growth factor protein was delivered directly to the site of neuronal injury or when a plasmid incorporating the gene was applied to the extracellular matrix.

Impressed with these results, Gillian saw an opportunity to establish a collaboration: Neurion's growth factor entering clinical trials at NIH. She consulted her scientific director about the project and was pleased that he was very interested. Alan and his partners in

Neurion were equally excited when she made that suggestion to them. Alan and Gillian quickly drafted a protocol for human trials, which was favorably received by Gillian's laboratory chief and scientific director, as well as by the venture capital group. After Gillian signed some of Neurion's forms, Neurion sent large amounts of good medical practice-grade materials for Gillian to use at NIH. The process of establishing the study appeared to be on the fast track to success.

Soon thereafter, while reviewing the final animal study data Alan had provided, Gillian noticed two things Neurion had missed. First, the rabbits in the "control" group (those given only blank plasmid) had no noticeable neuronal growth—that is, the number of nerve endings was unchanged with the injection of the plasmid—but they seemed to be improved in terms of muscle movement and strength. Upon closer examination of the rabbits, she found that the original injured nerve endings had in fact regrown. In contrast, all of the rabbits that received the gene had completely new nerves growing in addition to the original ones, and all the rabbits that received nothing had no neuronal stimulation at all. Something in the plasmid appeared to have activity. Second, she noticed that those rabbits receiving the gene had exuberant growth of neurons, even in regions in which all the original neurons were dead.

Alan was naturally excited to hear about these observations but told Gillian to keep them quiet just long enough so that Neurion could file a patent application. Reluctantly Gillian agreed; however, she quietly sent samples of the plasmid, with and without the gene, to John Rogers, M.D., a colleague of hers at Smallville, for careful analysis of the plasmid's sequence. The clinical trials began, and over the following weeks the pair began collecting data.

Then the major problems began. Alan and Gillian continued to prepare the manuscript for the paper disclosing Gillian's discoveries, but Neurion insisted Alan delay his efforts, telling Gillian that the delay was needed because the patent application was not yet ready. This created a problem for Gillian, who was obligated to publish her results as soon as possible. Then, while on a visit to Alan's offices at Neurion, Gillian saw some documents indicating that a patent application had already been filed by Neurion describing her discoveries, but she was not named as an inventor. Furious, Gillian quickly finished the rough draft and submitted the manuscript immediately. Upon learning of this act, Neurion demanded that Gillian retract the publication, return all remaining stores of the gene, and terminate the study, but Gillian refused.

To make matters worse, 10 subjects in the clinical trial were experiencing something very strange. The regions of tissue receiving the gene were experiencing hypersensitivity, to the point of severe pain. Histological analysis of the tissue revealed that the neurons were growing far more exuberantly in humans than in either rats or rabbits. The stimulating factor was out of control. As if matters were not bad enough, John used Gillian's sample plasmids to generate a large quantity of gene-bearing plasmid, which he had injected into 10 undergraduate volunteers at Smallville College without securing approval from the institutional review board, acquiring informed consent, or even controlling the quality of the materials he had injected. Six of these students have experienced the neuronal hyperplasia.

Gillian is now being sued by Neurion for breaches of their contracts, misappropriation of trade secrets, and patent infringement. Although the injured patients and students are suing Neurion for making the dangerous materials, Neurion has asked the court to order Gillian to pay Neurion's legal bills and any judgment associated with that product liability suit on the grounds that Gillian had agreed to do so in her various contracts. The media, having heard of the Smallville incident, has placed the whole story on national news. Congress has issued subpoenas to her entire lab, asking why NIH is sponsoring secret clinical trials of unproven, dangerous genes in our nation's children. The scientific director personally has asked her to resign. Finally, Gillian's attorney has told her the Assistant U.S. Attorney is investigating whether to charge her with criminal sanctions.

What went wrong, and how could the tools of technology development have helped avoid these problems? By unraveling the complicated mess, and reviewing each piece, we will illuminate the traps and show the tools that would help avoid them.

3. THE FIRST AND BIGGEST MISTAKE: SIGNING THE AGREEMENTS

3.1. Contract Execution in General

By the time most people have reached adulthood, they have been scolded to read all contracts before signing them, no matter how long and confusing the fine print may be. Indeed, in many cases, the documents we are asked to sign are so complicated and difficult to read that common sense demands hiring a lawyer. Nonetheless, because hiring lawyers is expensive and time-consuming, and because many of us are unaware of the actual risk of something going wrong,

we ignore that risk and sign—often without even reading—happy to have saved the time and money. Only later, when we need the lawyer's equivalent of a root canal, do we ruefully ask for help to clean up the mess.

Yet even if the document is simple and the person being asked to sign it has taken the time to read it, major pitfalls still lurk. For example, if something goes wrong, who is on the hook? As a general rule, a person who signs a contract is promising to fulfill the terms of the contract.[3] That means Gillian likely will be liable if the promises in the contracts she signed are not satisfied. This is especially dangerous if the agreement purports to make promises that the signer cannot keep, such as a promise to keep something secret that by law must be disclosed.

A bigger problem here is "agency," or the power to act on someone else's behalf. If the signer purports to bind another party (e.g., a company or institution) to perform a promise, the signer must, in truth, have authority from that party to bind it in order for the party to be bound.[4] Moreover, the authority must extend to the particular type of contract: If person A has limited authority to buy groceries for person B, A may not use B's money to buy investment bonds. Though these rules appear simple on their face, they are less simple in practice.

Although the owner of a private entity can bind that entity, generally, individual employees do not have authority to bind their employers. Certain employees, namely those who occupy key offices in a corporation (e.g., president or chief executive officer) or a university (e.g., provost or dean), typically have formal, written authority to bind their employers to the contracts they sign on their behalf.[5] The formal authority typically appears in charters, articles of incorporation, bylaws, or employment contracts. Other times, authority is expressly delegated in a memo or other writing, such as through a power of attorney. In government agencies, each statute passed by Congress that created each agency specify which offices can bind that agency, and actual authority below that level must be formally delegated in writing. In each case, this express grant of power is called "actual" authority.

Occasionally, authority to act as an agent reasonably can be inferred from the circumstances, even if no actual authority exists. If the general counsel, associate dean, or senior vice president of a company or university signs a contract, others might be justified in relying on the signature, even if the individual has no written delegation to display.[6] This is a narrow exception, however, and one cannot reasonably assume that any employee of a company (even a senior one) has authority to bind that company. Because Gillian did not have any indicia she had authority to bind her institute[7] (e.g., being the institute's director or technology development coordinator), Neurion had a poor basis for assuming her signature alone would bind anyone at NIH other than herself, and so would have weak grounds at best for asserting that the government breached any contracts.

This conclusion is cold comfort for Gillian. Normally, if an agent acts within the scope of the authority delegated by the principal, the agent will not be liable if the principal later breaks the contract.[8] That immunity, however, rests on whether the agent acted within the scope of the authority. Because Gillian's signature was not authorized by NIH, she will not be protected by the fact that she signed the agreements, even if she did it in an attempt to carry out her official duties.

Finally, even if a scientist who signs an agreement clearly lacked authority to bind the employer, the employer may still be placed in the position of facing an irate company. Two particularly high-profile cases highlight the problem.

According to an article published in *The Scientist*,[9] Dr. David Kern, a medical professor at Brown University, was asked by a local fabric company called Microfibers to consult on two cases involving a rare syndrome called interstitial lung disease. He discovered it was due to conditions in Microfiber's factories and also discovered cases in other employees of Microfibers working at two specific facilities. Immediately, he began the process of publishing his results. Microfibers, however, threatened to sue both Kern and his employer on the basis of certain nondisclosure agreements signed by students in Kern's department, who had come to Microfibers for a visit two years before on an unrelated matter. Apparently, neither Kern nor his employer had ever ratified the agreements, and it is unclear whether either was even aware of the agreements' existence. Even so, Kern's employer, placed in the highly awkward position of having to face litigation or restraining Kern, elected the latter.

Another high-profile example of an attempt to suppress research, reported in major newspapers,[10] occurred between the former Boots Pharmaceuticals[11] and the University of California at San Francisco (UCSF). In 1987, Dr. Betty Dong, a scientist at UCSF, signed Boots's research funding agreement personally in order to conduct a study on whether Synthroid (a synthetic drug for the treatment of hyperthyroidism) was superior to generic equivalents. The study, completed in 1990, indicated that the generics were bioequivalent to Synthroid. Dr. Dong handed copies of the data to Boots, but by 1995, Boots had not released any of the information, so Dr. Dong submitted a manuscript to the *Journal of the American Medical Association* (*JAMA*). Boots asserted the study was flawed and refused permission to publish, and the original research agreement stated that permission was required before the results could be made public. Despite the fact that the provision violated UCSF policy, UCSF's attorney told Dr. Dong that UCSF would honor the term, and if she wanted to publish on her own, she would have to defend herself against Boots's threatened litigation without UCSF support. Faced with this threat, Dr. Dong asked *JAMA* to halt the article. Only after intervention by Dr. Louis Sullivan, then Secretary of the U.S. Department of Health and Human Services, did Boots relent and allow publication,[12] but not before Boots had published a scathing critique, reinterpreting the data in a manner that cast a more favorable light on Synthroid.[13]

3.2. Scope of Actual Authority of Government Laboratories

In the context of government laboratories, there is an additional twist. For most people, laws are *disabling*: In other words, you can do whatever you want *unless* it is prohibited by law. For the government as an acting entity, with few exceptions, laws are *enabling*: A government agency (and its authorized representative) can do *only* what the law has expressly authorized. In the establishment of relationships between government agencies and nongovernment parties, this divergence of point of view often causes substantial problems. In particular, companies, nonprofits, and private universities, all accustomed to crafting essentially whatever terms their internal institutional policies will allow, simply do not understand why the agency says "No, we cannot do that."

The enabling character of law as it applies to government action stems from the Constitution—the very foundation for both federal and the state governments—which lists those specific things the government can do. Ultimately, the written authority for an agency to take a given action must be directly traceable from a provision in the Constitution, to a law passed by the legislative branch (or, occasionally, an order issued by the executive branch), through regulations promulgated by the secretary of the agency, and a written trail of delegations down the chain of command within that agency. At each delegation, the authority to act may be restricted further. The scope and meaning of these documents may be illuminated by opinions of courts, the Attorney General, and perhaps the general counsel of the agency. Finally, each agency may establish its own policies of implementation, which generally stem from the mission set out in the original legislation. As a consequence, even if a given person has the raw potential to receive authority to act on behalf of the agency, the scope of authority actually delegated may be severely circumscribed by these various layers of government. In certain circumstances, a particular office in an agency may want to take an action that is still within the law but exceeds existing delegations of authority. Unfortunately, circumventing a given authority may require so much review at so many levels, and may precipitate so much political fallout, that only an extraordinary case would justify the attempt.

Occasionally, the law also acts on agencies in a disabling way. For example, agencies of the government are directly forbidden to take an action that would incur upon the agency a debt that exceeds its appropriated budget, without express statutory authorization to do so.[14] Thus, in the Neurion scenario, NIH could not agree to protect Neurion from the product liability lawsuits brought by the injured students because the possible judgments against Neurion (not to mention Neurion's legal fees) might well exceed the agency's appropriated budget. At best, Neurion may feel cheated, having entered an agreement in good faith, and will be reluctant to enter future agreements with anyone at NIH. At worst, if any government employee purports to incur such a liability on behalf of the government—as Gillian did in the agreements she signed—the employee risks, in theory at least, going to jail.[15]

4. AGREEMENTS NOT TO DISCLOSE: TRADE SECRETS AND THE CONFIDENTIAL DISCLOSURE AGREEMENT

One political extreme holds the view that the government is engaged in the systematic suppression of information that the public has a need to know. The other extreme asserts that the government is not capable of keeping information secret without being forced to do so by law. Reality lies somewhere between these extremes. Since the passage of the Freedom of Information Act (FOIA), a lively debate has ensued over the proper balance between these two opposing positions. Sometimes, the government must reveal the information on which its actions and policies are based; other times, release of information in government possession would cause injury without providing any public benefit.

In the arena of scientific research, the debate is as strong as anywhere. Occasionally, government scientists need access to confidential information in the hands of private parties to do their jobs. By the same token, these same government scientists must publish their research results. The challenge is to find a way to accommodate the legitimate needs of industry to protect trade secrets and of individuals to protect their privacy, without giving a private party the power to restrict the government scientist's duty to publish results or the public's right to know.

The reach of FOIA is not limited to federal laboratories and offices. In 1997, Congress extended the reach of FOIA to nongovernmental researchers receiving federal funds.[16] Specifically, Congress ordered the Office of Management and Budget to amend Circular A-110 "to require Federal awarding agencies to ensure that all data produced under an award will be made available to the public through the procedures established under the Freedom of Information Act." Effective for all grants (new and continuing) awarded after March 16, 2000, data that are (1) first produced in a project that is supported in whole or in part with federal funds and (2) cited publicly and officially by a federal agency in support of an action that has the force and effect of law are subject to disclosure under FOIA.[17]

4.1. Background: Trade Secrets

As a general principle of trade secret law, a trade secret can be any piece of information that (1) is more or less exclusively known by the party claiming it (i.e., it is truly a secret in the field), (2) is protected by measures that are reasonable under the circumstances, and (3) has some economic value—either because the owner of the secret experiences a direct and tangible economic benefit (e.g., a cheaper way of making a formulation) or because the competitors of the owner would have to expend considerable resources to discover the secret through lawful means (e.g., by reverse engineering).[18] Classic trade secrets include methods of mass manufacture, detailed contact and pricing lists for each customer, industrial recipes, and inventions that are the subject of pending patent applications. A trade secret, however, could be anything.

If the basic criteria are met, the owner of a trade secret has grounds to ask a court to protect that secret against "misappropriation" by assessing money damages and sometimes by imposing an injunction.[19] A trade-secret lawsuit does not depend on the existence of a contract to be successful; "misappropriation" encompasses both the wrongful acquisition of a trade secret and the wrongful use or disclosure of a rightfully held trade secret.[20] Moreover, for as long as the information actually remains a secret, the legal right to protect the secrecy of that information continues.

The difficulty in trade secret litigation, typically, lies in proving that all the initial criteria are met. For example, assuming your confidante wrongly disclosed your secret, how do you prove that your information was actually a secret at the time you disclosed it to the confidante? Were the steps you took to keep your information secret "reasonable" (and will a jury agree)? Did the secret have commercial value? Was the information still a secret at the moment when the confidante publicly disclosed your information? These are difficult facts to prove, even in the best of conditions. Moreover, as a purely practical matter, the likelihood is low that an injured party will recover through the legal process the true value of what was lost when the secret was revealed, even if misappropriation has been proved.

Nevertheless, using some form of confidential disclosure agreement is a good idea for all concerned, for several reasons. First and arguably foremost, a signed agreement proves the people involved actually knew that a disclosure of a trade secret may occur, and merely putting a signature on paper often has the psychological effect of making those involved treat the terms of the written agreement more seriously than they would with a mere handshake. Second, clear terms can help avoid disagreements and ill will by clarifying which information should be treated as confidential, as well as what acts are or are not appropriate. Third, where a patent application has not yet been filed, a written confidentiality agreement reduces the risk that a patent office or court will later deem the prefiling disclosure to be a bar against patenting. Finally, even where there is a wrongful disclosure, if it is a minor disclosure, the party owning the trade secret still has a chance of getting legal protection for the information in the future because the party can point to the agreement as evidence that the party took every reasonable step under the circumstances.

4.2. Secrets and the Government

Under FOIA[21] (and its various state counterparts), all government records must be disclosed upon request, unless the government can demonstrate that the information in the record falls into a specific, narrow exception on a short list set out by Congress; even then, the government must disclose a redacted version if feasible. Of the exceptions on that list, five are routinely relevant to the federal government's biological and medical research. They are exceptions for trade secrets and "commercial and financial information,"[22] internal decision making,[23] personal information of a private

nature,[24] unfiled patent applications in which the government owns an interest,[25] and certain research information generated under a "Cooperative Research and Development Agreement" (CRADA) (discussed in more detail later).[26] This arrangement presents a dilemma for NIH.

On the one hand, from a scientific perspective, data should be meticulously collected, organized, and carefully analyzed before drawing conclusions; releasing preliminary conclusions could be irresponsible if they have not been grounded in properly collected data, particularly if the conclusions have not undergone some substantive review. This is especially true where the premature release of unsifted information would be misleading. Furthermore, NIH acknowledges that private research facilities have a legitimate need to protect their trade secrets and individuals have the right to privacy; NIH understands that these parties will not cooperate with NIH if the confidentiality of their information will not be protected.

On the other hand, even apart from the commands of FOIA, NIH has strong reasons to support disclosure of all research results as quickly as possible. For example, the bedrock mission of NIH is to uncover new knowledge that will lead to better health for everyone. NIH depends on the rapid communication of research results to advance that mission. Also, because the most talented scientists cannot advance their careers if impediments block their ability to publish important results in a timely manner, they will instead work in a more publication-friendly environment. For these and other reasons, NIH is strongly committed to the principle that scientific advancement relies on the unfettered and rapid dissemination of information. NIH will never approve any agreement in which a private entity has substantive control or veto power over the research publication of one of its scientists, lest valuable information that was developed by taxpayer funds be stifled to further private interests. On this point, NIH will not negotiate and encourages the academic community to follow its lead.

As a compromise, NIH strives to draw a line between the information provided to NIH and the research results derived from that information. NIH will work with collaborators to protect legitimate trade secrets from inadvertently being disclosed in publications. Specifically, NIH will delay disclosures enough to give collaborators a reasonable opportunity to file patent applications on discoveries. Also, NIH will seriously consider any requests by collaborators to redact or edit manuscripts and other disclosures before they are made public. Nonetheless, NIH must retain final authority to decide whether to go ahead with a given disclosure.

4.3. Anatomy of a Confidential Disclosure Agreement

A normal Confidential Disclosure Agreement (CDA)* addresses four major points,[27] in one form or another. First, it identifies the information to be disclosed. Second, it names the parties. Third, it states how the confidential information will be handled. Finally, it specifies the duration. Occasionally, some agreements discuss rights to intellectual property—that which exists prior to any disclosure under the agreement and that which is discovered because of the disclosure, should any arise—but this is not a necessary term.

The information to be disclosed defines scope and reach of the agreement. Consequently, this is the single most important part, and a well-crafted CDA will clearly identify the information to be disclosed. Unfortunately, there is a tension between the "provider" of the information, who typically wants the definition to be as broad as possible, and the "recipient," who wants it as specific as possible. Also, the provider will not want the CDA's description of the information to include the confidential information itself.

Nevertheless, some description should be fashioned that will make clear to the recipient exactly what the provider expects the recipient to keep confidential. Note that the agreement can accomplish this task in one of two ways, either by identifying the nature of the information with specificity (e.g., "the investigator's brochure for company's study drug") or by obliging the provider to mark all documents with the legend "Confidential" and reducing oral disclosures to writing (and marking them) within a set time. Although providers may dislike agreeing to accept the duty to mark, doing so is in their interests: As a rule, courts will not impose a greater duty on a recipient to identify and segregate a provider's confidential information than a provider imposed on itself.[28] In other words, if the information were truly valuable, a provider would have marked it.

Also, as a matter of reasonableness, the agreement should specify those situations in which information ostensibly provided under the agreement will not be deemed confidential, such as (1) information that is or becomes public through no misdeed by the recipient; (2) information that the recipient lawfully receives from a third party, that the recipient already knows, or that the recipient independently creates; and (3) information that must be disclosed by force of law.

Next, identifying the parties is simple, yet surprisingly often it is botched by making the individual receiving the information sign as the party, rather than the individual's employer. One reason this is a mistake is the question of agency: Providers have essentially no protection if they ask individuals to sign agreements on behalf of their recipient-employers, unless the individual's authority to do so is starkly apparent. Even if agency is not an issue, another problem lies in the hidden trap that caught Gillian Niher when she signed Neurion's CDA in her personal capacity: She breached her CDA merely by telling her scientific director and lab chief about Neurion's information—not to mention by telling John Rogers at Smallville—and any remedies specified in the CDA could be invoked against her.

How the parties will handle the confidential information is usually where the most substantial negotiations occur because the possibilities are virtually endless. For example, what measures will be taken to control who at the recipient's lab will have access to documents? When the agreement ends, what will be done with the documents, and for how long will the provider's rights survive? If the recipient wants to publish, what steps will the recipient have to take to ensure the publication does not contain the provider's confidential information? What will the provider's rights be if the recipient is ordered by a court to disclose the confidential information? Each of these issues could be negotiated, within the policies of the parties.

Finally, the agreement should have a clear, specified ending point. Some providers ask for (and sometimes receive) promises to keep information confidential indefinitely. Nevertheless, as Benjamin Franklin once wrote in *Poor Richard's Almanac*, "three can keep a secret, if two are dead;" in other words, the more who know a secret, the shorter its secret status will be. In addition, the dizzying pace at which biomedical technology is advancing strongly implies that the commercial value to a piece of confidential information depreciates rapidly, even if competitors never learn the secret. Consequently, a reasonable term to keep a secret should reflect the true life of the secret and little more. This is particularly important in the academic world, in which the act of dissemination is the source of value for information. NIH policy is to keep information given to it as confidential for three years, which can be extended for an additional two years upon request—subject, of course, to the limitations imposed by the Freedom of Information Act and other laws. Even for nongovernment parties, only in the most unusual circumstances is it even meaningful to promise to maintain a secret for more than five years.

---

*The CDA goes by other descriptions, including "nondisclosure agreement" and occasionally "secrecy agreement."

Intellectual property is only occasionally a true issue. Most parties appreciate the unlikelihood that the recipient will invent something immediately and directly upon seeing the provider's confidential information. Others, comfortable with the strength of their background patent position, do not concern themselves with what might happen if someone improves on the technology. In both of these cases, the agreement will state (at most) that patent law will govern ownership of patentable discoveries, and no licenses are promised.

Still, some providers (usually small companies having a single core technology in a competitive market) will insist that they be promised certain rights in anything invented by the recipient as a direct consequence of learning the confidential information. Companies and universities may, in the circumstances of the moment, decide that the benefit is worth the risk and agree to such a term. The government can never do so under a CDA. With the singular exception of a CRADA (discussed later), any term in an agreement that purports to promise rights in future government inventions, including even the option to negotiate a license, lacks authority under the law.

5. AGREEMENTS TO TRANSFER MATERIALS

5.1. The Basic Material Transfer Agreement

5.1.1. Background

A widely acknowledged axiom of academia is that the widest possible circulation of research materials is crucial to maintaining the pace of research. For years, and even today, little more than packing documents, cover letters, or even bills of lading provide the only evidence of transfers of materials. NIH has long searched for constructive methods of transferring materials without any kind of documentation, or at least to minimize the amount of paperwork required.[29]

Companies and a few universities, however, believe profits might be reaped by controlling the flow of the unique and useful things they have made. Others, moreover, have realized their vulnerability to product liability lawsuits (not to mention accusations of theft of trade secrets and patent rights, in addition to theft of the material). Accordingly, agreements to document the transfer of materials have begun to proliferate tremendously. For the foreseeable future, the Material Transfer Agreement (MTA) is here to stay.

Fundamentally, a routine MTA should be a simple, innocuous agreement, essentially promising that the recipient will not do anything unethical or stupid with the transferred material. Occasionally, the unique nature of the material to be transferred genuinely demands special treatment. Other times, the value of the material to the provider will justify added consideration. Nonetheless, the MTA should be an easy agreement to establish, even taking care to avoid the major pitfalls and accommodate the needs of an unusual case.

In principle, each pending MTA represents a set of experiments that are not being done because of paperwork. In practice, MTAs can get bogged down by posturing, by the overburden of negotiators, by unrealistic expectations of one of the parties, or perhaps by the sluggishness of a provider who is cooperating only out of courtesy and cannot be bothered to hurry. Perennially, delays caused by MTAs are the single most common complaint by scientists about technology transfer. Still, no matter how tempting cutting corners or bypassing procedure may seem, a failure to take care can create problems such as those suffered by Gillian.

5.1.2. Anatomy of the Material Transfer Agreement

A normal MTA will address the following separate topics: (1) identifying the provider and recipient; (2) identifying the material; (3) how the material will (or will not) be used; (4) how confidential information regarding the material, passed to the recipient incidental to the material transfer, will be maintained; (5) recipient's rights with respect to the material; (6) the term of the agreement; (7) indemnification and warranties; and (8) inventions derived from the use of the material. The MTAs now in circulation have particular terms that range from the truly innocuous to the truly outrageous. Each has its pitfalls for the unwary.

5.1.2.1. Parties

As with every agreement, the MTA should identify everyone involved, namely, the providing and receiving organizations. Many MTAs also name the provider's scientist and/or recipient's scientist, but where this is done, the MTA should clarify that the scientists are not the actual parties to the agreement. Again, this serves the very clear purpose of specifying who has agreed to be bound by the agreement and who is responsible if it is not carried out. Thus, when Dr. Niher signed Neurion's MTA in her personal capacity, she was personally bound by whatever terms Neurion had demanded, reasonable or unreasonable.

Increasingly, providers of material are demanding that all people who will handle the provided materials must actually sign an agreement in their personal capacities. To be sure, there is some wisdom in requiring that the recipient scientist acknowledge, in writing, having received the MTA, having read it, and having understood the terms under which the materials were transferred. Even so, in the overwhelming majority of cases, forcing the recipient scientist to be bound personally is pointless overkill because the recipient scientists are already bound by employment agreements, because other tort-based remedies exist regardless of whether the recipient scientist signed the MTA, and because the maximum damages for the breach of a contract such as this rarely will rise anywhere near a lawyer's litigation fee.

5.1.2.2. Materials

The MTA must also specify the materials to be transferred. Although this also is obvious, not all descriptions of materials are created equal. For example, some MTAs define the "materials" to include all "derivatives," regardless of whether the derivative incorporates any part of the original material. If the original material is a plasmid and the derivative is the plasmid incorporating an inserted oligonucleotide, this term may be understandable, but what if the original material is a cell line to be used to screen candidate drugs? Arguably, any drugs discovered or designed using the screening cell line could be construed as a "derivative." Everyone should watch for this subtle attempt to reach into future inventions (i.e., defining the "material" as including anything invented with it). Government labs must be particularly careful because rights to future inventions cannot be promised under the MTA; such a "back-door" transfer of invention rights would be unlawful.

One issue of particular concern to government laboratories is the status of the materials: Are they for sale? The MTA is authorized for the purpose of enabling research and no other purpose. If a private party recipient can buy a particular material from the marketplace, the recipient should pay for it; NIH is not a manufacturer or retailer, let alone a free supplier of commercial materials. Likewise, if NIH scientists can buy materials from competitive retailers, the use of MTA to circumvent the procurement laws and regulations would be inappropriate (and possibly illegal).

5.1.2.3. Uses

The MTA should include a brief research plan and clearly state prohibited activities—in particular, that the research materials should not be used in humans. Essentially, these provisions serve two purposes;

namely, they put the provider on notice of the nature of experiments the recipient plans to do, and they instruct the recipient not to do anything else. If Gillian had sent the plasmid to John Rogers under a formal MTA (assuming she was not prohibited from doing so by a prior MTA with Neurion), then she would have had a clear, easy answer to the congressional inquiry: John agreed in writing not to test the plasmid in humans; if he broke the agreement by doing just that, Congress should be asking him why he did it.

5.1.2.4. Confidentiality

In certain cases, confidentiality should be addressed, but rarely does this present a problem. If documents containing trade secrets about the material are transferred with the material, and to the extent the material constitutes a trade secret, confidentiality should be preserved; if the provider is still worried, the provider simply should not send those documents. Occasionally, however, companies will insist that certain limitations be placed on the recipient's ability to publish results. These limitations vary, from a mere 30-day delay (but only to permit the filing of patent applications on discoveries) at one end of the spectrum, to the right to review and redact in the middle of the spectrum, to the absolute right to prohibit any disclosures of any kind in perpetuity at the far end. Although private parties may negotiate whatever terms match their policies, NIH has a strict, essentially nonnegotiable policy never to permit any private party to control or limit the NIH scientist's prerogative to publish. Because NIH wants to collaborate, however, NIH will seriously consider any comments collaborators have and will accommodate any reasonable request to redact confidential information not absolutely necessary to publish.

5.1.2.5. Rights in the Materials

As a general principle, the typical MTA creates, in legal terms, a "bailment." In other words, the relationship between the parties, the scientists, and the materials is analogous to the relationship between a restaurant, the restaurant's coat-check host, a guest, and the guest's coat. If the guest, five minutes later, demands the coat back, the host cannot refuse to deliver it. The host may not do with the guest's coat as the host sees fit, regardless of whether the guest paid for the coat-check service, and even if the host's actions confer a benefit to the guest. Likewise, the recipient of research materials under an MTA may hold the materials, must return or destroy the materials upon demand, and may use the materials only as the provider says the recipient may use them. The recipient under an MTA does not have any ownership rights in the physical material transferred, even after the provider has asked the recipient to destroy the material.

The bailment relationship should be (and normally is) detailed in a term in the MTA. This term usually states that the recipient will have a limited license to use the materials, but that the provider retains title. The MTA often will state that the recipient will keep control over the materials and will not permit anyone to handle or use the materials other than those under the recipient's direct supervision. The MTA should state that the recipient will not transfer the materials to any third party without the written consent of the provider. All of this is routine and recommended even if not required.

5.1.2.6. Termination

Every contract should have a clear ending point. That event could be mutual consent, unilateral request by provider, the delivery/consumption of goods, the creation of a joint work product, or a simple expiration date. This is purely a matter of practicality. It addresses, for example: how long information must be kept confidential; how long the recipient has to track the MTA; and which rights, if any, continue after the material has been consumed, and if some do, for how long. Although parties certainly can agree to make an MTA last indefinitely, the absence of a formal termination event could cause bad feelings if each party's understanding is inconsistent with the other's. This is especially important where materials may sit in storage for years, long after the original recipient scientist (who understood the limitations imposed on the provider's materials by the MTA) has moved on to another position elsewhere. A recent version of the U.S. Public Health Service (PHS)[30] model MTA states simply that the recipient of materials will protect confidential information relating to the materials for a term of three years, which may be extended by another two years upon written request by the party providing the materials.

5.1.2.7. Warranties and Indemnification

Often, private parties to contracts make certain promises to each other that are beyond such matters as quantity, delivery date, price, etc. Promises such as these can constitute warranties and indemnification. These terms should be approached with great caution and under the advice of an attorney because such terms can create liability beyond the "four corners" of the agreement.

A warranty is a special promise, above the promises normally included in a contract, that a certain relevant fact is true.[31] In the ordinary sale of retail products, for instance, the merchant provides the consumer with the promise that the product in the box is what the label on the box says it is (called a "warranty of merchantability") and does what the merchant claims it will do (a "warranty of fitness for a particular purpose"). The warranty may be expressly stated, implied by the context, or imposed by law. If not forbidden by a law, parties may agree to waive certain warranties that ordinarily would apply automatically. In the absence of a warranty, if the merchant breaches a contract, the other party gets the cash value of the contract as damages—you get your money back—and no more. If a warranty is provided, and the promised fact turns out not to be true, the warrantor may be held liable for all foreseeable, consequential damages above the value of the contract, provided the damages can be shown to have been caused by the breach of warranty.[32] In essence, a warranty is an agreement to shift risk.

Research-related contracts often disclaim any warranty of merchantability and fitness for any particular purpose. These warranties were created to protect consumers against shady merchants selling shoddy goods. Such warranties, however, are rarely necessary to protect researchers handling materials of unknown properties and hazards—researchers are normally expected to be careful with such items. Also, agreements in the research arena routinely disclaim any warranty that materials being transferred do not infringe some third party's intellectual property rights. Sometimes, however, a provider of material will insist that the recipient warrant such things as that the investigator will comply with the laws of a certain country (other than the recipient's own) or that the terms of the MTA do not conflict with any other agreement entered by the recipient. Facts such as these may be impossible to ascertain, and so a warranty regarding these facts could be disastrous.

Indemnification essentially is a promise in the other direction: The customer promises the merchant that, if the customer does something stupid with the product and injures someone who then sues the merchant, the customer will "step into the shoes" of the merchant for the purposes of defending the litigation, including paying lawyer's fees, as well as paying any judgments against the merchant if the merchant loses. Indemnification essentially is another way parties can shift risks. Suppose in Gillian's case, for example, when she signed Neurion's agreements, she agreed to indemnify Neurion against any third party lawsuit concerning the materials she got from Neurion or arising from her use of them. If so, then even though she did not manufacture the materials, and even though she did not tell anyone that the materials were safe or would work properly, she could be forced to pay any judgments imposed on Neurion for making an unsafe product.

Indemnification creates a particular problem for agencies of the federal government and of many states. Companies and universities routinely acquire liability insurance specifically to cover litigation expenses, and though individuals often do not do so, they can; government agencies, in contrast, cannot indemnify anyone unless the law expressly states otherwise. Under the Adequacy of Appropriations Act[33] and the Antideficiency Act,[34] for example, a federal agency may not incur a debt or liability greater than the amount of money Congress has appropriated to that agency. Indemnification is an open-ended promise to pay whatever is assessed, even if that assessment exceeds the agency's budget. In the worst case, any federal employee purporting to incur such a liability on behalf of the federal government could be subject to criminal sanctions.[35] At best, when a company that thought it had secured indemnification from the government learns the truth, the company may believe that the scientist and the government negotiated in bad faith.

5.1.2.8. Inventions: "Reach-Through" Rights

The terms in MTAs relating to intellectual property are often the most nettlesome because they directly address the diverging views regarding how research material should be treated. Generally, a consensus has arisen that the clinical uses of materials (i.e., diagnostic, prognostic, or therapeutic applications) may be restricted by those who invented them to enable the inventor to recoup its investment and perhaps make a profit. For example, if a new, patented chemical is found to treat a disease, the inventor/patent owner should be able to control who can sell this new drug. The question is the extent to which pure research uses should be similarly restricted. In other words, if the new drug were being used to explore the mechanism of action of a cellular process unrelated to the condition the drug was invented to treat, should the inventor/patent owner be entitled to extract large royalties for each experiment or perhaps claim rights in discoveries made from those experiments?

Industry traditionally views all of its creations as things that required a capital investment and that can provide a source of revenue. Some even believe that all discoveries made using the creation, which could only have been made using the creation, are really part and parcel to the original creation. In various forms, some in industry now ask for so-called "reach-through" rights. Specifically, in exchange for the use of the materials, the provider would get some kind of rights in anything the recipient invents. Sometimes, the provider asks merely for an "option" to a license, to be negotiated later; other times, the provider asks for a prenegotiated license, often royalty-free, occasionally exclusive (i.e., no one can develop the invention but the provider). A few ask for total assignment of any inventions.

Academia views inventions as the practical consequence of theoretical discoveries, and that the former should serve the latter, not the other way around. Otherwise stated, any use of an invention that serves purely to investigate facts should be free and unfettered. Exorbitant fees or powerful reach-through rights, therefore, create barriers to research and learning, to the free flow of ideas. If a particular road to the development of a technology contains too many toll booths, the researcher will be forced to search for other, probably less efficient routes and so may miss important discoveries.

Additionally, at least from academia's point of view, the mere fact that someone has asked for reach-through does not necessarily mean granting it would be fair or reasonable. Put to the absurd extreme, if person A sells person B a screwdriver, should A be allowed to claim ownership of every piece of equipment, and perhaps every building, B builds with it? Aggressive reach-through by industry creates an even larger barrier for government researchers because the government is extremely limited in its authority to grant license rights, even when the grant is appropriate. In fact, the only mechanism now existing for a government laboratory to promise a private party present rights to the laboratory's future inventions is through a CRADA (discussed later).

5.2. The Uniform Biological Material Transfer Agreement

In the early 1990s, various nonprofit research organizations, universities, and NIH together realized that the MTA was an annoying, bureaucratic nuisance. All agreed on the major principles governing the transfer of materials among each other; all agreed not to do anything unethical or stupid with each other's research materials. So, they wondered, why must every MTA be renegotiated? To avoid the unnecessary extra paperwork, the academic community created the Uniform Biological Material Transfer Agreement (UBMTA)[36]—a "treaty," for lack of a better description—to which any nonprofit organization or university could become a member. Under the UBMTA, any signatory could transfer materials to any other signatory using a prenegotiated form that could be signed directly by the scientists doing the transfer rather than an administrator. The UBMTA is not mandatory, so if the provider has a special interest in the transferred materials (e.g., because the technology is exclusively licensed to a company), the provider could revert to the standard MTA process.

To the extent it has been utilized, the UBMTA process has dramatically streamlined the process and decreased the time needed to arrange for the transfer of materials among members. Unfortunately, the UBMTA has not been used as much as it might be. Part of the reason appears to be a lack of awareness that the mechanism exists, and another part seems to be that the UBMTA, crafted as a compromise among a committee of diverse parties, is confusingly written. The most visible part, however, appears to be the fact that universities and nonprofit organizations are marketing their technologies more aggressively, signing exclusive arrangements with companies more often, and thus finding that the UBMTA is not adequate. Still, it remains a valuable tool.

5.3. The Clinical Trial Agreement

Obviously, Gillian Niher could not have brought Neurion's materials to NIH under the MTA because MTAs expressly prohibit using transferred materials in humans. To address this limitation in the MTA, some of the NIH institutes and academic institutions have developed a variant, which would permit them to use received materials for clinical purposes. The Clinical Trial Agreement (CTA) is, at its heart, an expanded MTA. In addition to all the topics arising under the MTA, the CTA addresses other issues specific to clinical trials. A well-crafted CTA should reflect, at a minimum, special consideration relating to protocol drafting, regulatory filings, interactions with regulatory agencies, use of data, and how the agreement might be terminated in the middle of the clinical trial without endangering the patients enrolled in the trial.

Because the provider does not have to participate in research under a CTA, the CTA should make clear the provider's role. Some providers are pleased to be passive, particularly those who have little or no experience in running clinical trials or interacting with the Food and Drug Administration (FDA); other providers want at least an equal role as the recipient in drafting, reviewing, and approving any protocols, as well as in analyzing the data. NIH is flexible, provided that no outside party has the authority to command NIH personnel, restrict NIH research, or veto NIH publications.

Additionally, the CTA must clearly state who will be responsible for filing any regulatory documents with the FDA, such as an Investigational New Drug Application (IND), necessary to enable the research to begin. Because INDs are expensive and complicated,

companies often are happy to let NIH bear responsibility for filing the IND if the NIH is so inclined. If NIH is going to accept that responsibility, however, the provider should agree to send NIH the necessary formulation data or, at least, the provider must give NIH access to a drug master file.

As a matter of law, the holder of the IND is responsible for reporting adverse events,[37] for monitoring the conduct of the trial,[38] and for participating in any direct interactions with the FDA.[39] When NIH holds the IND, some providers want to participate in this process, and some do not; the term is negotiable. If the provider holds the IND, however, NIH must have the right to file its own adverse event reports and must be permitted to participate in any meetings with the FDA. This is to ensure that information negatively affecting the product being tested will be timely disclosed to the FDA in proper context. Almost all companies would never suppress such data, but the temptation for a company, which may be depending on the success of the product, to put a misleading spin on damaging information can be enormous. Physicians who are participating in the trial have a legal duty to report adverse events; the failure to do so could lead to administrative, or even criminal, penalties.[40] Consequently, NIH would rather risk insulting a company and insist on retaining this right.

Normally, a CTA will state that each party will share with the other all raw data generated under the clinical trial, except to the extent necessary to protect the confidentiality of the patients. Furthermore, each party normally has the right to use the data for its own purposes (reserving to each party, of course, the right to file patents on the inventions of its own employees). The parties may, if they like, agree to publish jointly; however, NIH will always reserve the right to publish independently if the provider declines to join in a particular publication. For studies involving the NIH or one of its grantees, if the provider of a study drug must have direct access to identifiable private information of any study subject, then the provider may inadvertently become regulated under the Common Rule.[41] If so, the parties would be well advised to seek legal help in navigating this situation.

Finally, some term should address what happens if one or both of the parties determines that the agreement should be terminated before the protocol has been fully carried out. As a matter of medical ethics, a doctor should not be forced to abandon a viable course of therapy already being administered to a patient due solely to a provider's refusal to continue providing the therapeutic agent. On the other hand, providers do not want to be forced to continue squandering significant resources on a project they have determined will not be profitable. Fortunately, there are several mechanisms to protect both parties' needs. For example, the provider could agree to provide enough agent at the beginning of the trial to supply the entire protocol. Alternatively, the provider could give a license, plus information on the manufacture of the materials, to hire a contractor to make enough agent to complete the trial (if the recipient cannot make the materials). The mechanism is negotiable, even if the principle is not.

5.4. Other Specialized Material Transfer Agreements

5.4.1. Materials in Repositories

The point of a repository is to enable researchers to access samples of research materials, typically biological materials, from a centralized source. Some of the institutes at NIH maintain repositories of biological materials, including transgenic animals, cDNA clones, and viruses. The National Cancer Institute (NCI), in addition, maintains a special repository of natural products collected from sites around the world. Private entities, such as the American Type Culture Collection and Jackson Laboratories, maintain repositories for public access.

Use of repositories raises one common issue relating to MTAs, specifically relating to "background rights." When the creator of the materials places a supply in the custody of a repository, the creator may have filed patent applications on the materials and may demand that the repository put restrictions on the further distribution of the materials. Normally, these restrictions are similar to those that would appear in a standard MTA (i.e., do not do anything stupid or unethical with the materials). Occasionally, the creator demands that the repository extract reach-through rights from any recipient for the benefit of the creator. Those who would access a private repository should be vigilant for such terms.

The NCI natural products repository has a unique twist, which is serving as a model for transnational research in other arenas. NCI's authority under the law to control what happens to materials it sends out of a repository is severely limited. Because most of the materials were collected from developing countries, NCI negotiated agreements with these countries, trying to find ways within U.S. law to ensure that a significant portion of any economic benefits derived from materials collected would flow back to the country of origin. Ultimately, NCI established a memorandum of understanding with each source country, which has resulted in the favorable cooperation of, and even collaboration with, the local scientists and universities in these countries.

5.4.2. Software Transfer Agreements

Suppose a scientist at NIH wants to work on software now under development. If the software was written by a potential collaborator, can an MTA be used to allow the collaborator to transfer the software? Alternatively, what about transferring the software out? The answer to both is a qualified "yes."

On a superficial level, the use of an MTA should be legally sufficient to permit the transfer of the physical magnetic media containing the code. On a deeper, more theoretical level, the issue is somewhat more complicated. Specifically, it is not clear whether NIH's authority to transfer biological materials[42] includes the intangible essence of software code (separated from the physical media on which it is written).

Regardless, an agreement to transfer software must always conform to all laws and NIH policies, such as that the software is not commercially available and that the provider does not demand reach-through to NIH inventions. The NIH Office of the General Counsel has approved use of various software transfer agreements by some of the institutes, many of which have been streamlined into a simple "click-wrap" version (i.e., clicking on the "accept" button before downloading software is sufficient to create a legally binding agreement).

6. COLLABORATION AND INVENTIONS: THE COOPERATIVE RESEARCH AND DEVELOPMENT AGREEMENT

6.1. Background

Uncounted collaborations occur every year that are never formally documented and that are never embodied in any kind of contract. Obviously, some kind of written agreement is required when the collaboration becomes complicated, the nature of the research requires the employers of the collaborating scientists to commit significant materials, or one or both parties is worried about how rights to inventions will be handled. For private parties, the possible terms are essentially limited only by each party's policies and available resources. For the government, matters are not so simple.

When a government employee invents something, the employee must assign ownership rights in that invention over to the government.[43] Yet, the core mission of NIH is to conduct research to improve the public health, not to sell products and make profit.

Therefore, when someone at NIH discovers a new prognostic/diagnostic tool or a new therapy, NIH is unable to commercialize products embodying the invention (i.e., engineer mass production, tap distribution channels, market, and sell)—only private parties can do that. The law requires the government to offer the opportunity to license government inventions to all interested parties in open competition. In a sense, the public owns each government invention, so everyone (the public) should have fair access to every opportunity to acquire rights in each invention.

This arrangement is appropriate for NIH inventions made purely by NIH personnel working exclusively with NIH personnel at NIH-owned facilities, but what about inventions through a collaboration with someone outside NIH? Indeed, these laws made companies nervous about collaborating with government scientists or laboratories because the companies had no assurance that they would have rights in inventions their work enabled. For example, a company probably would be reluctant to collaborate with the government on an improved analog to the company's main drug if it feared the government would license the analog to another company to increase competition. In particular, small companies worried that larger companies could outbid them, even though the small companies' collaborative contributions made the invention possible.

Therefore, in 1987,[44] and through updates in the ensuing years,[45] Congress further authorized government laboratories to enter a "Cooperative Research and Development Agreement," or "CRADA," which provided the laboratories a measure of flexibility in arranging such collaborations. For this purpose, each institute of the NIH constitutes a "laboratory." Currently, the CRADA is the only legal mechanism by which a government laboratory can, in the present, promise a collaborator certain rights in inventions yet to be created by the government as a consequence of the collaboration. The CRADA discussed in this chapter, therefore, is unique to government–private collaborations (although the principles involved may have applicability beyond this particular scope).

6.2. CRADA Basics

Foremost, the keystone of a CRADA is collaboration.[46] Each party must contribute some intellectual effort toward a specific research project. That collaboration drives the process of developing the agreement, and, in turn, that process is designed to authorize the negotiation of terms in the agreement suitable to enable the project.

Under a CRADA, the government laboratory may:

- contribute physical resources to a collaborator;
- dedicate staff time to a project;
- permit a collaborator's staff to work in government facilities without requiring that staff member to assign all inventions to the government (as is usually required);[47] and
- promise the collaborator an exclusive option to elect an exclusive or nonexclusive license (collaborator's choice) in any government rights in any invention that will be conceived or first reduced to practice in the conduct of research under the CRADA.

The CRADA is not a grant, procurement contract, or other "funding mechanism;"[48] in other words, the government laboratory is prohibited from transferring congressionally appropriated funds to a CRADA collaborator under any circumstances.

Under a CRADA, the collaborator may:

- contribute resources to the government laboratory;
- dedicate staff time to a project;
- permit government researchers to perform their CRADA-related research in the collaborator's facilities; and
- transfer funds to the government for the laboratory's use in carrying out the CRADA.

In addition, essentially all the issues pertinent to CDAs, MTAs, and CTAs can arise in the negotiation of a CRADA. Finally, the CRADA has some additional, administrative twists unique to the nature of the agreement, which will be discussed in more detail later.

As is obvious, the CRADA involves resolution of a wide variety of important issues. Consequently, an understanding of what a CRADA comprises can smooth the process greatly. The fastest NIH can establish a CRADA is approximately one month. Complicated cases have required a year of negotiations and occasionally more. A rough estimate for the time needed to establish a new CRADA is between four and eight months, depending in large measure on the speed and flexibility of the collaborator's review process. For the NIH, the major stages include selecting a collaborator, negotiating the agreement, institutional review of the agreement, and, finally, execution by the parties—each of which is discussed in turn.

6.2.1. Selecting the Collaborator

In the vast majority of cases, the selection of a CRADA collaborator is one of the simplest of the four main phases. Occasionally, however, this process presents serious hurdles. These hurdles can be classified as either fair access or conflict of interest.

By law, a federal laboratory must provide every possible collaborator "fair access" to any opportunity to enter a CRADA.[49] In the vaguely related context of selecting contractors to perform a service or selecting merchants to sell goods to the government, the Federal Acquisition Regulations thoroughly specify the procedure for ensuring that any interested party can apply for the opportunity. For CRADAs, in contrast, this process is not so well-defined, with good reason. In the overwhelming majority of cases, a given research collaboration can only be done with a single collaborator. For instance, a CRADA to develop the collaborator's patented new drug cannot be done by anyone but the owner (or licensee) of the patent. In such cases, no purpose would be served by opening the selection process to a competitive bid. Still, the government is not permitted to pick collaborators in an arbitrary or capricious way—the selection must always be reasonable under the circumstances.

As a general rule, if research under a CRADA genuinely depends on access by the government to a prospective collaborator's proprietary technology, unique expertise, or unique facilities, "fair access" is deemed satisfied without any effort having been made to find someone else (because no one else would suffice). This is not as beneficial for collaborators as it might appear at first blush, however, because the CRADA research would be circumscribed by that uniqueness. The laboratory would be free to initiate CRADAs on similar themes utilizing other technologies—provided, of course, that the laboratory can satisfy all the requirements of each CRADA, and that the research plan of each CRADA does not overlap any other. For instance, a laboratory having a new cDNA library may initiate one CRADA with a gene array maker using its propriety chip technology and another CRADA with a company with unique protein analysis technology to create an expression profile for this cDNA library. Indeed, in principle, if the research plans were written specifically enough and the research carefully segregated, the laboratory could engage in more than one CRADA to analyze different proteonomic aspects of the library, limiting each CRADA to research utilizing that collaborator's unique technology.

For those cases in which access to a particular technology is not a necessary prerequisite, the laboratory may announce to the world that a CRADA opportunity exists and permit anyone interested to submit a research proposal. Again, unlike the Federal Acquisition Regulations, the law governing CRADAs provides no formal guidance or specific mechanism for making such announcements. At a minimum, publication in the *Federal Register* should suffice, but there is no limit to the types or number of venues that may be used for announcing. Thereafter, if one collaborator is selected on the basis of a proposal submitted under that announcement, others who did bother to respond would have little grounds for complaining they did not have fair access.

A question often arises in the selection of collaborators, namely, whether a federal laboratory can enter a CRADA with either a nonprofit entity or a company based outside the United States. The answer to this question is "yes" for both kinds of collaborators, with certain caveats. For example, in a collaboration with a nonprofit entity, particularly universities, the parties must consider how the products that might be developed under the CRADA will be commercialized. Also, unlike private parties, the federal laboratory has limited authority to control the flow of money, which makes sharing royalties a tricky endeavor. These are issues the nonprofit entity should consider before embarking on the negotiation for a CRADA because the terms will have to be carefully crafted. For a foreign company, the law governing CRADAs requires only the following: (1) if two parties apply for the same opportunity, and if one is a U.S. company and the other is a foreign company, the federal laboratory must give preference to the U.S. company,[50] and (2) collaborators promising to substantially manufacture in the United States any products embodying subject inventions licensed to the collaborator will receive preference over those who do not so promise.[51]

Assuming the collaborator is appropriately selected under "fair access" principles, the other hurdle to cross before negotiations can begin is to confirm that the NIH's Principal Investigator (PI) will not have a conflict of interest. For example, if the PI owns stock in the prospective collaborator, or is in the process of negotiating employment with the prospective collaborator, the PI's independence could be questioned, even if not actually compromised.[52] To avoid such problems, NIH has designed a questionnaire for its PIs to complete and submit to their ethics officers for review. This process protects the PIs from accusations of unfairly steering opportunities to favored companies. Furthermore, the review uncovers subtle problems in the selection process before the negotiations become too involved, usually in time to address them to the satisfaction of everyone.

For clinical projects, NIH has implemented two variations on its standard CRADA, one for "intramural" studies (human studies to be conducted exclusively within NIH) and one for "extramural" studies (some or all to be conducted at grantee or contractor sites). For both types, as with the CTA, a clinical CRADA normally should reflect, at a minimum, special consideration relating to protocol drafting, regulatory filings, interactions with regulatory agencies, use of data, and how the agreement might be terminated in the middle of the clinical trial without endangering the patients enrolled in the trial. Unlike the CTA, however, the collaborator will always participate in a clinical CRADA, contributing intellectual effort to portions of the research, if not to all of it.

6.2.2. Negotiating the Agreement

Once the collaborator has been appropriately selected, the negotiations may begin. A complete CRADA should have at least three parts: (A) the research plan, which includes specific commitments of particular actions by each party; (B) the commitment of specific resources by each party; and (C) the terms' provisions that make the agreement operational under the law. Other items can be included if the parties see fit. At NIH, in order to make the review process more efficient, these three parts are written as separate documents that are attached to the back of a copy of the unmodified PHS model CRADA (called the "boilerplate") as appendices rather than integrating them into a single document.

6.2.2.1. Appendix A: The Research Plan

The research plan (RP) should serve three functions. First and foremost, it should lay out exactly what each party will do. The more specific these allocations of work are, the less likely confusion over responsibilities will be. Second, it should circumscribe the activities so that activities "outside" and "inside" the scope of the RP can be readily distinguished; these, in turn, define which inventions are governed by the agreement and which are not. For example, if the RP contemplates incorporating an antigen into a vaccine, the inadvertent discovery that the purified antigen makes a wonderful shoe polish probably would not be a subject invention. Third, if NIH invents something and the collaborator elects the option to a license, the collaborator is entitled under the law[53] to a prenegotiated field of use; NIH's normal prenegotiated field of use is "the scope of the RP."

Although not absolutely required, an RP may also incorporate additional information to serve other functions. For example, the RP presents a useful opportunity to explain the background of the technology, to highlight the experience and interests of the parties' PIs, and to explain in detail why the selected collaborator is particularly suited to the project. Also, the RP can contain an agreed abstract for public release, which each party understands up front may be freely disclosed to the public at any time by the other. Having such an abstract is especially important for NIH, which must often answer regular FOIA requests for routine data relating to CRADAs. Companies also appreciate

the reduced risk offered by such an abstract because they no longer have to worry about reviewing every proposed disclosure for these routine FOIA requests. Finally, the RP can include such other useful information as the parties deem appropriate, such as a list of the most relevant publications, background patents owned by each party, and any pertinent prior agreements between the parties.

6.2.2.2. Appendix B: Financial and Material Contributions

In NIH CRADAs, Appendix B contains the commitment of physical and financial resources. Specifically, this part of the CRADA spells out exactly what materials, facilities, equipment, and staff will be committed by each party and the funds (if any) that the collaborator will provide to NIH. Each Appendix B is unique; there is no requirement that every CRADA involve the commitment by either party to any particular one of these items. Ultimately, the resources to be committed by each party will depend on the research that each party wants to perform. If, for example, the collaborator wants NIH to perform an experiment using a particular piece of equipment neither party owns, the collaborator may choose to buy the equipment and loan it to NIH, to hire a contractor with the equipment to run the experiment, or to give the NIH laboratory money to buy one; alternatively, the NIH lab may decide to purchase the equipment directly. If neither the collaborator nor the NIH laboratory can afford the project's cost, or if each could pay but is unwilling to bear the expense for other reasons, the RP would have to be modified or scaled back.

The funding aspect of CRADAs offers a particularly useful source of opportunities to government laboratories. First, funds transferred by the collaborator to the government may be used to hire personnel who will not be subject to the hiring ceilings otherwise imposed by law. Second, unlike appropriated money, funds transferred to the government under a CRADA may be kept by the laboratory for the duration of the CRADA, and it will never revert to the U.S. Treasury. Third, subject to routine ethics review, the funds can be used for the travel-related expenses of government researchers in carrying out the CRADA. Furthermore, receipt of CRADA funds and materials allows the PI to explore additional, perhaps costly experiments that would not otherwise be supported by the laboratory's budget. Of course, the laboratory must regularly account to the collaborator how the funds are spent, the funds may be used only to pay for CRADA-related materials or activities, and any unobligated funds at the end of the CRADA must be returned to the collaborator.

The funding aspect of the CRADA also benefits companies. For example, a collaborator can leverage a relatively small contribution into a scientific project of far larger value. In addition, the CRADA presents a way for a company to support particular government research that is of interest to the company, without running afoul of the ethical concerns implicated in the gift process. Also, companies that do not have large budgets may be able to fund CRADA research with money received under a federal grant, such as the Small Business Innovative Research program. As long as the research project of the CRADA is distinct from the research project under the grant, such grant money can be used in this manner. In exchange, the company receives a wealth of expertise not available from any other source in the world—not just in a particular scientific field but also in regulatory filings.

With respect to this funding aspect of the CRADA in particular, one point should be clearly reemphasized: The foundation of every CRADA is intellectual collaboration. Although the CRADA mechanism offers NIH laboratories the opportunity to supplement the resources they receive through routine channels, this aspect should not dominate the CRADA. If the only reason a laboratory has for entering a CRADA is the material support, the use of the CRADA mechanism is inappropriate. Reciprocally, if the CRADA collaborator is only interested in acquiring a "pair of hands" for the collaborator's benefit, and has no interest in the intellectual contributions of the NIH scientists, there is no collaboration and the CRADA is not appropriate, even if the laboratory is willing to assist the collaborator.

6.2.2.3. Appendix C: Modifications to the CRADA Language

Appendix C contains changes to the CRADA boilerplate language. Some of the language in the boilerplate is little more than a restatement of existing law. For example, the mandatory government licenses to collaborator's subject inventions derive from a specific congressional command;[53] these cannot be removed. Others reflect NIH policy and can only be modified in consultation with the appropriate NIH offices. An example of this is the mechanism for licensing NIH inventions; because all NIH patents are licensed through the centralized NIH Office of Technology Transfer (OTT), individual institutes may not significantly change the process of licensing without confirming with OTT that it is willing and able to abide by those new terms. The remainder of the terms can be, and often are, negotiated to accommodate the unique needs of each collaborator.

6.2.3. NIH Review of the Agreement

Once the conflict of interest and fair access questions have been resolved, the scope of the research has been clearly identified in the RP, resources have been promised, and legal language has been hashed out, the complete agreement must be reviewed by NIH. Overall, this process currently requires at least nine separate formal approvals: four within the institute, four at the level of the NIH, and, after all these have been secured, final execution by the director of the institute.

First and foremost, the NIH PI must review the agreement as a whole because that individual will be ultimately responsible for doing what the CRADA promises. In addition, the PI's laboratory chief must approve, both because the CRADA represents a commitment of lab resources and as a first substantive review of the science behind the research plan. Next, the technology development coordinator for the institute must review the agreement to determine whether it complies with the institute's policies. Then, the scientific director must review the agreement to determine the merits of the project both on its own and in relation to the mission of the institute as a whole.

Once the institute has approved the package, it moves to NIH-wide review. Formally, the party to the agreement is the Department of Health and Human Services, however, the scope of the promises in the document extend only as far as the Institute or Center within NIH that signs the agreement on behalf of the Department. Even so, the law provides that NIH, on behalf of the Department, may disavow CRADAs within 30 days of execution, rendering them void.[55] To avoid this event, NIH requires review at four levels.

The first level of review is at OTT. OTT has been delegated the exclusive authority to prosecute patent applications and negotiate patent licenses for all the institutes of NIH. OTT reviews the CRADA for issues relating to the handling of intellectual property, such as modifications to the procedure by which any inventions under the CRADA will be licensed or the prenegotiated field of use for those inventions. Next, the NIH Office of the General Counsel (OGC) reviews the CRADA for legal sufficiency. Any modifications to the boilerplate, and any legally binding terms appearing anywhere else, will be scrutinized for whether they conform to, and are authorized by, the laws. Thereafter, the PHS CRADA Subcommittee* examines it for policy issues spanning the PHS, and in particular, it

*The CRADA Subcommittee belongs under the PHS Technology Transfer Policy Board.

reviews the CRADA for compliance with NIH policies and for conflicts with other CRADAs by other institutes.[56] Although the Subcommittee does not review the merits of a particular scientific project, and does not consider whether the commitment of particular resources by each party is "fair" or "wise," it does consider the precedential impact of an institute's decision to accept particular terms. Finally, the agreement is reviewed by the NIH Office of the Director. If this final review reveals no problems, the clearance of the CRADA by the NIH Office of the Director constitutes an assurance that the CRADA will not be disavowed after execution.

6.2.4. Execution by the Parties and the Effective Date

By its terms, the CRADA becomes effective on the day when the last signature is inked. Could the parties agree that the agreement will be effective on a date after final signature? Certainly. How about making the agreement retroactively effective—in other words, setting the effective date to a point before the final signature? By itself, this is apparently not authorized by the law; NIH cannot promise intellectual property rights without anything having been signed by the collaborator and the institute. Unfortunately, this inability to make CRADAs retroactive put prospective collaborators and NIH in a quandary; because CRADA negotiations take months, and because the NIH approval process takes weeks (sometimes more than one month), either the scientists must remain idle or the collaborator must risk losing rights to any NIH inventions that are invented just before the CRADA is signed. Several CRADA opportunities were lost because of this problem.

To solve it, NIH developed the Letter of Intent (LOI). The LOI is a simple promise that if a CRADA is signed, its effective date will be retroactive to the effective date of the LOI. Unfortunately, the mechanism has certain limitations. First, because the LOI is not a promise that a CRADA will ever be signed, some collaborators are unwilling to begin a project under an LOI. Also, some projects depend on the transfer of funds to begin; however, no funds may pass to NIH under an LOI because it is not a promise that the full CRADA will be signed. Furthermore, because the LOI was originally intended solely to allow research to begin while the paperwork is completed, it is limited to a short, six-month life, which may be extended for cause. Nonetheless, many collaborators are satisfied with this mechanism, and the LOI has proven to be a valuable mechanism for facilitating collaborations.

6.3. Possibilities

CRADAs have enabled a large, and growing, number of exciting projects. NIH laboratories and companies have been able to study therapies for rare diseases, new (perhaps high-risk) uses of existing drugs for new indications, and therapies and vaccines for diseases primarily occurring in poor countries—technologies most companies would consider too high a financial risk to invest resources developing—by pooling their resources and expertise. Beyond this, NIH laboratories have been able to access manufacturing channels and unique research materials, often which would be prohibitively expensive to procure without the CRADA, especially for the smaller institutes. Companies, in turn, have found they can access a unique source of expertise and can tap a research entity whose bedrock interest is to help successful products reach the bedside, without having to rely on the assistance of a competitor. In one specific and successful example, when NCI needed a tool to perform microdissection of cells for clinical pathology of cancerous tissue, NCI and Arcturus Engineering agreed to enter a CRADA to develop one. Laser capture microdissection was created, and it is now on the market.

In the case of Gillian Niher, a clinical CRADA would have enabled her project and protected her interests in publishing, receiving material and financial support, and handling regulatory filings. It would also have guaranteed NIH's interest in protecting the patients enrolled in the clinical trial. Additionally, it would have protected Neurion's interest in ensuring compliance by Gillian with the terms of their agreement and perhaps secured rights in Gillian's invention involving the bare plasmid. In short, a clinical CRADA would have established the ground rules by which the parties would act, ensured no one operated on a misconception, and authorized them to do what they wanted to do.

7. PROPRIETARY MATERIALS: THE MATERIALS CRADA

Assume that Gillian had not proceeded on her own and wanted to acquire Neurion's gene legally to run *in vitro* and *in vivo* tests of her own, though she did not want to collaborate. Assume further that the company has a supply and is willing to provide some to NIH for free, although no one at Neurion is interested in collaborating with Gillian. The gene is protected by patents and pending applications, but the company is worried about improved formulations or some other discovery that in combination might make the original technology even more valuable. Accordingly, the company refuses to release the gene or permit NIH to work with it, unless NIH promises the company rights in any related inventions Gillian creates during the project. Unfortunately, as previously noted, the keystone of a CRADA (the only mechanism by which such rights could be promised) is collaboration, of which there is none in this example. What can be done to get the materials to NIH?

A possible solution appears upon realizing that the intellectual property underlying the unique materials can be treated as the intellectual contribution of a collaboration. If this is sufficient, the CRADA can be modified to reflect that situation. Many standard CRADA provisions would no longer have meaning, such as those that govern sole inventions by the collaborator and the role of the collaborator's PI; these now can be removed and the agreement streamlined. In this way, NCI pioneered what has become the Materials CRADA, which the PHS Technology Transfer Policy Board has adopted as an officially approved mechanism.

Because there are limited situations in which the Materials CRADA would be appropriate, the Materials CRADA may be used only to transfer into NIH patented materials, or unpatented proprietary materials that are not available commercially.[57] No other materials or physical resources may be committed by either party. A collaborator may contribute limited funds towards the project, but that money may not be used to hire personnel. Finally, unless the agreement is unmodified or the modifications are minor, it will be treated like a normal CRADA for the purposes of NIH-review.

The greatest challenge to the Materials CRADA arises where the likely invention, if any, would be a "research tool." Although it is difficult to define exactly what constitutes a research tool, a good start is to state that a research tool is something that has a primary utility of enabling or enhancing scientific research, as opposed to utilities focused on diagnostic, prognostic, or therapeutic embodiments. Suppose the material to be transferred is a compound that dramatically improves the chances of success in making transgenic animals having whatever trait is desired. Transgenic animals have virtually no possible direct use in a clinical setting; rather, they are useful as tools to study other things, such as biological mechanisms and pharmacological activity.

A bedrock policy of NIH is that research tools should be made as widely available as possible. If the collaborator provides the materials under a Materials CRADA, the collaborator would be entitled to elect an exclusive license and, through it, could have the power to determine who would have access to any research tools. If the collaborator issues an ultimatum, demanding exclusive rights to research tool inventions in exchange for its material, should NIH hold ground and deny its researchers access to this exciting and scientifically rewarding opportunity or compromise its policy and risk allowing the collaborator to restrict research? Although many people have strong opinions—especially scientists, who need access to opportunities such as these to develop their careers—no easy answer exists.[58]

8. TRADEMARKS AND COPYRIGHTS FOR THE GOVERNMENT SCIENTIST

Occasionally, a research scientist encounters one of the two other main forms of intellectual property rights—copyright and trademark. Each form rarely has any direct impact on the scientist's ability to perform the responsibilities of employment, but whenever one becomes applicable, a minimal understanding of how they work can help the scientist determine what needs to be done.

8.1. Copyright

A copyright is the exclusive right to control the "copying" of an original "work of authorship" that has been "fixed in a tangible medium."[59] Works of authorship include such traditional things as books, articles, television shows, plays, music, photographs, sculpture, and computer software. It also applies to things that people normally do not think about, such as e-mail, cartoons, clip-art, flyers, and other advertisements, as well as the selection and arrangement of data, such as the telephone yellow pages,[60] the statistics on a baseball card,[61] and the pagination of a compiled work.[62] "Fixed in a tangible medium" refers to any physical embodiment. Thus, a videotape of a performance could be the subject of a copyright, but the live performance itself is not. Copyright exists the moment the work has been fixed—the moment the ink has dried or the software has been saved on a medium such as a hard drive.

Formal registration is not a prerequisite to acquiring, licensing, or transferring a copyright, although it does provide additional rights, such as the right to sue in federal courts for damages and injunctive relief. Also, infringing acts that occur prior to registration nevertheless infringe the copyright and can be stopped once the work has been registered. Regardless of registration, every work should bear the symbol "©," the name of the copyright owner, and the year in which it was created. This puts people on notice of the claim of ownership and deters unauthorized copying.

Under U.S. law,[63] the owner of a copyright has the exclusive right to do the following: (1) directly copy the work, (2) create a derivative work, (3) distribute the work,[64] (4) put the work on public display, (5) publicly perform the work, and (6) import copies of the work into the United States. These rights are circumscribed by the "fair use" exception, which allows limited copying and use of copyrighted materials in specific circumstances (e.g., academic research, legitimate commentary and criticism, education, and parody).[65] Even so, the "fair use" exception is neither broad nor particularly well defined, so particular questions should be brought to the attention of the institute's technology development coordinator or the NIH Office of the General Counsel before they become problems.

Although this collection of rights may seem straightforward, it becomes complicated when applied to the arena of digital information. There is no doubt that copyright applies to software, e-mail, Web pages, digitized music, and articles posted on the Internet. The question is, what can the recipient do with such electronic works? First, the wise course is to assume that everything is protected by copyright, unless it is expressly dedicated to the public domain. Second, it is reasonable to assume that trivial copies (e.g., loading a Web page into a computer's temporary memory or saving a copy on the hard drive) are either tacitly licensed by the person who put the work on the Internet or else at least a "fair use." Further distribution, however, should be done only with permission or great caution. For example, a simple, in-house clipping service—which scans for relevant articles in major news sites that do not charge access fees, and which distributes abstracts to a small, restricted group— probably is fair. Even if it were to be deemed not fair, chances are exceptionally remote that anyone will be injured enough to care. In contrast, a service that reposts fee-based articles on Internet bulletin boards, which can be accessed by an unlimited number of people for free, probably would not be fair.

Certain works of authorship, specifically those that were created by employees of the federal government as a part of their official duties, are not entitled to copyright protection.[66] Thus, articles written by NIH scientists may be freely copied by anyone. The journal in which the article was published may have some minimal rights to stop photocopying of the article, particularly if the journal contributed some original layout, used a creative typeface, or placed its own artwork on the same pages as the article. If the journal did not contribute substantively, however, it has no right to stop someone from transcribing the original article word-for-word.

Almost all scientific journals are aware of the exception for works by employees of the federal government, but occasionally, upon approving a manuscript for publication, the journal will send the NIH author a request to "assign" the copyright. Obviously, the author has nothing to assign, and the journal probably did not notice the affiliation. If an NIH scientist receives something like this, the scientist should simply call the journal and remind it of the author's affiliation; the journal will usually send a modified request that does not require assignment. If any confusion remains, the author should contact the technology development coordinator for the author's institute or the NIH Office of the General Counsel.

One copyright-related issue has begun to arise with increasing frequency—collaborations to write software. As a rule, when two authors create a single, integral work jointly, each owns a 50% share of the entire work, and when two authors contribute discrete parts that are linked but that can be easily distinguished (e.g., chapters in a book), the copyright to each discrete portion vests 100% with the author of that portion. If one of the authors is employed by the federal government and the contribution is within the employee's official duties, ownership of copyright is apparent only if the contributions of each author are clearly distinguishable. Unfortunately, the law relating to joint works that are integral is not clear. So what about collaborative research projects that involve writing software?

The Federal Circuit Court of Appeals has ruled that software can be the subject of a patent if the inventive idea behind the software otherwise meets the requirements for obtaining a patent.[67] Consequently, the collaborator would be well advised to enter a CRADA, if only to protect against the possibility of an invention arising from the project. As for copyright, the law authorizing CRADAs clearly permits each party to transfer property, including intellectual property, to each other. Accordingly, a copyright in a work created under a CRADA could be transferred to the government by the collaborator and licensed back, or else the copyright could be licensed to the government or the government could take no license (other than that needed for the government to continue to use the CRADA software). The term should be broadly negotiable. To avoid the conundrum of the existence of copyright in a jointly made, integral work, the CRADA RP should clearly identify who will write each portion.

One other issue occasionally arises relating to copyright, namely royalties. NIH has no statutory authority to keep royalties for copyrights assigned to it. Consequently, the main reason NIH might want to own the copyright in a work is to control the integrity of the work as it is distributed and recast. As for the author, if the author created the work as a part of his or her official government duties, receipt of royalties would be an actual conflict of interest (not to mention odd, given that there is no copyright). If the author created a work outside of official duties,[68] and if the author's ethics counselor has reviewed the situation, the author could receive royalties.

8.2. Trademarks

Occasionally, a research program finds itself in the position of offering a service to the public, perhaps even providing specific, tangible materials containing health-related information. In order to help the public become aware of the program, the program develops a name for the service or materials. As the program grows and becomes well-known, the program eventually will become concerned that other groups might try to piggyback on the reputation of the program, perhaps by falsely claiming endorsement by the program, claiming false information came from the program, or otherwise pawning off its materials as if they came from the program. The program can protect itself by registering the name of the materials as a mark in the U.S. Patent and Trademark Office.

A mark is any word, phrase, logo, graphical design, number, letter, scent, sound, or combination of these that serves to identify the source of goods/services and to distinguish the goods/services of the mark owner from similar offerings by other parties. A mark can fall into four categories. A *trademark* identifies goods, such as Ivory® soap and Forbes® magazine. A *service mark* identifies services such as United® airlines and Verizon® telephone. A *collective mark* identifies the provider as being a member of a select group, such as the Sunkist® fruit growers. A *certification mark* certifies that the goods or services of a provider have met the minimum requirements of quality or included features, such as the UL® mark, which appears on electronic products that have been tested as safe by Underwriter Laboratories. In some limited circumstances, a mark can appropriately be registered in more than one of these categories (such as the AAA® mark, which is both a collective mark and a service mark) or for an entire family of products (such as the wide range of Procter & Gamble® products that fill retail pharmacy shelves).

As with copyrights, registration of a mark is technically not a prerequisite to having rights in the mark, but registration provides important additional rights, and the sooner it is registered, the better. Marks that are registered should be identified with the "®" symbol; unregistered marks may be claimed by the "TM" or "SM" symbol for goods or services, respectively. Merely claiming and using a mark, however, is not always enough to earn the right to stop others from using it; the mark must, in fact, be distinctive from all other marks in use for related goods/services in order to fulfill its function. Thus, proposed marks that are confusingly similar to existing, registered marks will not be entitled to protection.[69] Also, marks that are generic references to the product or service (e.g., Fruit Stand for a roadside fruit vendor)[70] or that are purely descriptive of the product/service (e.g., Bed & Breakfast Registry for a lodging registration service)[71] will not be given any force by the courts or the U.S. Patent and Trademark Office.[72] To be reasonably assured of finding a successful mark, the owner should try to be as creative as possible, perhaps by creating a coined term (e.g., Kodak®) or using an arbitrary association of a word with the product (e.g., Apple® computers) or a fanciful term that has no descriptive quality whatsoever (e.g., Guess?® jeans).

Although the federal government may own a trademark, license its use, and seek injunctions to stop misuse, the NIH has no authority to keep royalties on the use of a trademark by another. Nonetheless, the protection to the reputation of an NIH-sponsored program remains a viable reason to acquire registration of a mark. Indeed, NIH already has several registered marks, including the NCI Comprehensive Cancer Center®, Wise Ears®, Back To Sleep®, and PDQ®. Queries about existing or new marks should be sent to the NCI Technology Transfer Branch or the NIH Office of the General Counsel.

9. CONCLUSION

Over the centuries, the intrepid trailblazers mapped rivers, built monuments, and explored new terrain using tools such as a compass, sextant, and telescope. Today, they map genes, build new devices, and explore new ideas using, among other things, the tools of technology development. Properly utilized, these tools help avoid the dangers and reveal the best that the new landscape has to offer. Vast opportunities await those who have the vision to seize the tools along with the moment.

References and Notes

1. In this chapter, "government" means an agency of either the federal government or a state government, including, where applicable, a university acting in its capacity as an agent of the state.

2. "Technology transfer," as the term is normally used, usually encompasses issues focused on acquiring and licensing patents, which are discussed in another chapter, in addition to the various research-related transactional agreements that are discussed in this chapter. Accordingly, to avoid confusion between the two topics, the tools discussed in this chapter are grouped in a subcategory called "technology development."

3. See, for example, RESTATEMENT (2D) AGENCY, § 322 (an agent who fails to disclose existence of agency or identity of the principal is personally liable) and § 329 (an individual lacking agency authority may be liable for breach of warranty of agency).

4. See, for example, id. § 1 (definition of agency, principal, and agent), § 26 (creation of actual agency), § 140 (principal liability for the acts of an authorized agent), and § 159 (principal liable for acts of agent with apparent authority).

5. See, for example, id. § 1 comment c (attorney at law) and § 14C (although individual members of the board of directors are not agents of the corporation, officers hired to conduct the company's business are). In theory, actual authority does not have to be written, see id. § 26 (creation of agency relationship may be oral) and § 27 (creation of agency by apparent authority may be by oral statements of principal), but as a matter of practical reality, agency relationships based on oral statements are difficult to prove.

6. Id. § 27.

7. NIH has 27 subdivisions under its aegis, each of which is either an "institute" or a "center." The institutes and centers of the NIH, together with the Food and Drug Administration and the Centers for Disease Control and Prevention (see infra, n. 47), are all referred to as "institutes" for simplicity's sake.

8. Id. § 320.

9. Gwynne P. Corporate collaborations: Scientists can face publishing restraints. The Scientist, May 24, 1999, pp. 1, 6.

10. Weiss R. Thyroid drug study reveals tug of war over privately financed research. Washington Post, April 16, 1997, p. A03; King R, Jr, Bitter pill: How a drug firm paid for university study, then undermined it. Wall Street Journal, April 15, 1997, p. A01; Rennie D. Thyroid storm [Editorial]. JAMA 1997;277(15): 1238–1243.

11. Boots Pharmaceuticals was purchased by BASF AG in April 1995.

12. Dong BJ, et al. Bioequivalence of generic and brand-name levothyroxine products in the treatment of hypothyroidism. JAMA 1997;277:1205–1213.

13. Mayor GH, et al. Limitations of levothyroxine bioequivalence evaluation: Analysis of an attempted study. Am J Ther 1995;2:417–432. Dr. Mayor was also an associate editor of this journal at the time.

14. The Adequacy of Appropriations Act, 41 U.S.C. § 11, and the Antideficiency Act, 31 U.S.C. § 1341.

15. 31 U.S.C. § 1350: "An officer or employee of the United States Government or of the District of Columbia government knowingly and willfully violating section 1341(a) or 1342 of this title shall be fined not more than $5000, imprisoned for not more than 2 years, or both." Based on the fact that the author is

unaware of any case in which the U.S. Department of Justice has even attempted to prosecute anyone for this crime on the basis of an unauthorized indemnification clause appearing in a research-related agreement, jail appears to be an extraordinarily remote possibility.

16. Omnibus Budget Reconciliation Act, Public Law 105–277 (1997).

17. See 65 Fed. Reg. 14406 (March 16, 2000).

18. See, for example, RESTATEMENT (2D) TORTS § 757 comment b; Roger Milgrim, Milgrim on Trade Secrets § 101 (discussing the Uniform Trade Secret Act); cf. Economic Espionage Act of 1996, 18 U.S.C. § 1839(4) (1997) (definitions). Each state in the United States has its own trade secret law. In addition, the federal government recently enacted the Economic Espionage Act, which is intended to complement existing state laws without preempting them. As a result, there are many overlapping definitions and rules concerning trade secrets. Specific matters should be addressed by attorneys who have particular familiarity with the laws of the jurisdiction in question.

19. Milgrim, § 16.01[7].

20. Economic Espionage Act, 18 U.S.C. §§ 1831–32; Milgrim, § 13.03.

21. 5 U.S.C. § 552.

22. Id., § 552(b)(4). Information generated by a government scientist under a CRADA is also exempt, provided the information is such that it would be deemed a trade secret if it had been given to the government by the collaborator. 15 U.S.C. § 3710a(c)(7).

23. Id., § 552(b)(5).

24. Id., § 552(b)(6). See also Privacy Act of 1974, 5 U.S.C. § 552a.

25. 35 U.S.C. § 205. This exemption only applies for a "reasonable time in order for a patent application to be filed."

26. 15 U.S.C. § 3710a(c)(7). Of particular note, subparagraph (7)(B) extends the "trade secret" exemption of the Freedom of Information Act to cover data generated by government scientists under a CRADA, provided that the data so generated would qualify as a trade secret if it had been provided by the CRADA collaborator. However, this extra exemption only lasts five years from the development of that data.

27. Depending on the parties negotiating the agreement, it often, but not always, contains some additional terms. Examples of such provisions include those that specify the law of the agreement (e.g., "Federal law shall govern"), certification provisions (e.g., certification by signer of authority to bind the party), indemnification provisions, and disclaimers of warranties. An attorney should be consulted before any of these provisions are accepted. Although these terms may be common, they do not necessarily have to appear in an agreement to make the agreement valid and binding.

28. See, for example, Weigh Systems South, Inc. v. Mark's Scales & Equipment, Inc., 68 S.W.3d 299 (Ark. 2002) (although the extent of measures taken to guard secrecy of information is only one of the factors a court will consider in determining its status as a trade secret, it is a prominent factor); Tyson Foods, Inc. v. ConAgra, Inc., 79 S.W.3d 326 (Ark. 2002) (where the employer did not restrict access to secret documents, stamp documents "Confidential," or notify staff which data the company considered to be trade secret, the company cannot rely on broad, nonspecific secrecy obligations in employment contracts and employee handbooks); Capsonic Group, Inc. v. Plas-Met Corp., 361 N.E.2d 41, 44 (1st Dist. Ill. 1986), cert. denied, 505 N.E.2d 353 (1987) (lack of guards or secure zones, no controls over nonemployees visiting the plant, and failure to mark documents or lock them away all suggest company does not consider its know-how a

valuable trade secret); *ConAgra, Inc. v. Tyson Foods, Inc.,* 30 S.W.3d 725, 729–30 (Ark. 2000) ("If Tyson did not consider it necessary to preclude the dissemination of pricing information by its customers, why should this court on *de novo* review enforce the secrecy of that same information?"); *Engineered Mechanical Svcs., Inc. v. Langlois,* 464 So.2d 329, 335–37 (La. App. 1984), *cert. denied,* 467 So.2d 531 (1985) (despite employment contract terms against disclosure of trade secrets, company failed to communicate to employees which data or documents it considered secret, thus no data were really secret); *Electro-Craft,* 220 U.S.P.Q. 820–21; *MBL (USA) Corp. v. Diekman,* 445 N.E.2d 418, 424–25 (1ˢᵗ Dist. Ill. 1983) (despite employment agreement requiring that employees maintain trade secrets, the fact that the company failed to label which documents contained secrets—and that employees were unaware of which was which—undercut claims of trade secret); *Dynamics Research Corp. v. Analytic Sciences Corp.,* 400 N.E.2d 1274, 1287 (Mass. App. 1980) (the fact that company information is commercially valuable does not make it a trade secret, so the company's failure to distinguish its secret information from information of general knowledge voids claims of trade secret, despite an agreement promising blanket nondisclosure of trade secrets). Cf. *Tele-Count Engineers, Inc. v. Pacific Tel. & Tel. Co.,* 168 Cal.App.3d 455, (1ˢᵗ Dist. Cal. App. 1985) (in "breach of confidence" tort, plaintiff bears burden of proving that the defendant knew with particularity which information is secret).

29. Federal Register Notice published on May 25, 1999 (64 FR 28205).

30. Although the Public Health Service no longer functions as a discrete subunit of the Department of Health and Human Services, the name still serves to identify the National Institutes of Health, the Centers for Disease Control and Prevention, and the Food and Drug Administration as a group.

31. *Black's Law Dictionary,* 6th ed., pp. 1586–1589 (1990). See also Corbin A. *Corbin On Contracts,* § 14 (single-volume edition).

32. Williston S. *Williston On Contracts 3d* § 1364C (buyer's consequential damages under the Uniform Commercial Code), § 1394 (general consequential damages for breach of warranty).

33. 41 U.S.C. § 11.

34. 31 U.S.C. § 1341.

35. 31 U.S.C. § 1350.

36. See the Web page of the Association of University Technology Managers at www.autm.net.

37. 21 C.F.R. § 310.305 and § 312.32.

38. 21 C.F.R. §§ 312.50 (general duties of sponsor), 312.53 (selecting investigators and monitors).

39. See, for example, 21 C.F.R. § 312.47(meetings with FDA), § 312.50 (general duties of sponsor), § 312.58 (FDA inspection of sponsor's records), and § 312.68 (FDA inspection of records of sponsor's clinical investigator).

40. 21 U.S.C. §§ 335a, 335b.

41. 45 C.F.R. Part 46, Subpart A.

42. 42 U.S.C. § 282(c) ("substances and living organisms").

43. See Executive Order No. 10096 (1952), as amended.

44. The Federal Technology Transfer Act, P.L. 99–502 (1986) (amending 15 U.S.C. § 3710a).

45. The National Technology Transfer and Advancement Act, P.L. 104–113 (1995) (amending 15 U.S.C. § 3710a).

46. To be sure, not every agency of the U.S. government views the minimum degree of "collaboration" equally. The only case pertaining to this point is *Chem Service, Inc. v. Environmental Protection Agency,* 12 F.3d 1256 (3d Cir. 1993). In this case, the Third Circuit suggested that the CRADA statute must be viewed together with procurement and grant statutes, such that if the primary purpose of the interaction is to procure goods for the benefit of the government, the government must use a procurement contract, not a CRADA. Implicitly, then, the CRADA is appropriate where the primary purpose is collaborative research and development.

47. See, for example, NIH Policy Manual No. 2300-320-03 (the NIH Visiting Program).

48. Some confusion occasionally arises between a "cooperative agreement" (15 U.S.C. § 3706), which is a mechanism analogous to a grant by which federal funds can be legally transferred to a private party, and a "cooperative research and development agreement," which is not a funding mechanism.

49. 15 U.S.C. § 3710a(c)(4).

50. *Id.* § 3710a(c)(4)(B).

51. *Id.*; see also 35 U.S.C. § 204.

52. Although rarely exercised, in instances posing an apparent (but not actual) conflict of interest, the NIH institute has the power to elect to waive that conflict if the research is of overriding importance to the institute and no other PI could carry out the research.

53. 15 U.S.C. § 3710a(b)(1).

54. 15 U.S.C. § 3710a(b)(1, 2).

55. 15 U.S.C. § 3710a(c)(5).

56. Because OTT and OGC have representatives on the CRADA Subcommittee, issues that OTT and OGC have about a CRADA are typically raised as an integral part of the Subcommittee's review, which increases the efficiency of resolving those issues by airing them all in a single forum.

57. On a case-by-case basis, the Subcommittee will consider Materials CRADAs for materials that are commercially available but that are so exorbitant that they are effectively unavailable without the promise of intellectual property rights. Such requests are not reviewed favorably, but some have been approved.

58. The final NIH policy on research tools appeared in the *Federal Register* on December 23, 1999 (64 FR 72090), but it does not entirely answer this conundrum.

59. 17 U.S.C. § 101 (definitions), § 102 (subject matter of copyright), § 103 (compilations and derivative works), and § 106 (core rights of copyright owner).

60. *BellSouth Advertising & Publ. Corp. v. Donnelley Information Publ., Inc.,* 999 F.2d 1436 (11ᵗʰ Cir., 1993), *cert. denied,* 501 U.S. 1101 (1994).

61. *Kregos v. Assoc. Press,* 937 F.2d 700 (2ᵈ Cir., 1991); *Eckes v. Card Prices Update,* 736 F.2d 859 (2ᵈ Cir., 1984).

62. *West Publishing Co. v. Mead Data Corp.,* 799 F.2d 1219 (8ᵗʰ Cir., 1986).

63. 17 U.S.C. § 106 (core rights of copyright owner), § 106A (rights of attribution and integrity), and §§ 601–603 (importation).

64. A major exception to this right is the "first sale" doctrine. In essence, if I buy a book from a store, I can do whatever I want with that book, including sell it to someone else. However, assuming I have a license from the copyright owner to make copies of the book, that license does not automatically include the right to distribute the duplicates. 17 U.S.C. § 109.

65. 17 U.S.C. § 107.

66. 17 U.S.C. § 105. The only twist to this rule is that the government may accept assignment of a copyright from a private party, but this is rarely done.

67. *State Street Bank & Trust Co. v. Signature Financial Gp., Inc.,* 149 F.3d 1368 (Fed. Cir., 1998), *cert. denied,* 119 S.Ct. 851 (1999).

68. For example, a chapter in a medical textbook that broadly teaches about an area of health might be a legitimate outside activity for an NIH physician, but a chapter on the particular research in which the physician is engaged probably would not. The ethics counselor for each institute must review such projects.

69. See *Id.*; TMEP § 1207.

70. *Trademark Manual of Examining Procedure* § 1209.01(c) (TMEP).

71. TMEP § § 1209.01(b).

72. See 15 U.S.C. § 1052. Other marks that cannot be protected include those that are deceptively misdescriptive, are purely geographical references, are a mere use of a surname, are official government insignia or flags, or are offensive and scandalous. *Id.*

23

Technology Transfer

JACK SPIEGEL

Office of Technology Transfer, Office of the Director, National Institutes of Health, Rockville, Maryland

1. WHAT IS TECHNOLOGY TRANSFER?

Technology transfer does not have a universally accepted definition. In its broadest aspects, it relates to a process of sharing knowledge. As with many broad concepts, technology transfer takes different forms according to one's motivations and desired outcomes. Government agencies, academic institutions, and private industry invoke the term to elicit remarkably disparate intents. This polymorphism extends to variants within each group. Technology transfer may have a very different look and flavor at the National Institutes of Health (NIH) compared to NASA or a Department of Defense agency. Likewise, small biotechnology companies and large pharmaceuticals may reveal strikingly different colors when technology transfer light travels through their respective prisms of commercial interest.

We need to refine this broad concept as a starting point in our understanding of technology transfer at NIH. Consider technology transfer as the exchange of information, materials, or intellectual property rights between and among government, academic, or industry laboratories to facilitate further research and commercialization. Much of this definition is familiar to scientists in a research environment. NIH scientists are experienced and comfortable exchanging information and materials with colleagues in varied institutions, including industry. They engage in such exchange in furtherance of research on a regular basis through publication, meetings and symposia, material transfer agreements, informal material sharing, formal and informal collaborations, as well as myriad collegial communications.

The exchange of intellectual property rights to facilitate further commercialization is the element of the definition that may appear foreign to many NIH scientists. At first blush, such endeavors may appear both alien and offensive to an investigator's instincts to share basic science. Yet, this aspect of technology transfer may be as critical to the mission of advancing public health as more traditional modes of sharing knowledge. Indeed, obtaining intellectual property rights to further commercialization may well be the defining step that transforms good science to a public health benefit. A goal of this chapter is to support this proposition.

Toward this end, this chapter explores the esoteric world of patents. It provides insight into the purpose of patents in our commercial society. It leads us to a realization that patents are a tool and, like many other powerful tools, can be used for noble or lesser purposes. This chapter aims to educate and, it is hoped, reassure NIH researchers in the use of this tool to advance this organization's goals and mission. Finally, the chapter introduces the many faceted ways patents are used in NIH technology transfer and what to expect when patents are employed to advance your scientific discoveries.

2. PATENTS AS INTELLECTUAL PROPERTY

Patents, trademarks, copyrights, and trade secrets are the four types of intellectual property protection that may be applied to inventions. Each of these protects different aspects of intellectual property, and each is obtained and enforced under distinct sets of laws.

Patents and copyrights are controlled solely by federal law, whereas trademarks are governed by both federal and state law. Trade secrets are the antithetical alternative to patents and are controlled by state law.

Patents will be developed in this chapter as the intellectual property tool used for technology transfer at NIH. Copyright protection is not available to cover the work developed by federal employees at NIH. Trade secrets are not compatible with the operation of federal facilities, nor with the open scientific philosophy and mission of NIH. Trademarks do make a small contribution to technology transfer at NIH and were presented in Chapter 22. Trademarks, however, have very limited applicability to promote commercial transfer of our early stage inventions toward the goal of developing products for the public health.

Patents are a tool used to protect and exploit certain categories of new and useful inventions. That protection and exploitation takes form as an enforceable legal right to exclude others from making, using, selling, or importing the patented invention. Similar to real property, a patent right may be assigned, licensed, sold, bought, and willed. There is no natural right to patents in the way that there is a natural right to life, liberty, and the pursuit of happiness. Rather, patent rights are derived from and issued by national governments according to their national laws. Most countries issue and enforce patents, including all industrialized nations. The U.S. Patent and Trademark Office (USPTO) in Alexandria, Virginia, issues patents in this country. The USPTO is part of the Department of Commerce. Patent rights are not enforceable outside a country's national borders. Efforts are under way, however, to lessen this territorial nature and harmonize different national patent laws. For example, European countries are striving to establish a single European patent enforceable in all countries belonging to the European Patent Community.

It is important to remember that patents confer an exclusionary intellectual property right. Patents do not give inventors a per se right to make, use, or sell their inventions. There are circumstances that can preclude a patent owner from working a patent. One example is very common in the biomedical arena. A drug requiring Food and Drug Administration (FDA) regulatory approval cannot be used merely because it is patented. A second common example of this principle occurs when the practice of one invention is restricted by a patent to another inventor. The patent laws prohibit two patents to the same invention, but it is possible to have patents of different scope that overlap one another. The rationale permitting such overlapping patents is discussed later as part of the rules governing patentability.

Another important characteristic of patents is that the exclusionary right only lasts for a definite and limited period of time. The length of patent protection varies according to national patent laws. In a few countries, patent term is calculated from the time the patent issues. This was the case in the United States for patent applications filed prior to June 8, 1995. Such patents expire 17 years from the date they issue. United States law was changed as part of the General Agreement on Tariffs and Trade (GATT) to harmonize certain aspects of our patent laws with the rest of the industrialized world. Thus, as in most industrialized countries, patents issued on U.S. patent applications filed after June 8, 1995, now expire 20 years from their filing date. The 20-year term of U.S. patents is subject to limited adjustments and extensions of time based on certain delays at the Patent Office and in seeking regulatory approval from the FDA. When the patent term expires, the invention enters the public domain and the patent owner's exclusionary rights end.

Scientists who are uncomfortable associating NIH research with patents will not be assuaged by this thumbnail characterization of patent rights. It is reasonable to ask why NIH should embrace a tool designed to exclude others from making or using the science from our laboratories, and why our government should issue a tool to promote monopolies in the marketplace.

3. RATIONALE FOR USING PATENTS

3.1. Different Research Outcomes

Apprehension about linking our science and institutional philosophy to a system of exclusionary rights is not misplaced. Patents should have nothing to do with the vast majority of good science coming from NIH laboratories. Most of our scientists' work product comprises scientific knowledge elucidating fundamental mechanisms and pathways of disease. This knowledge is often an incremental advance in the existing knowledge base and, occasionally, is a breakthrough and enabling discovery. Additionally, a multitude of biological materials come from our labs. Most of these materials are tools useful in advancing research. Both these tools and knowledge need to be distributed and shared with colleagues as quickly as possible. Traditional avenues of technology transfer, such as publication, material transfer, and other modes of open disclosure, are well suited for this purpose. Notably, patents do not add value to this type of technology transfer and may not only slow the transfer process but also stifle it.

Another genre of work product occasionally comes from basic research efforts. These technologies still contribute legitimately to the knowledge base when transferred via traditional means. However, their maximum value in advancing health outcomes requires further research and development. Such technologies typically take the form of potential vaccines, therapeutics, diagnostics, and devices. These technologies impact health dramatically when they are successfully developed into publicly available products. In many regards, these products are the pinnacle achievements of research goals. They are the outcomes that much of routine research seeks to stimulate and support. Nonetheless, despite their potential importance, these technologies remain early stage and are many years away from their final form and from wide distribution to the general patient population. The further work to develop these technologies into final form suitable for public distribution will not be done in the laboratory where it originated. In all likelihood, that development is not appropriately done anywhere at NIH.

3.2. Product Development in Private Industry

Indeed, history informs us that this special category of technology has little to no chance of being developed further into publicly available health products if disclosed to the scientific community by traditional publication alone. Private biotechnology, diagnostic, and pharmaceutical industries are the province for bringing research and development of such early stage technologies toward publicly available products. Furthermore, most of these products require some level of regulatory review and approval at the FDA. The probability of any candidate making it to a final product in the marketplace is very small, and the cost associated with bringing such products to market can easily run in the hundreds of millions of dollars.

Technology transfer of these special technologies is not about dissemination of information and research results to inform the scientific community. The object is to transfer these technologies into the hands of private companies willing, able, and committed to moving them forward into the marketplace. Many biotechnology companies advance products part way down the development road before passing them on to larger pharmaceutical companies. Therefore, the pathway toward product launch may involve the subsequent transfer of the technology from one company to another.

The basic research community embraces incremental advancement built on prior research from colleagues. Such incremental advances are adequately rewarded through publication and career advancement. By contrast, pharmaceutical or vaccine developers seek rewards from sales of their developed products. Those sales must underwrite the enormous research and development costs to launch the products, including obtaining any necessary FDA or other regulatory approval. It is critical to sell the developed products in sufficient volume and at the high enough price to support those costs and return a fair profit. Competition in the marketplace reduces market share and drives down the price of products. It is not surprising that the preferred business model is a monopoly market for each product.

3.2.1. Eliminate Competition

Success in a market attracts competitors. This is particularly true if a competitor can enter a market more cheaply than the pioneer. Generic drugs enter a market significantly faster and cheaper than the first-to-market pioneer drug because the copycat generic does not have to reproduce all the development work of the pioneer (e.g., clinical trials necessary to obtain regulatory approval). In other words, the generic piggybacks on the development paid for by the pioneer. Having reached the market at reduced cost compared to the pioneer, the generic can undercut the pioneer's product price.

Eliminating competition in this market scenario is a twofold proposition. The first goal is to establish a dominant position in a market. This can be accomplished by being the first to market. The second goal is to maintain a monopoly position by restricting subsequent entry of competitors into the market. A simple and effective way to accomplish both goals is through patent protection. A patent on the product provides a clear path to be first to market and prevents immediate entry of competitors. Until a patent expires, it creates the perfect market monopoly. Rather than relying on slow and costly market dynamics to eliminate competition and recoup developmental costs, a patent owner need only obtain an injunctive court order against infringers enforcing the exclusionary right.

3.2.2. The Drug Development Model

Industries such as pharmaceuticals are built on the strength of their patent protection. There are many more new drug candidates than resources to pursue their development. In an environment of drug candidate excess, companies only pursue those drugs having strong patent protection. The necessity for an exclusive patent position is nonnegotiable in the drug development industry. This paradigm is not altered by the intercession of intermediate players such as biotechnology companies. Such intermediate participants must also satisfy their financial sources (e.g., venture capitalist) and the future development partner. None of these players are willing to accept the risk inherent in nonexistent or weak patent protection.

The pharmaceutical drug development model is extraordinary in our economy. It exemplifies a disciplined rigorous use of patent laws to drive progress in an industry characterized by extreme financial, regulatory, and social pressures. The drug development industry flourishes in high-risk ventures by exploiting patent monopolies on their products.

The severest critic of patent regimes should now appreciate the necessity of NIH seeking patent protection on those inventions requiring significant corporate research and development to bring important health products to the public. Comfort follows from confidence that such patent filings neither undermine nor jeopardize commitment to basic research and the unencumbered dissemination of scientific discovery to the biomedical community. Inventors of technologies chosen for patent filings can take pride not only in the scientific merit of their inventions but also in the public health benefits that may arise from their commercial transfer to private industry.

3.3. Inventor Interaction and Communication

Successful commercial technology transfer at NIH requires ongoing interaction and communication between inventors and the Office of Technology Transfer (OTT) at NIH (see Chapter 22). It is critical that the attorney drafting the patent application tap a scientist's insight into the science, diagnostic, and therapeutic potentials surrounding the inventions. Obtaining a patent is not a simple bureaucratic registration. Patent applications undergo rigorous examination at the USPTO and foreign patent offices, often taking several years to complete. Deciding that an invention is patentable and determining the appropriate scope of patent protection involves iterative communications with a patent examiner. These communications are formal documents relating the invention to various patent law requirements. Each legal requirement must be satisfied before a patent can be issued and more often than not this involves an assessment of the invention and published work related to the invention. Inventors are copied on these communications, and scientific input from inventors can be critical with regard to an NIH patent attorney and the patent examiner agreeing on the proper application of the patent laws to the invention.

Inventor input may also be critically important when OTT seeks commercial partners and negotiates licenses related to the patent rights on behalf of NIH and its inventors. That input helps OTT assess the commercial value of the technology, appropriate companies in the marketplace, appropriate benchmarks and milestones for the development of the technology, and assess the scientific merits of statements from license applicants about their capabilities and technology development plans.

The rest of this chapter is a primer designed to familiarize NIH and other inventors with basic concepts of patent law, USPTO patent examining procedure, specific NIH patenting and licensing policies, and basic OTT patenting and licensing processes. The purpose is twofold. First, better appreciation of the technology transfer process should increase the likelihood scientists will seek OTT's opinion (in the case of NIH-funded research) regarding the potential commercial value of their research outcomes. Second, this information should improve inventors' communications and interactions with patent attorneys during preparation and prosecution of their patent applications.

4. HISTORICAL BEGINNINGS OF PATENTS

Patent systems exist in all industrial countries. The philosophical foundation of our patent system extends back centuries, with the first formal patent statute enacted in Venice in 1474. Concepts of intellectual property were important to the rise of industrialization in Europe. Intellectual property concepts spread to the American colonies based largely on British practice.

The importance of developing intellectual property systems was realized by our founding fathers. Article 1, Section 8, of the Constitution provides Congress authority to enact laws embodying patents and copyrights. In a single sentence, the Constitution sets out the fundamental principle underpinning these two intellectual property modalities. Congress shall have power "to promote the progress of science and useful arts by securing for limited times to authors and inventors the exclusive right to their respective writings and discoveries." The terms *science*, *authors*, and *writings* refer to what evolved to be copyrights, whereas *useful arts*, *inventors*, and *discoveries* refer to what evolved to be patents. It is interesting that two centuries ago the domains of literature, music, and art were associated with the term "science" and what we think of today as science was referred to as "useful arts." The concept of securing for inventors an exclusive right to their discoveries for a limited time is the fundamental property right the government bestows with a patent. The first phrase of the sentence establishes another extremely important concept about patents. The exclusive right to a discovery for a limited time is granted in return for something. The exclusive patent grant must promote the progress of the useful arts. In other words, there is a quid pro quo between the patent owner and society. Unless society receives its benefit, there is no basis to grant the inventor a limited exclusionary property right. The Constitution struck a bargain between the inventor and society. Whereas the Constitution distinctly defined the benefit granted to the inventor, it left to Congress the responsibility to define what the inventor must do to obtain that benefit.

Throughout the years, Congress has promulgated patent laws in satisfaction of the previously discussed constitutional charge. The patent laws are codified in Title 35 of the United States Code (35 U.S.C.), and the implementing administrative regulations are found in Title 37 of the Code of Federal Regulations. Since federal law establishes and controls patents, these laws are interpreted and adjudicated by various federal courts, including the Court of Appeals for the Federal Circuit and the U.S. Supreme Court. An important set of patent laws establish the requirements for patentability. Three sections of these patentability requirement laws (Sections 101, 102, and 103) establish that a patent must be new, useful, and unobvious. Section 101 in Title 35 of the U.S. Code addresses the concepts of "useful" or utility and one aspect of being "new." Another aspect of being new, known in patent terminology as "novelty," is found in 35 U.S.C. Section 102. 35 U.S.C. Section 103 introduces the concept of obviousness.

5. 35 U.S.C. 101: CONCEPTS OF NEW AND USEFUL

Section 101 states, "Whoever invents or discovers any new and useful process, machine, manufacture, or composition of matter, or any new and useful improvement thereof, may obtain a patent therefore, subject to the conditions and requirements of this title." Patent law thus sets forth statutory categories of inventions eligible for patent protection. The "process" category includes both methods of making and methods of using. Manufacture refers to things made in industry (i.e., the proverbial widget). Compositions of matter usually involve chemical compositions. The law states that inventions within these categories must be new and useful. This concept of new excludes that which naturally and always exists. Thus, products of nature,

natural phenomena, and scientific principles are part of the public domain and cannot be patented. For example, Newton and Einstein were the first to identify and describe scientific principles always existing in nature; they did not invent them. The U.S. patent system does not confer an exclusive monopoly on the first person to identify, understand, or describe a law of nature. However, although a scientific principle may not be patented, new and useful processes applying that principle are eligible for patent protection.

Advancements in the scientific landscape and evolution in judicial interpretation influence when certain discoveries qualify as patentable subject matter under Section 101. There have been dramatic shifts in this area during the past 25 years. The advent of recombinant DNA technology raised the question of whether genetically modified organisms are not patentable as products of nature. The landmark Charkabarty Supreme Court decision in 1980[1] declared that such inventions are patentable. The Court viewed recombinant organisms as not previously existing in nature. The new organism arose through the industry of the inventor and, therefore, did not remove from the public domain that which was always there. That Court decision established the principle that "new" under Section 101 encompasses "anything under the sun made by the hand of man." Simple extension of this principle has led to patenting naturally occurring genes and gene sequences by claiming them in a form not normally found in nature (i.e., in an isolated or purified form). This interpretation of Section 101 has had profound impact on the development and growth of the biotechnology industry.

The Charkabarty principle has had important ramifications in the patent and commercial world. It has been extrapolated through more recent judicial decisions to other categories of invention historically thought not to be patentable. The application of algorithms to software and the inclusion of methods of doing business into the ranks of patentable subject matter are recent examples causing concern in a number of industries. Whereas patents are territorial, industries are global. The patent laws of the major industrial nations vary, but they tend to revolve around similar basic concepts. Seismic eruptions in the fundamental patent laws of a major economic player cause shock waves throughout the international patent and business communities. Anxieties and rhetoric rise in various commercial, financial, political, legal, and academic venues as national courts interpret patent laws and national legislatures adjust their patent laws and philosophies.

The second prong of Section 101 requires that patentable inventions must be useful. As usual, the meaning of this statutory term has been interpreted by the courts through numerous litigations. That case law deems a utility must be credible, substantial, and specific in order to satisfy the usefulness requirement of Section 101.

5.1. Credible Utility

Credible utility historically has been a low threshold requirement employed to weed out inoperative inventions. The USPTO does not have laboratory facilities to test inventions. Consequently, patent examiners accept the scientific and utility statements of applicants unless there is a compelling reason to question them. For engineering inventions, this usually involves challenging inventions that disobey the laws of physics, such as perpetual motion machines. Patent examiners resolve this problem by having applicants provide evidence or a working model demonstrating that the invention is operable.

Interpretations vary in certain technology areas as to what constitutes a proper threshold requirement for credible utility. Such was the case in the pharmaceutical and gene therapy fields. For a period of time in the 1980s through the mid-1990s, many patent examiners consistently rejected the utility of therapeutic inventions in areas such as cancer and gene therapy as incredible under Section 101. Citing publications critiquing the available *in vitro* and *in vivo* animal models of cancer, as well as conflicting court decisions about unpredictability in this area, these patent examiners resolved that evidence for therapeutic utility short of positive phase II/phase III clinical trials was not credible. Applicants argued against those criticisms and availed themselves of administrative procedures, keeping related applications pending for years. The prosecution histories of these cases are marked by endless rounds of "no it isn't," "yes, it is" repartee. Demonstrating choreographic precision putting Balanchine to shame, applicants ended this "dance of the intransigent examiner" by submitting clinical trial evidence in anticipation of their new drug application filings at the FDA. The patent soon issued, providing applicants 17 years of market exclusivity coordinated around the same time they gained FDA approval to market the drug.

Section 2 of this chapter described a patent law change in 1995 whereby patent term changed from 17 years from issue of the patent to 20 years from filing of the application (or its earliest parent application) from which the patent issued. As the GATT implementation rambled toward reality, it became evident that the next ballet season needed a new dance program. The USPTO solicited input from the patent bar and interested parties, held hearings, and published a new set of utility guidelines. Those new guidelines supported a low threshold—minimum barrier approach to the Section 101 credibility requirement of utility for therapeutic inventions. Patent examiners were reminded that the patent office is not the FDA. Appreciative pharmaceutical and biotechnology communities rose for a rousing standing ovation. The USPTO reveled in the glorious curtain call.

5.2. Substantial Utility

The substantial utility requirement provides that the proposed use of the invention be a "real-world" utility. This requirement is designed to avoid two problems. Occasionally, applicants seek a patent on an invention they believe may be or may lead to something important, but they do not really know what their invention actually does or where it might lead when they file the patent application. Since Section 101 requires them to identify some utility, applicants proffer an insignificant "throwaway" possibility that is not incredible on its face (i.e., it obeys the laws of physics), but it is not very specific, meaningful, or relevant. For example, the inventor might make a knockout mouse but not know yet how the genetic deficiency impacts the animal. The inventor wants a patent on the mouse, not on how to use it. The applicant tries to avoid the issue by declaring the mouse is useful as snake food. Nice try, but no patent. Snake food would not be considered a substantial real-world use for a genetically engineered knockout mouse. Were the scenario changed such that the knockout caused the mouse to be digestible to a species of snake incapable of digesting normal mice, then a proffered utility as food for that species of snake would be acceptable. There is now a real-world relationship between the nature of the invention and the proposed utility.

The other situation in which the issue of substantial utility arises is the case of "research utility." For example, an inventor isolates and purifies a cell surface receptor from embryonic brain tissue that is not expressed in the adult. Analysis of the domain structure of the protein leaves no doubt that it must function as a transmembrane receptor. Unfortunately, the inventor does not know what the receptor binds to. Its differential expression implies it may be important to brain development. The patent application proffers the receptor is useful for screening embryonic brain tissue for morphogenetic factors in development. This would be deemed an unsubstantial research utility under Section 101 because the object of the utility is to do research on the invention to determine its real function.

The concept of a research utility must be distinguished from a utility for research. As discussed previously, a research utility performs research on the invention. In contrast, a utility for research involves a tool useful for doing research on something else. Sephadex is a tool useful for separating molecules based on molecular size. It is known that Sephadex functions by molecular exclusion. It has a legitimate patentable utility even though you may not know the identity of the molecules being separated. Many research tools are patentable inventions. The previous receptor example would have been better served to pass as a research tool were it known that it bound serotonin. The utility for research could be to screen for serotonin agonists in developing brain.

5.3. Specific Utility

The third requirement for utility is that it must be specific. Problems arise when the utility of an invention is described only by generalized characteristics of a large heterogeneous group to which it belongs. The key is that applicant is not able to identify any utility that specifically applies to and defines the specific invention as opposed to the generic group to which it is thought to belong. Consider, for example, the case of a particular expressed sequence tag (EST) where the identity of the associated gene is unknown. The applicant enumerates a laundry list of generalized utilities traditionally associated with ESTs, such as probes for full-length genes, chromosome markers, and forensic probes. None of these generalized utilities, which are common to all ESTs, distinguishes the special and specific function of the applicant's invention, the particular EST. At least one specific activity associated with that EST must be identified. Where the EST is used as a gene probe, one must know to what gene or larger sequence it specifically binds or hybridizes. Even if its utility is as broad as a chromosome marker, one must at least know which chromosome it can specifically distinguish from all the chromosomes in the cell. When an invention is defined merely by generalized function, it ultimately reduces to being a research utility as described previously. When one uses an EST as a generic gene probe, one is actually conducting research on the EST to identify its real specificity. This contrasts to applying the specificity of the EST to probe for the known corresponding gene in a diagnostic assay for the gene.

Both the specific and the substantial requirements for utility advance the premise that at least one legitimate patentable utility must exist in a currently available form. This requirement does not preclude learning new uses for the invention at a later time. Those new uses may be distinct, separately patentable inventions. The patent monopoly is granted for successfully providing a useful new deliverable to the American people. Paraphrasing the Supreme Court in *Brenner v. Manson*, a patent is granted for the prize, not for the hunt.[2]

6. 35 U.S.C 102: CONCEPT OF NOVELTY

The law does not permit patents for that already in the public domain. To do otherwise would remove something from the public for a period of time. It matters not whether the subject matter became part of the public domain as a gift of nature or through human industry. Section 102 extends the concept of "new" introduced in Section 101 beyond things already in the public domain by the grace of nature. Section 102 establishes the concept of "novelty" to exclude from patent protection things introduced into the public domain by others or through certain prohibited actions by the inventor.

Section 102 is divided into seven subsections, (a) through (g), defining different circumstances or events resulting in a loss of novelty and forfeiture of the right to patent protection. Novelty may be lost when an invention is disclosed to the public or exploited (e.g., sold) by the inventor before engaging the patenting process. Engaging the patenting process is defined in different subsections of 35 U.S.C 102 with respect to when the subject matter is invented or when the application for patent is filed. This distinguishes U.S. patent law from that of the rest of the world, which defines novelty solely in relation to the date an application is filed.

6.1. 35 U.S.C. 102(b)

Four subsections—102(b), 102(c), 102(d), and 102(f)—set forth activities that absolutely bar an inventor from seeking a patent. Section 102(b) denies a patent if

> the invention was patented or described in a printed publication in this or a foreign country or in public use or on sale in this country, more than 1 year prior to the date of the application for patent in the United States.

This complex subsection identifies a number of issues, but they all relate to events occurring more than one year before the patent application is filed in the United States. The first issue is that the invention cannot be described in another issued patent or published in the literature anywhere in the world. If so patented or published, the invention is considered to be in the public domain and not patentable. Inventors' own publications are included in this prohibition. With the advent of other publication media, printed publication is interpreted to include any indexed form of information storage reasonably available to an interested party. Patents and literature relating to inventions are referred to in patent terminology as "prior art." If the invention is described in the prior art anywhere in the world less than one year before filing the patent application in the United States, the issues are controlled under the provisions of Section 102(a).

Section 102(b) also identifies certain public and commercial activities that cannot be conducted in the United States. The public use or sale of the invention may take place outside the United States as long it does not involve a patent or publication, as indicated previously. Public use in the United States does not have to be for commercial purposes. It merely needs to take place in such a way that the public is aware of the completed invention operating for its intended purpose. In appropriate circumstances, public use occuring in the presence of a single person can initiate the 102(b) bar to a patent. The "on sale" provision of this subsection does not require a consummated sale or signed contract. Certain offers for sale can initiate the bar as well. The public policy and court interpretations are very clear: Do not publicly use or try to commercialize your invention in this country more than one year before you file for a patent.

6.2. 35 U.S.C 102(c)

Section 102(c) is a rarely invoked provision indicating a patent is barred if the inventor abandons the invention. The public policy behind this provision requires inventors to be diligent in seeking patent protection once they make an invention. Inventors, of course, are free to maintain an invention as a trade secret. If an inventor takes that route and later decides to file for a patent, the resulting patent is in jeopardy of being unenforceable due to this subsection of 35 U.S. C. 102. Evidence of the inventor's abandonment of the invention comes in the discovery process of interference or litigation proceedings by another who independently invents the same invention and diligently seeks a patent or by an infringer seeking to invalidate the patent rather than being excluded by it, respectively.

6.3. 35 U.S.C. 102(d)

Our patent laws set out circumstances and rules whereby inventors can file for patents in foreign countries and subsequently file for the same invention in the United States. Section 102(d) is a provision of the novelty laws designed to impress diligence on inventors who first file patent applications abroad. For example, it provides that a patent will be barred if an application for the invention is filed in the United States by the same inventor more than one year after it issues as a patent anywhere else in the world. This circumstance rarely arises.

6.4. 35 U.S.C. 102(f)

Section 102(f) denies issuance of a patent if the applicant did not invent the subject matter sought to be patented. This arises when an inventor derives the invention from someone else. Although it is rare for scientists to seek patents on inventions stolen from others, rejections based on this section appear at times when a patent examiner cites publications from the inventor's laboratory. These references include authors who are not inventors on the application. Such rejections are unfortunate because different authorship does not imply or provide evidence that the inventor derived the invention from the other authors. Indeed, there are more appropriate ways for the patent examiner to resolve such publications. Regardless, the issue is resolved in a technical manner that does not imply fraudulent behavior by the inventor.

The four subsections of the 35 U.S.C 102 novelty law (described previously) constitute bars against the issuance of a patent. If the patent examiner accurately applies the facts to these subsections of Section 102, the bar is not arguable. It may be possible to avoid a 102(b) bar based on prior art by amending the invention so the cited reference no longer applies.

6.5. Date of Invention/Reduction to Practice

The remaining three subsections of the novelty law—102(a), 102(e), and 102(g)—relate to the date of the invention. The date of invention is the date the invention is completed or reduced to practice. There are two ways to reduce an invention to practice under U.S. patent law. As a matter of patent law, an invention is constructively reduced to practice when an application for it is filed in the U.S. Patent Office. Therefore, the filing date is also its constructive reduction to practice date. Prior to the constructive reduction to practice, an invention may be actually reduced to practice by physically making or practicing the completed invention.

6.6. 35 U.S.C. 102(a)

Section 102(a) states that a person shall be entitled to a patent unless "the invention was known or used by others in this country, or patented or described in a printed publication in this or a foreign country, before the invention thereof by the applicant for patent." The prior art portion of this section applies if the patent issued or the reference published before the date of invention. When applying Section 102(a), the patent examiner takes the date of invention to be the filing date of the application (its constructive reduction to practice date). However, the applicant can overcome 102(a) prior art by showing evidence of an earlier actual reduction to practice to be the date of the invention. This can be done by submitting a particular form of declaration to the patent examiner providing evidence of the earlier actual reduction to practice. This evidence may be excerpts from laboratory notebooks.

Another significant element of Section 102(a) is the "by others" concept. An earlier discussion under Section 102(f) described a type of prior art reference from the inventor's laboratory having additional authors. Such a reference is legitimate prior art under Section 102(a) because, on its face, it represents invention by others. Section 102(a) prior art can be overcome by providing evidence that it is not the work of "others." Evidence of this kind again is submitted via a special type of declaration to the Patent Office, which has the effect, for patentability purposes, of removing the "others" from the prior art (e.g., coauthors from a publication). Viewed now as only the work of the inventors, the reference is no longer appropriate prior art under this section of 35 U.S.C 102.

A very important and distinctive feature of U.S. patent law derives from analyzing the relationship between Sections 102(a) and 102(b). Any prior art published more than one year before the filing date [the 102(b) date] is a statutory bar under Section 102(b). Prior art published between this critical 102(b) date and the filing date of the application is prior art under 102(a). We just discussed that a reference authored only by the inventors, published during this 102(a) period, is not considered the work of others and cannot be used to deny a patent under Section 102(a). Consequently, inventors have a one-year grace period from the time they publish or disclose their invention before they must file an application on their invention in the United States to avoid a 102(b) bar. This is because during that year grace period their own publication/disclosure is not prior art against them under 102(a). This is a significant benefit provided by the U.S. patent system. The value of this benefit must be balanced against the fact that other countries do not have similar grace periods. Most of the industrialized world operates under an absolute novelty system in which any disclosure prior to filing is a bar to getting a patent. Therefore, an applicant taking advantage of this grace period in the United States forfeits patent rights throughout the rest of the world.

6.7. 35 U.S.C. 102(e)

The next subsection of the novelty law relating to the date of invention is Section 102(e). The prior art effect of patents under Sections 102(a) and 102(b) is determined against the date those patents issue. Subsection 102(e) of 35 U.S.C. 102 bestows a preferred prior art status to U.S. patents. Section 102(e) bases the prior art effect of U.S. patents on the filing date of the patent application. This is analogous to viewing a literature reference as prior art as of the date the manuscript was received by a single special publisher rather than by its publication date. Consequently, an invention is not novel under 35 U.S.C. 102(e) if a U.S. patent describing the same invention had a filing date prior to the date of invention sought by the patent applicant. In a manner similar to 102(a), this conflict can be overcome by showing evidence of an actual reduction to practice predating the filing date of the prior art patent.

The same U.S. patent may constitute prior art against an invention both under 102(a), based on its issue date, and under 102(e), based on its filing date. Both attacks on the novelty of the invention are defeated by the same evidentiary showing of an earlier reduction to practice. The 102(e) prior art effect, however, is markedly more difficult to overcome. This follows from the fact that the filing date of a patent may be years earlier than its issue date. This makes U.S. patents potentially powerful prior art tools, and it illustrates the advantage/preference provided by U.S. patent law to U.S. patents compared to foreign patents. This advantage is sometimes exploited by using early filed U.S. patents, containing voluminous disclosures of numerous potential applications and embodiments of the invention (including prophetic ones), as a defensive publication against future competitor patents.

Recent changes in U.S. patent law permit U.S. patent applications to be published 18 months after filing. Once a patent application publishes, it becomes eligible as prior art under Section 102(e) against other patent applications. Again, the prior art effect of the published application is measured against its filing date.

6.8. Sections 102(a) and 102(e) Relate to Disclosure, Not to Claims

The novelty defeating property of patents under 35 U.S.C 102(a) and 102(e) depends on their disclosures describing the same invention. Patent applications contain a specification portion that provides a detailed description of the invention as well as background information about the subject area. The patent culminates with a claim, or set of claims, that sets out the boundaries of the invention protected by the patent. Patent rights relate to the embodiments defined in the claims of an invention. The description and teachings in the specification often are broader than patent rights defined in the claims. If the claims of a prior art patent define the same invention claimed in the patent application seeking a patent, then resolution of the conflict requires additional consideration. It is not permissible to overcome a Section 102(a) or 102(e) prior art patent claiming the same invention by showing evidence of an earlier actual reduction to practice. Otherwise, two patents would exist claiming the same invention. This is not permitted. The application will be denied a patent if the filing date is more than 6–12 months (depending on the complexity of the technology area) later than the 102(a) or 102(e) prior art patent claiming the same invention. If the two filing dates are within this range, the USPTO resorts to 35 U.S.C. 102(g), the final subsection of the novelty law, to resolve the conflict.

6.9. 35 U.S.C. 102(g) and Interference Proceedings

Section 102(g) instructs that the applicant is entitled to a patent unless

> before the applicant's invention thereof, the invention was made in this country by another inventor who had not abandoned, suppressed, or concealed it. In determining priority of invention under this subsection, there shall be considered not only the respective dates of conception and reduction to practice of the invention, but also the reasonable diligence of one who was first to conceive and last to reduce to practice, from a time prior to conception by the other.

This very complex subsection of the novelty law introduces a new consideration—that is, conception.

Conception relates to the formation in the mind of the inventor of a definite and permanent idea of the complete and operative invention as it is thereafter to be applied in practice. Conception is established when the invention is made sufficiently clear to enable one skilled in the art to reduce it to practice without the exercise of extensive experimentation or the exercise of inventive skill. Since conception is a mental process, there must be some documented record or evidence of the idea that took place in the mind of the inventor—some type of corroboration of the idea. For example, an inventor A might have conceived of a compound A and asked another person to synthesize compound A after drawing him the chemical structure of compound A.

Documentation is critically important to resolution of Section 102(g) issues. Up to now, all communications at the USPTO were between the applicant and the patent examiner. This is referred to as *ex parte* prosecution. Under ex parte rules, evidence of actual reduction to practice, etc. is submitted under oath, and the examiner accepts its authenticity accordingly. The resolution of issues under 102(g), however, involves comparing evidence between two different parties using much more stringent rules of proof. To accomplish this, the USPTO sets up a special *inter partes* proceeding known as an "interference" to determine the earliest date of invention (i.e., who invented the invention first) under Section 102(g). A panel of three administrative patent judges at the USPTO's Board of Patent Appeals and Interferences handles interferences. Simple submissions of evidence under oath are not sufficient in an *inter partes* environment. Interference evidence must comply with the Federal Rules of Evidence used in federal litigations. Indeed, interferences resemble small-scale litigations.

Interference rules require evidence related to conception, diligence, and actual reduction to practice of the invention be corroborated and authenticated. This places severe requirements on laboratory notebooks to be of probative evidence. Generally, this involves paper laboratory notebooks being hardbound, consecutively numbered/dated pages, and the entries witnessed by a noninventor capable of appreciating the data. Records kept in a haphazard fashion and "lack of diligence" (i.e., unexplained and unreasonable gaps in time in preparing the invention for patenting) can also present problems. Interferences are difficult and expensive propositions (one or two years and approximately $1 million dollars) that NIH avoids when possible. Important inventions, however, tend to be pursued competitively at the Patent Office, as well as in the laboratory and marketplace. Furthermore, NIH is often involved with corporate partners who rely on our effective cooperation and participation in such interfering cases. It is not unreasonable to expect that important inventions arising in active competitive fields may occasionally become involved in interference. NIH inventors working in areas of this nature that may lead to commercially important inventions should consider contacting the Office of Technology Transfer or their institute or center technology development coordinator regarding guidance in this regard sooner rather than later.

Among the U.S. patent laws, 35 U.S.C. 102(g) and interference practice epitomize the concepts of date of invention, rewarding the first to invent, and diligence in bringing inventions to the Patent Office. The interference process also reveals a recurrent theme in our patent laws giving preference to U.S. inventions and inventors. The requirement that "the invention was made in this country" severely limits foreign inventions in the interference process. They generally are limited to their constructive reduction to practice date (filing date) as the best date of invention in this country because evidence of conception, actual reduction to practice, and diligence are performed outside this country.

7. 35 USC 103: CONCEPT OF OBVIOUSNESS

Development through the courts of the concept of novelty relative to the prior art led to an important realization. In order to defeat an invention under Section 102, a prior art reference must anticipate every element of the claimed invention. Any element of an invention not recited in or inherent in (e.g., if a reference describes mixing NaOH and HCl, it inherently describes producing NaCl) the prior art reference renders the invention, viewed in its entirety, novel relative to that prior art. Patent attorneys are a clever species capable of tweaking claim language subtly to avoid prior art without unduly limiting the invention. Additionally, every element of the claimed invention must be found within the teaching of a single reference. The teachings of two individually deficient prior art teachings cannot be combined into a hypothetical "super reference" that anticipates every element of the invention.

What if there was a difference between what a prior art reference described and the claimed invention, but that difference was minor or insignificant? The patent system struggled for a long time with various concepts of obviousness and how to cope with obvious differences between claimed inventions and the prior art.

In 1952, the patent laws were amended to introduce 35 U.S.C. 103 to state the following:

> A patent may not be obtained though the invention is not identically disclosed or described as forth in Section 102 of this title, if the differences between the subject matter sought to be patented and the prior art are such that the subject matter as a whole would have been obvious at the time the invention was made to a person of ordinary skill in the art to which said subject matter pertains.

The landmark *Graham v. John Deere Company* Supreme Court decision[3] in 1966 established the following factual inquiries for determining obviousness:

(1) determine the scope and content of the prior art, (2) ascertain the differences between the prior art and the claims in issue, (3) resolve the level of ordinary skill in the pertinent art, and (4) evaluate evidence of secondary considerations of nonobviousness.

Other court decisions refined these inquires and helped focus the basic considerations of this obviousness concept and the frequent pitfalls encountered applying them. A common problem in obviousness determinations is a tendency to fragment claimed inventions into isolated parts and apply art against the various parts of the invention instead of the complete invention. The courts have consistently cautioned that the invention must be considered as a whole when applying prior art. It is important that references be viewed without benefit of impermissible hindsight vision afforded by knowledge of the claimed invention. Many excellent inventions seem obvious once we are taught about them, and we integrate the invention into our knowledge base. The challenge is to analyze prior art based on what they teach, not what we want them to mean to defeat the invention. References may be combined for their respective teachings in making a single obviousness argument. When references are combined, however, there must be a motivation for making the combination. That motivation to combine must be suggested by the teachings of the references and cannot arise from knowledge of the invention gained from reading the application. Obviousness is meant to be viewed through the eyes of a hypothetical person of ordinary skill in the art. That mythical figure has been variously described as one who knows all (is aware of all relevant prior art) but has no imagination (cannot extend the teachings of the prior art beyond what it says). The courts have cautioned that there can be additional factors that militate against an invention being considered obvious. These are referred to as secondary considerations of nonobviousness, and they include unexpected results, commercial success, long-felt need, failure of others to solve the problem, copying by others, and skepticism of experts that the invention would not solve the problem. It is interesting that some secondary considerations relate to events and information obtained after the invention is made and filed. For example, evidence of commercial success of an invention in the marketplace undoubtedly comes after the invention is made and usually after the patent application is filed.

The concept of secondary considerations helps explain how patents may encompass overlapping inventions. Section 2 of this chapter discussed the possibility of two patents having claims of overlapping scope. This can happen even though two patents cannot issue to the same invention. An example of overlapping claims arises when a patent issues to a species of invention after a prior patent claiming the generic invention. The generic patent is said to dominate the species and may exclude the species patent holder from working the species invention. Likewise, the species patent holder may exclude the generic patent from working the species within the scope of the generic invention. The generic patent holder, however, is free to exercise exclusionary rights regarding all other species within scope of the claims. The question may arise as to how a later discovered species can issue in view of a prior generic disclosure of the invention. Shouldn't the species be obvious in view of the generic disclosure? In many cases, species are deemed obvious when they appear to possess all the distinguishing characteristics of the genus. If an otherwise obvious species demonstrates unexpected results (i.e., secondary considerations of nonobviousness) compared to other members of the genus, however, it may be a basis to issue a patent to that now nonobvious species within the scope of the genus. This provides an important concept in patent law that distinguishes the patentability of invention (satisfying all the patentability statutes to obtain a patent) from phenomena such as dominance that prevent a patent right from being enforced.

One the other hand, a generic invention is anticipated and not novel in the face of a prior art species. Such prior art species force an applicant for a generic invention to limit the scope of the genus so as to exclude or avoid the previously known species.

Obviousness is a conclusion of law reached after a determination of relevant facts (e.g., the *Graham v. Deere* factual inquires). It is remarkable that two patent attorneys viewing the same facts seldom reach the same legal conclusion regarding obviousness (unless they work for the same client). Obviousness determinations involve much subjective argument that inundates patent prosecution histories and litigations. This certainly is the situation in biotechnology areas such as obviousness issues related to DNA sequences. Current case law attempts to treat DNA sequences in a manner similar to theories developed for chemical patent practice. It will be interesting to see how the legal system evolves obviousness to deal with the informational nature of DNA, as well as issues of homology and polymorphism.

8. MORE IS NEEDED TO ESTABLISH THE QUID PRO QUO

The quid pro quo scorecard arguably still seems to favor the patent owner. The utility requirement is a positive start, but Section 101 issues provide only a minimal threshold entry barrier to patentability. Was it ever a serious concern that entrepreneurs would abuse the patent system by flooding it with useless patents? Only a small percentage of exclusionary patent rights are actually enforced in commerce. The vast majority of the patents that issue have little value in the marketplace. It could be argued that the landscape of enforced patents would look fairly similar today if the utility requirement did not exist.

The "new" requirement of Section 101, the elaborate "novelty" law of Section 102, and the "obviousness" of Section 103 serve as public guardians of the system. They help keep applicants from receiving inappropriate patents on things already in the public domain. Fortunately, there is another set of patentability requirements that may help balance the deal.

9. 35 U.S.C. 112 AND THE NEED TO KNOW

There are a group of requirements set forth in Section 112 of Title 35 of the U.S. Code defining the sufficiency of an invention disclosure (i.e., defining what an applicant must satisfy before a patent can issue). The public policy is that society deserves to be informed about the invention. It is important that society knows how to make and use the invention so it can effectively exploit it once the patent expires and the invention enters the public domain. While the patent monopoly is in force, it is important for society to know exactly what the patent excludes. This enables society to get out of the way of the protected area (i.e., not infringe the patent claims) and to be able to exploit and develop the technological field from the boundary of that protected area outward. Thus, knowledge of how to make and use the invention is necessary to engineer around the invention and to make improvements on it even during the enforceable life of the patent. Remember, improvements are a statutory category of invention. Improvements are eligible for further patents and may displace the original invention in the marketplace. Both these activities advance the field by introducing new approaches and better mouse traps. The quid pro quo then becomes a limited time monopoly in return for an enabling disclosure allowing society to fully understand the new and useful invention. This enabling knowledge is an incentive to innovate on and around the patented invention, and it enables society to fully exploit it when the patent expires.

9.1. The First Paragraph of Section 112

35 U.S.C. 112 has a number of parts organized in separate paragraphs of text. This chapter discusses

only the first and second paragraphs of this section of the patent law. The first paragraph of Section 112 states,

> The specification shall contain a written description of the invention, and of the manner and process of making and using it, in such full, clear, concise, and exact terms as to enable any person skilled in the art to which it pertains, or with which it is most nearly connected, to make and use the same, and shall set forth the best mode contemplated by the inventor of carrying out his invention.

This first paragraph of Section 112 makes three separate requirements, which are referred to as the written description, enablement, and best mode requirements.

9.1.1. The Written Description Requirement

The written description requirement ensures that the applicant provides a full description of the invention. The requirement instructs the applicant that the description must be clear, concise, and in exact terms. It is directed toward skilled artisans in the field of the invention. This is a clear instruction not to wordsmith an obfuscated exposition that keeps the real invention secret or unclear. The courts have interpreted this requirement as providing evidence the applicant invented the claimed invention and was in possession of the invention at the time of filing.

The written description requirement has taken on heightened significance in gene patenting. A number of court decisions during the past 15 years developed the principle that a gene is a chemical composition defined by its structural and physical properties. Patent case law regarding chemical compositions indicates that a composition must be described by its physical properties, not by its function alone. Knowledge of at least one function associated with the composition is necessary to establish patentable utility, but functional knowledge must correlate to a physical structure. One cannot claim to be in possession of a chemical composition merely by describing its function. Genes can be defined by a distinguishing combination of physical properties (e.g., size, restriction patterns, and melting temperature). Nucleotide sequence is the typical way the structure of a gene is defined. Possession of a gene composition similarly demands evidence of being in possession of its physical structure. Therefore, written description is not satisfied absent disclosure of the nucleotide sequence or some other set of physical properties that distinguish the structure of the gene. Importantly, the written description requirement is not satisfied merely by describing a gene by its function.

9.1.2. The Enablement Requirement

Section 112 places another legally distinct requirement on the description of the invention. The disclosure must be sufficient to enable those working in the field of the invention (skilled in the art) to make and use the invention. This is referred to as the enablement requirement. Whereas the written description requirement aims at ensuring the inventor was in possession of the invention at the time of filing, the enablement requirement aims at ensuring that society is in possession of the invention when the patent issues. As indicated previously, that possession, in the form of knowledge about the invention, may be an incentive for others to invent around and improve upon the excluded invention during the term of the patent. Ultimately, that enabling knowledge should be sufficient to ensure possession of the invention within the public domain once the patent term expires.

An important question is how to judge whether any particular disclosure is sufficient to establish enablement. The courts have interpreted this requirement to mean that the skilled artisan should not have to engage in undue experimentation in order to make and use the claimed invention based on the description in the application. To aid in determining what constitutes undue experimentation, the federal court and the USPTO have provided a set of eight illustrative factors to be considered:[4]

1. The nature of the invention
2. The state of the prior art
3. The relative skill of those in that art
4. The amount of additional experimentation required
5. The amount of direction and guidance provided by the application
6. The presence or absence of working examples in the application
7. The degree of unpredictability in the art
8. The breadth of the claims

Analyzing the interplay of these factors provides guidance in establishing the proper balance between the sufficiency of the enabling disclosure and the scope of the claims (patent rights). The more unpredictable the art associated with the invention, the more direction, guidance, and working examples are required to support any particular breadth of claim scope. Ultimately, the scope of claims seeking patent protection should be commensurate with the enabling disclosure teaching how to make and use that breadth of invention. An important factor in determining the scope of claims is the nature of the prior art. Broader claims increase the chance that the invention will impinge the prior art under Sections 102 and 103. Crowded mature technology fields tend to force new patents to have narrower claim scope. This is independent of whether the disclosure teaches how to make and use a broad scope of invention. Pioneering patents in new technology areas tend to have broad claims because their scope is dependent only on the sufficiency of the enabling disclosure.

9.1.3. *The Best Mode Requirement*

The third requirement of the first paragraph of Section 112 is for the applicant to disclose the best mode of the invention. The policy behind this requirement is that the applicant should not disclose a less preferred way of making and using the invention to gain market exclusivity while reserving the best mode as a secret. In return for the patent monopoly, society deserves knowledge of the best way of making and using the invention known by the inventor when the application was filed.

9.2. The Second Paragraph of Section 112

The second paragraph of Section 112 states, "The specification shall conclude with one or more claims particularly pointing out and distinctly claiming the subject matter which the applicant regards as his invention." The second paragraph of Section 112 requires the patent application delineate at least one claim separate from the specification of the application that is defined and controlled by the written description, enablement, and best mode requirements of the first paragraph of this section. The claims are the portions of the patent that define the property lines of the invention receiving the patent right. Claims set out the boundaries or metes and bounds of the invention. Claims must employ clear and distinct language to accomplish this goal as compared to real property that can rely on land surveys and fences to define property boundaries. The language of the claims must be sufficiently clear to determine if the scope of the claimed invention is commensurate with the enabling written description in the specification. The language of the claims also must be sufficiently clear to permit those working in the field of the invention to know if they are infringing or "trespassing" on the claimed invention.

10. PROCEDURES FOR PROSECUTING A PATENT APPLICATION AT THE USPTO

The statutes described in this chapter represent the main patentability requirements. There are many additional statutes, regulations, and guidance that control formal requirements of the patent application and the examination procedures of the patent office. These can be viewed at the USPTO website, www.uspto.gov, with special attention to the *Manual of Patent Examining Procedure*. As indicated previously, inventors are copied on all major actions and responses involving the patent office. Inventors may be requested to comment and provide scientific assistance toward responding to such actions. A brief description of the major administrative action–response chains is provided here to place communications from and to the USPTO in context.

Soon after receiving the application, the patent office will notify the applicant of receipt and any formalities regarding the filing that may be deficient. Once the formal matters have been satisfied, a patent examiner eventually takes up the application. The examiner may issue a restriction requirement action that indicates the application claims more than one invention capable of supporting a patent. For example, claims to a composition, a method of making, and various methods of using the composition may each support a separate patent. Examiners are not required to examine more than one patentable invention in a single application. The restriction requirement forces the applicant to choose or elect claims corresponding to one of the indicated patentably distinct inventions for examination in that application. The claims to nonelected inventions are said to be restricted, and they are withdrawn from consideration in that application. The applicant is free to file one or more additional applications, called divisions, seeking examination on the withdrawn claims. Division applications cannot change the written description of the invention in the specification. Although each division is a separate application with its own serial number and filing date, divisions receive benefit of the filing date of the original application for purposes of applying prior art under Sections 102 and 103 (and for purposes of calculating patent term).

After analyzing (examining) the elected invention, the patent examiner sends a first Office Action on the Merits discussing the invention relative to each section of the patentability laws. If any of these statutes is not satisfied, the examiner rejects the claims. The Office Action sets a six-month statutory deadline to respond. Failure to respond to the Office Action within this time period results in abandonment of the application. In the response to the Office Action, the applicant may amend the claims and specification to satisfy and overcome the criticisms and rejections. Changes to the specification must be formal (e.g., correct a spelling error) and cannot add new matter in support of or that changes the nature of the invention.

In addition to, or in place of amendments, the response can argue why a rejection is improper based on the facts of the case or the patent examiner's interpretation of the law.

The patent examiner again examines the application in view of the applicant's response. If all the rejections and criticisms are overcome, and no new ones are proffered, the patent examiner mails a Notice of Allowance. Again, the applicant has a statutory period to pay an issue fee and satisfy any outstanding formal matters to have the patent issue. More likely, however, the patent examiner maintains some or all of the previous rejections and will send out another Office Action. If the new Office Action contains any new ground of rejection not necessitated by the applicant's amended response, this second Office Action is sent out under the same ground rules as the first Office Action on the Merits. On the other hand, if the new Office Action maintains the same rejections of claims and/or adds new rejections necessitated by the applicant's amendment, the new Office Action is made final. A Final Rejection closes examination of the application. The Final Rejection has a six-month statutory period for response during which time the applicant may again submit amendments and arguments in an After Final Response to overcome the rejections. Since examination is closed by the Final Rejection, there is no requirement on the patent examiner to enter into the record any amendment or argument that raises new examination considerations or that does not satisfy all the outstanding rejections so as to place the entire application into condition for allowance. If the After Final Response is not entered into the record or does not place the case in condition for allowance, the patent examiner notifies the applicant via an Advisory Action. The Advisory Action indicates the disposition of the After Final Response and any claims remaining under rejection. It also advises that the statutory time period set in the Final Rejection continues to operate. In other words, After Final Responses that do not place all claims in condition for allowance do not stop the statutory clock of the Final Rejection.

At this point, the applicant has several options. The applicant can allow the application to go abandoned by not responding before the end of the Final Rejection statutory deadline. The applicant can submit another After Final Response. This follows the same rules and time issues as the previous After Final Response. Namely, it has no right of entry, and the statutory clock on the Final Rejection continues to run. Another option is to file a Request for Continuing Examination. This request, together with a fee equivalent to a new filing fee, stops the Final Rejection clock, reopens examination, and requires the patent examiner to enter into the record any previously nonentered After Final Responses. The patent examiner once again examines the claims in view of all the responses now on the record and issues a new Office Action. The cycle of amended response, Final Rejection, and After Final practice may be repeated.

A final option for responding to a Final Rejection or Advisory Actions is to submit a Notice of Appeal. This notice stops the Final Rejection statutory clock and begins a new statutory deadline to file an Appeal Brief. This Appeal Brief and a corresponding Examiner's Answer are transmitted to the USPTO Board of Patent Appeals and Interferences. The appeal is reviewed and ruled on by a panel of three administrative patent judges. Decisions of this board of appeals affirming the patent examiner's rejections can be appealed further to the Court of Appeals for the Federal Circuit. Appeal from this federal appellate court is to the U.S. Supreme Court. For NIH, appeals to the federal courts are handled by the Department of Justice.

11. OBTAINING FOREIGN PATENTS

Commercial partners, requiring U.S. patent rights as an incentive to invest in product development of NIH technologies, often desire foreign patent protection as well. Products may have commercial value worldwide, and many of the market forces described previously exist in all major industrial countries. Foreign patent rights, therefore, can enhance the value of NIH technologies to commercial partners. Foreign patent laws are complex and vary by nation. Even a superficial survey of them is beyond the scope of this chapter. Filing and prosecution of these cases are handled by foreign associates of the NIH contract law firms responsible for handling the corresponding U.S. applications. Since patent issues in the various countries often track U.S. prosecution, domestic inventors seldom are burdened with foreign prosecution events. However, there are some basic foreign filing concepts that are useful for U.S. inventors to understand in order to follow the commercialization of their technologies.

Two important considerations about foreign filing have been mentioned previously. Patents are territorial, so each country issues its own patents, and national patent rights have no effect outside individual country borders. It has also been discussed that nearly all foreign countries award patents to the first to file rather than the first to invent. As a result, these countries operate under an absolute novelty system that does not permit a grace period on disclosure before a patent must be filed. Despite this decentralization and first to file requirement, there are two mechanisms of cooperation between all industrialized countries to lessen the burden of worldwide filings.

11.1. The Paris Convention

The first of these mechanisms is the Paris Convention of 1883. This is a treaty administered by the World Intellectual Property Organization (WIPO). WIPO is an agency of the United Nations. All industrialized countries that have joined this treaty recognize filings made in other member countries. The nature of this recognition extends a one-year priority benefit to patent applications earlier filed in any other member country. For prior art purposes, this treaty treats the filing date in the later-filed country as if it were the first-filed country. Thus, prior art published in the intervening period between filing in the first country and the subsequent filing in the second country is shielded. This allows an applicant to first file in his or her home country and then wait up to one year to file elsewhere without jeopardizing any rights in the foreign countries.

11.2. The Patent Cooperation Treaty

The second mechanism to facilitate foreign filing is the Patent Cooperation Treaty (PCT), which is also administered through WIPO. The PCT took the benefits accorded by the Paris Convention and significantly extended and advanced them. Again, all industrialized countries have joined this treaty. The PCT permits a single international patent application to be filed by member countries. This PCT application can be filed at the end of the Paris Convention year and extends the Paris Convention benefit up to an additional 18 months. Consequently, it is possible to file in your home country and not have to file individual national applications elsewhere in the world for 30 months. The PCT application establishes an international filing date used to determine patent term in later-filed national patents. If a country's patents expire 20 years from filing, then applications entering that country via PCT expire 20 years from their PCT international filing date.

PCT applications are searched for prior art and optionally examined by patent examiners in the U.S., Japanese, or European patent offices. However, no patent issues from the PCT process. The PCT application is published, and the results of the search and examination are provided to any national patent offices entered via PCT.

11.3. The European Patent Convention

The European community has organized a European Patent Convention (EPC) with a European Patent Office (EPO) that advances the spirit of the Paris Convention and the PCT by developing a regional European patent. The EPO grants a European patent that can be converted into national patents throughout most of Europe without the time, expense, and effort of further search and examination in each country. The European patent, however, has no enforcement rights in the EPC countries. The EPO is a designated country in the PCT. Therefore, it is possible to enter the EPO at 30 months after first filing and have the invention examined in English. The benefits afforded by the Paris Convention, PCT, and EPO permit the NIH, for example, to preserve foreign patent rights in much of the industrial world economically and almost effortlessly for a significant period of time. This time permits the U.S. partner to better realize the commercial value of the technology and to seek commercialization partners to develop the technology into products.

12. THE NIH PATH TO FILING PATENT APPLICATIONS

The patent filing path typically pursued by NIH involves initial filing of a provisional patent application in the USPTO. Provisional applications are not examined but serve as placeholder applications for 1 year. Provisional applications automatically expire at the end of 1 year. They are placeholders in the sense that they provide priority benefit for prior art purposes similar to the Paris Convention, but they do not count against the 20-year term of any eventual U.S. patent. On the anniversary of the provisional application filing, it is converted into another patent filing. There are then two options. In the event there are no foreign patent rights available (i.e., there was a disclosure prior to filing the provisional application destroying the absolute novelty requirement of foreign countries), the provisional application is converted to a regular U.S. patent application. This application is examined as described previously. In the more typical situation in which potential foreign rights still exit, the provisional application is converted into a PCT application. The PCT application is filed back into the United States as a national filing at the end of the 18-month PCT process. This provides 30 months from the initial provisional application filing date to evaluate the technology and seek commercial partners before having to prosecute the application in the USPTO. When NIH desires to preserve and pursue foreign rights world-wide, EPO and other selected national patent applications are also filed at the end of the 18-month PCT process.

The path OTT takes in deciding to file for patent protection is guided by the NIH patent policy. The policy is applied to inventions reported in employee invention reports (EIRs) coming to OTT from the laboratories via technology development coordinators or offices in each institute/center. OTT cooperates with institute/center technology transfer personnel to evaluate inventions relative to potential prior art, commercial potential, and NIH patent policy. Prior art and commercial potential issues vary with each technology. The patent policy is consistent and clear. The foundation of that policy is that NIH seeks patents where further investment is needed to develop a product. The corollary of this proposition is that NIH does not seek patent protection for inventions that clearly are research tools. Our policy appreciates the purpose of the patent system to stimulate innovation in return for public disclosure. However, that quid pro quo is not what drives our decision process toward filing patents. NIH scientists do not require the incentive of exclusive patent rights to encourage their ingenuity and industry. Neither NIH scientists nor their intended audiences rely on patents to obtain their knowledge about NIH science outcomes. That knowledge will be communicated in an enabling fashion to the public much more rapidly and effectively through traditional publication than through the patent process.

NIH files for patent protection when patents will be a necessary incentive for commercial partners to invest in the technologies and develop them into products to improve the public health. Markets such as therapeutics, vaccines, and some diagnostics operate in environments of extreme risk. Players in these markets require strong patent protection before they will consider investing in developing a product. It is necessary for NIH to seek patent protection on such inventions so they may reach their fullest potential for advancing public health. The NIH patent policy and invention review processes reflect these realities.

All entrepreneurs manage risks in their respective businesses. Most entrepreneurs desire monopoly status in their markets and will employ all legal tools and business practices to attain it. It is not surprising, then, that most companies seeking NIH technologies prefer exclusive patent rights to maximize their advantage over competitors. Market forces, vagaries, and expediencies in our economy, however, cause the contribution, significance, and need for intellectual property in diverse business sectors to diffract across a broad spectrum. Part of our challenge in transferring NIH technology to the commercial world is determining the best wavelength along that spectrum to encourage competition without stifling the incentive to develop our product.

Rarely must a single enterprise operate simultaneously near both ends of this spectrum. Such is the fate of technology transfer at NIH. Much of NIH's research outcomes benefit from free and open dissemination unencumbered by intellectual property issues. Some of these research outcomes rely on rigorous patent protection and its exclusivity to realize its maximum potential for advancing our mission. It is relatively easy to discriminate candidates belonging solely at one dipole or the other. Prudence dictates that NIH deals with each end of this dipole appropriately. NIH must not disadvantage, prejudice, or compromise one mode of technology transfer because it coexists alongside the other.

The challenge is what to do with research outcomes that do not neatly sort into one of these distinct technology transfer modes. Many NIH inventions are early stage discoveries with multifaceted components and potentials. Some of those components and potentials may be diagnostic or therapeutic in nature and would require various amounts of additional research and development to realize their benefit. Some components may be characterized as research tools useful in aiding or stimulating further basic or applied research. The markets related to these diagnostic, therapeutic, or applied research tool inventions may range from niche to expansive. Some inventions are so early stage that markets and market players are not evident. It is seldom easy, and sometimes impossible, to predict which potential will pan out scientifically or will resonate in the marketplace.

A preferred course of action would allow the technologies to mature, and then take appropriate intellectual property action as the uncertain potentials crystallize and reveal themselves. As indicated in this chapter, patent systems do not encourage such a deliberate and measured approach to seeking intellectual property protection. The patent system forces an early commitment if meaningful patent protection is contemplated. This translates into making "now or never" patent filing decisions. Consequently, the technology transfer process must make rapid decisions on pursuing patent protection for early stage inventions. The general policy is to err on the side of caution and file for patents in these gray areas. Once filed, one should rigorously seek the broadest possible patent protection. There is a mechanism to introduce incremental improvements to an earlier invention via a special application called a Continuation-in-Part. What is needed is a mechanism to enforce the ensuing exclusionary patent rights in ways that are complementary to the spectrum of NIH research (for example) and commercialization goals in technology transfer. Much effort is directed toward ensuring the emerging intellectual property is transferred in the most advantageous way to the private sector.

13. THE NIH LICENSING PROCESS

The tool employed to transfer NIH patent rights to its commercial partners is the license. A license is a legal agreement that grants permission to engage in an activity that is otherwise prohibited. As already indicated, patents create the right to exclude others from making, using, selling, or importing inventions described by the patent claims. NIH licenses are legal agreements by which NIH agrees not to exercise its patent right to exclude the licensed party (licensee) from making, using, and selling the invention.

13.1. Flexibility Provided by Licenses

There is significant flexibility in negotiating the terms of licenses. The patent owner (licensor) may license the patent right exclusively to a single party. This contractually binds the licensor not to license the patent right to anyone else. Even though an exclusive licensee does not own the patent, the licensee contractually is the only party that can operate free of its exclusionary rights. This effectively transfers the ability to establish a monopoly position in the marketplace to an exclusive licensee. Provisions of the exclusive license permit the licensee to enforce the patent right against competitors. The size and nature of a market sometimes are such that two parties are willing to invest in developing an invention and then compete in the marketplace. This permits the licensor to coexclusively license to the two parties.

The licensor may forgo exclusive licensing and choose to license its patent rights nonexclusively to many parties. Nonexclusive licenses grant licensees freedom to operate in the marketplace, but they have to compete with any number of other licensees of the invention. The licensor retains the right to exclude others who do not take a nonexclusive license.

Licensors may exercise additional flexibilities in the licensing process. Different parts of patent rights, for example, can be parsed in the license. In this way, the license may be limited to certain fields of use. If a company does not desire, or is not able, to develop all

potential fields of use, agreement may be reached to limit the scope of the license. Patent rights to a cancer drug, for example, may be exclusively licensed to one party for treating breast cancer and licensed to another party for treating prostate cancer. This permits both health problems to be addressed. Otherwise, products directed to only one may be developed.

When there are foreign patents, those territorial rights may be bundled into a single license or licensed independently. This may facilitate NIH strategies for transferring technologies for neglected diseases to companies interesting in making products available in developing country markets. Finally, the licensor may parse a license to distinguish the right to make and use an invention from the right to sell. This permits NIH to give out licenses for research purposes or internal use but deny the right to commercialize or sell the invention. Alternatively, the NIH as licensor can grant an exclusive commercial license that reserves the right to grant other licenses for research purposes. Patent rights to a monoclonal antibody, for example, potentially could be licensed exclusively for therapeutic uses, coexclusively for *in vivo* diagnostic uses, nonexclusively for *in vitro* diagnostic uses, and nonexclusively for research purposes only.

License agreements permit licensors to include benchmark provisions to ensure diligent development of the invention. If a benchmark requirement is not satisfied in an exclusive license, it can be a basis to terminate the license. This would free the patent rights and make the technology available to another party better able to develop the commercial product.

13.2. NIH Licensing Policy

NIH has developed an official licensing policy aimed at exploiting the flexibilities of the licensing process to adapt its patent portfolio to coincide with its institutional philosophy and goals relative to NIH technologies. Application of this licensing policy becomes the mechanism to reconcile and compensate miscalculations precipitated by the need to rush to patent filing. This licensing policy becomes the mechanism to calibrate and fine-tune the best practice of our patent rights as the technologies and markets mature. The application of this licensing policy is the tool that transforms a one-dimensional right to exclude into a multidimensional means to advance our public health mission.

The NIH licensing policy instructs to license nonexclusively where possible and exclusively when necessary. When engaging in exclusive licensing, provisions should be included and care taken to ensure appropriate scope in the fields of use and territory and to ensure expeditious development of the invention. The licensing policy takes special notice of NIH responsibility not to encumber the research process and to ensure the continuing availability of NIH research tools and materials.

The OTT is responsible for commercial technology transfer at NIH. OTT has developed a number of license models and procedures to advance the NIH licensing policy. In addition to models for commercial exclusive and commercial nonexclusive licenses, there are other models to meet particular goals of the licensing program. A commercial evaluation license (CEL) model allows companies to test the invention to determine if it serves their commercial purposes. This nonexclusive license grants the right to make and use the invention for a limited period of time. This license prohibits sale or further distribution of the invention. At the end of the evaluation period, the company can apply for a commercialization license or another special license for internal use. Similar to the CEL, an internal commercial use license permits the licensee to make and use, but not to sell, the invention. Unlike the CEL, however, the internal commercial use license is not time limited. It is designed to permit the company to use the invention as an internal tool within its research and development programs to produce other products.

NIH scientists occasionally collaborate with colleagues at academic institutions in making inventions. Patent rights to inventions arising from such collaborations are co-owned by NIH and the academic institution. Each co-owner of a patent right has an undivided right to the invention in the entirety. This means the co-owners each can license the invention independently. It is wasteful and potentially embarrassing for each party to file separate patent applications for the same invention or for one party to exclusively license its rights while the other nonexclusively licenses its rights. Therefore, it is advantageous that one party take the lead in patenting and licensing such co-owned inventions. NIH has developed a series of model licenses to establish and govern such interinstitutional relationships. The reader is invited to visit the OTT website at www.ott.nih.gov to view copies of these models, as well as ones for Cooperative Research and Development Agreements, Material Transfer Agreements, and Biological Material License Agreements.

14. OTHER OTT FUNCTIONS

Evaluating and transferring NIH technologies to the private sector is a complex enterprise.[5] In order to maximize the licensing program, effective strategies have been developed to market early stage NIH inventions.[6] OTT is responsible for developing policy for both intramural and extramural technology transfer. OTT has developed and advanced significant policy positions regarding sponsored research agreements, research tool guidelines, and best practices for licensing of genomic inventions.[7,8] OTT has initiated a program in international technology transfer to transfer relevant technologies and enhance capacity building in developing countries.[9] This program has had marked success in transferring NIH technologies associated with malaria, dengue, pertussis, AIDS, rotavirus, typhoid fever, and meningitis to public and private institutions in India, Mexico, Brazil, Korea, Argentina, Egypt, China, and South Africa. Monitoring NIH licensees to ensure they are diligent in the development of the technologies and their financial responsibilities is an important and expanding program at OTT.[10] The reader is directed to the OTT website and the cited references of this section for additional information about OTT, the patenting and licensing processes, and these other aspects specifically related to NIH technology transfer.

15. CONCLUSION: THE MEASURE OF NIH TECHNOLOGY TRANSFER IS ITS SUCCESS

OTT is proud of its success in advancing technology transfer during its brief lifetime in this endeavor. The OTT website elaborates statistics regarding various aspects of OTT patenting and licensing activity since 1995. Reflective of its level of activity are data from fiscal year 2005, when OTT received 388 invention disclosures (EIRs), executed 307 licenses, and received $98.2 million in royalty income from licensees. In accordance with its royalty policy, $8.9 million of that income was shared with 916 inventors in recognition of their inventive contributions. The remainder was distributed to the institutes/centers to underwrite their technology transfer operations and support new scientific research. The NIH OTT is proudest of the roster of FDA-approved products to which NIH inventions contributed and were licensed to product developers. These include Synagis, Videx, Velcade, Taxol, Kepivance, Taxus Express 2, Gardasil, Prezista, Hivid, Fludara, RotaShield, Havrix, Twinrix, Zevalin, Zenapax, Sporanox, NeuTrexin, Certiva, Vitravene, Thyrogen, LYMErix, AcuTect, and NeoTect. The arrays of NIH technologies currently in clinical trials evoke confidence that the next decade's roster of FDA-approved products will eclipse this decade's list. These past and future products never would exist to benefit large numbers of patients were it not for the inventiveness of NIH scientists and the application of its commercial technology transfer process to the outcomes of that research endeavor. The compendium of FDA-approved products improving patients' lives underscores the importance of the technology transfer process to the NIH mission.

Acknowledgments

The opinions expressed are those of the author and do not necessarily represent the views of OTT, NIH, or HHS. I express gratitude to my wife, Carol, for editing the manuscript and thwarting my assault on the English language.

References

1. *Diamond v. Charkabarty*, 447 U.S. 303 (S.Ct. 1980).
2. *Brenner v. Manson*, 383 U.S. 519 (S.Ct. 1966).
3. *Graham v. John Deere Co.*, 383 U.S. 1 (S.Ct. 1966).
4. In re *Wands*, 858 F.2d 731, 8 USPQ2d 1400 (Fed Cir. 1989).
5. Ferguson SM. Products, partners & public health: Transfer of biomedical technologies from the U.S. government. *J Bio Law Business* 2002;5(2):35–39.
6. Ramakrishnan V, Chen J, Balakrishnan K. Effective strategies for marketing biomedical inventions: Lessons learnt from NIH license leads. *J Med Marketing* 2005;5:342–352.
7. Ferguson SM, Kim JP. Distribution and licensing of drug discovery tools—NIH perspectives. *Drug Discovery Today* 2002;7(21):1102–1106.
8. Gupta R, Kim J, Spiegel J, Ferguson S. Developing products for personalized medicine: NIH research tools policy applications. *Personalized Med* 2004;1(1):1–9.
9. Salicrup LA, Fedorkova L. Challenges and opportunities for enhancing biotechnology and technology transfer in developing countries. *Biotechnol Adv* 2006;24:69–79.
10. Keller GH, Ferguson SM, Pan P. Monitoring of biomedical license agreements: A practical guide. *Pharm Dev Regul* 2003;1(3):191–203.

24

Writing a Protocol

ROBERT B. NUSSENBLATT

Laboratory of Immunology, National Eye Institute, National Institutes of Health and Office of Protocol Services, National Institutes of Health Clinical Center, Bethesda, Maryland

Doing clinical research is the goal of so many interested in clinical medicine. We are constantly urged to bring observations from the laboratory into the clinic as quickly as possible. We are anxious to know if there is any human consequence to these findings. If the phrase "the devil is in the details" has relevance, it certainly does when it comes to clinical research. Although an idea for a clinical study may be very good, the challenge is putting it into a format and structure that has a good chance of yielding clinically valid information. The process of converting an idea into an infrastructure that results in clinically valid information is in essence creating a clinical protocol. There is a more formal definition for a protocol: "a complete written description of, and scientific rationale for, a research activity involving human subjects."[1] This chapter reviews the general guidelines of protocol writing in the United States.

1. TYPES OF PROTOCOLS

There are two major categories of clinical protocols. The first category is natural history protocols, which can be retrospective reviews of cases/histories. Case reviews are usually not initially performed with patients but with their data. Follow-up with questionnaires or further testing can certainly come from the initial review. Natural history studies can also follow what happens to patients with a specific disorder. The second major protocol category is an interventional study or clinical trial designed to change the course of a disease through a therapy or the use of instrumentation. Interventional protocols exist in four phases.

Phase I studies initially are performed in the evaluation of new drugs and are designed to determine safe doses of a drug. Phase II studies search for evidence of efficacy and provide further safety testing; they usually are conducted using 20–100 people and may or not involve a placebo. A phase III study is carried out when previous experience has identified the degree of activity of a drug, its approximate dose, and possible side effects. In a phase III study, one writes a protocol to compare a new intervention with a currently standard practice or placebo. Phase III studies are performed when a drug is being considered for Food and Drug Administration (FDA) approval for a license of a new indication and usually involves hundreds or thousands of participants. Finally, phase IV, or postapproval, studies may be performed and are designed to monitor safety in thousands or millions of subjects. For those interested in multicenter studies as well as their ethical aspects, several texts can be consulted.[2–5]

2. WRITING A PROTOCOL

Often, an idea for a protocol comes about after your research group or clinical colleagues have seen and discussed a particularly interesting case that might stimulate you to think about a specific question. Usually, there is a real enthusiasm about the prospect of doing a study, and that is good. However, once the enthusiasm begins to ebb, the difficult work begins (Table 24-1). Generally, the principal investigator of the study will assume the responsibility of coordinating

TABLE 24-1   Basic Elements of Body of Protocol

Précis of 400 words
Table of contents/outline
Introduction
    Study background
    Animal/human research
    Describe new techniques
    Will an Investigational New Drug request be sought?
Objectives
Study design and methods
Inclusion and exclusion criteria
    Women and children
Monitoring subjects and criteria for withdrawal of subjects from study
    Define end points and criteria for withdrawal
Human subjects protection
    Subject selection
    Benefits and risk/discomforts
    Compensation?
    Adverse events
Protocol consent and assent
Appendices
References

these considerations. Someone must write the initial protocol, and it is the principal investigator who will begin with the help from the coinvestigators of the study. What is the question you wish to ask? Is this something that can be asked in the context of a clinical study? Are patients available to be evaluated? Often, there is a very good idea, but the types of patients needed may not be available. Would the study being considered put the patients at any risk? Every study carries some degree of risk, but whatever the risk, the risk must be justified by the possible benefit. If there are adverse events, how will these be handled? There are important practical considerations as well. Once studies begin, there is a need for coordination of visits, collection of data, and the handling of phone calls or other inquiries to the study. All these need to be considered before beginning a study.

It usually makes sense to include all persons who are involved in the protocol as associate investigators on the protocol. This has become more important as concerns about conflict of interest in clinical research are being concretized in a more formalized evaluation (see Chapter 11). The principal investigator should use the associate investigators' expertise in the development of the protocol. Each associate investigator should review the protocol, with special attention to the part related to his or her expertise. The principal investigator must collect the comments of her or his associate investigators and create the final protocol, which is then submitted for review.

3.   WHAT HAPPENS TO YOUR PROTOCOL?

Before going to the institutional review board (IRB), many institutions will have a pre-IRB committee that will review protocols for scientific quality and potential cost. Other regulatory agencies, such as the FDA, may need to be informed as well. For some protocols, such as randomized, masked studies, a data and safety monitoring board may be constituted to review the data and safety of a study as it progresses. The protocol is sent to this group for its concordance and suggestions.

4.   ESSENTIAL ELEMENTS TO THE PROTOCOL

The elements to a clinical protocol may vary to some degree between academic institutions, and what differs are the added features required by an institution. An interventional trial usually requires more information and will be scrutinized very carefully.

4.1.   Précis

The body of the protocol begins with the précis, a short (400 words or less) description of the study. The précis should describe the objectives, study population, design, and what outcomes will permit you to evaluate the study.

4.2.   Introduction

The introduction describes the background of the study that is being proposed, often with a description of the disorder under study and the general study design. If the protocol is a clinical trial using a new drug or technique, these should be described. An outline of the research, both human and animal, that has been done to date should be provided as well as a justification of the dosing in this study that is different from that of other studies already performed. The mechanism of action of a new drug or how a new device works should be included.

4.3.   Objectives

The objectives section can be short and succinct. It describes what will be accomplished with the study and often is divided into primary and secondary objectives.

4.4. Study Design and Methods

The study design and methods describe how the objectives will be achieved. The following questions should be addressed: Is the study a clinical trial or a natural history study? What type of patients will be recruited? What type of disease will the patients have? Will the patients be followed as clinic patients, inpatients, or both? Will this be a follow-up to a study that was recently completed? What is the patient recruitment plan? The IRB will seriously review the number of patients and proposed period of time for the study. A power analysis to determine the number of patients required should be presented. Many studies are stopped because patient recruitment is far below what the investigators claimed they would be able to obtain. Special consideration needs to be given to "vulnerable" patient populations, such as children and those with mental infirmities. A detailed description of what will be done should be provided. Will patients receive a new medication and, if so, will it be compared to a standard therapy? Will patients be randomized and, if so, how? What is the role of the pharmacy? An outline of the number of visits and the tests planned is essential. This is called a protocol timeline. (Table 24-2). A description of the methods and procedures and how potential complications will be managed is required. Projection of the need for special resources, such as research bloods (or other fluids), should be provided. The impact of the protocol on standard of care requirements for the hospital should also be outlined.

TABLE 24-2 Example of a Protocol Timeline

Scheduled Visit Week (See protocol for handling of delayed or missed treatments.)	B[1]	0[1]	2	4	6	8	10	12	14	16	18	20	22	24	26
General Assessments															
Medical history	X														
Brief body systems review and examination	X	X[1]			X			X				X			X
AE assessment[2] and current meds[2]	X	X	X	X	X	X	X	X	X	X	X	X	X	X	X
Quality of life determinations	X							X							X
Vital signs (BP, respiration, temperature)	X	X	X	X	X	X	X	X	X	X	X	X	X	X	X
Visual System Exams															
Manifest refraction[3]	X							X							X
Visual acuity	X[3]	X[1]	X	X	X	X	X	X[3]		X		X		X	X[3]
Slit lamp exam and tonometry	X	X[1]	X	X	X	X	X	X		X		X		X	X
Dilated fundus exam	X	X[1]	X	X	X	X	X	X		X		X		X	X
Inflammation grades	X	X[1]	X	X	X	X	X	X		X		X		X	X
Substudy evaluations (site-specific)		(X)						(X)							(X)
Study Therapy															
Open-label therapy[4]		X	X	X	X	X	X	X	X	X	X	X	X	X	X
Laboratory															
CBC with differential	X	X[1]	X	X	X	X		X		X		X			X
Hematology,[5] LFTs,[5] and urinalysis	X			X		X		X		X		X			X
Pregnancy test for females	X							X				X			X
Serum test drug and anti-antibodies		X[6]	X[6]	X[6]	X[6]										X[6]

[1]The baseline (week B) visit may immediately precede the initial (week 0) treatment if all requirements for enrollment have been met and are documented. Listed evaluations (marked with [1]) under week 0 should be repeated only if > 5days had elapsed since the initial baseline visit, or if medically indicated.

[2]Adverse events should be reported at any time between scheduled visits as necessary. At each visit, a review with directed questions is performed with the patient regarding adverse events in the interval since the last visit, including an assessment of the injection site(s). Current medications are recorded at each visit.

[3]Manifest refractions must be performed when scheduled and repeated as indicated, including whenever a drop in BCVA ≥ 10 ETDRS letters (≥ 0.20 logMAR) occurs within a 12-week period.

[4]Continuation of study therapy will occur unless a safety or withdrawal study end point is reached. If an end point has been reached, the participant will exit the trial.

[5]Hematology and liver function tests include sodium, potassium, chloride, CO_2 (total), creatinine, glucose, urea nitrogen, alkaline phosphatase, ALT/GPT, and AST/GOT.

[6]Serum test drug and anti-idiotypic antibody levels will be performed at all phase II sites. The default intervals will be at days 0*, 14*, 24, 35*, and 182*. Up to three additional intervals may be specified to obtain pharmacokinetic (PK) samples from participants during the early induction phase. (The * means trough value.) Participating sites will be shipping out coded specimens to a sponsor-designated central laboratory.

4.5. Inclusion and Exclusion Criteria

A careful description of inclusion and exclusion criteria is necessary for successful subject recruitment to a protocol. A clear and succinct list of particulars needed for a patient is necessary. For example, for a study involving patients with diabetes, what type of diabetes will be studied? Will it be necessary for patients to be on a specific insulin regimen? Will patients likely have certain medical complications requiring prolonged hospitalization? Similar information is required for the exclusion criteria. Are there age limitations? Not only are there inclusion and exclusion criteria related to the disease under study but also there are exclusion criteria because patients are unable to undergo certain tests. For example, one possible exclusion could be hypersensitivity to an imaging dye required for testing. Another might be prior drug use or stage of a disease. If a particular category of patients are to be excluded from the study, a thoughtful justification for the exclusion is necessary.

4.6. Women, Children, and Minorities

Phase III clinical studies and natural history studies should be designed so that the results can be stratified to establish whether or not benefits occur in men, women, children, and minority populations. This may not be possible for phase I clinical trials, in which a limited number of patients will be studied.

4.7. Monitoring Subjects and Criteria for Withdrawal of Subjects from Study

The risk/benefit ratio is carefully considered by IRBs. The protocol should describe criteria to minimize harm. Expected minor and serious adverse events criteria for detecting and reporting adverse events need to be described in detail. The protocol needs to define event end points and when a patient will be withdrawn because of adverse events. Examples might be worsening of an underlying medical condition or a patient's poor compliance with the protocol. Of course, a patient has the right to withdraw from the protocol at any time. If a patient is terminated or withdraws from a protocol, details of what type of follow-up is necessary should be provided. If the study is using a new intervention, which most often is performed under an Investigational New Drug request, then a separate section is needed.

4.8. Compensation

Payment of clinical research subjects has been an established policy for quite some time.[6] The amount of remuneration varies depending on the type of research protocol. For some studies, details of travel and subsistence provisions are required in the protocol.

4.9. Protocol Consent and Assent

Informed consent is one of the most important elements of a protocol. Required elements of the consent document include the following:

1. A statement that the study involves research.

2. An explanation of the purpose of the research, an invitation to participate, an explanation of why the subject was selected, and the expected duration of the subject's participation.

3. A description of procedures to be followed and identification of which procedures are investigational and which might be provided as standard care. Use of research methods such as randomization and placebo controls also needs to be described.

4. A description of any foreseeable risks or discomforts to the subject, an estimate of their probability and magnitude, and a description of what steps will be taken to prevent or minimize them, as well as acknowledgment of potentially unforeseeable risk.

5. A description of any benefits to the subject or to others that may reasonably be expected from the research and an estimate of their likelihood.

6. A disclosure of any appropriate alternative procedures or courses of treatment that might be advantageous to the subject.

7. A statement describing to what extent records will be kept confidential, including examples of who may have access to research records, such as hospital personnel, the FDA, and drug sponsors.

8. For research involving more than minimal risk, an explanation and description of any compensation and any medical treatments that are available if subjects are injured through participation, where further information can be obtained, and whom to contact in the event of a research-related injury.

9. An explanation of whom to contact for answers to questions about the research (include the name and telephone number of the principal investigator) and the research subject's rights.

10. A statement that research is voluntary and that refusal to participate or a decision to withdraw at any time will involve no penalty or loss of benefits to which the subject is otherwise entitled.

11. A concluding statement indicating that the subject is making a decision whether or not to participate, and that his or her signature indicates that he or she has decided to participate, having read and discussed the information presented.

12. If the subject is or may become pregnant, a statement that the particular treatment or procedure may involve risks, foreseeable or currently unforeseeable, to the subject or to the embryo or fetus.

13. A description of circumstances in which the subject's participation may be terminated by the investigator without the subject's consent.

14. Any costs to the subject that may result from participation in the research.

15. The possible consequences of a subject's decision to withdraw from the research and procedures for orderly termination of participation.

16. A statement that the principal investigator will notify subjects of any significant new findings developed during the course of the study that may affect the subjects and influence their willingness to continue participation.

17. The approximate number of subjects involved in the study.

18. If the investigator is not planning to return results to the subjects, a statement should be included that explains the reasons for planned nondisclosure and recognizes the subject's right to that information under the Privacy Act. The following language has been recommended for use in protocols at the National Institutes of Health Clinical Center:

> The investigators conducting this study do not plan to provide you with the results of any medical tests or evaluations or other information pertaining to you, or other research data or results because (the results will be preliminary) (the results will require further analysis) (the results may reveal unwanted information about family relationships) (further research may be necessary before the results are meaningful). (If meaningful information is developed from this study that may be important for your health, you will be informed when it becomes available.)
>
> By agreeing to participate in this study, you do not waive any rights that you may have regarding access to and disclosure of your records. For further information on those rights, please contact Dr. _____ (principal investigator).

Consent forms should be written in simple language, at a sixth- to eighth-grade level, always trying to use short terms. The consent document should outline what the patient is to expect during the study. What are the tests to be done? What should the patient expect, and are there any adverse effects the patient might suffer? Are you giving a new medication or new surgical technique? The investigator needs to outline why he or she is considering this approach, the possi-

ble problems, the possible advantages, and also what alternative therapies are available. Will blood be taken? Define how much blood will be obtained and put the amount in terms that can be easily understood, such as teaspoons or tablespoons. If a new medication is provided by a drug company, this should be stated in the informed consent. Conflicts of interest of investigators need to be addressed (see Chapter 11). The informed consent should describe whether researchers participating in the protocol have a relationship with the drug company. The consent document also should outline whether or not the investigators will receive royalty income if the study is successful.

Remember that the consent process is an evolving process and institutions may have specific requirements. For example, one possible addition is HIV testing. Specific wording is available to cover many of the required and suggested concepts that were mentioned previously. A translated consent document is required when English is not the primary language of the population you are planning to recruit.

4.10. Child Assent

The assent is how you obtain a child's agreement to participate in a clinical research study. The form is usually shorter and simpler than the consent document, but you still need to outline what you plan to do and what the child should expect. This, of course, is a complicated subject since the child's age, maturity, and psychological state will effect whether the child is even capable of understanding what will happen. You need to consider this and the IRB will determine whether it agrees with the approach outlined.

The written consent document does not substitute for a detailed oral explanation of the protocol to the patient. Patients should have an opportunity to ask and receive good answers to all questions in order to be sure they are fully informed.

References

1. National Institutes of Health. *Protomechanics*. Available at http://intranet.cc.nih.gov/protomechanics.
2. Meinert C. *Clinical Trials, Design, Conduct and Analysis*. New York, Oxford University Press, 1986.
3. Piantadosi S. *Clinical Trials: A Methodologic Perspective*. New York, Wiley, 1997.
4. Levine RJ. *Ethics and Regulation of Clinical Research*, 2nd ed. Baltimore, Urban & Schwarzenberg, 1986.
5. Emanuel EJ, Crouch RA, Arras JD, Moreno JD, Grady C (eds.). *Ethical and Regulatory Aspects of Clinical Research*. Baltimore, Johns Hopkins University Press, 2003.
6. Grady C. Payment of clinical research subjects. *J Clin Invest* 2005; 115:1681–1687.

25

Evaluating a Protocol Budget

MARGARET A. MATULA

Clinical Research Management Branch, National Institute of Allergy and Infectious Diseases, National Institutes of Health, Bethesda, Maryland

1. INTRODUCTION

Preparing and evaluating a budget for a protocol or clinical trial can be done for different purposes. It could be to obtain financial support for your own research through a grant application, a pharmaceutical company may have offered to pay you to conduct a protocol at your site, or maybe it is the reverse—you are the one who is going to be paying other researchers to conduct a protocol at their sites. From any of these perspectives, it is important to gather complete information in order to sufficiently prepare a protocol budget. With that in mind, preparing a budget should focus on evaluating the protocol to determine what the requirements of the protocol are and what the resource requirements will be at the clinical trial site where the protocol will be implemented. Once you have established those requirements, you will be able to develop a comprehensive budget. This chapter describes the different requirements that should be considered and then a method for establishing a protocol budget.

2. REQUIREMENTS

In order to evaluate the protocol requirements, you need to determine exactly what is going to be done as part of the protocol. Once that is completed, you should be able to determine what resources the site will need in order to implement the protocol. The items in Table 25-1 are discussed in this section since they are the most common requirements that should be included when you evaluate the protocol and site resource requirements, especially if it is for a clinical trial.

2.1. Duration

In terms of the duration, you will need to know how long the protocol is going to last—days, weeks, months, or years. It is also important to consider whether or not duration is based on the accrual rate, if it is divided into steps or phases (e.g., a treatment phase and a follow-up period), and, if it has different steps or phases, whether they are of equal intensity. There is a difference between a duration that is 48 weeks versus a duration that is 48 weeks after the last subject is enrolled. If it takes 24 weeks to enroll that last subject, the protocol duration has just changed from 48 weeks to 72 weeks. Increasing the duration by 50% will impact your budget remarkably.

2.2. Subjects

The research subjects to be enrolled in the protocol are considered in both the protocol requirements and the site resource requirements. In terms of the protocol, you will want to consider the following:

- How many will be enrolled?
- How many will you need to screen to achieve the target enrollment?
- How long will it take to achieve the target enrollment?
- What are the eligibility criteria? How is that going to impact your recruitment?
- What are the subjects signing up for? How many study visits are required? What other commitments or special procedures are required that might impact your recruitment?

TABLE 25-1   Requirements

Protocol	Site Resources
Duration	Subjects
Subjects	Clinical
Screening	Laboratory
Clinical	Study product
Laboratory	Data management
Study product	Travel
Toxicity management	Personnel
Data management	Equipment
Monitoring	Supplies
Travel	Regulatory
	Subcontracts
	Indirect costs

From the perspective of site resources, the focus of areas to consider related to the research subjects is different, although it is based on the protocol requirements. It will include consideration of the following:

• How large is the site's population base?
  Does it represent the population affected by the disease? Does it include racial or ethnic minorities? Women? Children? Other special populations (e.g., intravenous drug users, pregnant women, and prisoners)?
  Based on eligibility criteria, how many people in the pool of patients are potentially eligible?
  How many people will need to be screened for every enrollment?
• What will you need to do to recruit subjects for enrollment, and how will you reach out to them?
  Do you have a strategy or plan to recruit subjects? This is especially critical for recruiting women, minorities, and other special populations.
  Will you need to advertise your protocol to recruit for it? This may include such things as public service announcements, newsletters, and informing primary care providers, public health clinics, or hospitals.
  Do you need to get institutional review board (IRB) approval for your recruitment methods?
• What will you need to do to keep your subjects in the protocol? Recruiting subjects is not enough; you also have to consider retention.
  Do you have a strategy or plan to retain subjects once they have been enrolled? If it is a long-term study, this is especially important.
  Does your plan contain financial incentives? It is not uncommon for subjects to be reimbursed for travel expenses, receive food or formula, be provided with child care during visits, or receive compensation for painful procedures.
  Would these incentives be considered coercive?

  Do you need to get IRB approval for your retention methods?
• Do you plan to involve the community, and what do you expect the involvement to be? Community involvement may contribute to the success of your recruitment and retention, whereas lack of community support may contribute to the failure of recruitment and retention, so this needs to be carefully considered.
  What support will you need to provide to the community so that it is more knowledgeable about your protocol (e.g., information about the disease and education about research)?
  How will you address a community that may have knowledge and experience ranging from the layman to technical expertise?
  In what kind of forum will you do this?
  Do you need to provide any financial support for community involvement? Refreshments, free parking, and printed materials may be necessary.
  What do you expect to get out of your community (e.g., input into protocol and assistance with recruitment and retention)?

2.3. Screening

Before a subject can be enrolled into a protocol, screening is usually required to determine that the person is eligible. Screening may range from interviewing the person about his or her history to specific clinical and laboratory evaluations. The protocol may require special clinical or laboratory tests, and it may also have a time constraint on when the screening data are obtained in relation to enrollment in the protocol (e.g., within 45 days prior to enrollment). If a laboratory test or procedure is done outside a protocol-required time constraint, it may need to be repeated. Who will bear that cost? If screening involves tests and examinations that are considered to be part of routine care, then the investigator may be expected to bill the person's insurance for reimbursement. If the person is uninsured or the insurance company denies it, who will bear that cost? It is essential to know who will be paying for the costs related to screening, what it will include, and if it will cover the costs of screening everyone or only those who are successfully enrolled in the protocol. Again, you will need to closely estimate how many potential subjects you will need to screen to achieve the target enrollment. The budget related to screening is extremely important because it may have a major impact on your resources. If you have to screen a large number of people to achieve your targeted enrollment, it could be very expensive.

2.4. Clinical

If the protocol involves a clinical trial, then you will need to consider the budget in terms of both the protocol requirements and the site resource requirements. In terms of the protocol, you will need to know what evaluations are required. The protocol will specify how many study visits are to occur and what is to occur during those visits. However, the protocol may not specify how long each visit will last and who will need to conduct the visits. This is something you will need to determine based on the protocol requirements in order to adequately budget for it. You will also need to consider if each visit is the same or if they have varying intensity. Typically, study visits are more intense at the beginning of a study, especially in multiyear studies. Year 1 may have more intense visits and more frequent visits than the following years. Thus, a visit that lasts only 30 minutes versus one that lasts several hours will have a different impact on the cost. Visits conducted by a physician, nurse, midlevel practitioner, or some combination of care providers will also impact the cost. In addition, you will need to consider if there are any special procedures (e.g., lumbar puncture, biopsy, magnetic resonance imaging, computed tomography scan, and timed blood sampling for pharmacokinetic studies) that will need to be done as part of the protocol so they can be incorporated into the budget as well.

After determining the clinical protocol requirements, you will need to assess what site resources are required to fulfill those requirements. There are several possible features of site resources related to the clinical aspect of the protocol that will impact the budget. The first one is location. This includes considerations such as if it is a single-site or multisite trial; if it will be done in urban or rural areas; if it will be done in one country or multiple countries; and whether it will involve university, private, or public facilities.

The second feature is infrastructure. You will need to know where you will see subjects for their study visits and what infrastructure you will need to perform the evaluations. Can the study visits be conducted in an office or do you need an examination room? Do special services or facilities, such as inpatient care, radiology, special procedures, and emergency or resuscitation equipment, need to be accessible? The costs associated with this variety of clinical resources will differ. Other considerations include the following:

- Is the existing space sufficient?
- If multiple sites are involved, is the infrastructure consistent across the sites?

- Can the laboratory support both routine and special laboratory tests?
- Are any alterations or renovations required?

If you are conducting clinical trials in resource-poor countries, an accurate assessment of infrastructure is very important because it does not exist as we think of it in many places. For example, does the clinical area have an area for hand washing or can a hand sanitizer be used? Does the pharmacy have air conditioning and humidity controls as well as secure doors and windows? If you have to build infrastructure first, it can become a major expense.

2.5. Laboratory

Another requirement you may need to consider is the protocol and site resource components related to the laboratory. In terms of the protocol, you will need to know what laboratory tests are required. This includes the number and type of laboratory tests to be done at each visit, whether they are routine safety tests or special research tests, if any serial studies will be required, the type of specimens (serum, plasma, cells, and tissue), what will need to be done with the specimens when they are collected (processing, shipping, and storage), if the tests will be done in real time or later in batches, and what laboratories will be used (local, commercial, and research). Again, the laboratory testing may vary in intensity, so you will need to carefully evaluate the requirements for each visit.

Once the protocol requirements are determined, the budget will need to reflect the site resources that are necessary to support the laboratory protocol requirements. This means it will be necessary to evaluate the site's capacity related to laboratory. If there are any limitations to that capacity, determine how they will be corrected and the cost of doing so. This includes obtaining specimens, processing specimens, preparing specimens for both storage and shipping, and whether the laboratory participates in certification or quality assurance programs to ensure that validated results are being obtained. Costs associated with obtaining specimens include those for needles, gloves, tubes, and possibly personnel time. The budget for processing specimens will depend on what will be done at the site and could include a centrifuge, reagents, supplies, equipment, and support for equipment maintenance contracts. If a site is to ship specimens, the budget will need to include funds for packing materials; ice packs, dry ice, or liquid nitrogen; transportation; and training for shipping hazardous materials. If the shipping is international, the budget may also need to include the use of special couriers to refresh dry ice and escort packages through customs. If the site is to store specimens, then the site will need to have access to freezer space and an ability to track and retrieve specimens. Therefore, the budget may include refrigerators, freezers, a computer, backup generators, and alarm systems.

2.6. Study Product

If the protocol involves the use of a study product (drug, vaccine, microbicide, device, etc.), you will need to determine the protocol requirements with regard to what type of product it is, how many there will be, the route of administration (intravenous, injection, oral, topical, etc.), the frequency, and the duration. The site resources required for managing study product will depend on those answers. Site resource requirements may include access to a pharmacy and pharmacist; distributing the study product to the site and also to the subjects, determining how much of a supply will be dispensed at each interval, and determining if any supplies are needed to do this; storing the study product in recommended conditions (temperature and humidity) at the site, pharmacy, or the subject's home; maintaining accountability of the study product once it has been received; and providing education, training, and counseling to the subjects.

2.7. Toxicity

If a protocol involves study products, you will need to know how toxicities will be managed, and the budget might include a combination of both protocol and site resource requirements. Based on the protocol, what can you expect in terms of toxicities or adverse experiences? The protocol may specify the treatment and evaluation of common or expected toxicities, and that could involve additional visits, laboratory tests, or special procedures to evaluate the event and follow it through to resolution. However, you should also expect to budget for the costs of more serious or unexpected toxicities. Costs related to a rash versus renal failure are very different. You will also need to know if the protocol needs to meet FDA reporting requirements, the IRB reporting requirements for toxicities, and what impact those reporting requirements will have on site resources, most commonly personnel time to prepare reports and respond to queries.

2.8. Data

Data for protocols are usually collected on case report forms (CRFs) or questionnaires; therefore, you will need to know who is developing them and if this needs to be included as part of your budget. Statistical support to analyze the data may also be necessary. Computer programming for the CRFs and statistical analysis may need to be included as well. Evaluating the protocol requirements will also include determining how much data will be collected at each visit. This includes the number of CRFs and if the individual forms consist of multiple pages. A CRF consisting of 5 pages may take longer to complete than a single-page CRF, implying a higher cost. In addition, the protocol may require that the data be collected and submitted within a certain time period. There may also be costs associated with transmitting the CRF data from the site to a data management or statistical center, depending on what the protocol requires for data disposition.

Site resources for data collection and management usually consist of several components. Staff to complete CRFs, to verify that CRFs are complete and accurate, and to resolve errors and data queries may be the most expensive component since these activities can be very time-consuming. Other site resources related to data management may include equipment such as computers, printers, facsimile machines, and copiers; supplies such as paper and toner; telecommunication and Internet access; maintenance contracts for equipment; and technical support. Another aspect of data management at the site is maintaining the source documents and regulatory files related to the protocol. The budget to support maintaining records such as these will be based on how files are organized and what is needed to provide storage that is secure and has limited access, both during the study and after the study (if necessary).

2.9. Monitoring

The term monitoring may have different meanings. It may refer to clinical (adherence to the protocol, regulations, and good clinical practices), safety (toxicities, adverse experiences, and end point achievement), and data monitoring (source documents and CRFs); it may be done by someone who is part of the site staff or someone independent of the site; and it might also refer to auditing done by the sponsor, IRB, FDA, Office for Human Research Protections (OHRP), or other regulatory body. Each of these consumes staff time, whether it is to conduct the monitoring or to work with a monitor when he or she is at the site, so this requirement should be incorporated in the budget.

2.10. Travel

Protocol and site resource requirements for travel may include meetings or trainings associated with a protocol, scientific presentations, visiting other sites, performing outreach or recruitment, and as part of retention efforts. Once you determine what travel is required, you will need to determine the type, number, length, location, and the staff (investigator, coordinator, nurse, etc.) to be included in the budget. Then, you will also need to determine the costs related to travel. These may include transportation (air, train, mileage, taxis, etc.), parking, lodging, meals, and incidental expenses.

2.11. Personnel

As you evaluate the various requirements related to the protocol and site resources, many of them include staffing or personnel time and effort. So when you examine the site resources requirements for personnel you will want to consider the different types of personnel, their experience, and their other commitments. Personnel may include those who specialize in research and others who do not.

The first group could include the principal investigator, subinvestigators, coordinators (study, project, clinical, and administrative), research nurses, laboratory scientists and technicians, data staff (entry, analyst, and manager), and statisticians. The latter group may include physicians, midlevel providers (nurse practitioner and physician's assistant), nurses, pharmacists, specialists, consultants, social workers (case manager and outreach coordinator), monitors, quality managers, regulatory affairs specialists, and administrators (fiscal and secretarial). It is not unusual to see professional or higher paid staff doing work that could be done by support staff. Therefore, it is important to carefully evaluate who is needed to do the various tasks related to the protocol. Is it more cost effective to have someone transcribing data onto CRFs or to have the research nurses enter the data onto the CRFs? How will either method affect costs related to quality assurance and data management? Who will transport specimens from the clinic to the lab? Who is going to do all the copying or preparing of subject files? You will need to determine how your budget will be most efficiently utilized by the types of personnel you support.

The experience of the personnel involved is also important and will affect the budget. Someone who is less experienced or more "junior" will usually cost less to support than an expert; however, that must be balanced with the need to have personnel with sufficient expertise and qualifications to implement the protocol. In resource-poor countries, it can be a challenge to hire the appropriate staff. Also, the roles and responsibilities of personnel in other countries can vary compared to those of similar positions in the United States. If access to experienced staff is limited, the budget may need to include costs for training or recruiting.

The availability or other commitments of the personnel involved is important to consider since you want to make sure that the personnel will give you the time and effort that you are supporting financially. For example, if an investigator is also committed to doing other research, serving on the faculty of a university, and serving as an attending physician at a hospital, will he or she have sufficient time to focus on your protocol? You will want to negotiate a realistic commitment to support with your budget.

2.12. Equipment and Supplies

A variety of equipment and supplies may be needed to fulfill the various site resource requirements. This could include items for the clinic (ECG machine, scale, blood pressure cuff, exam gloves, needles, and hazardous waste containers), laboratory (freezer, centrifuge, reagents, plastic disposables, syringes, tubes, and dry ice), office (computers, furniture, filing cabinets, copier, paper, toner, file folders, and binders), mailing and shipping (postal and courier service), and telecommunications (phone, facsimile, and Internet). Some of these could be specific items in the budget and others could be included as part of another cost. For example, reagents could be part of the fee for specific lab tests, or clinic supplies could be included as part of a fee for using the clinic space.

2.13. Regulatory Issues

An increasingly common part of budgets is the cost of the regulatory burden associated with doing research. This refers to both the staff resources and fees associated with fulfilling IRB requirements, complying with state and federal (FDA and OHRP) regulations, complying with National Institutes of Health policies, meeting good clinical practice expectations, establishing policies and procedures, and complying with other sponsor requirements. The ability to track and comply with this assortment of regulatory requirements can consume a significant amount of staff time, and many IRBs now charge fees to the investigator for each protocol that is submitted. With regulatory requirements continuing to increase, you will want to ensure this is considered as part of the site resource requirements for your budget.

2.14. Subcontracts

If the site does not have the resources in place to fulfill the protocol requirements, subcontracts with other organizations or individuals may be necessary. It is not uncommon to outsource for services in the areas of monitoring, pharmacy, laboratory, data management, computer support, and record storage. When this needs to be done, restrictions related to salary structure, limits on funding levels, and types of costs covered may need to be incorporated into the budget.

2.15. Indirect Costs

Another aspect of site resource requirements that may be part of the budget is the indirect cost rate. Indirect costs are also referred to as overhead or facilities and administration costs. These are fees charged by an institution and may include such items as space, utilities, cleaning, maintenance, administrative support, and equipment. Indirect costs are usually a negotiable rate that is based on a percentage of the budget. If your budget includes indirect costs, it is important to clarify what the rate is based on and if there are any limitations. For example, is it based only on personnel costs, or does it include equipment, supplies, travel, and clinical costs? You also need to know what the indirect costs will specifically cover so that you can ensure you are getting what you are purchasing. For example, do utilities include phone and Internet? Do you need to budget for a security guard if it includes building security? Another aspect to consider is if you can make any changes to lower your indirect cost rate, such as moving to a different location. That consideration will need to be balanced with other costs you may then need to support instead, such as rent, access to the research subjects, laboratory support, or other site resource requirements.

3. ESTABLISHING A PROTOCOL BUDGET

Now that the different protocol and site resource requirements have been described, this section focuses on establishing a protocol budget. To prepare a protocol budget, you will need to determine on what your costs will be based. Your institution may have an established price list, you may need to use industry standard pricing, or you may need to base it on your previous protocol budgeting experiences. If you are conducting a multicenter protocol or using an industry standard, you may need to compare prices across practices, institutions, cities, and countries and, possibly, vary those prices to reflect any differences.

In addition, for some aspects of the budget, there are different ways to calculate how you will charge the costs. For example, rates for personnel may be charged at a flat rate per study visit, at an hourly rate, or as a percentage of full-time effort (a full-time equivalent equals 100%). Another example is laboratory tests. Testing can be done after the specimen has been collected (real time) or specimens can be saved for analyses until a number of samples have been collected (batched). Batching is sometimes done on all of the samples from a subject, the site, the study, or some other grouping. Batching is usually more cost efficient even when you include the costs of storage until analysis.

The budget you prepare should cover the full spectrum of protocol implementation. That means, in addition to the conduct of the protocol, your budget will need to include the costs associated with the startup activities or all of the regulatory and administrative work necessary to initiate a protocol; screening subjects to determine eligibility; follow-up for six to eight weeks following study completion, if required; and close-out activities or all of the regulatory and administrative work necessary to end a protocol.

When preparing your protocol budget, it is important to remember that you want to determine the real or actual cost of conducting the protocol; however, there is no single right way to do this. Therefore, you want to make sure you do it correctly; otherwise, you might overspend your budget before you complete the protocol. It is also reasonable to assume that you will need to establish a balance between conserving costs and preserving subject safety and the scientific integrity of the protocol. Suggestions for doing this include the following:

- Only schedule tests and collect data that are necessary. It is very easy to keep adding things that would be "interesting" or "might" be analyzed in the future; however, this can increase your costs dramatically.
- Negotiate or shop around to find the best rates for your needs.
- Change the study visit schedule.
- Change laboratory and research test schedules.
- Change periods of recruitment, sample size, or duration of study.
- Consider alternative sites to conduct the protocol.
- Outsource or subcontract services that are too expensive to perform on your own.

Two commons ways to establish your budget are to determine a cost per subject and to record a cost for each resource on a line of your budget. Either method is effective, especially if you consider all the costs that go into conducting the study. The following example uses a spreadsheet to establish a cost per subject. In setting up a spreadsheet, you can make it as simple or complex as you like. For illustration purposes, the spreadsheets displayed here are fairly basic.

In preparing a spreadsheet, it is useful to group types of costs together so you can examine how much money is budgeted for each of the categories or groups. Depending on the protocol, categories may include personnel, subject reimbursement or incentive, clinic supplies, office supplies, clinical procedures, radiology, and laboratory. These categories can also be separated further. For example, personnel could include physician, nurse, pharmacist, data manager; this could be subdivided even further into the types of duties. Laboratory costs could be divided into hematology, chemistry, immunology, microbiology, and virology. The rows in the spreadsheet will list the categories, and the columns will indicate the parameters to be applied, such as the number of visits, hours per visit, and cost per hour. The final column should be a total for that row with a formula inserted to automatically calculate it. Figure 25-1 is an example of a spreadsheet to calculate personnel costs per subject, and Figure 25-2 is an example of a spreadsheet to calculate laboratory costs per subject. So where does the data come from in order to complete the spreadsheets? One source is the protocol's schedule of evaluations. Figure 25-3 illustrates a schedule of the clinical and laboratory evaluations from a sample protocol.

The information from Figure 25-3 is used to determine some of the protocol and site resource requirements described in the previous section and, therefore, can be used to establish your protocol budget. You will notice that the schedule includes evaluations required for screening, entry, each study visit after entry (weeks 2–48 and after 48 weeks), if virologic failure occurs, if the subject enters step 2 (this is only for subjects who develop resistance to one of the study drugs), and when the subject is discontinued (D/C) from the study. Looking at this schedule, you can determine some of the personnel and laboratory requirements.

The schedule in Figure 25-3 indicates that clinical assessments are required at screening; entry; weeks 2, 4, 8, 12, 16, 24, 36, and 48 and every 12 weeks after week 48; step 2; and D/C. The sample spreadsheet in Figure 25-1 indicates a total of 11 visits on the physician rows. The first row under the physician category is for the initial visit. This is a separate line for the physician, nurse, and pharmacist categories because the initial visit usually requires more time. Note that the "Hours/Visit" column indicates 1 hour for this type of visit and the number of visits is only "1." In our example, this will account for the screening visit. The second row under the physician category is for the remaining study visits; however, it is important to note that it specifies year 1. In fact, column A, row 1 indicates this whole worksheet is only for direct costs in year 1. That means the costs associated with the scheduled visits after week 48 are not reflected in this worksheet. For budgeting purposes, it is usually easier to

	A	B	C	D	E	F
1	DIRECT COSTS YEAR 1	NO. of	NO.	HOURS/	COST/	
2		Prescriptions	VISITS	VISIT	HOUR	TOTAL
3						
4	PERSONNEL TIME					
5	Physician					
6	Initial Visit		1	1.0	$75	$75
7	On Study, Year 1		10	0.5	$75	$375
8	Registered Nurse					
9	Initial Visit		1	3.0	$40	$120
10	On Study, Year 1 (Physical, Draw & Aliquot Blood, Charting)		10	2.0	$40	$800
11	Adherence Assessment		4	0.3	$40	$40
12	Questionnaire Administration		2	0.2	$40	$16
13	Pharmacist					
14	Initial Set-up		1	0.4	$60	$25
15	Time per visit per prescription	4	10	0.2	$40	$320
16	Administrative					
17	Regulatory Affairs Specialist		11	0.5	$35	$193
18	Quality Assurance Specialist		11	1.0	$30	$330
19	Secretary		11	1.0	$15	$165
20	Data Management					
21	Manager		11	0.2	$30	$66
22	Technician		11	0.5	$15	$83
23						
24	TOTAL PERSONNEL COSTS					$2,607
25						

Total Year 1 Direct Costs \ **Personnel** / Lab /

Ready   CAPS

FIGURE 25-1 Personnel spreadsheet.

	A	B	C	D	E
1	DIRECT COSTS YEAR 1		NO.	COST/	
2		Shipments	UNITS	UNIT	TOTAL
3	LABORATORY EVALUATIONS				
4	Hematology				
5	CBC with Differential		8	$18	$142
6	Platelet Count		8	$17	$132
7	Chemistries & Liver Function				
8	Blood Chemistry Package		8	$24	$192
9	CPK		1	$10	$10
10	Lipase, Serum		1	$29	$29
11	Lipid Panel		4	$57	$227
12	Liver Function Tests ONLY		11	$20	$220
13	Immunology				
14	CD4+/CD8+		7	$165	$1,155
15	Miscellaneous				
16	Pregnancy Test (Urine)		9	$40	$360
17	Pharmacology				
18	NVP pk		1	$50	$50
19	Virology				
20	HBsAg		8	$47	$378
21	HIV-1 plasma RNA		9	$104	$936
22	PBMC Store Only		3	$41	$123
23	Plasma/Serum Store Only		8	$25	$200
24	Sequencing (Genotypic Analysis)		1	$258	$258
25					
26	SPECIMEN SHIPPING				
27	Dry Ice to sites outside USA	32		$500	$16,000
28	Dry Ice to USA	32		$1,500	$48,000
29	Liquid Nitrogen to sites outside USA	32		$1,500	$48,000
30	Liquid Nitrogen to USA	32		$3,000	$96,000
31					
32	TOTAL LAB COSTS			$	212,412
33					

Total Year 1 Direct Costs / Personnel \ **Lab** /

FIGURE 25-2   Laboratory spreadsheet.

separate the costs for different years onto separate worksheets. Returning to the schedule in Figure 25-3, after screening occurs there are 9 clinical assessments scheduled for year 1 of the protocol. However, our sample spreadsheet (Fig. 25-1) indicates the number of visits is 10. One extra visit was included in the budget to account for the possibility that the subject may require a step 2, D/C, or unscheduled visit in year 1. In this same row, note that the hours per visit have been changed to 0.5. This is because the amount of physician time for these visits is not as intensive as it is for the initial visit.

Also note that in Figure 25-1 the other personnel categories have been subdivided to reflect different types of personnel or different intensities in visits. Under nursing, besides the initial and on-study visits, there are additional rows for adherence assessment and questionnaire administration. These activities do not occur at every study visit, but they do increase the duration of the visit, so listing them in a separate row is one way to account for the added intensity of these visits.

Accounting for laboratory costs is usually more straightforward. When you compare the schedule with the spreadsheet in Figure 25-2, the calculations are

done in a similar manner. Hematology and blood chemistry tests are required seven times during the first year (including screening); however, eight tests are included in the spreadsheet for each of those categories. Again, this allows for the possibility that a step 2, D/C, or unscheduled visit may occur in year 1. There are a few laboratory tests on the schedule (pregnancy, HbSAg, lipase, CPK, and lactate) that are only required if necessary. This means that you will need to determine how many of these tests should be budgeted, and you may have to justify how you arrived at that decision. You may have noted that one section of the laboratory spreadsheet includes a category that is not on the schedule. This is specimen shipping. To determine the costs and frequency of specimen shipping to include on your spreadsheet, you will have to refer to the protocol to determine the requirements for shipping (dry ice or liquid nitrogen) and the frequency of shipments (real time vs. batched).

4. SUMMARY

As described in this chapter, evaluation of a protocol budget begins with determining the requirements

Evaluation	Screening	Entry Week 0	Weeks after Entry								After 48 weeks	Virologic Failure	Step 2 Entry	D/C
			2	4	8	12	16	24	36	48				
Documentation of HIV	X													
Medical/Medication History	X													
Concomitant Medications/ Treatment Modifications		X	X	X	X	X	X	X	X	X	q 12w		X	X
Clinical Assessments	X	X	X	X	X	X	X	X	X	X	q 12w		X	X
Hematology	X	X		X		X		X	X	X	q 12w		X	X
Blood Chemistries	X	X		X		X		X	X	X	q 12w		X	X
Lipid Levels		X						X		X	q 48w		X	X
Liver Function Tests	X	X	X	X	X	X	X	X	X	X	q 12w		X	X
Pregnancy Testing	X	X	Repeat if Indicated											
HbSAg		X	Repeat if Indicated											
Lipase			Perform for symptoms suggestive of pancreatitis											
CPK			Perform for sx suggestive of myositis											
Lactate			Perform for sx suggestive of lactic acidosis											
CD4+/CD8+	X	X				X		X	X	X	q 12w		X	X
HIV-1 RNA, real time	X	X				X		X	X	X	q 12w	X	X	X
HIV-1 RNA, batched				X			X							
Plasma for Resistance Testing		X										X	X	
Stored Plasma		X		X	X	X		X	X	X	q 12w	X	X	X
Stored PBMC		X						X		X	q 24w	X		
PK Sampling, Arm 1A only			X	X										
Adherence Assessments					X		X	X		X	q 24w			
QOL/RU Assessments								X		X	q 24w			

FIGURE 25-3 Protocol schedule of evaluations.

of the protocol and site resources. Establishing a protocol budget is affected by the cost base you are using and how the costs are calculated. Depending on the protocol and site resource requirements, your spreadsheets may include a variety of cost categories and can be as simple or complex as necessary. Finally, it is important to remember that you will need to be prepared to justify how you arrived at your budget, and performing a detailed evaluation should provide you with that justification.

26

Data Management in Clinical Research: General Principles and a Guide to Sources

STEPHEN J. ROSENFELD

MaineHealth, Portland, Maine

1. INTRODUCTION

Data management is at the heart of the clinical research process. Although good data management practices cannot make up for poor study design, poor data management can render a perfectly executed trial useless. This book is about clinical research; the failure of a study to produce generalizable knowledge because of bad data management practices carries both resource and ethical costs. The investigator or sponsor can be held accountable for the former, but nothing can undo the harm to which subjects may have been exposed. It is incumbent upon every investigator to understand the statistical rules of study design, the practical steps that can be taken to protect the integrity of data, and the rules and regulations that govern the submission of data and the protection of subject privacy.

2. THE DIMENSIONS OF DATA

There are several ways of looking at data, all of which are valid, and all of which have consequences for data management practices. These "dimensions" of data are as a surrogate measurement for a complex state, as an objective recording of an observation, and as a "snapshot" of a point in time.

2.1. Data as a Surrogate Measurement

Typical measurements made in the course of clinical research reflect complex functional states or are simple endpoint assessments that mask a great deal of complexity. For example, hematocrit is often used as an indicator of the oxygen carrying capacity of blood. It can also be used as an indicator of the health of the bone marrow, where red blood cells are produced, or, in the right context, as an indicator of response to an acute intervention such as a red blood cell transfusion. The actual number reported as hematocrit, however, is none of these things. It is the percentage of the blood volume taken by red blood cells after blood is centrifuged in a heparinized microcapillary tube under controlled forces and for a prescribed period of time.

Similarly, endpoint assessments that seem straightforward, such as mortality or the occurrence of an event, mask many assumptions that have consequences for the statistical interpretation of data and for the subsequent understanding of study outcomes. Hidden in the simple classification of survival are decisions regarding how to statistically handle patients who died from causes unrelated to the condition or treatment being studied, how to account for patients who are lost to follow-up, how to correct for normal age-associated mortality, etc. These decisions need to be identified at the time of study design so that the data necessary to make appropriate corrections are collected. It is also easy to lose track of the larger context of the study and the assumptions underlying the choice of data to collect, especially when these assumptions are part of the prevailing wisdom at the time of the study's design.

These observations lead to practical recommendations for data management: recognize when observations are really surrogates for outcomes of interest; for outcome data, plan to capture all the data needed to make necessary statistical corrections; and for data that are surrogates for outcome, make explicit and then question all the steps that lead from the surrogate to the outcome of interest.

2.2. Data as an Objective Recording of an Observation

"Objective" has several meanings, at least two of which are relevant to the conduct of research. The first meaning is "not influenced by personal feelings or opinions in considering and representing facts."[1] This is obviously the goal of observations recorded for clinical research and emerges naturally from the requirement that all scientifically established findings are reproducible. Although mechanically assuring objectivity in a particular measurement or outcome assessment may seem straightforward, such determinations often contain hidden assumptions and biases. The act of measuring a particular physiologic parameter may be objectively reproducible, but the fact of recording that observation in the file of a subject who has been assigned to a control arm versus a treatment arm means that the observation inherits context from outside the act of measurement. Similarly, when assessing functional outcome, a researcher may be unconsciously biased to persevere longer in tracking down lost subjects in the treatment arm than in the control arm, leading to the same subject being assigned a status of "living" versus "lost to follow-up" depending on the arm to which he or she was assigned. The possibility and subsequent impact of unrecognized bias cannot be overemphasized. Clinical researchers go to extraordinary and very expensive efforts to systematically minimize the likelihood of bias, usually through study design. This is the justification for the "gold standard" of design for interventional trials—the double-blind study.

Reproducibility has two other implications. First, it implies that limits to the precision of data should be explicitly recognized, especially if the data are used for further calculation. Second, it implies that data should be collected in as standard a fashion as possible. Unless

there are compelling scientific reasons to do otherwise, an investigator should use well-characterized and generally accepted tests, instruments, and measures. This extends not only to the values recorded but also to the names of tests, diagnoses, symptoms, etc. The aim of a research publication is to communicate new knowledge, and this can only be done well if the writer and reader share a common vocabulary and understanding of the problem.

The second definition of "objective" is "not dependent on the mind for existence; actual."[1] Although there may be philosophic debate about whether this can be realized or not, the practical implication is that an investigator should avoid, whenever possible, using interpretation as primary data. If such use is necessary (e.g., in assessing a subject's mental state), the criteria for the interpretation should be explicit and the interpretation explained in reference to these criteria.

2.3. Data as a Snapshot in Time

Whether data are "actual" or not, the recording of observations made during the course of a clinical research study stand in, at the time of analysis, for the real events and states of research subjects. It is the data, not the subjects' experience, that are used to argue for or against the null hypothesis. It is the data, not the subjects, that are examined for evidence of bias. And it is the data, not the subjects, that will be subjected to the regulatory measures established to ensure research quality and subject privacy.

The recognition that data are observations collected at particular points in time, and that these points of observation will later represent the entire research study, makes explicit that it is up to the investigator to choose not only what to measure but also when and how frequently to measure it. Both of these characteristics of a measurement, the "what" and the "when," must be chosen to best reflect the underlying physiology and also to be most amenable to statistical analysis. Statistics is the art of collecting data in such a way that its expected distribution under the null hypothesis is known. In many cases, this distribution is independent of most aspects of the underlying physiology and is rather a function of the internal relationships of the data. It is essential, therefore, that the investigator consider the statistical methods that will be used to analyze the data at the same time that the choice is made of what and when to measure.

3. KINDS OF DATA

Obviously, data include experimental observations. However, there are also other types of data that need to be collected by the clinical researcher. Most of these are used either to validate or establish the context of the experimental observations. Some examples of this type of data include the research protocol document, the subject consents, the statistical analysis plan, the credentials of the investigators, the files documenting who entered/modified electronic data and when, and the privacy policy of the research institution. The guidelines and regulations for data management apply to these types of data as well as to data documenting traditional research end points.

4. GENERAL CONSIDERATIONS

As noted previously, data will stand in for the actual observations made during a clinical research project. The circumstances of its recording must be clear, including who made the observation, whether the record of the observation was amended or deleted, and when and for what reason such amendments and deletions were made. These requirements are very much like those that are legally required for medical records, for similar reasons. Legal requirements on data arise because they may be used as evidence in an argued court case. Similarly, every research publication is an "argument" for the particular interpretation advanced by its author(s), and the "rules of evidence" are similar. In the case of paper records, such requirements are met by striking through, rather than erasing, the original data when making changes, and signing and dating each change. In electronic environments, the need to reconstruct the history of any datum means that systems must be designed not simply to store data elements but also to record all changes or deletions and keep the original observation available. This is usually achieved through a combination of security rules that limit access and activity to authorized individuals and that treat a username and password (or other identification mechanism, such as a fingerprint) as equivalent to a signature, combined with a log of what records were accessed and what changes were made, by whom and when. Implicitly, such a system only works if there are administrative controls that forbid the sharing of usernames and passwords and define penalties for unauthorized use. In the current environment, in which most databases are on computers that are directly or indirectly connected to the Internet, there need to be additional safeguards to ensure that access to and alteration of data are prevented from outside the institution. These safeguards typically take the form of hardware or software that can block or detect unauthorized access. Just as important paper records often are stored in fireproof vaults and have

physical copies maintained, so electronic databases also must be regularly "backed up," preferably with at least one set of backups stored at a secure off-site location.

In addition to considerations of physical storage and maintenance, a researcher should collect and store data so that they are as reusable as possible. Data are lost far too frequently when an investigator leaves an institution, taking his or her computer password or knowledge of how he or she collected data with him or her and leaving data that are inaccessible or uninterpretable. Loss of data represents loss of an investment of institutional time, and it also means that to answer questions that could potentially have been answered with existing data, new subjects will need to be recruited and exposed to risk. As a corollary to the need to preserve data, it should also be collected in as standard a form as possible, and "metadata," or definitions of what each data element means and how it was defined and collected, should be stored alongside the data. Storing data for reuse is a general good, but it can be difficult to enforce in an academic environment that creates incentive for individual investigators to control and limit access to research data. Such a structural environment limits the long-term value of any particular piece of information and blocks the research community from taking advantage of the ability of computers to store and combine very large amounts of data. It will be one of the major challenges of the coming years to reconstruct the academic research environment to recognize and reward researchers for contributing data to repositories outside their personal or programmatic control. The habitual accumulation of data has the potential to transform clinical research from an activity focused on the individual study to an enterprise concerned with the evolving interpretation of a growing body of knowledge.

5. DOCUMENTS THAT GOVERN DATA MANAGEMENT

The conduct of clinical research is heavily regulated for several reasons. The primary reason is that all clinical research is conducted on human subjects. Participation in clinical research is seen by some as a right and is generally agreed to be subject to the demands of equity. The population of research subjects is fundamentally vulnerable, in that subjects are not expected to bring the same level of understanding to participation in research that the investigator may have, and they may have a different kind of interest in its outcome. The largest body of regulations addresses the protection of human research subjects. These protections are largely covered elsewhere in this book. However, the increasing awareness of the value of information, and the complementary value of privacy, which is an individual's right to control information about him- or herself, has led to human subjects protection legislation that is specific to data management. This regulation is embodied in the Health Insurance Portability and Accountability Act of 1996 (HIPAA).

Research is also regulated because life-and-death decisions are made based on its findings. Where other kinds of research have to prove their findings in the marketplace or through scientific debate, no one is willing to allow these forces alone to determine whether a particular medical intervention is safe or effective. Research involving new drugs or devices is subject to regulation by the Food and Drug Administration (FDA), which imposes standards for research conduct, study oversight, human subjects protections, and the submission of materials for regulatory approval of new agents. These regulations and standards are published in the Code of Federal Regulations, which is the compendium of rules and regulations issued by the executive departments and agencies of the federal government of the United States. The Code of Federal Regulations is divided into 50 volumes, each one dealing with a particular area of the government. The relevant sections for clinical research are 21, Food and Drugs, and 45, Public Welfare, which contains regulations issued by other agencies of the Department of Health and Human Services.

Research is also very expensive, both in terms of resources and in terms of potential exposure of human subjects to harm. There is a general interest in clinical trials that are widely interpretable so studies do not need to be repeated. This is particularly true in the international venue, where countries may have different statutory requirements for research. The desire to share research results across borders, without concern that quality and process differences may make trials incomparable, led to the establishment of the International Conference on Harmonisation of Technical Requirements for Registration of Pharmaceuticals for Human Use (ICH). The stated purpose of ICH is

> to make recommendations on ways to achieve greater harmonisation in the interpretation and application of technical guidelines and requirements for product registration in order to reduce or obviate the need to duplicate the testing carried out during the research and development of new medicines.[2]

ICH publishes guidelines in four areas: quality, safety, efficacy, and multidisciplinary, identified by the letters "Q," "S," "E," and "M," respectively. The efficacy guidelines contain those relevant to conduct of research on human subjects.

5.1. HIPAA

HIPAA was designed to simplify health care transactions and lower costs by encouraging filing of insurance claims and other documents related to reimbursement electronically using standard formats. Because of fears that sensitive information in electronic form would be susceptible to theft and disclosure, HIPAA also contained the requirement that Congress or the Secretary of Health and Human Services develop rules to safeguard the privacy of "protected health information" and that such rules address "(1) the rights that an individual who is a subject of individually identifiable health information should have; (2) the procedures that should be established for the exercise of such rights; (3) the uses and disclosures of such information that should be authorized or required."[3]

The resulting Privacy Rule went into effect in April 2003. Although it was designed to safeguard information collected from patients in a health care setting, the fact that much of clinical research is conducted in this setting, and that research largely depends on the kind of information safeguarded by the rule, means that the collection, storage, and disclosure of research data are profoundly affected. The rule explicitly applies to entities that transmit health information for a broad range of financial transactions.[4] Although an independent researcher might not be considered a "covered entity," for practical purposes most of the clinical data collected for research are collected within the environment (a hospital, clinic, or doctor's office) of a covered entity and are therefore covered by the rule. In addition, whereas the original HIPAA statute applied to information stored or transmitted electronically, the Privacy Rule applies to all protected health information, regardless of the media on which it is recorded.

The general impact of the Privacy Rule on research is to require much more specific consent from individual subjects for the uses of protected information. The rule defines the required components of the consent and what it can cover and then goes on to discuss the detailed conditions under which the requirement for consent can be modified or waived. Many of these conditions are attempts to balance the public good, including research, against the individual right to privacy, and they are both specific and prescriptive.

The following discussion should be considered an introduction and orientation to HIPAA and the Privacy Rule. These requirements carry the force of law, and their detailed application to a particular research study and research institution must be well understood at the time of study design.

A research authorization must contain the following:

> (i) A description of the information to be used or disclosed that identifies the information in a specific and meaningful fashion.
>
> (ii) The name or other specific identification of the person(s), or class of persons, authorized to make the requested use or disclosure.
>
> (iii) The name or other specific identification of the person(s), or class of persons, to whom the covered entity may make the requested use or disclosure.
>
> (iv) A description of each purpose of the requested use or disclosure. The statement "at the request of the individual" is a sufficient description of the purpose when an individual initiates the authorization and does not, or elects not to, provide a statement of the purpose.
>
> (v) An expiration date or an expiration event that relates to the individual or the purpose of the use or disclosure. The statement "end of the research study," "none," or similar language is sufficient if the authorization is for a use or disclosure of protected health information for research, including for the creation and maintenance of a research database or research repository.
>
> (vi) Signature of the individual and date. If the authorization is signed by a personal representative of the individual, a description of such representative's authority to act for the individual must also be provided.[5]

In short, the authorization must contain what information will be used, for what purpose, by and to whom it will be disclosed, and whether or not the authorization to use the information has an expiration date. Covered entities are, in general, prohibited from requiring consent to release information as a condition of treatment, but there is a specific exemption of this provision that allows an entity to refuse to provide research-related treatment in the absence of authorization to use the resulting data for research purposes.

There are a number of exceptions or modifications of these authorization requirements, some of which are specifically applicable to research. These exceptions include[6]

- A waiver of authorization by an IRB or a privacy board
- Work "preparatory to research"
- Research on deceased subject

The Privacy Rule describes in detail the composition of the privacy board but defers to other cited regulations governing the structure of IRBs. An approved waiver must contain the following assurances:

> (A) The use or disclosure of protected health information involves no more than a minimal risk to the privacy of individuals, based on, at least, the presence of the following elements: (1) an adequate plan to protect the identifiers from improper use and disclosure; (2) an adequate plan to destroy the identifiers at the earliest opportunity consistent with conduct of the research, unless there is a health or research justification for retaining the identifiers or such retention is otherwise required by law; and (3) adequate written assurances that the protected health information will not be reused or disclosed to any other person or entity, except as required by law, for authorized oversight of the research study, or for other research for which the use or disclosure of protected health information would be permitted by this subpart;
>
> (B) The research could not practicably be conducted without the waiver or alteration; and
>
> (C) The research could not practicably be conducted without access to and use of the protected health information.[7]

The researcher must provide evidence of the critical need for access to the data, whether for work in preparation of future research and/or for research on deceased subjects. Additional documentation is required for each exception. Other exceptions are made for certain public health reporting, including adverse event reports[8] to the FDA and reports as part of required postmarketing surveillance.[9]

The HIPAA Privacy Rule erects a new structure of barriers around the use of identified or identifiable personal health information in research. In addition to requiring new authorizations and board reviews, it disallows future use of data under blanket authorizations. Thus, for example, a covered research institution could not create an archive of data for a particular study and then go back and "mine" this information for other than its original stated purpose. The research subject explicitly controls identifiable information, and any future use requires a new authorization or waiver.[10] Fortunately, the Privacy Rule contains a broad exception that can be exploited for much preliminary research and, in the right circumstances, for entire studies. This exception is based on the idea that it is the association of the data with an identifiable individual that is protected, not ownership of the data by the subject. So if this association is broken, the data are no longer considered protected health information and are free from the constraints of the rule. There are two accepted ways to break this association and "de-identify" such information. The first is to remove or obfuscate the data and then show, through a well-documented statistical analysis, that the risk of identification is very small.[11] The other method is to remove all of a set of 18 identifiers (Table 26-1).[12]

Although the researcher may associate a code with such a de-identified record so that the subject's identity can be reestablished, the code cannot be derived from identifying information in such a way that it can be used by itself to reestablish identity. Sharing the rules for re-identifying a subject based on such a code is tantamount to sharing protected health

TABLE 26-1   Data Elements That Must Be Removed for De-identification

Names
Geographic subdivisions smaller than a state
All dates specifiers more specific than year, and all specific ages
  over 89 (to be specified as 90 or older)
Telephone numbers
Fax numbers
E-mail addresses
Social Security numbers
Medical record numbers
Health plan numbers
Account numbers
Certificate/license numbers
Vehicle identification numbers and license plate numbers
Device identifiers and serial numbers
URLs
IP addresses
Biometric identifiers
Images that could identify the individual subject
Any other uniquely identifying number or characteristic

information and is covered by all the provisions of the Privacy Rule.[13]

The explicit de-identification methodology of the Privacy Rule allows the creation of large research repositories free of the legal and regulatory considerations imposed by HIPAA and free of the uncertainties surrounding privacy protections that existed before the establishment of the rule. Ubiquitous de-identification will allow the sharing of data in ways that are likely to be a major boon to population-based research.

5.2. 21 CFR 11: FDA Regulations for Electronic Records

There are many FDA regulations that directly affect data management, but most of them are specific to particular circumstances and not part of general data management guidelines. Such specific regulations are published in the Code of Federal Regulations. They include the requirements for an Investigational New Drug Application (21 CFR 312), Applications for FDA Approval to Market a New Drug (21 CFR 314), and Application for FDA Approval of a Biologic License (21 CFR 601). A researcher who anticipates requiring any of these approvals should be very familiar with the applicable regulations because these will dictate both the content and the form of much of the data that are collected.

The FDA regulations regarding electronic records and electronic signatures apply across all these regulated activities and specify the data management standards for any data that are collected or stored

electronically for submission. Unlike the HIPAA Privacy Rule, these regulations cover only information in electronic form. They are intended to define the steps to be taken to make electronic information equivalent, from a regulatory perspective, to information in paper files authenticated with a handwritten signature.[14] Most of these requirements reflect technical controls and procedures to ensure that a history of the data can be reconstructed, including its original authorship and any subsequent changes. The specific requirements are:

(a) Validation of systems to ensure accuracy, reliability, consistent intended performance, and the ability to discern invalid or altered records.

(b) The ability to generate accurate and complete copies of records in both human readable and electronic form suitable for inspection, review, and copying by the agency. Persons should contact the agency if there are any questions regarding the ability of the agency to perform such review and copying of the electronic records.

(c) Protection of records to enable their accurate and ready retrieval throughout the records retention period.

(d) Limiting system access to authorized individuals.

(e) Use of secure, computer-generated, time-stamped audit trails to independently record the date and time of operator entries and actions that create, modify, or delete electronic records. Record changes shall not obscure previously recorded information. Such audit trail documentation shall be retained for a period at least as long as that required for the subject electronic records and shall be available for agency review and copying.

(f) Use of operational system checks to enforce permitted sequencing of steps and events, as appropriate.

(g) Use of authority checks to ensure that only authorized individuals can use the system, electronically sign a record, access the operation or computer system input or output device, alter a record, or perform the operation at hand.

(h) Use of device (e.g., terminal) checks to determine, as appropriate, the validity of the source of data input or operational instruction.

(i) Determination that persons who develop, maintain, or use electronic record/electronic signature systems have the education, training, and experience to perform their assigned tasks.

(j) The establishment of, and adherence to, written policies that hold individuals accountable and responsible for actions initiated under their electronic signatures, in order to deter record and signature falsification.

(k) Use of appropriate controls over systems documentation including: (1) adequate controls over the distribution of, access to, and use of documentation for system operation and maintenance and (2) revision and change control procedures to maintain an audit trail that documents time-sequenced development and modification of systems documentation.[15]

In addition, in the common circumstance in which technical control of a system is separated from the individuals who are responsible for the contents of the records it contains, data must be further protected through the use of document encryption and digital signatures.[16]

The issue of accurately and legally mimicking the physical signing of a paper document to attest to authorship, approval, etc., on an electronic system is significant enough to warrant a separate set of guidelines.[17] These guidelines include both technical and administrative elements. Technically, an electronic signature must be based on a biometric (a physical characteristic of an individual that is unique, measurable, and persistent over time, such as a fingerprint) or on at least two separate components, such as a name and password. Signatures must be linked to the "signed" electronic document in a way that cannot be removed, falsified, or falsely associated. There must be administrative controls in place to make sure that no signature is assigned to more than one individual, that the association between an individual and an electronic signature is periodically rechecked, that lost or stolen signatures are de-authorized in the affected systems, and that system checks are in place to detect any unauthorized attempts to use signatures. Lastly, the legal equivalence between electronic and handwritten signatures for a system must be certified and submitted on paper to the FDA.

5.3. ICH Guidelines

The most important ICH guideline for general data management principles is E6: Guidelines for Good Clinical Practice.[18] E6 and 21 CFR 11 overlap in several areas because both are largely concerned with the validity of documents used in regulated clinical trials, but E6 enunciates more general principles, most of which apply to both paper and electronic documents. In particular,

> All clinical trial information should be recorded, handled, and stored in a way that allows its accurate reporting, interpretation, and verification.[19]
>
> The confidentiality of records that could identify subjects should be protected, respecting the privacy and confidentiality rules in accordance with the applicable regulatory requirement(s).[20]
>
> Any change or a correction to a CRF (case report form) should be dated, initialed, and explained (if necessary) and should not obscure the original entry; this applies to both written and electronic changes or corrections.[21]

With regard to electronic systems:

> (a) Ensure and document that the electronic data processing system(s) conforms to the sponsor's established requirements for completeness, accuracy, reliability, and consistent intended performance (i.e., validation).
>
> (b) Maintains SOPs (standard operating procedures) for using these systems.

> (c) Ensure that the systems are designed to permit data changes in such a way that the data changes are documented and that there is no deletion of entered data (i.e., maintain an audit trail, data trail, edit trail).
>
> (d) Maintain a security system that prevents unauthorized access to the data.
>
> (e) Maintain a list of the individuals who are authorized to make data changes (see 4.1.5 and 4.9.3).
>
> (f) Maintain adequate backup of the data.
>
> (g) Safeguard the blinding, if any (e.g., maintain the blinding during data entry and processing).[22]

E6 gives the following guidance to individuals or organizations assigned to monitor a clinical trial—guidance that is useful to researchers because it sets the standard of what is expected. Records must be sufficient to verify that the protocol has been followed, that informed consent was obtained, that only eligible subjects were enrolled, that source documents and trial records are accurate and well maintained, and that the data in source documents and case report forms correspond.[23]

E6 also contains a list of documents that are considered essential for the conduct of a trial.[24] These documents are described in detail and categorized by the phase of the trial with which they are associated. Other relevant ICH guidelines are E3: Structure and Content of Clinical Study Reports, and E9: Statistical Principles for Clinical Trials. The former elaborates on E6, whereas the latter gives general statistical guidance, including requirements for validation of computer hardware and software used in statistical analysis.[25]

6. SUMMARY

Every investigator should consider data management before beginning a clinical research project, whether it is an interventional clinical trial or a natural history study. The anticipated use of the data will drive the HIPAA Privacy Rule authorization process or possibly justify a waiver of authorization. Establishing standard operating procedures for data management will simplify future audits and ensure that adequate data are collected for planned statistical analyses. During the conduct of a clinical protocol, the researcher should put him- or herself in the role of someone trying to discredit the study's conclusions and question every procedure that could cast doubt on the accuracy, validity, or relevance of the data collected. Every research conclusion is an argument, and the conclusions will only stand if the data stand. Data management needs to be forward looking. If data is collected in a way that it will never be examined when the original study is closed, it is realizing a fraction of its usefulness. Data should be collected using standard methods and instruments, and the definitions and explanations that define the data and provide context should be stored along with it. Lastly, it should be routine practice to maintain data in a de-identified state so that it can be used for future studies and for hypothesis generation.

This chapter provided an overview of data management principles, but there are several indispensable resources on the World Wide Web that should be familiar to the clinical researcher:

The ICH guidelines available at www.ich.org.
The FDA good clinical practice guidelines available at www.fda.gov/oc/gcp/regulations.html.
Information on the HIPAA statute, including the Privacy Rule, available at www.hhs.gov/ocr/hipaa. Note that each title of the Code of Federal Regulations is updated annually, and that references or excerpts cited on sites other than that of the Government Printing Office may not be current.
Information on HIPAA and research, available at http://privacyruleandresearch.nih.gov.
The Code of Federal Regulations, including regulations pertinent to FDA and other HHS entities, available at www.gpoaccess.gov/cfr/index.html.

References

1. *New Oxford American Dictionary*, 2nd ed, electronic edition. Apple Mac OS 10.4, 2005.
2. International Conference on Harmonisation of Technical Requirements for Registration of Pharmaceuticals for Human Use (ICH). Available at www.ich.org. Accessed January 10, 2006.
3. 104th Congress. Public Law 104-191, Health Insurance Portability and Accountability Act of 1996. August 21, 1996. SEC 264(b).
4. 104th Congress. Public Law 104-191, Health Insurance Portability and Accountability Act of 1996. August 21, 1996. SEC 1173(a)(2).
5. Code of Federal Regulations Title 45, Part 164, Section 508, Paragraph (c), p. 762. October 1, 2005.
6. Code of Federal Regulations Title 45, Part 164, Section 512, Paragraph (i), p. 769. October 1, 2005.
7. Code of Federal Regulations Title 45, Part 164, Section 512, Paragraph (i)(2)(ii), p. 769. October 1, 2005.
8. Code of Federal Regulations Title 45, Part 164, Section 512, Paragraph (b)(iii)(A), p. 765. October 1, 2005.
9. Code of Federal Regulations Title 45, Part 164, Section 512, Paragraph (b)(iii)(D), p. 765. October 1, 2005.
10. *HIPAA and Research Repositories*, NIH Publication No. 04-5489, p. 6. January 2004.
11. Code of Federal Regulations Title 45, Part 164, Section 514, Paragraph (b)(1), p. 773. October 1, 2005.
12. Code of Federal Regulations Title 45, Part 164, Section 514, Paragraph (b)(2), p. 773. October 1, 2005.
13. Code of Federal Regulations Title 45, Part 164, Section 514, Paragraph (c), p. 773. October 1, 2005.
14. Code of Federal Regulations Title 21, Part 11, Section 1, Paragraph (a), p. 109. April 1, 2005.
15. Code of Federal Regulations Title 21, Part 11, Section 10, p. 111. April 1, 2005.
16. Code of Federal Regulations Title 21, Part 11, Section 30, p. 111. April 1, 2005.
17. Code of Federal Regulations Title 21, Part 11, Sections 100–300, p. 112. April 1, 2005.
18. International Conference on Harmonisation of Technical Requirements for Registration of Pharmaceuticals for Human Use. ICH Guideline E6: Good Clinical Practice: Consolidated Guideline, 1997.
19. International Conference on Harmonisation of Technical Requirements for Registration of Pharmaceuticals for Human Use. ICH Guideline E6: Good Clinical Practice: Consolidated Guideline, 1997, p. 9, paragraph 2.10.
20. International Conference on Harmonisation of Technical Requirements for Registration of Pharmaceuticals for Human Use. ICH Guideline E6: Good Clinical Practice: Consolidated Guideline 1997, p. 9, paragraph 2.11.
21. International Conference on Harmonisation of Technical Requirements for Registration of Pharmaceuticals for Human Use. ICH Guideline E6: Good Clinical Practice: Consolidated Guideline, 1997, p. 18, paragraph 4.9.2.
22. International Conference on Harmonisation of Technical Requirements for Registration of Pharmaceuticals for Human Use. ICH Guideline E6: Good Clinical Practice: Consolidated Guideline, 1997, p. 21, paragraph 5.5.3.
23. International Conference on Harmonisation of Technical Requirements for Registration of Pharmaceuticals for Human Use. ICH Guideline E6: Good Clinical Practice: Consolidated Guideline, 1997, p. 26, paragraph 5.18.
24. International Conference on Harmonisation of Technical Requirements for Registration of Pharmaceuticals for Human Use. ICH Guideline E6: Good Clinical Practice: Consolidated Guideline, 1997, p. 41, section 8.
25. International Conference on Harmonisation of Technical Requirements for Registration of Pharmaceuticals for Human Use. ICH Guideline E9: Statistical Principles for Clinical Trials, 1998, p. 27, paragraph 5.8.

27

Getting the Funding You Need to Support Your Research: Navigating the National Institutes of Health Peer Review Process

OLIVIA T. BARTLETT* AND ELLIOT POSTOW**

*Research Programs Review Branch, National Cancer Institute, National Institutes of Health, Bethesda, Maryland
**Division of Biologic Basis of Disease, Center for Scientific Review, National Institutes of Health, Bethesda, Maryland

Once an investigator has selected a general research topic, determined what work has already been done by others, identified promising new research areas, planned a specific new project, and estimated the budget required for it, the investigator will need to obtain funding to perform the study. New developments in biomedical technology have created more opportunities than ever for clinical researchers and raised interest in new diagnostic and therapeutic approaches to many diseases.

Although a large number of public and private organizations support laboratory and clinical biomedical research and clinical trials, most of the biomedical research conducted at research centers and academic institutions in the United States is supported by the National Institutes of Health (NIH).

Recent studies at NIH show that the overall success rate of applications submitted by physicians is slightly higher than the success rate for applications submitted by basic scientists, but that the overall success rates for applications proposing clinical research and clinical trials are lower than the success rate for applications proposing basic or laboratory-based research. The lower success rates were not due to the review panel assignment, the composition of the review committee, the cost of the proposed research, or whether clinical applications were reviewed in the same review group as basic research applications.

Although NIH is continuing to study why the success rate for applications proposing clinical research is lower, it is likely that the applications are just not as well prepared. Translational and clinical research can be broader in scope than laboratory research and more difficult to plan, describe, and carry out. In addition, most clinician scientists receive little training in grant writing.

Rejected grant applications can challenge the ego, as well as result in loss of research opportunities. This delay is especially critical if the planned research project is linked to ongoing clinical trials.

A thorough understanding of the peer review process that NIH uses to inform funding decisions will help both new and established clinical investigators be more competitive in applying for research funds. The purpose of this chapter, therefore, is to provide (1) an overview of the mission and organization of NIH and the NIH peer review process, (2) suggestions for writing more competitive grant applications, (3) brief descriptions of some of the many different types of NIH grant programs that are available for clinical researchers at various stages of their careers, (4) an overview of some new directions at NIH, and (5) suggestions about how to find current information about the NIH grants process.

1. OVERVIEW OF NIH

1.1. Mission and Organization of NIH

NIH is a federal agency that is part of the Department of Health and Human Services (DHHS) (Fig. 27-1). The mission of NIH is to improve people's health by increasing understanding of the processes underlying human health and by acquiring new knowledge to help prevent, detect, diagnose, and treat disease. NIH accomplishes this mission by:

1. Supporting research in universities, medical schools, hospitals, small businesses, and research institutions in the United States and abroad

2. Conducting research in its own laboratories and clinics

3. Supporting training for promising young researchers

4. Helping to develop and maintain research resources

5. Identifying research findings that can be applied to the care of patients and helping to transfer such advances to the health care system

6. Promoting effective ways to communicate biomedical information to scientists, health practitioners, and the public

7. Developing and recommending policies related to the conduct and support of biomedical research

NIH consists of 19 research institutes, the National Center for Complementary and Alternative Medicine, the National Center for Research Resources, the National Center for Minority Health Disparities, the National Library of Medicine, the Fogarty International Center, the Center for Scientific Review (CSR), the Center for Information Technology, and the Clinical Center (Fig. 27-2). Most of the 19 research institutes have both intramural programs, with laboratories and clinics operated by NIH employees, and extramural programs, through which research is supported in institutions worldwide. Although most of NIH is located in or near Bethesda, Maryland, the National Institute of Environmental Health Sciences is located in Research Triangle Park, North Carolina; the intramural program of the National Institute on Aging is located in Baltimore, Maryland; and some research components of other NIH institutes are located in other areas of the United States.

It is important to note that although each NIH institute has specific scientific areas of primary interest, there are many areas of interest that are shared between institutes. For example, asthma is a shared interest of the National Institute of Allergy and Infectious Diseases (NIAID) and the National Heart, Lung, and Blood Institute (NHLBI); NIAID and the National Institute of Arthritis and Musculo-Skeletal Disorders share interest in autoimmune diseases; and the National

U. S. Department of Health and Human Services

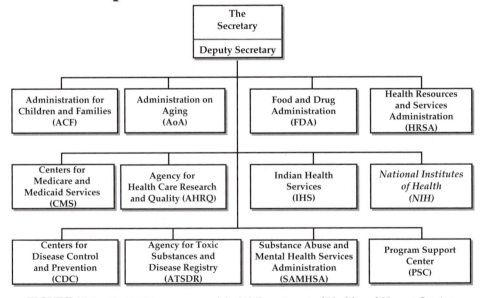

FIGURE 27-1 The NIH is an agency of the US Department of Health and Human Services.

National Institutes of Health

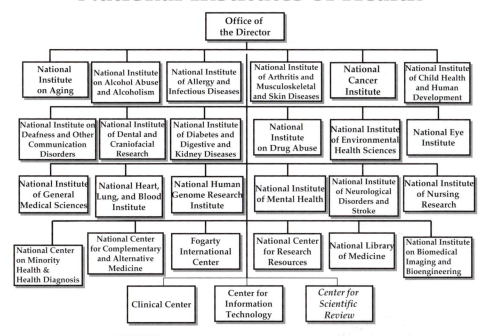

FIGURE 27-2 Structure and organization of the NIH.

Cancer Institute (NCI) and NIAID are both interested in transplantation biology and the life cycle of oncogenic viruses. Prospective applicants should discuss potential research projects with program staff in all relevant NIH institutes/centers before submitting a grant application.

The total budget of the NIH was approximately $28.5 billion in fiscal year 2005, the last fiscal year for which complete data are available (Fig. 27-3A). Approximately 85% of the NIH budget is for support of extramural research through a series of different types of grants and contracts. In a typical year, approximately 85% of that extramural budget is used to support various types of research grants, approximately 4% is for grants for training and fellowships, and approximately 10% is used to support contracts for research and research support (Fig. 27-3B).

An institute's extramural program is generally organized into a number of specific scientific areas, each of which may provide research funding through grants, contracts, and/or cooperative agreements.

1.2. Responsibilities of NIH Staff

An NIH scientist, the program official or program director, directs scientific management of the extramural research program in each of NIH's scientific areas. Grants management and contracts management staff provide financial stewardship and oversight of an institute's extramural research programs. Each institute also has a review office that manages the scientific peer review of contract proposals and highly mission-oriented grant applications, as explained in more detail later. As shown in Table 27-1, the review, program, and grants and contracts management staff of the NIH each have important, but separate, responsibilities.

1.3. Types of NIH Extramural Research Support

1.3.1. Grants, Contracts, and Cooperative Agreements

There are three different mechanisms though which NIH supports extramural research and development: grants, contracts, and cooperative agreements. The relationship between NIH and the awardees in each of these general funding mechanisms is summarized in Table 27-2. Links to specific funding opportunities and guidelines for some of the most commonly used types of NIH research, training, career development, small business, and specialized grant programs can be found at http://grants1.nih.gov/grants/oer.htm.

In general, an investigator who applies for a grant through an applicant institution is responsible for developing the ideas, concepts, methods, and

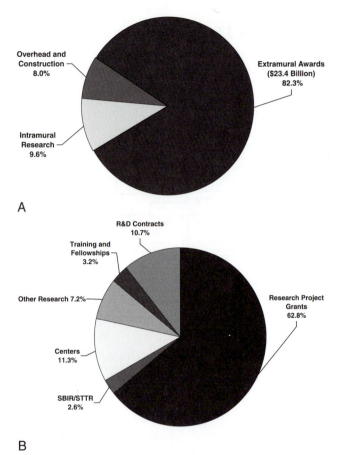

A

B

FIGURE 27-3 **A,** Fiscal Year 2005 NIH Budget; **B,** Fiscal Year 2005 NIH Budget for Extramural Research.

TABLE 27-1 Roles and Responsibilities of NIH Extramural Staff

NIH Staff	Role and Responsibilities
Scientific review administrator (SRA)	In Center for Scientific Review and in each NIH institute/center scientific review office Organizes, manages, conducts, and reports scientific peer review of grant applications and/or contract proposals Liaison between applicants and reviewers
Program officer/ director	In NIH institutes and centers Manages a portfolio of awarded grants/contracts Monitors scientific progress made on grants/contracts
Grants/contracts management officer	In NIH institutes and centers Fiscal stewardship of portfolio of awarded grants/contracts Negotiates fiscal aspects of awards Monitors financial progress made on grants/contracts

TABLE 27-2 NIH Extramural Award Mechanisms

Mechanism	NIH Role	NIH Provides
Grant	Patron	Assistance, encouragement
Cooperative agreement	Partner	Assistance, with substantial program staff involvement
Contract	Purchaser	Direction

approaches for a project. In contrast, a contract is the mechanism by which NIH purchases a service or product, and the awarding NIH institute is responsible for establishing the plans, parameters, and detailed requirements for projects supported by contracts. Although most grant applications are "investigator-initiated" rather than "solicited," contract proposals are almost always solicited through requests for proposals. Other differences between contracts and grants involve variations in the submission and review processes, review criteria, mechanism for reimbursement of costs and administration, extent of involvement of the funding institute, and delivery of the end product. Because most NIH support of extramural research is via grants, we do not address contracts in detail. For more information about NIH support of research and development contracts, see http://ocm.od.nih.gov/contracts/contract.htm.

Cooperative agreements are similar to grants in that they are awarded by NIH to assist and support research or related activities. They differ from traditional grants, however, in that they include a substantial program-

matic (i.e., scientific and/or technical) role by NIH staff. This may involve cooperation and/or coordination to assist the awardee in carrying out the project or may require review and approval by NIH staff of certain processes/phases in the project. Policies and procedures for application, review, and administration of cooperative agreements are similar to those for grants. An important difference, however, is that cooperative agreement applications are usually solicited via a specific Request for Applications (RFA), which describes the program, functions, or activities that will be supported by the agreement as well as the nature of NIH staff involvement.

1.3.2. Funding Opportunity Announcements in Grants.gov

From December 2005 through the fall of 2007, NIH is transitioning from submission of paper applications on the paper PHS 398 application form to electronic submission of applications on the Standard Form 424 Research and Related (R&R) application form through the federal Web portal Grants.gov.

(See http://era.nih.gov/ElectronicReceipt/files/ Electronic_receipt_timeline_Ext.pdf for the transition schedule.) The Grants.gov portal allows potential applicants to search all federal grant programs through the "Find" option and to apply for grants through the "Apply" option.

As each NIH grant mechanism is transitioned to Grants.gov, all applications for that type of grant—including those that traditionally were considered "investigator initiated"—must be submitted in response to an NIH Funding Opportunity Announcement (FOA). The NIH will post a general (or "parent") FOA for its most common grant mechanisms in Grants.gov. Specific FOAs will be published for each RFA, Program Announcement, and NIH institutes/center-specific grant mechanisms. Each active FOA (except Small Business Innovation Research [SBIR]/Small Business Technology Transfer [STTR] FOAs) will allow submission of only one type of grant mechanism.

Each FOA will have an opening date and a submission deadline. Many FOAs posted by NIH will include multiple submission/receipt dates and will be active for several years. Therefore, the FOA will show an open/close period that spans the entire active period of the announcement. Applicants should read the entire FOA carefully to determine the specific submission/receipt deadlines. Applications submitted after a posted submission deadline will be returned to the applicant for submission during the next submission period.

This change affects both how NIH does business and how applicants do business with NIH. First, all applicant organizations and institutions must register with Grants.gov. In addition, applicant institutions must register their project directors/principal investigators (PD/PIs) in the NIH electronic research administration (eRA) website called the NIH eRA Commons. The PD/PI must hold a PI account in the NIH eRA Commons and be affiliated with the applicant organization. The registration process can take several weeks, especially close to submission dates when registration volume peaks. Start early to avoid delays. NIH recommends starting 4 weeks ahead of your target submission date.

The electronic grant submission process through Grants.gov involves several steps that must be completed in sequence:

1. Search for and identify a FOA on Grants.gov.
2. Download the specific grant application package for that FOA.
3. Complete the application package. (Be sure to save a local copy.)

4. The applicant organization submits the application package through Grants.gov.
5. NIH eRA software retrieves the application from Grants.gov and checks the application against NIH's business rules.
6. NIH notifies both the PI and signing official (SO) for the applicant institution by e-mail to check the NIH eRA Commons for results of the NIH validations check.
7. The eRA Commons will show if the grant application passed or failed the NIH validations check.

> If it passed, there wll be a grant image in the eRA Commons. The PI and the SO have two weekdays in which they may review the application image via the eRA Commons before the application is processed for referral and review. If the application is not accurate, the PI and SO may reject the application in the eRA Commons during those two weekdays. If this happens, the PI and the SO must make the necessary changes, and must submit the entire changed/corrected application again via Grants.gov.

> If the application failed NIH validation checks, a list of the errors and warnings will appear in the eRA Commons. The PI and SO must fix the errors and submit the entire corrected application again through Grants.gov.

8. eRA Commons saves the data and grant image, and NIH begins processing the application.

It is important to note the following about the transition to the SF424 (R&R) forms:

- As soon as a grant mechanism is transitioned to the electronic mode and the grant opportunity is posted on Grants.gov, applicants will be able to download and begin working on the application package. However, you cannot submit the application until the funding opportunity's open date. For example, if an FOA has an open date of November 7 for a December 1 submission deadline, you can submit an application electronically to Grants.gov any time between November 7 and December 1.
- Until a grant mechanism is transitioned, all applications submitted for that grant mechanism should be submitted on paper PHS 398 forms. NIH systems will not be ready to receive those applications electronically until the transition date.
- Once a grant mechanism is transitioned, paper applications and applications on the old PHS 398 form will no longer be accepted for that grant mechanism. Applications for the transition submission date and thereafter must utilize the

SF424 (R&R) form and be submitted electronically through Grants.gov.

- Read all instructions carefully. Each NIH grant application package in Grants.gov will include both SF424 (R&R) components and NIH-specific PHS 398 components. Each application package will include an application guide. This document is critical to submitting a complete and accurate application to NIH. For example, some fields within the SF424 (R&R) components that are not mandatory on the federalwide form are required by NIH. NIH-specific instructions for such fields are clearly identified in the application guide by the NIH logo.
- Early submission is encouraged. Allow time for correction of errors and resubmission through Grants.gov if necessary.

There are a number of resources on the transition to the SF424 (R&R) forms on the NIH electronic receipt website (http://era.nih.gov/ElectronicReceipt), including

SF424 (R&R) application guides, sample application packages, and related resources
How to Complete an Application Package on Grants.gov
eRA Commons registration training
Videotaped presentation, "A Walk through the SF424 (R&R)"
End-to-end demo facility for applicants to "practice" the entire process from finding an opportunity in Grants.gov to viewing a submitted application in the eRA Commons.

1.3.3. Requests for Applications and Program Announcements in the NIH Guide

The NIH may invite submission of grant applications to address areas of special interest to an awarding institute by issuing an RFA or program announcements (PAs). RFAs and PAs are published in the *NIH Guide for Grants and Contracts*, which can be found at http://grants.nih.gov/grants/guide/index.html and are also listed as FOAs in Grants.gov. Each RFA and PA announced in the NIH guide will have a link to the corresponding FOA in Grants.gov for the convenience of potential applicants.

Table 27-3 summarizes the key features of RFAs and PAs. Issuance of an RFA FOA generally indicates that the issuing institute has set aside funds specifically to support research, training, or infrastructure on a given topic. RFA FOAs also list NIH staff contacts, and it is a good idea to discuss your potential project with them to ensure that you meet all responsiveness and eligibility criteria. Scientific peer review of applications

TABLE 27-3   Key Features of NIH Program Announcements and Requests for Applications

Program announcement (PA)	Addresses a relatively broad field/category of research
	Usually no set-aside I/C budget
	Usually submit on regular receipt dates
	Regular review criteria for type of applications requested
	Frequently more than one NIH institute involved
	Applications may be reviewed by CSR or the issuing NIH institute
Request for applications (RFA)	Addresses a well-defined area of research
	Set-aside I/C budget for RFA applications
	Submit on special, one-time-only receipt dates
	Often special eligibility and/or review criteria
	Often special application format and/or submission instructions
	Usually only one NIH institute involved
	Applications usually reviewed by the issuing NIH institute

responding to an RFA is usually managed by the review office in the issuing institute or center. In contrast, a PA usually indicates an NIH institute or center's interest in supporting research in a broad area without a specific set-aside budget. Applications responding to PAs are usually reviewed along with other "investigator-initiated" applications on similar topics through the usual channels in study sections organized by the Center for Scientific Review (see later).

Both RFA and PA FOAs may have special eligibility requirements, application preparation procedures, receipt dates, and/or conditions of award, so it is important to read the announcement carefully before preparing an application.

1.3.4. Multiple PIs

Beginning with research grant applications submitted for February 2007 receipt dates, NIH will allow applicants and their institutions to identify more than one Principal Investigator (PI). NIH expects the availability of the Multiple PI option to encourage interdisciplinary and other team approaches to biomedical research. NIH will extend the Multiple PI option to most research grant applications as they transition to electronic submission through Grants.gov using the SF424 R&R application package. Some paper applications submitted on PHS 398 application forms also will allow inclusion of more than one PI, but only when the multiple PI option is clearly specified in the soliciting Request for Applications (RFA) or Program Announcement (PA). Some types of applications including individual career awards, individual fellowships, Dissertation Grants, Director's Pioneer Awards, and Shared Instrumentation Grants will not accommodate more than a single PI. The restriction to a single PI will be described in FOAs for those programs. The decision to apply for a single PI or a multiple PI grant is the responsibility of the investigators and the applicant organization. Those decisions should be consistent with and justified by the scientific goals of the project. Applications that involve more than one PI must include a Leadership Plan that describes the roles, the responsibilities, and the working relationship of the identified PIs (see later). More information about the multiple PI option is available on the Multiple Principal Investigator website at http://grants.nih.gov/grants/multi_pi/index.htm.

2. THE NIH PEER REVIEW PROCESS FOR GRANTS

Because of the magnitude, diversity, and complexity of its research mission and its pursuit of excellence, NIH draws on the national pool of scientists actively engaged in research for assistance in evaluating the tens of thousands of grant applications that it receives annually. These scientists, who are scientific "peers" of the applicants, advise NIH which applications are the most meritorious and the most promising in each field.

2.1. The NIH Dual-Review System

The cornerstone of NIH peer review of grant applications is the "dual-review system," which consists of two sequential levels of review with different functions (Fig. 27-4). The first level of review is carried out by panels of experts organized according to scientific discipline or research area for the primary purpose of

First Level of Review:

Scientific Review Group

Provides initial scientific merit review of grant applications

Rates applications and recommends appropriate level of support and duration of award

Second Level of Review:

Advisory Council/Board

Assesses quality of SRG review of grant applications

Makes recommendations to Institute staff on funding

Evaluates Program priorities and relevance

Advises on Institute policy

FIGURE 27-4 The NIH dual review system.

evaluating the scientific and/or technical merits of grant applications. These panels are referred to as scientific review groups (SRGs) and are also commonly called "study sections" or "review committees." Each SRG is managed by a scientific review administrator (SRA), who is a health scientist administrator with experience and expertise in the areas of science addressed by the SRG. A cluster of SRGs chartered as a single entity under the Federal Advisory Committee Act and responsible for the review of grant applications in scientifically related areas is called an integrated review group.

The second level of review is performed by an institute's or center's national advisory board or council, hereafter referred to as a council, which is composed of both scientific and public representatives who are noted for their expertise, interest, or activity in matters related to the mission of the institute for which it serves. Council recommendations are based not only on considerations of scientific merit, as judged by the SRGs, but also on the relevance of the proposed study to an institute's mission, programs, and priorities.

The NIH dual-review system therefore separates the scientific assessment of proposed projects from policy decisions about scientific areas to be supported and the level of resources to be allocated.

2.2. NIH Review "Cycles"

The NIH Center for Scientific Review processes all grant applications submitted to the PHS for more than 150 award mechanisms. During fiscal year 2005, the CSR processed approximately 80,000 applications. To handle this load, each type of grant application has a designated receipt date(s) that will be indicated in the FOA in Grants.gov. There are currently three receipt dates per year for most types of grant applications. In order to spread workload in the research administration offices at applicant institutions, which must now submit all applications through Grants.gov and eRA systems; NIH announced new application receipt dates effective in January 2007. The grant application receipt dates are listed on the NIH website at http://grants.nih.gov/grants/funding/submission-schedule.htm. Note that RFAs and some other specific FOAs may have special receipt dates. Table 27-4 shows the three overlapping review cycles for grant applications that result from these receipt dates. The review cycle for a grant application begins when an investigator submits an application to NIH, generally through a research organization, and concludes when the applicant organization and the principal investigator are notified of the recommendation of the council (Fig. 27-5).

TABLE 27-4   NIH Grant Application Receipt, Review, and Award Cycles

Application Receipt Dates	Scientific Review Group Meetings	National Advisory Council Meetings	Earliest Possible Award Date
January 25–May 1	June–July	September–October	December 1
May 25–September 1	October–November	January–February	April 1
September 25–January 2	February–March	May–June	July 1

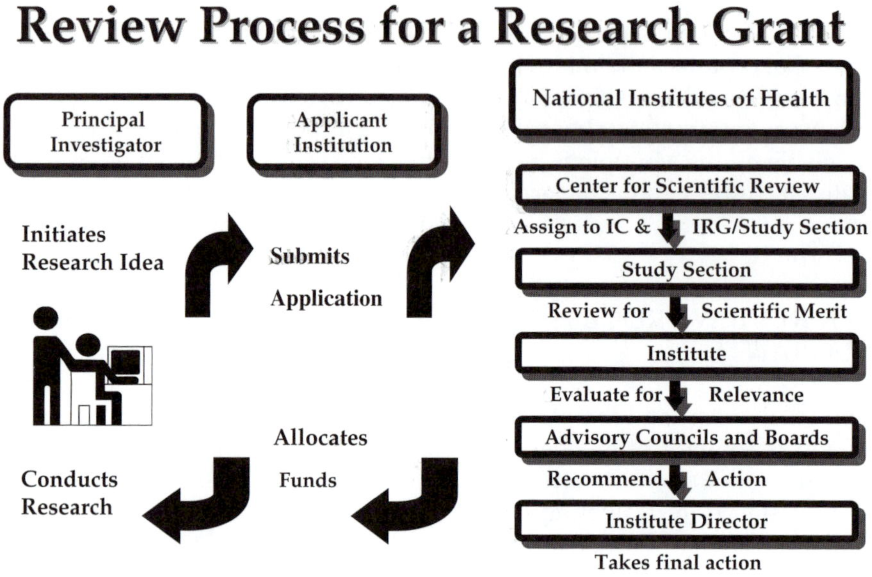

Review Process for a Research Grant

FIGURE 27-5   Flow of a typical grant application from applicant through the NIH peer review and funding process.

2.3. Assignment of Applications to a Review Group and a Possible Funding Institute

All grant applications submitted to NIH are processed centrally by the CSR Division of Receipt and Referral, where it is determined if the application is appropriate for NIH and where it should be reviewed. Based on the type of application and the written guidelines describing the mission areas of the various NIH components, an application is assigned to a potential awarding institute/center (IC) and to a CSR study section or an IC review committee for scientific and technical merit review.

The rosters and scientific areas of the CSR study sections and links to other review committees are on the NIH website at http://era.nih.gov/roster/index.cfm. The project director/principal investigator of an application is encouraged to provide suggestions about appropriate review groups and/or scientific expertise areas needed to evaluate the application in the cover letter attachment that accompanies the application. If specialized expertise is required to review an application, additional temporary members may be invited by the SRA to serve as reviewers. If the research objectives and approaches of an application or group of applications cannot be reviewed appropriately by an existing review group, a special emphasis panel may be constituted for the review. Applicants are notified via the NIH Commons (https://commons.era.nih.gov/commons) of the review group and the SRA within 6–8 weeks after receipt of the application.

Assignment to a funding component is based on the institute's congressionally mandated program responsibilities. If the subject matter of an application is pertinent to the program responsibilities of two or more institutes, a dual, or multiple, assignment may be made. The CSR has no responsibility for either decisions about funding or the management of grant programs.

2.4. How Are Members of Scientific Review Groups Selected?

The primary requirement for serving on an SRG is demonstrated competence and achievement as an independent investigator in a scientific, technical, or clinical discipline or a biomedical or biobehavioral research specialty, based on the quality of research accomplished, publications in refereed professional journals, and other significant scientific or clinical research activities, achievements, and honors. Service also requires mature judgment, balanced perspective, objectivity, ability to work effectively in a group context, commitment to review assignments, and personal integrity to ensure the confidentiality of applications and discussions and the avoidance of real or potential conflicts of interest. NIH also considers such factors as geographic distribution, institutional affiliation, and adequate representation of ethnic minority and female scientists in the selection of SRG members.

Candidates for regular membership on an SRG are nominated by the SRA. The director of NIH makes final appointments to SRGs and advisory committees. The secretary of DHHS makes appointments to councils, except for the National Cancer Advisory Board and the President's Cancer Panel, whose members are appointed by the president of the United States. Appointments are usually made for 4 years, with staggered terms so that approximately one-fourth of the membership of any given SRG changes each year. Several NIH institutes also include lay members (patient representatives or advocates) on SRGs reviewing clinical research applications, especially those involving clinical trials. Patient advocates bring expertise in the impact of the disease on patients and their caregivers, an understanding of strategies and approaches likely to succeed in patient recruitment and follow-up, and important quality of life issues. Patient advocates are full voting members of the review panels.

2.5. How Does the Review Proceed?

SRGs ("study sections") normally meet three times a year for 1–3 days each time, depending on the number and types of grant applications to be reviewed. Typically, a CSR study section is responsible for the review of 60–100 research project grant applications at each meeting.

2.5.1. Reviewers Receive Applications and Instructions for Preparing Their Reviews before the Meeting

The SRA assigns each application to two or more members of the SRG, who provide detailed written reviews and present their evaluations orally at the SRG meeting. Each member may be asked to prepare detailed written critiques for 5–10 applications. In addition, each member is assigned as a discussant

(reader) on additional applications. Four to 6 weeks before the SRG meeting, the SRA sends SRG members CDs with images of the applications to be reviewed at the meeting, appendices and other supporting materials, and instructions for preparing the reviews, or makes these items available to the reviewers on a secure NIH website.

2.5.2. The SRA Is Your Liaison during the Review Process

If reviewers need additional information from the applicant, they will ask the SRA to obtain the required materials. Reviewers may not contact an applicant directly, and applicants may not contact the reviewers. The SRA is the official representative of NIH and must handle all communications with applicants during the review process. For some applications, an SRA may also seek opinions from additional experts in the field via mail or may invite the expert to be a temporary member of the SRG and participate in the meeting either in person or via telephone.

2.5.3. Who Runs the Study Section Meeting?

The SRA, who, under rules of the Federal Advisory Committee Act, is the designated federal official in charge of the meeting, and the chairperson, who is one of the members of the SRG, conduct the meeting. During the review portion of the meeting, the chairperson calls on the assigned reviewers and discussants to present their critiques. After these presentations, the chairperson moderates a discussion among all SRG members. Other members of the SRG often question the assigned reviewers about the application or their evaluation of it. Members score each application on the basis of their own assessment of its merit. Members whose assessment of an application is significantly different from that of the majority must explain their views.

2.6. Review Criteria for Research Project Grant Applications

The review criteria for research project grant applications are summarized in Table 27-5. The goals of NIH-supported research are to advance our understanding of biological systems, improve the control of disease, and enhance health. In their written critiques, reviewers are asked to comment on each of the following criteria in order to judge the likelihood that the proposed research will have a substantial impact on the pursuit of these goals. They were updated in 2005 to address interdisciplinary, translational, and clinical projects more effectively (see http://grants.nih.gov/

**TABLE 27-5 NIH Review Criteria for Research Project
Grant Applications**

Review criteria:
  Significance
  Approach
  Innovation
  Investigator
  Environment
Types of applications:
  Unsolicited R01s, R03s, R21s, P01s
  R01s, R03s, R21s, P01s for PAs
  Most R01s, R03s, R21s, P01s for RFAs

grants/guide/noticefiles/not97010.html) and in late
2006 to address applications with multiple principal
investigators (see later and http://grants.nih.gov/
grants/guide/notice-files/NOT-OD-07-017.html).

Significance: Does this study address an important
  problem? If the aims of the application are
  achieved, how will scientific knowledge or clinical
  practice be advanced? What will be the effect of
  these studies on the concepts, methods,
  technologies, treatments, services, or preventative
  interventions that drive this field?
Approach: Are the conceptual or clinical framework,
  design, methods, and analyses adequately
  developed, well integrated, well reasoned, and
  appropriate to the aims of the project? Does the
  applicant acknowledge potential problem areas
  and consider alternative tactics? For applications
  designating multiple PD/PIs, is the leadership
  approach, including the designated roles and
  responsibilities, governance and organizational
  structure consistent with and justified by the
  aims of the project and the expertise of each of the
  PD/PIs?
Innovation: Is the project original and innovative? For
  example, does the project challenge existing
  paradigms or clinical practice or address an
  innovative hypothesis or critical barrier to progress
  in the field? Does the project develop or employ
  novel concepts, approaches, methodologies, tools,
  or technologies for this area?
Investigators: Are the PD/PI(s) and other key
  personnel appropriately trained and well suited to
  carry out this work? Is the work proposed
  appropriate to the experience level of the PD/PI(s)
  and other researchers? Do the PD/PI(s) and the
  investigative team bring complementary and
  integrated expertise to the project (if applicable)?
Environment: Do(es) the scientific environment(s) in
  which the work will be done contribute to the
  probability of success? Do the proposed studies

benefit from unique features of the scientific
environment(s), or subject populations, or employ
useful collaborative arrangements? Is there
evidence of institutional support?

RFA FOAs may list additional elements relating to
the specific requirement(s) of the RFA under each of
these criteria and/or may have additional review
criteria.

Each of the review criteria is considered in assigning
the overall score, with reviewers weighing them as
appropriate for each application. Note that an application does not need to be strong in all categories to be
judged likely to have major scientific impact and thus
deserve a high priority score. For example, an investigator may propose to carry out important work that
by its nature is not innovative but is essential to move
a field forward.

Reviewers also provide an overall evaluation—a
one-paragraph summary of the most important
strengths and weaknesses of the application in terms
of the five review criteria. The reviewers recommend
a priority score reflecting the overall impact of the
project on the field, weighing the review criteria as
appropriate for each application.

2.6.1. Additional Review Criteria

In addition to these criteria, the following items will
also be considered in the determination of scientific
merit and the priority score for research project grant
applications:

Protection of human subjects from research risk: The
  involvement of human subjects and protections
  from research risk relating to their participation in
  the proposed research is assessed. This includes
  plans for monitoring data and safety of all
  proposed clinical trials.
Inclusion of women, minorities, and children in research:
  The adequacy of plans to include subjects of both
  genders, from all racial and ethnic groups (and
  subgroups), and children as appropriate for the
  scientific goals of the research is assessed. Plans for
  the recruitment and retention of subjects is also
  evaluated.
Care and use of vertebrate animals in research: If
  vertebrate animals are to be used in the project,
  reviewers assess the plans for use and care of the
  animals.

2.6.2. Additional Review Considerations

The reasonableness of the proposed budget and the
requested period of support in relation to the proposed

research are considered. Reviewers are instructed that the priority score should not be affected by the evaluation of the budget.

In addition, for renewal (formerly called competing continuation) applications, reviewers evaluate progress during the past project period.

For resubmission (formerly called amended or revised) applications, reviewers also address progress since the previous submission and changes made in the research plan in response to the critiques in the summary statement from the previous review. These would indicate whether the application is improved or the same as or worse than the previous submission.

2.7. Hazardous Research Materials and Methods

The principal investigator and the applicant institution are responsible for protecting research personnel, the public, and the environment from hazardous conditions. If reviewers identify special hazards, concerns about the adequacy of safety procedures are included in a special biohazard note in the summary statement. An award may not be made until all concerns about hazardous conditions have been resolved to the satisfaction of NIH. Serious shortcomings in the plans to protect against biohazards may be addressed under the "approach" review criterion.

2.8. Sharing Research Data

NIH has had a long-standing policy to promote the sharing of research data and to make available to the public the results of the projects that it funds. All applications with direct costs greater than $500,000 in any single year are expected to address data sharing. Applicants are encouraged to discuss their data sharing plan with their program contact at the time they negotiate an agreement with the IC staff to accept assignment of their application as described at http://grants.nih.gov/grants/guide/noticefiles/NOTOD02004.html. In some cases, an FOA may request data sharing plans for applications that request less than $500,000 in direct costs in all years. Reviewers will evaluate the appropriateness of the proposed data sharing plan, but this is not factored into the determination of scientific merit or priority score. Program staff is responsible for overseeing the data sharing policy and for assessing the appropriateness and adequacy of the plan in each awarded application. For more information on data sharing, see http://grants.nih.gov/grants/policy/data_sharing.

2.9. NIH Model Organism Sharing Policy

NIH encourages the sharing of research resources developed with NIH funding to make them available for further research, development, and application. At the same time, NIH recognizes the rights of grantees and contractors to elect and retain title to subject inventions developed with federal funding pursuant to the Bayh–Dole Act. Model organisms include both mammalian models, such as the mouse and rat, and non-mammalian models, such as budding yeast, social amoebae, round worm, fruit fly, zebra fish, and frog. (See NIH Model Organism for Biomedical Research, located at www.nih.gov/science/models, for information about NIH activities related to these resources.)

All NIH applications are expected to include a plan for sharing and distributing unique model organisms generated using NIH funding or provide appropriate reasons why such sharing is restricted or not possible. Investigators may request funds to defray costs associated with this sharing. Investigators are encouraged to confer with their technology transfer office and/or office of sponsored programs for guidance.

Reviewers will assess the adequacy of plans for sharing model organisms, and the assessment will be included in an administrative note. The assessment will generally not influence the overall priority score. However, for some special initiatives and grant programs specifically directed toward the development of model organisms, reviewers may be asked to integrate their evaluation of the plan for sharing with other review criteria and factor their assessment into the overall evaluation of scientific merit.

2.10. Review of Research Project Grant Applications from New Investigators

New investigators are encouraged to submit research project grant (R01) applications and to identify themselves by checking the "new investigator" box on the PHS 398 Cover Page Supplement to the SF424 form so that reviewers can evaluate these applications in a manner appropriate for the experience of the new investigator.

When reviewing applications from new investigators, reviewers consider the experience of and the resources available to the applicant. The five review criteria are applied in a manner appropriate to the expectations for, and problems likely to be faced by, a new investigator. Specifically,

Approach: More emphasis is placed on demonstrating that the techniques and approaches proposed are feasible than on the presentation of preliminary results.

Investigator: More emphasis is placed on the applicant's training and research potential than on his or her track record and number of publications.

Environment: There should be some evidence of institutional commitment in terms of space and time to perform the research.

2.11. Research Involving Human Subjects

Almost by definition, a clinical research project will involve human subjects, either living persons with whom you will interact directly or specimens from them that can be identified. Safeguarding the rights and welfare of human subjects involved in research activities supported by DHHS is primarily the responsibility of the institution that is accountable to DHHS for the funds awarded to support the research activity. DHHS will fund research involving human subjects only if the applicant institution has filed an appropriate assurance with the DHHS Office of Human Research Protections (OHRP) and has certified that the research has been approved and is subject to continuing review by an institutional review board (IRB).

In 2004, OHRP issued new guidance on research involving coded specimens. The NIH definition of human subjects research includes all research involving the use of human organs, tissues, and body fluids from living individuals, as well as graphic, written, or recorded information derived from living individuals. The exception to this definition is that human subjects are not considered to be involved if (1) the research uses only coded private information or specimens and (2) this information meets the conditions that (a) the private information or specimens are not collected specifically for the proposed research and (b) the investigator(s) cannot identify the individual(s) providing the coded private information or specimens because either the key to decipher the code has been destroyed or a formal agreement exists prohibiting release of that key to the investigators during the lifetime of the subjects.

Before preparing an application, potential applicants should see the instructions in the SF424 (R&R) Application Guide (http://grants2.nih.gov/grants/funding/424) and the OHRP website (www.hhs.gov/ohrp) for guidance on and decision charts for research involving human subjects.

NIH does not require IRB approval at the time of application submission, but it must be submitted to the awarding NIH component before an award can be made. NIH also requires education on the protection of human research participants for all investigators involved in research involving human subjects.

The application must include a plan for monitoring data and safety for all clinical trials. Large-scale (phase III) trials must have a data and safety monitoring board. NIH expects investigators to share these plans with IRBs.

2.11.1. Peer Review of Plans for Use of Human Subjects

Applications that do not adequately address research on human subjects may be returned without review, or the review or award may be delayed. SRG members evaluate the proposed use of human subjects, the risks to the subjects, the plans to protect them from risks, and whether the risks are reasonable in relation to the anticipated benefits to the subjects. They also evaluate the importance of the knowledge that may be expected to result from the research.

Views expressed by the SRG regarding the adequacy of protection of human subjects are included in the summary statement in a special resume section. No award may be made unless all concerns have been resolved to the satisfaction of NIH, and the applicant institution has given OHPR an acceptable assurance of compliance with all relevant regulations.

2.11.2. Inclusion of Both Genders, Minorities, and Children as Research Subjects

NIH policy requires that applicants who propose clinical research include minorities and both genders in study populations so that research findings can benefit all persons at risk of the disease, disorder, or condition under study (see the NIH guide, Vol. 23, No. 11, March 18, 1994, available at http://grants.nih.gov/grants/guide/noticefiles/ not94100.html). Applicants must describe and justify the gender and racial/ethnic composition of the proposed study population in terms of the scientific objectives of the study. Reviewers evaluate whether the representation of minority groups and both genders is appropriate and, if not, whether the justification provided by the investigator is adequate. Cost is not an appropriate justification for limited representation. It is not sufficient to state that no one will be excluded on the basis of gender or race. The application must include an explicit plan to recruit and retain members of minority groups and subjects of both genders, unless it is not appropriate scientifically, as in gender-specific conditions such as prostate cancer, ovarian cancer, pregnancy, and pelvic inflammatory disease.

It is also NIH policy that children (defined as individuals younger than the age of 21 years) be included in all clinical research conducted or supported by NIH, unless there are scientific or ethical reasons not to include them (http://grants.nih.gov/grants/guide/noticefiles/not98024.html). This policy applies to all NIH-conducted and NIH-supported clinical research, including research that is otherwise "exempt." If children will be excluded from the research, a justification for their exclusion must be provided.

If representation is limited or absent and the scientific justification for the proposed study population is inadequate, this is considered a scientific weakness in the study design and is reflected in the priority score. The SRG's views on these matters are included in special sections of the summary statement under the headings "Gender," "Minorities," and "Children."

2.12. Research Involving Vertebrate Animals

PHS Policy on Humane Care and Use of Laboratory Animals requires that applicant organizations proposing to use vertebrate animals file a written Animal Welfare Assurance with the Office of Laboratory Animal Welfare (OLAW), establishing appropriate policies and procedures to ensure the humane care and use of live vertebrate animals involved in research supported by PHS. The applicant organization bears responsibility for the humane care and use of animals in PHS-supported research. PHS policy defines *animal* as "any live, vertebrate animal used or intended for use in research, research training, experimentation or biological testing or for related purposes."

NIH no longer requires approval of the proposed research by the Institutional Animal Care and Use Committee (IACUC) before scientific peer review of an application (http://grants.nih.gov/grants/guide/noticefiles/NOTOD02064.html), but it must be submitted to the awarding NIH component before an award can be made.

Potential applicants whose research involves vertebrate animals should see the appropriate instructions in the SF424 (R&R) form and the OLAW website (http://grants.nih.gov/grants/olaw/olaw.htm) before preparing an application. Failure to adequately address issues about vertebrate animals may result in delayed review or award, or return of the application without review.

Views expressed by the SRG regarding the adequacy of protection of vertebrate animals are included in the summary statement in a special resume section. No award may be made unless all concerns have been resolved to the satisfaction of NIH, and the applicant institution has given OLAW an acceptable assurance of compliance with all relevant regulations.

2.13. Streamlined Review Procedures

To use the limited time available for the review meeting most effectively, many SRGs use a streamlined review process. Briefly, only those applications judged to have scientific merit in approximately the upper half of all applications reviewed by the SRG are discussed at the SRG meeting. The rest of the applications are not discussed and are not scored at the meeting. To carry out this process most effectively, the upper and lower halves of the applications are tentatively identified by the reviewers before the meeting, and the list is confirmed at the beginning of the SRG meeting. Nonconcurrence by any member of the SRG who is not in conflict with the application is sufficient to bring an application to full discussion at the meeting.

The PD/PIs of unscored applications receive a summary statement including the reviewers' comments, but the unscored applications are normally not reviewed by council. However, these applications are nevertheless considered to be favorably recommended, unless the summary statement explicitly states otherwise, and, in very rare circumstances, may be recommended for council consideration by program staff.

2.14. Merit Ratings for Applications: Priority Scores and Percentiles

In rating applications, reviewers are instructed to:

Base their opinions strictly on thoughtful and objective considerations of the review criteria, not on emotional or institute programmatic or budgetary considerations;

Judge the merit of each application independently of other applications and according to the "state of the science" in the research area; and

Score according to their own judgment and evaluation of the application.

Each reviewer who is not in conflict with an application assigns a score ranging from 1.0 (best) to 5.0 (worst). After the meeting, the individual reviewers' numeric ratings for each scored application are averaged and multiplied by 100 to provide a three-digit rating called the priority score.

In addition to the priority score, percentile ranks are also calculated for most research project grant (R01) applications. The percentile represents the relative position or rank of each priority score (on a 100.0 percentile band) among the scores assigned by the SRG, with scores in the first percentile being the best and scores in the 100th percentile being the worst. Percentiles are calculated against a reference base of R01 grant applications reviewed by a study section at three consecutive meetings. Percentile ranking is currently the primary factor used by most NIH institutes in deciding which applications to fund, although each may set a different percentile "payline" for applications.

2.15. Possible SRG Actions

Scientific review groups have several possible options for each application that is discussed:

Score: If the application has been discussed and the SRG members have sufficient information to make a final recommendation about the application, they will score it.

Unscore: If, after a brief discussion, the SRG decides that the application is of poorer quality than those that were not discussed, it may, by unanimous vote, decide to not score the application. The summary statement for such an application will include a resume and summary of discussion but no score or percentile rank.

Deferral: In the rare circumstance that an SRG cannot make a recommendation without additional information, it may defer the application to the next review cycle. The SRA will contact the applicant to obtain the necessary information, or, if the information can be obtained only by discussion with the applicant or by direct observation, a telephone conference with the applicant or a project site visit may be scheduled. Deferred applications are not presented to councils and are usually reviewed again by the same SRG during the next review cycle. In general, deferral is not an option for review of applications in response to a one-time RFA or FOA.

Not recommended for further consideration: In reviews that are not streamlined, applications may be "not recommended for further consideration" if they lack significant and substantial merit or if they involve procedures that are gravely hazardous or are otherwise unethical. Priority score ratings are not given to these applications, they are not considered by the advisory councils, and they may not be awarded.

2.16. Summary Statements: How You Find Out What the Reviewers Thought About Your Application

The summary statement is the official document describing the outcome of the review. Immediately after the SRG meeting, the SRA prepares a summary statement for each application documenting the deliberations of the SRG. Summary statements generally include the reviewers' essentially unedited written comments. For applications that were discussed, the SRA also prepares a resume and summary of discussion to convey the highlights (i.e., major strengths and weaknesses) of the discussion at the review meeting and explains how the SRG arrived at the final rating. Summary statements for scored applications may also include budget recommendations, including how many years of support are recommended. The summary statement may also include administrative notes regarding special points or aspects of an application outside of scientific or technical merit that the SRA or SRG considers important enough to be brought to the attention of the applicant, institute, or council, such as concerns about research involving human subjects or animals or potential overlap with other ongoing projects.

Summary statements for unscored applications contain only the reviewers' written comments and are generally not presented to council.

An applicant should expect to be able to access his or her summary statement through the NIH eRA Commons within 4–8 weeks after the review meeting. The summary statement is also forwarded to the assigned institute for consideration by its national advisory council or board.

Summary statements have several important uses:

1. Advisory council members use summary statements as the main source of information about applications and as the primary basis for their recommendations.

2. Institute staff use summary statements in making funding decisions, as a basis for discussions with councils and applicants, and as guides in the future management of any resulting awards.

3. Unsuccessful applicants use summary statements in reassessing, adjusting, or revising their research projects.

4. Summary statements may provide background information to reviewers evaluating resubmission (formerly called a revised or amended), revision (formerly called supplemental), or renewal (formerly called competing continuation) application.

2.17. Review by National Advisory Councils and Boards

The second level review for grant applications is by institute advisory councils or boards, which assess the quality of the scientific merit review by the SRG, consider the relevance of the proposed research to the institute's programs and priorities, and advise the institute on policy issues. With the exception of individual fellowship applications and some grant applications with recommended direct costs less than $50,000 annually, grants cannot be awarded without consideration by a council or board. Generally, councils review only scored applications. Although councils may not change the score or percentile ranking of an application, they may recommend, usually on the basis of high or low "program relevance," whether an application should be funded and in what order. For most applications, councils concur with the recommendation of the SRG. These applications are usually acted on as a group (en bloc). If council disagrees with an SRG recommendation because of a perceived flaw in the review process, it may recommend deferral for rereview. In addition, council may advise the institute that a particular application, based on the relevance of the project to the institute's mission, should receive more favorable or less favorable consideration for funding than would be indicated by the priority score and/or percentile rating.

2.18. What Determines Which Applications Are Awarded?

Awards are made based on the scientific and technical merit of the application, as reflected in the priority score and/or percentile rating it received, the programmatic relevance of the application, and the availability of funds. Each NIH institute generally sets a payline for each of the different types of applications, and these paylines may differ considerably among the institutes, depending on their overall budgets, their portfolio of award mechanisms, and the advice of their advisory councils about portfolio balance. Paylines may also differ from program to program within an institute.

2.19. Confidentiality and Conflict of Interest

Protection of the confidentiality of review proceedings is essential to maintain the integrity of the peer review system. Under no circumstances may reviewers advise applicants or others of recommendations, nor may they discuss the review proceedings outside of the SRG meeting. The SRA in charge of the SRG handles all inquiries from applicants and from reviewers. In addition, review group members may not independently solicit opinions or reviews of particular applications, or parts thereof, from experts outside the SRG. If a reviewer believes that additional expertise is needed to review an application, the reviewer informs the SRA, who obtains additional scientific input. Review materials and the proceedings of review meetings are privileged communications prepared for use only by reviewers and NIH staff. Conflict of interest in scientific peer review is defined as a situation in which a reviewer has a personal, professional, or financial interest in an application. A conflict of interest is also assumed when an application involves a close relative or a close professional associate of the reviewer. Close professional associates include colleagues with whom the reviewer does research or with whom the reviewer is closely associated professionally.

The SRA for the review identifies conflicts of interest among the reviewers before the review. Before the review meeting, reviewers sign a document stating that they will not participate in the discussion of any application with which they are in conflict and listing those applications with which they are in conflict. At the beginning of each SRG meeting, the SRA explains the NIH confidentiality and conflict of interest policies. During the meeting, review staff keeps a record of which members leave the room because of conflicts of interest. At the end of the meeting, the SRA requests a second written certification from all members that they have not participated in the review of any application where their presence would have constituted a conflict of interest, and that they will maintain the confidentiality of the review process, materials, and information. Reviewers must also leave all review materials with the SRA at the conclusion of the review meeting. After the meeting, the CDs are destroyed, and paper materials are shredded.

3. HINTS FOR PREPARING BETTER GRANT APPLICATIONS

After you have decided on your research project, the most important element in increasing the chances that you will be successful in getting funding for your project is a well-prepared grant application. The reviewers assigned to your application will be scientists working in the general area of your research project. Consider them "informed strangers." The application must convey a large amount of information to them and also generate excitement about the project for them. After reading the application, they must understand the rationale for, and the objectives of, the project; see where the project fits in the "big picture" and where it will lead in the future; and feel confident that you can actually design, carry out, and interpret the experiments proposed to move the field forward. A well-prepared application leads the reviewers through the project logically and says much about you as the principal investigator, particularly that you "think like a scientist." The process for preparing an application for NIH support therefore requires a significant amount of time, a high level of organization, and attention to detail.

3.1. Planning Your Application

The following are key points to remember before, during, and after writing your application:

1. In October, 2006, NIH announced new receipt dates for grant applications, effective January 1, 2007. There is now a specific submission deadline for each type of grant application (http://grants1.nih.gov/grants/funding/submissionschedule.htm.). There are three receipt dates per year for most types of grant applications.

2. The submission, review, and award process for applications usually takes 8–10 months (see Table 27-4).

3. There are special deadlines for applications related to AIDS and for applications submitted in response to RFAs and some PAs.

4. The application should be complete for review as submitted. Submit your very best application because reviewers expect you to have taken the time needed to think it through before submitting.

5. NIH frequently updates policies, procedures, and application requirements. Do not rely on "hearsay" from colleagues who have submitted applications in the past. Visit the NIH website for the latest information.

6. NIH uses a "just-in-time" approach to the receipt of certain types of information (e.g., current "other support") that is not required until an award is about to be made. If a grant will be awarded, the NIH component making the award will request the information from you.

3.2. Allow Sufficient Time to Prepare the Application

Establish deadlines for the preparation of the grant application, particularly when collaborating investigators are involved. Be aware of administrative deadlines within your institution that could delay internal processing, and leave enough time to correct validation errors and resubmit through Grants.gov, if necessary. Allow time for equipment failures, personnel shortages, etc. If possible, prepare your application sufficiently early so that objective experts (e.g., successful grantees or an institutional panel) can review your application and provide extremely

frank feedback and criticism. Revise the application based on their critique. Friends and close associates are rarely as critical as the reviewers on an NIH study section.

3.3. Get Help

If possible, find someone in your institution who can help you understand the NIH process and completing your application. Ask your colleagues for copies of successful NIH grant applications to get a more concrete idea of what each section should include. Incomplete applications are returned without review. Do not feel inhibited about requesting technical assistance from the funding agency or your institution. Talk to the program representative(s) from the NIH institute(s) with interests in your research area to obtain advice on scientific and technical issues and to the grants management specialist to obtain advice on administrative issues. Representatives from your institutional grants office can also be of assistance regarding Grants.gov and NIH administrative and application format requirements. Talk to them and find out how they can help you. Investigate any special research priorities of funding agencies and ascertain from the program representative whether your project falls within the scope of an existing initiative (RFA or PA) or an area of special emphasis.

3.4. Follow the Instructions Carefully

Before you begin writing your grant application, read the SF424 (R&R) instructions carefully to become familiar with all the requirements and certifications necessary. If you are submitting your application in response to a specific PA or RFA, read the announcement in detail for special eligibility requirements, formatting instructions, and/or submission deadlines.

3.5. Submit a Complete and Carefully Prepared Application

If several people are contributing to writing the application, decide who will do the final editing. Reread your application. Have someone else read it. Proofread it again before submission. NIH receives and processes approximately 80,000 applications for PHS grants every year. You, and only you, are responsible for making sure your application is written with good grammar, that the references and figure legends are accurate, and that the flow of experiments is clear. Grants.gov and the NIH eRA Commons will validate that all required sections of the application are submitted in the required format and will electronically collate the sections and components of the SF424 (R&R) application and the required PHS 398 components into a complete application. As discussed previously, you will be notified if your application does not pass required validation checks. You also have two weekdays to view the application image in the eRA Commons to ensure that it was assembled accurately. Once the application is referred to a study section, NIH cannot "change pages." However, if you discover serious mistakes in the application after the two weekdays expire, call the SRA to find out if you can send corrections before the review.

3.6. Resubmission Applications

When resubmitting a revised application, address all reviewer concerns mentioned in the earlier summary statement in the "Introduction" section of the PHS 398 Research Plan Component. Regardless of how you feel, do not insult the reviewers. Try to convince the reviewers of your point of view courteously; do not ignore a comment because you disagree with it. In addition to responding to specific reviewer concerns, review all other aspects of the application to determine whether updating or improvement is possible. Just because something was not criticized is no guarantee it will not be criticized in the review of the revised application. NIH allows resubmission of an application up to two times; see http://grants2. nih.gov/grants/funding/phs398/instructions2/ p3_revised_nih_policy_revised_application.htm.

3.7. General "Do's" and "Don'ts"

- Observe application guidelines strictly, particularly if you are responding to an RFA or PA.
- Use good English. Avoid jargon and spell out all acronyms when used initially.
- Type single-spaced and use 1-inch margins on all sides of the page.
- Observe type size and page limitations strictly.
- Graphs, diagrams, charts, and tables should be legible (be consistent with formats). Label these items carefully.
- Make sure that all cited figures are included in the application and that all included figures are cited in the text.
- Have an outside reader review the application for clarity and consistency.
- Proofread carefully by reading aloud. Do not rely on computer spell-check programs to point out mistakes.
- Be consistent with terms, references, format, and writing style.

- Do not use a small font to get more information into your application or exceed page limits noted in the SF424 (R&R) instructions because this will result in your application failing the validation checks in Grants.gov.

3.8. Hints and Suggestions for Preparing Each Part of Your Application

This section should be used in conjunction with the Grants.gov Application Guide for the SF424 (R&R) forms package. The items discussed here are important parts of the application on which reviewers focus; many first-time applicants have problems with them. [Note that they are not listed here in the order in which they are attached to the SF424 (R&R) application.]

3.8.1. SF424 (R&R) Project Summary/Abstract

The purpose of the Project Summary/Abstract (formerly called the "Description") is to convey succinctly every major aspect of the proposed project, except the budget. It must contain a summary of the proposed project suitable for release to the public. A separate, detailed Research Plan component is required for NIH and other PHS agency applications.

The Project Summary/Abstract is a very important part of your application. It is used in the application referral process, along with a few other parts of the application, to determine what study section is most appropriate to review the application and to what NIH institute it is most relevant. Members of the review committee who are not primary reviewers may rely heavily on the Project Summary/Abstract to understand your project.

The Project Summary/Abstract should include a brief background of the project, a concise statement of the specific aims or hypotheses, the unique features of the project, the methodology (action steps) to be used, the expected results and how you will evaluate them, and the significance of the proposed research, including how your results will affect related fields. The following are suggestions for writing the SF424 (R&R) Project Summary/Abstract:

- Be complete but concise.
- Mention the short- and long-term objectives, the specific aims, and the types of methods (i.e., genetic, immunologic, genomic, proteomic, population surveys, etc.) that will be used in the project.
- Do not exceed the space allotted or your application may fail the validation check in Grants.gov.
- View the Project Summary/Abstract as your one-page advertisement.

- Write the Project Summary/Abstract last so that it reflects the entire project.

Remember that the Project Summary/Abstract will have a longer shelf life than the rest of the application and may be used for purposes other than review, such as to provide a brief summary of an awarded grant in annual reports or NIH programmatic presentations or in response to requests from top management at NIH, Congress, or the public. If an award is made, the Project Summary/Abstract will be available to the public, so it should not contain any proprietary information.

3.8.2. PHS 398 Specific Research Plan Component

Table 27-6 summarizes the key features of the PHS 398 Specific Research Plan component of successful applications. The research plan describes the "what," "why," and "how" of the proposed project. This is the most important part of the application and will be evaluated by the reviewers with particular care. The "what" is Section 2: Specific Aims. The "why" is addressed in Section 3: Background and Significance. Section 4: Preliminary Studies/Progress Report, and Section 5: Research Design and Methods constitute the "how." You will also address use of human subjects and vertebrate animals in the research plan. The reviewers' assessments of the research plan will largely determine whether the application receives a favorable merit rating and is recommended for funding. The maximum length of Sections 2 through 5 of the research plan for R01 applications is currently 25 pages, except

as specified in specific PAs or RFAs. (Note: Section 1: Introduction to Application will be used only for a resubmission or revision application and is not counted in the 25-page limit.)

The Research Plan as a whole should answer the following questions:

- What do you intend to do?
- Why is this worth doing?
- How is it innovative?
- What has already been done in general, and what have other researchers done in this field?
- What will this new work add to the body of knowledge in this and related fields?
- What have you (and your collaborators) done to establish the feasibility of what you are proposing to do?
- How will the research be accomplished? By whom? On what schedule?
- How will the results be obtained, analyzed, and interpreted?

The following are general suggestions for preparing the research plan:

- Make sure that all sections (2–5) are internally consistent and that they support each other. One person should revise and edit the final draft. The thought processes behind the project should be clear to the reviewers.
- Lead the reviewers through the research plan. Avoid "dense," difficult-to-read sections that may give reviewers a negative impression of your organizing skills or clarity of reasoning. Use a numbering system and/or subheadings to make all subsections of the application easy to find. Use diagrams for complex processes, relationships, or organizational schemes.
- Demonstrate understanding of recent key literature and explain how the proposed research will further what is already known or fill gaps in knowledge.
- Emphasize how some combination of a strong hypothesis, important preliminary data, a new experimental system, and/or a new experimental approach will enable important progress to be made.
- Emphasize biological mechanisms in your hypotheses, experiments, and interpretation of results as much as possible.
- Make sure the project aims are focused; avoid a "fishing expedition."
- Anticipate results, both positive and negative. Discuss alternative approaches that will be pursued if the proposed approach is not successful.
- Establish the credibility of the proposed principal investigator and the collaborating researchers.

TABLE 27-6   Key Features of Successful Research Grant Applications

Hypothesis
   A meaningful hypothesis *and* a means of testing it
   A sound rationale for the hypothesis
   A set of related aims focused on the hypothesis
Preliminary data
   Shows proper training for the research proposed and the ability to interpret results
   Include alternative interpretations of results and address limitations of methods
Well-organized research plan
   Aims focused, not diffuse
   Rationale for methods proposed, with problems and alternatives addressed
   Research priorities clearly indicated
   Sufficient experimental detail
   Emphasize mechanism—avoid "descriptive data gathering"
   Clear data analysis plans, with alternative interpretations addressed
   Access to key reagents, patients, specimens, facilities, etc., well documented

3.8.2.1. Section 2 of the PHS 398 Specific Research Plan: Specific Aims

The purpose of the Specific Aims section is to describe concisely and realistically what the proposed research is intended to accomplish. The recommended length of the Specific Aims section is one page. The Specific Aims should cover broad, long-term goals (e.g., the hypothesis to be tested) and specific time-phased research objectives. Generally, the Specific Aims section should begin with a brief narrative describing the long-term goals of the project and the framework in which it fits, followed by a numbered list of aims. The following are suggestions for preparing the Specific Aims section:

• If your research project is hypothesis based, state the hypotheses clearly and definitively. Use terms such as "causes," ". . . is mediated by . . . ," or "results in . . ." when formulating your hypotheses. Make sure the hypotheses are understandable, testable, and adequately supported by citations in the background section and by data in the preliminary results section. Be sure to explain how the results to be obtained will be used to test the hypothesis.
• Show that the objectives are attainable within the stated time frame.
• Be as brief and specific as possible. For clarity, each aim should be conveyed in one sentence. Use a brief paragraph under each aim if detail is needed. Most successful applications have two to four specific aims.
• Do not "bite off more than you can chew." A small, focused project with a feasible timetable is generally better than a diffuse, multifaceted project.
• Be certain that all aims are related to the main focus of your project. Have someone read them for clarity and cohesiveness.
• Focus on aims where you have good supporting preliminary data and scientific expertise.

3.8.2.2. Section 3 of the PHS 398 Specific Research Plan: Background and Significance

The purpose of the Background and Significance section is to set the stage for demonstrating that your proposed research is significant and/or innovative. Discuss the current state of knowledge relevant to the proposed project, including literature citations and highlights of relevant data, gaps that the project is intended to fill, and the potential contribution of this research to the problem(s) addressed. Discuss how the proposed project fits in the field and how it will extend or contribute to advancing the field. Show how your project is innovative, examines the topic from a fresh perspective, or develops or improves technology. The recommended length of this section for a typical R01 application is three to five pages; an even more detailed significance section may be needed for new projects or projects in emerging research areas.

The following are suggestions for preparing the Background and Significance section:

• Make a compelling case for your proposed research project. Why is the topic important? Why are these specific research questions important?
• Demonstrate familiarity with recent relevant research findings. Avoid citing outdated research. Use citations not only as support for specific statements but also to demonstrate familiarity with relevant publications and points of view. Your application is likely to be reviewed by someone working in your field, so include various points of view and findings from other laboratories. This section should be a critical analysis of the state of the field rather than an exhaustive tutorial.
• Make sure the citations are specifically related to the proposed research. Cite and paraphrase correctly and constructively.
• Highlight why research findings are important beyond the confines of the proposed project (i.e., how can the results be applied to further research in this field or related areas?).

3.8.2.3. Section 4 of the Research Plan: Preliminary Studies/Progress Report

The purpose of the Preliminary Studies/Progress Report section is to describe prior work by the applicants relevant to the proposed project. The recommended length of this section of an R01 research grant application is six to eight pages. In a new application, preliminary results are important to establish the experience and capabilities of the applicant investigators in the area of proposed research and to provide experimental support for the hypothesis and the research design. This section is not mandatory for new research grant applications or for applications for exploratory/developmental projects, pilot projects, or some career development awards, but it is extremely difficult to obtain a favorable review of an R01 research grant application without some preliminary data. This section should include enough preliminary data to give the reviewers confidence that your hypothesis has a reasonable chance of being correct, to show that the project is worth doing, and to show that you can, in fact, design, execute, and interpret experiments.

The most important type of preliminary data come from recent studies by the applicant that established the feasibility and importance of the proposed project.

It may also include a brief description of older published studies by the applicant that provides background information relevant to the proposed project and results of previous studies by the applicant not directly relevant to the proposed project if they are needed to establish the applicant's competence and experience with the proposed experimental techniques. In a competing renewal application, this section becomes a Progress Report describing the studies performed during the last grant period.

The following are suggestions for preparing the Preliminary Studies/Progress Report section:

- All tables and figures necessary for the presentation of preliminary results should be included in this section of the application.
- Do not dwell on results already published. Summarize the critical findings in the text and include links to the full articles. Up to three publications that are not available on free public websites or manuscripts accepted for publication can be included as appendix material.

3.8.2.4. *Section 5 of the Research Plan: Research Design and Methods*

The purpose of the Research Design and Methods section is to describe how the research will actually be carried out. This section is crucial to how favorably an application is reviewed. The recommended length of this section is 12–15 pages. It should include the following:

- An overview of the experimental design
- An explicit description of the specific methods that will be used to accomplish the specific aims, along with the rationale for choosing these particular methods
- A detailed discussion of the way in which the results will be collected, analyzed, and interpreted
- A projected sequence or timetable (work plan) for the experiments or sets of experiments
- A description of new methodology that will be used and why it represents an improvement over existing methods
- A discussion of potential difficulties and limitations and how these will be overcome or mitigated
- Expected results and alternative approaches that will be used if unexpected results are found or problems are encountered
- Precautions to be exercised with respect to procedures, situations, or materials that may be hazardous to personnel or human subjects

The following are suggestions for preparing the Research Design and Methods section:

- Number the sections in this part of the application to correspond to the numbers of the Specific Aims section.
- Give sufficient detail. Do not assume that the reviewers will know how you intend to proceed. For example, it is not sufficient to state, "We will grow a variety of viruses in cells using standard model systems." The reviewers will want to know which viruses, cells, and model systems; your rationale for selecting these particular viruses, cells, and model system(s) for achieving your aims; how the model systems will be used; and if you have ever done work like this previously.
- Avoid excessive experimental detail—this is not a "Materials and Methods" section of a publication—by referring to publications that describe the methods to be used. Publications cited should be by the applicants, if possible. Citing someone else's publication establishes that you know what method to use, but citing your own (or that of a collaborator) establishes that the applicant team is experienced with the necessary techniques.
- If relevant, explain why one approach or method will be used in preference to others. This establishes that the alternatives were not simply overlooked. Discuss not only the "how" but also the "why."
- If using a complex technology for the first time, take extra care to demonstrate familiarity with the experimental details and potential pitfalls of the methods. Consider adding a coinvestigator or consultant experienced with the technology.
- Document proposed collaborations and offers of materials or reagents of restricted availability with strong letters of commitment from the individuals involved.
- For clinical research applications involving patients, populations, or clinical trials, document that the patients and/or specimens are actually available for this project. If appropriate, include letters of collaboration discussing the specifics of what will be made available.
- Include a section on statistical analysis discussing both the study design and data analysis, with appropriate power calculations. This section should be reviewed by a biostatistician before the application is submitted. Especially in the case of clinical research, it is desirable to include dedicated effort of a statistician.

3.8.3. *PHS 398 Specific Research Plan: Human Subjects Sections*

The purpose of these sections of the application is to describe the procedures that will be used to ensure protection of the rights and welfare of individuals who participate in research projects, describe the demographics of the projected subject population, and provide information on how data and safety of all proposed clinical trials will be monitored. There is no page limit for these sections, but be succinct.

3.8.3.1. Section 6: Protection of Human Subjects

Provide a complete description of the proposed involvement of human Subjects as it relates to the work outlined in the Research Plan sections two to five. If an exemption has been designated in Item 1 of the SF424 (R&R) Other Project Information, enough detail must be provided to show how the appropriateness of the exemption was determined. This decision must be made by someone other than the principal investigator. If no exemption is claimed, there are four items that must be addressed in this section. A full description of these points can be found in Part II, Supplemental Instructions for Preparing the Human Subjects Section of the Research Plan, of the Grants.gov Application Guide for the SF424 (R&R) form. Be thorough in addressing these items since a human subjects "concern" expressed by the reviewers will bar award of a grant and may affect the merit rating of your application.

To help reduce workloads for IRBs, NIH policy allows submission and review of grant applications before they are approved by your IRB. If your application may be funded, you will be contacted by NIH to provide documentation of IRB approval. No work with human subjects can begin on an NIH grant without documentation of IRB approval.

3.8.3.2. Sections 7 and 8: Inclusion of Women and Minorities and Targeted/Planned Enrollment Table

All applications proposing clinical research must explicitly address the plans for including women, minorities, and children in the subject population. Remember that it is not sufficient to state that there will be no exclusions based on race or gender. You must provide a specific plan for the recruitment and retention of women and minorities in the study population. A justification is required if there is limited representation of women or minorities. Reviewers will consider this justification in evaluating your application. Failure to address this issue will result in an administrative bar to funding until all concerns raised by the SRG have been resolved. You must also complete the "Targeted/Planned Enrollment" table or the "Enrollment Report" table, if applicable, as described in the Supplemental Instructions for Preparing the Human Subjects Section of the Research Plan, to indicate racial/ethnic breakdown of your study population.

3.8.3.3. Section 9: Inclusion of Children

NIH defines "children" as persons younger than the age of 21 years. To determine if inclusion of children applies to your application, follow the instructions in the Supplemental Instructions for Preparing the Human Subjects Section of the Research Plan. A justification is required if there is limited or no representation of children. Reviewers will consider this justification in evaluating your application. Failure to address this issue will result in an administrative bar to funding until all concerns raised by the SRG have been resolved.

3.8.4. HS 398 Specific Research Plan Section 11: Vertebrate Animals

The purpose of this section is to document the humane treatment of live vertebrate animals in the proposed research. There is no specified length, but be succinct. Provide a complete description of the proposed use of vertebrate animals as it relates to the work outlined in the Research Plan section. Five points must be addressed in this section [see SF424 (R&R) Application Guide]. Be thorough in addressing these five areas. Failure to adequately address the use of vertebrate animals may result in an administrative bar to award. NIH policy allows submission and review of grant applications before receipt of IACUC approval for use of vertebrate animals. If your application may be funded, you will be contacted by NIH to provide IACUC approval. No work with vertebrate animals can begin on an NIH grant without documentation of IACUC approval.

The following are suggestions for preparing the Vertebrate Animals section:

- Most research and academic institutions have a multiple project assurance from OLAW. If your institution does not, contact OLAW as soon as possible to obtain a single project assurance.
- Be sure the number of animals is realistic and appropriate for the studies proposed.
- Justify all animal expenses for applications not prepared in the modular budget format.

3.8.5. Multiple PI Leadership Plan

For applications designating multiple PD/PIs, a new section of the research plan, entitled "Multiple PD/PI Leadership Plan" [Section 14 of the Research Plan Component in the SF424 (R&R) or Section I of the Research Plan in the PHS 398], must be included. The rationale for choosing a multiple PD/PI approach should be described. The governance and organizational structure of the leadership team and the research project should be described, including communication plans, processes for making decisions on scientific direction, and procedures for resolving conflicts. The roles and administrative, technical, and scientific responsibilities for the project or program should be delineated for the PD/PIs and other collaborators. If budget allocation is planned, the distribution of resources to specific components of the project or the individual PD/PIs must be delineated in the Leadership Plan. In the event of an award, the requested allocation may be reflected in a footnote on the Notice of Grant Award.

3.9. Other Important Parts of the Application

3.9.1. Budget and Justification

The purpose of the Budget section is to present and justify the costs requested to accomplish the project aims and objectives. For multi-institutional applications, there must be a separate subaward/consortium budget component for each subawardee or consortium organization that will perform a substantive portion of the project.

The application forms package associated with most NIH FOAs will include two budget components: the SF424 (R&R) Budget Component and the PHS 398 Modular Budget Component. Each NIH application will use one of these budget forms, not both.

The modular budget format is applicable for certain types of research grant applications requesting $250,000 or less per year for direct costs in all years. Consortium/contractual facilities and administrative costs are not included in this direct cost limit and may be requested in addition to the $250,000 per year limit. Modular budgets are simplified budgets, without detailed information about what is being requested in each budget category. Applicants estimate the total research budget required for each year in multiples of $25,000 (e.g., $125,000, $150,000, or $225,000) and do not itemize categories such as glassware, reagents, animals, equipment, and travel. The budget justification should specify the roles and person-months of effort proposed for each of the listed project personnel and explain any large costs, unusual items, or unapparent costs that contribute to the overall estimate for the first year of the project. The budget for future years of the project should be similarly estimated, and any increases or decreases in the number of requested budget modules should be explained. For grants with modular budgets, the award will not be increased by an inflation factor each year.

Most clinical research studies will require more than $250,000 in any year. Such applications must include a complete SF424 (R&R) Budget Component, with a detailed budget for each year of support requested. The SF424 (R&R) Budget Component includes three separate data screens. Read the instructions carefully and include all required fields. The form will generate a cumulative budget for the total project period. The budget should include costs for all personnel, consultants, equipment, supplies, travel, patient care, and other expenses (e.g., animal maintenance, equipment service contracts, and offsite space rentals). The Budget Justification attachment should explain the roles of the proposed personnel and the need for each item requested.

It is the policy in most research institutions that the sponsored projects office must approve the budget before submission of the application.

The following are suggestions for preparing detailed budgets for projects over $250,000:

- Be realistic. Both "padding" and deliberately under budgeting reflect naiveté or lack of appreciation of the scope of the work proposed, which will be recognized and viewed negatively by the reviewers.
- Provide brief descriptions of duties for all positions requested in the budget, with the person-months of effort requested each year and any anticipated fluctuations. Special skills or accomplishments of a designated person may be included if not discussed elsewhere.
- If possible, identify specific individuals for each position requested. "To be named" personnel are often deleted by reviewers.
- Justify all requested equipment. The proposed acquisition of major pieces of equipment is likely to be scrutinized very carefully. Details are important, especially for equipment that is not project specific, such as fax machines and computers.
- Break out supply costs into major categories (e.g., reagents, disposables, or animals). Provide justification for unusual expenses.
- Detail and justify travel costs. Make sure they reflect current fares and lodging costs and that the proposed travel is project related.
- Explain any year-to-year fluctuations in the budget, particularly the level of effort of personnel. Changes should parallel the research plan and project aims.
- If there is a coinvestigator at another institution who will require salary and/or supplies in order to work on the project, be sure to include her or him in your budget.

- Check indirect costs. Some institutions have on-campus and off-campus rates.
- Be complete but concise. There is no page limit for the Budget Justification section.
- Provide strong justification for the need to use outside consultants, if applicable.
- If applicable, provide documentation of institutional rates for animal maintenance and acquisition. Exceptionally large numbers of animals will need more detailed justification.
- Prorate service contracts to reflect the percentage of time equipment will be used for this project.

3.9.2. Senior/Key Personnel Profiles Component and Biosketches

This section of the application is your chance to showcase the expertise and experience of the senior/key personnel, other significant contributors, and consultants involved in your research project. Senior/key personnel are all individuals who contribute in a substantive, measurable way to the scientific development or execution of the project, whether or not salary is requested in the budget. Consultants should be included in this category if they meet this definition. Other significant contributors are individuals who have committed to the scientific development or execution of the project but who are not committing any specified measurable effort to the project; these are typically shown as "as needed" in the budget.

Reviewers look carefully at the biosketches to evaluate whether the proposed research team has the qualifications and experience to carry out the proposed work and overcome any problems that may arise. Reviewers evaluate the suitability of proposed personnel under the "Investigators" review criterion.

3.9.3. Facilities and Other Resources

The purpose of the Facilities and Other Resources section is to describe the resources, facilities, and institutional support available to the researcher. Use this section to show the reviewers that you have all of the necessary equipment and space, including clinic and clinical laboratory space, to perform the proposed project successfully. Do not assume that the reviewers will know what is in your institution or what is actually available for your use.

The following are suggestions for completing the Facilities and Other Resources section:

- Make sure this section addresses all of the requirements of the proposed research plan.
- Justify any reliance on resources external to your research laboratory. Include letters of collaboration from the providers of those resources.

- Make sure all subcontractors and consortium members have the capability to perform the tasks assigned to them.
- Make certain your resources and budget requests are consistent. Do not request funds for equipment listed in the Resources section as already available to you.

3.9.4. Appendix

In November 2006, NIH issued new guidance on materials allowed in the Appendix of grant applications (http://grants.nih.gov/grants/guide/notice-files/NOT-OD-07-018.html). Published manuscripts and/or abstracts that are publicly available in a free, online format should be referenced by the URL in the application. These publications may *not* be included in the Appendix. URLs or NIH PubMed Central (PMC) submission identification numbers should be included along with the full reference in the Bibliography and References Cited section (SF 424RR)/Literature Cited (PHS 398) section, the Progress Report Publication List section and/or the Biographical Sketch section. Applications requiring electronic submission on the SF424 (R&R) may include full color graphic images of charts, gels, micrographs, photographs, etc. in the Research Plan PDF; these images may *no longer* be included in the Appendix (except when part of a qualifying publication). See the SF 424 (R&R) Application Guide for guidance about size and resolution of images.

Applicants may submit up to three of the following types of publications.

- Manuscripts and/or abstracts accepted for publication but not yet published.
- Published manuscripts and/or abstracts *that do not have* a free, online, publicly available journal link available.
- Patent materials directly relevant to the project.

Any exceptions will be noted in specific FOAs. Other items that may be included in the Appendix include surveys, questionnaires, data collection instruments, clinical protocols, and informed consent documents.

For electronic submission using the SF 424 (R&R) forms and Grants.gov, Appendix materials must be submitted in PDF format.

For applications still submitted on paper PHS398 forms, the Appendix may include full-sized glossy photographs of material such as electron micrographs or gels that do not reproduce well in black and white; however, an image of each (may be reduced in size but readily legible) must also be included within the page limitations of the Research Plan. For paper submission using the PHS 398, Appendix materials may be submitted in paper format; five collated sets are needed;

however, applicants are encouraged to send Appendix materials submitted with paper PHS 398 applications on a CD in PDF format in lieu of the five collated sets. See application instructions for details on preparing CDs. Only a single CD need be sent.

For materials that cannot be submitted electronically or materials that cannot be converted to PDF format, (e.g., medical devices, prototypes, DVDs, CDs), applicants should contact the Scientific Review Administrator for instructions following notification of assignment of the application to a study section. If the SRA is listed in the FOA, he or she should be contacted in advance to address acceptability of Appendix materials.

3.10. Revising Unsuccessful Applications

Although the NIH budget increased significantly from 1998 to 2003, the number of applications submitted and the average cost of research grants also increased, and recent budget increases have been much smaller. Competition for NIH research and career development awards is tough, and it is common for applicants not to succeed on the first attempt.

Table 27-7 lists some of the most common reasons for unsuccessful applications. Although a rejected grant application can be hard on the ego, the reality is that most investigators have to resubmit applications before securing funding for their research. Revising an application provides an opportunity to rethink weaknesses in your design, approach, and methods and to address the reviewers' concerns. Current NIH policy allows two resubmissions of an application.

3.10.1. How to Decide Whether to Revise Your Application

Read and reread the summary statement. Look for the main problems identified by the reviewers. Discuss

TABLE 27-7   Most Common Problems with Unsuccessful NIH Grant Applications

Lack of new or original ideas
Hypothesis is not scientifically sound
Lack of scientific rationale for the project or proposed approach
Diffuse, superficial, or unfocused research plan
Questionable reasoning in experimental approach
Poor choice of experimental methods
Lack of adequate controls
Lack of sufficient experimental detail
Insufficient statistical power for clinical studies
Lack of knowledge of published relevant work
Unrealistically large amount of work
Uncertainty concerning future directions
Lack of experience in the essential methodology

the summary statement with the NIH program officer responsible for your application. If the reviewers thought the main idea or research question is worthwhile and important, then it is worth revising the application. If the review identified fundamental problems in the scientific rationale or the hypothesis, then it may be best to begin with a new idea and develop a new project.

Common fixable problems include poor writing or organization of the application, insufficient information about experimental details, insufficient preliminary data, diffuse aims, too much work for the project period requested, concerns about the experience of the proposed personnel, and insufficient attention to potential problems or how the data will be interpreted. More significant concerns, which may not be fixable, are that the project will add little to advance the field, the hypothesis is not sound, the work has already been done, or the design or methods are not appropriate for testing the hypothesis.

3.10.2. How to Revise and Resubmit Your Application

The key to successfully revising and resubmitting your application is to address the reviewers' concerns. Add preliminary data, an experienced collaborator, and additional details if needed. Delete weak and peripheral aims or experiments, and refocus diffuse projects tightly on the hypothesis. Change the approaches or methods that will be used if necessary. Rethink the design of a clinical trial to address concerns about statistical power. Ask a colleague who is experienced in your field and in grantsmanship, but who is not involved in your project, to read your application and the summary statement and to comment on your plans for changes in the research plan. Address each main concern noted in the summary point by point in an introduction to the resubmitted application. For R01 applications, the introduction may not exceed three pages. If you disagree with the reviewers, explain why and provide additional information if required. Indicate, by font changes, indenting, or a line in the margin, which sections of the application have been changed.

Even if you respond to all of the reviewers' comments, your resubmitted application may still not receive a fundable score. This may happen for several reasons. The summary statement is not meant to be an exhaustive critique, and some of the reviewers' concerns may not be highlighted in it. Also, when you make changes in the application, you risk introducing new problems. In addition, science "moves on," so a project with high significance when first submitted may not be as important by the time it is resubmitted and reviewed. Finally, the membership of review committees changes so that new reviewers with different perspectives may review your resubmitted application. However, you may still resubmit your application one more time; NIH allows up to two resubmissions of an application.

3.10.3. What if It Appears That the Study Section Was Inappropriate or Biased?

If it appears that there was not sufficient expertise on the review panel (e.g., a molecularly oriented study section reviewing a clinical trial), or you have reason to believe that there was a bias in the review, you should revise and resubmit the application and request a different study section for the review in the cover letter that accompanies your application. Real bias in the review is very rare. Reviewers are alert to potential bias among competitors on the review group and argue against it vigorously. SRAs are also alert to potential bias among reviewers.

4. NIH AWARD MECHANISMS FOR SUPPORT OF CLINICAL RESEARCHERS AT VARIOUS STAGES IN THEIR CAREERS

Although the R01 research project grant is the most well-known and popular of NIH's grant mechanisms, NIH has several award mechanisms specifically designed to support clinical researchers at various stages in their careers. In addition, other award mechanisms described here, such as career transition awards and small (R03) and exploratory (R21) grants, are available to any researcher; these awards are useful ways in which new clinical investigators can obtain the preliminary data and proof of concept that are needed to prepare a competitive R01 application. Finally, many NIH institutes also have special career development awards for researchers in specific scientific areas.

4.1. Individual Career Development ("K") Awards

Detailed information about career development awards can be found at the NIH K Kiosk at http://grants2.nih.gov/training/careerdevelopmentawards.htm. There are a number of different types of career development awards, and not all NIH institutes and centers participate in all of them. In addition, each participating NIH component may have its own guidelines and requirements for a particular career develop-

ment award to accommodate the career needs of researchers working in different fields. You should therefore contact the training and career development office in the NIH institute closest to your research interests before preparing an application.

At the time of award, candidates for most NIH career development awards must be citizens or noncitizen nationals of the United States or permanent residents; individuals on temporary or student visas are not eligible for these awards. The exception is the K99/R00 Pathway to Independence Award described later.

Note that the review criteria for career development awards are different from the review criteria for research project grants discussed previously. Review criteria for the various career development awards vary somewhat, but generally focus on the following:

- Qualifications of the candidate
- Qualifications of the sponsor(s) for mentored awards
- Career development plan
- Research project to be conducted as part of the career development plan
- Institutional environment in which the career development will take place

In addition, career development award applications require a plan for training the candidate in the responsible conduct of research. These plans must detail the proposed subject matter, format, frequency, and duration of instruction. No award will be made if an application lacks this component.

4.2. Mentored Career Development Awards

The candidate must identify a mentor with extensive research experience and must devote at least 75% effort to career development research activities during the period of the award. Former and current principal investigators on PHS research grants are not eligible. However, awardees may apply for a PHS research project grant during the period of the award.

4.2.1. Mentored Clinical Scientist Development Award (K08)

The Mentored Clinical Scientist Development Award (K08) provides support for clinical professionals who wish to develop into independent investigators. In general, K08 awards support more laboratory-oriented, translational, or preclinical research projects; clinicians who wish to pursue patient-oriented research training should see the section on the K23 award. There is substantial variability among the sponsoring NIH institutes in eligibility requirements, allowable costs, and application procedures. Applicants should contact the individual institutes for specific guidelines.

Candidates should hold a clinical doctoral degree and should have initiated postgraduate clinical training. The requested project period may be for three, four, or five years, depending on the candidate's prior research experience, additional experiences needed, and the policy of the awarding NIH institute. Awards are not renewable.

4.2.2. *Mentored Patient-Oriented Research Career Development Award (K23)*

The purpose of the Mentored Patient-Oriented Research Career Development Award (K23) is to support the career development of investigators who have made a commitment to focus their research on patient-oriented research (POR) and who have the potential to develop into productive, clinical investigators focusing on POR. For the purposes of this award, POR is defined as research conducted with human subjects (or on material of human origin, such as tissues, specimens, and cognitive phenomena) for which an investigator directly interacts with human subjects. This area of research includes (1) mechanisms of human diseases, (2) therapeutic interventions, (3) clinical trials, and (4) the development of new technologies.

Candidates must have a clinical degree or its equivalent: M.D., D.O.S., D.M.D., D.O., D.C., O.D., N.D. (doctor of naturopathy), and doctorally prepared nurses. In addition, individuals holding the Ph.D. degree may apply for the award if they have been certified to perform clinical duties, such as a clinical psychologist and clinical geneticist. Candidates must have also completed their clinical training, including specialty and, if applicable, subspecialty training, before receiving an award; however, they may submit an application before the completion of clinical training. Candidates may request three to five years of support, depending on their previous training and experience.

4.3. Career Transition Awards

4.3.1. *K99/R00 Pathway to Independence (PI) Award*

One of the most challenging transitions in any research career is that from postdoctoral trainee to independent scientist. In January 2006, NIH announced a new K99/R00 Pathway to Independence (PI) Award designed to help the most promising, exceptionally talented new investigators make the transition from trainee to independent investigator. Candidates must have no more than five years of postdoctoral research training experience. This new career transition award is part of NIH's ongoing effort to reverse the trend that researchers are receiving their first independent research awards at increasingly later stages in their careers.

The K99/R00 PI award will provide up to five years of support consisting of two phases. The initial mentored phase will provide support for salary and research expenses for up to two years to complete research, publish results, and bridge to an independent research position. As part of the application, the candidate must propose a research project that will also be pursued as an independent investigator during the second phase of the award. The candidate and mentors together will be responsible for all aspects of the mentored career development and research program. The individual must select an appropriate mentor with a track record of funded research related to the selected research topic and experience as a supervisor and mentor. The sponsoring institution must ensure that the candidate has the protected time needed to conduct the proposed research.

Following the mentored phase, the awardee may request up to three years of support to transition, as an independent scientist, to an extramural sponsoring institution/organization to which the individual has been recruited at the level of a tenure-track assistant professor or equivalent. This support will allow the awardee to continue to work toward establishing his or her own independent research program and prepare an application for regular research grant (R01) support. Support for the independent phase, however, is not automatic and is contingent on being accepted by an extramural institution and NIH programmatic review of progress during the mentored phase of the award.

The total cost per year for the initial mentored phase may not exceed $90,000. Salary is limited to $50,000, plus applicable fringe benefits, and up to $20,000 for research support costs for a 12-month budget period. Candidates must commit at least 75% effort to pursue their career development and research experience during the mentored phase. The total cost for the independent investigator phase may not exceed $249,000 per year, including salary, fringe benefits, research support allowance, and applicable facilities and administrative costs.

For more information about this program, see http://grants2.nih.gov/training/careerdevelopmentawards.htm.

4.3.2. K22 Career Transition Awards

Several NIH institutes support other types of K22 career transition awards. In general, this award is intended to facilitate the transition of investigators, particularly clinical investigators, from the mentored to the independent stage of their careers in research. There are two general types of K22 awards. One type is to provide "protected time" for newly independent investigators to develop their initial research programs completely in extramural settings in a research institution of the candidate's choice. The unique feature of this award is that individuals may apply without a sponsoring institution while they are still in a "mentored" position. The other type of K22 award involves up to two years of support to do research in an intramural program of one of the NIH institutes, followed by extramural grant support to develop a research program at a research institution of the candidate's choice. Because policies about the K22 awards differ markedly among NIH institutes, potential applicants should contact the training office in the NIH component most closely associated with their research interests before preparing an application.

4.4. Independent Scientist Awards

The Independent Scientist Award (K02) provides up to five years of salary support for newly independent scientists who can demonstrate the need for a period of intensive research focus as a means of enhancing their research careers and enabling them to expand their potential to make significant contributions to their field of research.

A candidate must have a doctoral degree and independent, peer-reviewed research support at the time the award is made; some NIH institutes and centers require the candidate to have an NIH research grant at the time of application, whereas others will accept candidates with peer-reviewed, independent research support from other sources. Scientists whose work is primarily theoretical may apply for this award in the absence of external research grant support.

The candidate must devote at least 75% of his or her full-time professional effort to conducting research and research career development during the period of the award. In addition, the candidate must be able to demonstrate that the requested period of salary support and protected time will foster his or her career as a highly productive scientist in the indicated field of research. Former principal investigators on PHS research grants are not eligible, and a concurrent PHS award may not be held. However, awardees may apply for a PHS grant during the period of the award.

The requested project period may be three to five years. Some NIH awarding units allow submission of renewal applications.

4.5. Midcareer Investigator Award in Patient-Oriented Research

The purpose of the Midcareer Investigator Award in Patient-Oriented Research (K24) is to provide salary support for clinician investigators to allow them protected time to devote to POR as defined previously for the K23 award and to act as research mentors primarily for clinical residents, clinical fellows, and/or junior clinical faculty. It is expected, for example, that investigators will obtain new or additional independent peer-reviewed funding as the PD/PI for POR and establish and assume leadership roles in collaborative POR programs; and that there will be an increased effort and commitment to mentor beginning clinician investigators in POR to enhance the research productivity of these investigators and increase the pool of future well-trained clinical researchers.

This award is primarily intended for clinician investigators who are at the associate professor or equivalent level and who have an established record of independent, peer-reviewed federal or private research grant funding in POR. Candidates must have a clinical degree or its equivalent, as discussed under the description of the K23 awards.

NIH will provide salary plus commensurate fringe benefits for up to 50% effort; at least 25% effort is required. The effort may overlap with the 50% effort required for the Loan Repayment Program awards (described later). NIH will also provide up to $50,000 per year for research development support, which can be used for (1) research expenses, such as supplies, equipment, and technical personnel for the principal investigator and his or her mentored clinical investigators; (2) travel to research meetings or training; and (3) statistical services including personnel and computer time.

4.6. Exploratory/Development Grant (R21) Applications

The R21 award mechanism is intended to encourage exploratory and developmental research projects in innovative new research areas by supporting early and conceptual stages of these projects, such as pilot projects or feasibility studies. The ideas may not be developed sufficiently to submit as a regular research project grant (R01) application. For example, such projects could assess the feasibility of a novel area of investigation or a new experimental system that has the potential to enhance health-related research, or they could propose the unique and innovative use of an existing methodology to explore a new scientific area. These studies may involve considerable risk but may lead to a breakthrough in a particular area or to the development of novel techniques, agents, methodologies, models, or applications that could have a major impact on biomedical, behavioral, or clinical research.

Many of the NIH institutes and centers accept investigator-initiated R21 applications. However, those that do not may solicit R21 applications to meet specific program needs by issuing specific FOAs. You should consult the NIH R21 website at http://grants.nih.gov/grants/funding/r21.htm for further information before preparing an application.

Applications for R21 awards should describe projects distinct from those supported through the traditional R01 research project grant. For example, long-term projects and projects designed to increase knowledge in a well-established area are not appropriate for R21 awards. Conversely, projects of limited cost or scope that use widely accepted approaches and methods within well-established fields are better suited for the NIH small grant (R03) mechanism.

The review criteria are the same as the review criteria for R01 research project grant applications described previously. However, because the research plan is limited to 15 pages, an R21 application need not have extensive supporting background information. Accordingly, reviewers will focus their evaluation on the conceptual framework, the level of innovation, and the potential to significantly advance knowledge or understanding. Because this type of grant is designed to support innovative new ideas, preliminary data as evidence of feasibility are not required. Justification for the proposed work can be provided through literature citations, data from other sources, or, when available, investigator-generated data. However, the applicant is still responsible for developing a sound research plan.

R21 grants are generally limited to a total budget request of $275,000 for 2 or 3 years of support and generally are not renewable.

4.7. Small Research Grant (R03) Applications

Small research grants provide research support that is limited in time (usually 1 or 2 years) and amount (usually $50,000–$100,000 direct costs per year) and are nonrenewable. R03s are generally for support of preliminary studies or short-term projects. The results of an R03 grant often provide the preliminary findings for an R01 grant application. Not all NIH institutes support R03 awards, and of those that do, different institutes have different objectives, guidelines, and requirements for their small grant programs. In several institutes, these applications are accepted only in response to specific PAs or RFAs. Therefore, applicants interested in R03 awards should contact the program officials in the institute(s) most closely associated with their research interests, or check the NIH Guide for Grants and Contracts and Grants.gov for current R03 FOAs, for more information before preparing an application. The review criteria for R03s are generally similar to those for R01s. More information on the R03 program can be found at http://grants.nih.gov/grants/funding/r03.htm.

5. OTHER NIH PROGRAMS FOR CLINICAL RESEARCHERS

5.1. Loan Repayment Program

The NIH Loan Repayment Programs (LRP) were initiated in 2002 to attract health professionals to careers in clinical, pediatric, health disparity, or contraceptive and infertility research. There is also a Loan Repayment Program for Clinical Researchers from Disadvantaged Backgrounds. For more information, see the LRP website at www.lrp.nih.gov.

In exchange for a 2- or 3-year (for intramural general research) commitment to a research career, NIH will repay up to $35,000 per year of your qualified educational debt. In addition, NIH will make corresponding federal tax payments for credit to your Internal Revenue Service tax account at the rate of 39% of each loan repayment to cover your increased federal taxes. NIH may also reimburse any increased state or local taxes and/or additional increased federal taxes (where the federal tax payments were not sufficient to fully cover your increased federal taxes) that you incur as a result of your LRP benefits.

Applicants must have a M.D., Ph.D., Psy.D., Pharm. D., D.O., D.D.S., D.M.D., D.P.M., D.C., N.D., or equivalent doctoral degree from an accredited institution. The D.V.M. degree is appropriate for all LRPs except the Clinical Research LRP or Clinical Research for Individuals from Disadvantaged Backgrounds LRP. Applicants must be a U.S. citizen, U.S. national, or permanent resident.

An applicant's research must be funded by a domestic nonprofit or U.S. government (federal, state, or local) entity. An NIH grant is acceptable but is not required. Salary support and/or research funding from your university department is acceptable if your employer is nonprofit. You must commit 50% of your time (at least 20 hours per week) for 2 years to the research.

Your educational debt—from qualifying types of student loans—must equal at least 20% of your "institutional base salary" paid by the institution where you are conducting research. NIH issues payments directly to lenders on a quarterly basis. To remain eligible for NIH LRPs, the student loans must remain segregated from all noneducational loans and loans held by another person, such as a spouse or a child.

5.2. Research Supplements to Promote Diversity in Health-Related Research

This NIH program provides administrative supplements to ongoing research grants to improve the diversity of the research workforce by supporting and recruiting students, postdoctorates, and eligible investigators from groups that have been shown to be underrepresented in biomedical science. All NIH awarding components participate in this program. Candidates eligible for support under this supplement program include individuals from groups that have been shown to be underrepresented in science, including individuals from underrepresented racial and ethnic groups, individuals with disabilities, and individuals from disadvantaged backgrounds. Awards under this program are limited to U.S. citizens or noncitizen nationals and permanent residents.

Staff of the awarding IC will review requests for these supplements using the following criteria:

- Qualifications of the candidate, including career goals, prior research training, research potential, and any relevant experience
- Evidence of educational achievement and interest in science, if the candidate is a student
- Strength of the description of how this particular appointment will promote diversity within the institution or in science nationally
- Plan for the proposed research and career development experiences in the supplemental request and their relationship to the parent grant
- Evidence that the proposed experience will enhance the research potential, knowledge, and/or skills of the candidate and that adequate mentorship will be provided
- Evidence that the activities of the candidate will be an integral part of the project

5.3. Research Supplements to Promote Reentry into Biomedical and Behavioral Research

NIH also supports a program for administrative supplements to research grants to support individuals with high potential to reenter an active research career after taking time off to care for children or to attend to other family responsibilities. This program will support full-time or part-time (at least 50% effort) research by these individuals to bring their existing research skills and knowledge up-to-date. By the completion of the supplement, the reentry scientist should be in a position to apply for a career development (K) award, a research award, or some other form of independent research support.

The parent grant should have at least two years of support remaining at the time of the proposed beginning date of the supplemental funding so that there is ample opportunity for the candidate to develop his or her research skills further. A maximum of three years of supplemental support can be awarded under this program. Because NIH institutes and centers may have varying policies in implementing the reentry program, potential applicants should consult the extramural training office in the NIH awarding component at the earliest possible stage to discuss his or her unique situation.

Candidates must have a doctoral degree, such as M.D., D.D.S., Ph.D., O.D., D.V.M., or equivalent, and they must have had sufficient prior research experience to qualify for a doctoral-level research staff or faculty position at the time they left active research. Candidates who have begun the reentry process through a fellowship, traineeship, or similar mechanism are not eligible for this program. Awards are limited to citizens or noncitizen nationals of the United States or permanent residents.

In general, the duration of the career interruption should be for at least one year and no more than eight years. The program is not intended to support additional graduate training and is not intended to support career changes from nonresearch to research careers for individuals without prior research training. Generally, at the time of application, a candidate should not be engaged in full-time paid research activities.

Staff of the awarding IC will review requests for these supplements using the following criteria:

- Qualifications of the reentry candidate, including career goals, prior research training, research potential, and any relevant experience
- Plan for the proposed research experience in the supplemental request and its relationship to the parent grant
- Evidence from the principal investigator that the experience will enhance the research potential, knowledge, and/or skills of the reentry candidate
- Evidence from the principal investigator that the activities of the reentry candidate are an integral part of the project

- Evidence of effort by the reentry candidate to initiate the reentry process, such as attending scientific meetings and keeping current with journals
- Evidence that proposed research will achieve the stated objectives of the reentry supplements
- Evidence that the principal investigator understands the importance of the mentoring component of this supplement and has prepared a mentoring plan

6. NEW DIRECTIONS AND INITIATIVES AT NIH

6.1. NIH Road Map for Biomedical Research

During fiscal year (FY) 2002 and 2003, with input from hundreds of outside scientists, members of industry, and the public, NIH developed an integrated vision, called the NIH Road Map for Biomedical Research, to deepen our understanding of biology, stimulate interdisciplinary research teams, and reshape clinical research to accelerate medical discovery and improve health. The purpose was to identify major opportunities and gaps in biomedical research that no single institute at NIH could handle alone but that the agency as a whole must address in order to make the largest impact on the progress of medical research.

The NIH Road Map has three main themes: new pathways to discovery, research teams of the future, and reengineering the clinical research enterprise (Table 27-8). NIH began issuing special initiatives in these areas in FY 2004.

The theme of reengineering the clinical research enterprise will be most relevant to readers of this

TABLE 27-8   Main Themes of the NIH Road Map for Biomedical Research

Theme	Topic Areas
New pathways to discovery	Building blocks, biological pathways, and networks
	Molecular libraries and molecular imaging
	Structural biology
	Bioinformatics and computational biology
	Nanomedicine
Research teams of the future	High-risk research
	Interdisciplinary research
	Public–private partnerships
Reengineering the clinical research enterprise	Clinical research networks
	Clinical research policy analysis and coordination
	Clinical research workforce training
	Dynamic assessment of patient-reported chronic disease outcomes
	Translational research

chapter. The following are some of the initiatives under this theme:

Clinical research policy analysis and coordination: This effort is intended to enhance the leadership and coordination of approaches to harmonize, standardize, and streamline federal policies and requirements pertaining to clinical research while emphasizing the integrity and effectiveness of federal and institutional systems of oversight. As part of its stewardship responsibilities, NIH is responsible for taking steps to foster the responsible conduct of high-quality clinical research.

Clinical research workforce training: This effort will address the career development of clinical researchers at multiple points in the educational pipeline, both to attract individuals to clinical research and to enhance the expertise and careers of those already engaged in clinical research. One initiative, the Multidisciplinary Research Career Development Program, will support the early career development of clinical researchers from a variety of disciplines, including patient-oriented research, translational research, small- and large-scale clinical investigation and trials, and epidemiologic and natural history studies. Another initiative, the National Clinical Research Associates (NCRAs), will create a cadre of qualified health care practitioner–researchers who are well trained to ensure responsible conduct of clinical research. The NCRAs will refer and follow their own patients in clinical research and disseminate research findings to the community, thus playing a critical role in advancing the discovery process.

Clinical research networks: The efficiency and productivity of the United States' clinical research enterprise will be enhanced by promoting clinical research networks capable of rapidly conducting high-quality clinical studies and trials in which multiple research questions can be addressed. This initiative involves testing the feasibility of establishing the National Electronic Clinical Trials and Research Network. This network will be developed through a phased planning and development process. The network will allow community-based clinicians from the NIH NCRAs to participate in important national studies, facilitate the sharing of data and resources, and augment clinical research performance and analysis.

Dynamic assessment of patient-reported chronic disease outcomes: There is a pressing need to better quantify clinically important symptoms and outcomes, including pain, fatigue, and quality of life. Through this effort, new technologies will be developed and tested to measure these self-reported health states and outcomes across a wide range of illnesses and disease severities.

Translational research: Translational Research Core Services will facilitate the translation of basic discoveries to early phase clinical testing by providing bench and clinical investigators with cost-effective core services, including the expertise needed to move projects through complex logistical and regulatory barriers and the technical services to synthesize chemical and biological agents for early phase clinical studies.

For more information about the NIH Road Map and specific initiatives in areas of your research interests, see the following website: http://nihroadmap.nih.gov/.

6.2. Multiple Project Directors/ Principal Investigators

Modern biomedical research increasingly requires multidisciplinary or interdisciplinary teams. Innovation and progress still depend on creative individual investigators, but increasingly, collaborative synergy is necessary to fully realize the promise of biomedical and biobehavioral research. Increasingly, health-related research involves teams that vary in terms of size, hierarchy, location of participants, goals, disciplines, and structure.

Therefore, NIH is in the process of adopting a model in which multiple PD/PIs may be designated if appropriate for the research proposed. Multiple PD/PIs will be allowed as grant mechanisms transition to electronic submission on the SF 424 (R&R) form. (Multiple PD/PIs will not be appropriate for some types of grants, such as fellowships and career development awards, even after transition to the SF 424 form.) Each of the listed PD/PIs will be designated by the grantee institution and will be expected to share responsibility for directing the project or activity supported by the grant. As with current policies, each PD/PI will be responsible for the proper conduct of the project or activity and accountable to the grantee institution and to NIH. To facilitate communication with NIH, the applicant institution will be asked to indicate a contact PD/PI at the time of application. As in current applications, peer reviewers will consider whether the designated PD/PIs have appropriate training and experience to carry out the proposed study.

Applications with more than one PD/PI will have to include a leadership plan to describe the roles and areas of responsibility of the named PD/PIs as well as the processes for making decisions on scientific direction, allocating resources, and resolving disputes that may arise. Consistent with current NIH practice, the quality of the leadership plan will be considered by peer reviewers during the assessment of scientific and technical merit. (For more information, see http://grants1.nih.gov/grants/guide/notice-files/NOT-OD-07-017.html.)

6.3. Pilot Studies on Shortening the Review Cycle

As described previously, the NIH peer review cycle typically takes approximately 8–10 months. Shortening the review cycle is as important to NIH as it is to the biomedical and behavioral research communities. NIH is also committed to career development and to supporting new investigators in their efforts to obtain research grant funding. Since new investigators by definition do not have NIH research grant support, expediting their ability to submit an amended application is particularly important for their careers. Therefore, beginning in the summer of 2006, the NIH began a pilot of a rapid turnaround process for revised (amended) R01 applications from new investigators. This pilot will allow new investigators to submit a revised application for the very next receipt date, saving approximately $4\frac{1}{2}$ months in the overall process.

For the purpose of review and funding, applicants are considered new investigators if they have not previously served as the PI on any PHS-supported research project other than a small grant (R03), an Academic Research Enhancement Award (R15), an exploratory/developmental grant (R21), or certain research career awards directed principally to physicians, dentists, or veterinarians at the beginning of their research career (K01, K08, and K12). Current or past recipients of independent scientist and other nonmentored career awards (K02 and K04) are not considered new investigators (see http://grants2.nih.gov/grants/guide/notice-files/not97-231.html).

A limited number of study sections in the NIH CSR are participating in the pilot. The results of the pilot will be analyzed to determine whether the shortened review cycle can be expanded to all R01 applications submitted by new investigators or to all R01 applications and also to consider if other grant mechanisms should be included.

Since NIH limits the number of resubmission applications to two, new investigators will need to consider carefully whether this option is appropriate for their situation. Consultation with NIH institute/center program staff and senior colleagues about the specific weaknesses cited in the summary statement, and whether they are amenable to a "quick fix," will be very important.

For more information about the pilot of a shortened review cycle, see the NIH website at http://grants.nih.gov/grants/guide/notice-files/NOT-OD-06-013.html.

7. HOW TO STAY INFORMED ABOUT THE NIH PEER REVIEW SYSTEM

Although the fundamental dual peer review process that is the foundation of the NIH grants system is likely to continue, specific policies, forms, and procedures regarding the peer review process are frequently revised and updated. Therefore, before you prepare a grant application, you should visit the NIH website to obtain the latest information and discuss current application and review procedures with program or review staff at NIH. The following are a few of the "starting points" for finding current information about NIH when you are ready to apply for research support.

7.1. The NIH website

The Internet address for the main NIH home page is www.nih.gov. From there, you can navigate to the Grants and Funding Opportunities page, which has information about NIH grants programs, policies, forms, etc. and links to the NIH Guide for Grants and Contracts and other important information.

7.2. NIH Institute Home Pages

Each of the NIH component institutes also has a home page. The general format for the Internet addresses is "www.Institute acronym.nih.gov." Therefore, the home page for NCI is www.nci.nih.gov, the home page for NHLBI is www.nhlbi.nih.gov, etc. Many programs within each NIH institute also have their own home pages. Each institute home page will have a way for you to find the contact person for each of the general areas of science that the institute supports, as well as the program office responsible for managing the institute's training and career development portfolio.

7.3. The CSR Home Page

Potential applicants are encouraged to visit the CSR home page for additional information about CSR and peer review. The address is www.csr.nih.gov. The home page includes telephone numbers and e-mail addresses for all CSR employees and the following:

1. Breaking news items and information on the activities of the NIH Peer Review Advisory Committee.

2. Information on peer review meetings, including schedules of CSR study section meetings, study section rosters, and review procedures.

3. Resources for applicants, including advice to investigators submitting clinical applications, developments in review (including information on electronic grant applications); particularly instructive for new applicants is a video of a mock study section that can be found at www.csr.nih.gov/Resourcesfor Applicants/InsidetheNIHGrantReviewProcessVideo. htm.

4. Organizational information

28

Clinical Research from the Industry Perspective

DENISE T. RESNIK

Medical Research Consulting Services, Yonkers, New York

Clinical trials are conducted to determine whether medicines are safe and efficacious in humans. Pharmaceutical company-sponsored clinical trials are conducted to create a body of research supporting an investigational new medicine prior to submission for approval to the Food and Drug Administration (FDA).

Currently, industry-sponsored clinical trials are being conducted on 2200 medicines in 800 disease conditions worldwide.[1] In the United States alone, approximately 1000 industry-sponsored clinical trials are currently in progress. This research leads to approval of 35–40 new medicines every year. In addition to industry-sponsored clinical trials, approximately 3000 government-sponsored studies at the National Institutes of Health (NIH) and other federal agencies are currently being conducted.[2] In order to fully understand clinical research from the pharmaceutical industry perspective, it is important to define the components of the industry and discuss the challenges under which the components operate.

The major components of the pharmaceutical industry are traditional pharmaceutical companies, stand-alone biotechnology companies, and "biopharmaceutical companies" that represent a melding of traditional and biotechnology approaches to discovery. In addition, there are medical device companies that, again, are either stand-alone entities or folded into larger pharmaceutical companies.

1. COMPONENTS OF THE PHARMACEUTICAL INDUSTRY

1.1. Traditional Pharmaceutical Companies

Traditional pharmaceutical companies have been in existence for more than 100 years. Early in their histories, many produced chemicals. In the years during and immediately following World War II, many companies perfected mass production techniques, and penicillin was produced on a large scale for the war effort. These companies became the large, traditional pharmaceutical companies that discover, develop, manufacture, and market small molecule prescription medicines.

Traditional pharmaceutical companies identify and develop new medicines by large-scale screenings of new chemical entities, which they have patented or licensed from other sources and which they predict will have medicinal activity. Upon identification of a compound exhibiting activity, the compound is submitted to preclinical pharmacology and preclinical safety testing. Compounds that exhibit desirable safety and efficacy profiles then undergo years of clinical testing to determine human safety and efficacy.

Attrition is a major issue in the development of new medicines. Although thousands of potential medicines (both new chemical entities and medicines seeking approval which are refinements of previously-approved medications) are screened for medicinal activity, only a few new chemical entities exhibit desirable properties and warrant continued investigation. Indeed, for every 5000 potential new medicines developed, only a handful of new chemical entities survive chemistry and preclinical studies (including animal research and *in vitro* studies), and of those, only 1 or 2 new chemical entities continue through clinical research and become approved for use. The entire process from new chemical entity to approved medicine typically takes between 11 and 15 years. The Tufts Center for the Study of Drug Development, which provides strategic information regarding drug development to drug developers, regulators, and policy makers, depicts the process of developing a new chemical entity from the chemist's bench through preclinical testing and, eventually, through clinical testing and regulatory approval as shown in Figure 28-1.[3]

The Tufts Center also estimates that the cost of bringing a new chemical entity to market is approximately $802 million.[4] This figure is based on its survey of 10 drug companies and includes the following costs:

- Out-of-pocket discovery and preclinical (animal and *in vitro*) development costs

- Out-of-pocket clinical costs
- Attrition rate (the pace at which a compound undergoing clinical trials will fail and be removed from the testing regimen at the various clinical phases)
- Clinical success rate (the probability that a compound undergoing clinical trials will result in an approved medicine)
- Development times and the cost of capital over those periods of time[4]

The FDA has documented the impact of attrition of new drugs (both new chemical entities and product refinements) in clinical research as they progress through their review and approval process, as shown in Table 28-1.[5]

1.2. Biotechnology Companies

The methods used by biotechnology companies to discover new medicines are different from those of traditional pharmaceutical companies. Biotechnology is a collection of technologies that capitalize on the functions of various cell components, including DNA and proteins, within a given type of cell or among different types of cells. Biotechnology tools and techniques are used to study the molecular basis of health

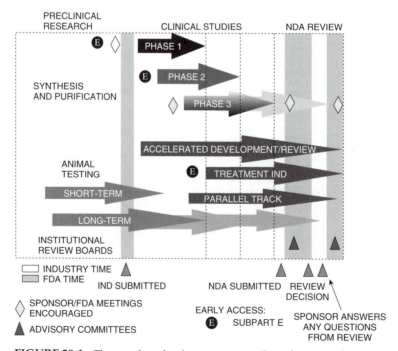

FIGURE 28-1 The new drug development process: Steps from test tube to new drug application review.

TABLE 28-1 FDA Approval Rates for Drug Testing in Humans

	Number of Patients	Length	Purpose	Percent of Drugs Successfully Tested*
Phase 1	20–100	Several months	Mainly safety	70 percent
Phase 2	Up to several hundred	Several months to 2 years	Some short-term safety but mainly effectiveness	33 percent
Phase 3	Several hundred to several thousand	1–4 years	Safety, dosage, effectiveness	25–30 percent

• For example, of 100 drugs for which investigational new drug applications are submitted to FDA, about 70 will successfully complete phase 1 trials and go on to phase 2; about 33 of the original 100 will complete phase 2 and go to phase 3; and 25 to 30 of the original 100 will clear phase 3 (and, on average, about 20 of the original 100 will ultimately be approved for marketing).

• Note: The data presented above include approval of new chemical entities as well as approval of medicines which are refinements of previously-approved medications.

and the changes that take place as a result of disease. This knowledge is resulting in improved and novel methods for treating and preventing disease. By exploiting the extraordinary specificity of cells and biological molecules in their interactions, biotechnology products can often solve specific cellular problems efficiently and with minimal adverse events. By using biotechnology research applications, insights are being gained into the precise details of cell processes, including the following:

• The specific tasks assigned to various cell types
• The mechanics of cell division
• The flow of materials in and out of cells
• The path by which undifferentiated cells become specialized
• The methods cells use to communicate with each other, coordinate their activities, and respond to environmental changes[6]

Biotechnology therapeutics have been approved by the FDA to treat many diseases, including anemia, cystic fibrosis, growth deficiencies, rheumatoid arthritis, hemophilia, hepatitis, genital warts, transplant rejection, and leukemia and other cancers. It is expected that biotechnology will continue to make possible improved versions of today's therapeutic regimes as well as treatments that would not be possible without these new techniques. Currently, there are more than 370 biotechnology vaccines, biologicals, and drug products being investigated in clinical trials, targeting more than 200 diseases, including various cancers, Alzheimer's disease, heart disease, diabetes, multiple sclerosis, immune suppression, immune stimulation (including AIDS), and arthritis. Biotechnology is also critical in many nontherapeutic areas, including medical diagnostic tests, food science, environmental science, industrial applications, and DNA fingerprinting used for criminal investigations and forensic medicine.[6]

The field of biotechnology has mushroomed since 1992. United States revenues increased from $8 billion in 1992 to $39.2 billion in 2003, and research and development (R&D) costs exceeded $17.0 billion in 2003.

The biotechnology product development and regulatory approval processes are similar to those of traditional drug companies and are shown in Figure 28-2.

The basic tools of biotechnology include

• Recombinant DNA technology—used to manufacture products such as human insulin and hepatitis B vaccine;
• Advanced methods in cell culture;
• Monoclonal antibody technology, which uses immune system cells that make highly specific proteins called antibodies;
• Proteomics—the systematic study of the structure, function, cellular location, expression, and interaction of proteins within and between cells; and
• Genomics and pharmacogenomics—analysis of gene structure, expression, and function to tailor therapeutics to the genetic makeup of individual patients with the goal of identifying genetic differences that predispose patients to adverse reactions to certain drugs or make them good subjects for other drugs.[6]

Current medical uses of recombinant DNA techniques, in conjunction with molecular cloning, include:

• Production of new medicines and safer vaccines;
• Treatment of some genetic diseases;
• Controlling viral diseases; and
• Inhibition of inflammation.

In addition, recombinant DNA technology is important in agriculture and food sciences, environmental sciences, and in developing biodegradable plastics.[6]

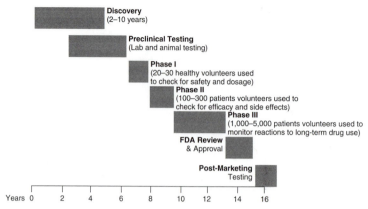

FIGURE 28-2 Overview of the drug discovery process. Reprinted with permission by Ernst and Young (2005).

Improvements in cell culture technology have increased our understanding of the molecular basis of the cell cycle. Scientists have found that the rigorously controlled sequence of steps in the cell cycle depends on both genetic and nutritional factors, and that a delicate balance exists between factors that stimulate cell division and those that inhibit it. Any disruption of this balance leads to uncontrolled cell proliferation (cancer) or cell death (apoptosis).[6]

Current monoclonal antibody research is being conducted to develop methods to selectively suppress the immune system in organ transplantation patients and those with autoimmune diseases. Additionally, a new generation of vaccines is being developed that consists of only the antigen, not the actual microbe. These vaccines are produced by inserting the gene that produces the antigen into a manufacturing cell, such as yeast. During the manufacturing process, each yeast cell makes a perfect copy of itself and the antigen gene. The antigen is then isolated and used as a vaccine without the risk of transmitting the virus.[6]

An ever-expanding knowledge base in proteomics and genomics is serving as the foundation for the following initiatives:

- Predictive tests for diseases that can be prevented with targeted interventions
- Fundamental changes in the way drugs are discovered, tested, and developed
- Therapies that are tailored to the specific genetic makeup of individual patients
- Therapies that address and sometimes correct the biochemical causes of a disease rather than only alleviating the symptoms[6]

For example, gene therapy is a promising technology that uses genes or related molecules such as RNA to block mutated genes and thereby to treat diseases.

Research is currently being conducted to determine whether instead of injecting patients with missing proteins to treat deficiencies, patients could be administered nondefective genes to enable the patient's body to produce previously deficient proteins and correct the genetic defects.[6]

Pharmacogenomics is the study of genome-derived data elucidating individual patient genetic variations to predict disease risk and progression and the response by individual patients or groups of patients to specific drugs. Following the 2003 completion of the human genome sequence, the expected impact of genetics and genomics on the diagnosis and treatment of disease seems endless, and it is predicted that pharmacogenomics will be crucial for successful discovery, development, and delivery of new medicines.

In order to study pharmacogenomics, appropriate biomarkers must be identified, and tools to measure biomarkers must be developed. A biomarker is a characteristic that is objectively measured and evaluated as an indicator of normal biologic processes, pathogenic processes, or pharmacologic responses to a drug. Traditional biomarkers have been used to ascertain efficacy and safety of medicines for large populations. One widely used traditional biomarker is the measurement of blood pressure as an indicator of cardiovascular health. For individualized therapy, pharmacogenomic biomarkers are being developed to identify individuals likely to benefit from a particular treatment as well as those individuals at risk from the same treatment.[7]

The economic impact of the use of biotechnology to develop medicines and biologicals is beginning to unfold. It is expected that the use of these techniques to test the safety and efficacy of medicines early in the drug development process will have a positive impact on the total development cost. For example, if, by using biotechnology techniques, compounds early in the development process are found not to possess expected attributes, work on these compounds can be halted in favor of other compounds with more promising profiles. Improved profitability might also be realized by shortening the product development process as a result of using a single technology at many steps in the process. For example, a small piece of DNA used to locate a gene might eventually become a component of a diagnostic test. Similarly, a monoclonal antibody developed to identify therapeutic leads might be used to recover and purify that therapeutic compound during scale-up.[6]

Biotechnology has already favorably impacted the costs of diagnostics. These diagnostics are not only less expensive than those produced by traditional methods but also more accurate and quicker than previous tests. These changes greatly improve a patient's prognosis by allowing for earlier diagnoses of disease processes.[6]

1.3. Biopharmaceutical Companies

The pharmaceutical industry trade group, Pharmaceutical Research and Manufacturers of America (PhRMA), coined the term *biopharmaceutical* industry.[8] Traditional pharmaceutical companies use this term to describe the incorporation of biotechnology principles. Pharmaceutical companies traditionally discover medicines by studying chemical reactions in the body, specifically searching for the effect on a specific disease target, whereas biotechnology develops methods that capitalize on the attributes of cells and use DNA and proteins to modify the cell functions as a way to combat disease. In recent years, traditional pharmaceutical companies have adopted many of these new technologies either by developing biotechnology groups within their own organizational structures or by partnering or purchasing biotechnology companies. This convergence of biotechnology and traditional pharmaceutics has led to the development of biopharmaceutical companies, which possess strengths from both disciplines.

1.4. Medical Device Companies

The process of developing a medical device is different from that of developing a new medicine, although developing a medical device is time-consuming (up to 15 years) and may be expensive (up to $350 million). Also, although there are three different classes of medical devices from a regulatory standpoint, the development of all classes of medical devices follows a stepwise process as outlined here.[2]

1.4.1. Basic Research

Primary to development of any medical device is the underlying basic research. Basic research—typically conducted by physicists, biologists, and mathematicians—provides the fundamental understanding of physical phenomena (e.g., gravity) and human physiology and the interaction of the two.[2]

1.4.2. Applied Science

Based on the fundamental research, scientists develop a prototype device believed to have a medical application. By using computer simulations with the prototype design, scientists can predict the feasibility of the design as well as projected safety and efficacy. In addition, during this stage, scientists can estimate the costs of "scaling up" the prototype to produce a commercially viable medical device.[2]

1.4.3. Engineering

During the engineering stage, fully operational products designed to meet clinical needs are developed from the prototype. Not only is the medical device developed but also all necessary hardware and software are developed and integrated into the medical device.[2]

The device is then tested in animals to evaluate the reliability of the product. When the device has proved to be reliable, it is then tested in human volunteers to evaluate safety, efficacy, and user acceptance.[2]

1.4.4. Commercialization Stage

The medical device is prepared for final use. Detailed clinical trials and testing are performed to secure final regulatory approval and to support product labeling. Marketing plans are developed to produce necessary training tools and to address pricing issues, and manufacturing capacity is established for commercial product production.[2]

1.4.5. Classification and Regulations Surrounding Medical Devices

There are three classes of medical devices defined by the complexity of the device and the amount of risk they present to the user:

Class I: These devices are not used to support or sustain life but do require general controls. Examples of class I medical devices are surgical gloves and tongue depressors.

Class II: These devices not only require general controls but also must meet performance standards. An example of a class II medical device is a hypodermic needle.

Class III: These devices sustain or save lives and require premarketing approval, similar to that of a new medicine. Examples of class III medical devices are ventilators, x-ray machines, and vascular stents.[2]

2. ISSUES IMPACTING INDUSTRY CLINICAL RESEARCH

Pharmaceutical company-sponsored clinical trials are conducted to support labeling claims made for a new medicine, biologic, or medical device. These trials are similar to those designed and conducted by any noncommercial entity such as the NIH or an academic medical center; however, there are additional clinical trials that pharmaceutical companies must conduct, either as necessary to support the new medicine in the marketplace or as required by regulatory agencies.

Recent additions to regulations regarding the design and conduct of clinical trials have created new challenges for the pharmaceutical industry. The number of subjects and length of therapy required for a New Drug Application (the dossier of clinical, preclinical, chemistry, and manufacturing information about a medicine seeking approval) has been steadily increasing. In addition, the FDA has required that measurements of safety and efficacy and the biostatistical analyses used to evaluate the clinical trials parameters be increasingly rigorous.

The issues impacting the industry's conduct of clinical research are interrelated. As regulations surrounding the development of new medicines and the clinical trials used to support claims of efficacy and safety become more encompassing, the costs incurred to develop new medicines increase accordingly. Therefore, as costs increase, pressures on the pharmaceutical companies by both the public who use and pay for the medications and their company shareholders impact on the conduct of pharmaceutical business.

2.1. Voluntary Postapproval Trials

Many clinical trials are conducted by pharmaceutical companies after new medicines are approved. Although it might seem that the approval of a new medicine would signal the end of clinical investigations, that is rarely the case. More often, the pharmaceutical companies conduct additional long-term safety and efficacy trials designed to answer questions that did not need to be addressed during phases I–III, including:

- Determination of the medicine's place in the array of medicines already available to treat the condition under study;
- Cost-effectiveness of the medicine;
- The effect of the medicine on patients' quality of life; and
- The safety and efficacy of the medicine on specific subpopulations of patients.

2.1.1. Clinical Outcomes Trials

Large-scale late phase III and phase IV clinical outcomes trials are being conducted with increased frequency by industry. These outcomes trials, which are not FDA mandated but are essential for a fuller understanding of a new medicine, are designed to measure the long-term safety and efficacy of a new medicine on large patient populations. These outcomes trials typically collect morbidity (including myocardial infarction and stroke frequencies) and mortality data from the use of a new medicine by patients for a period of up to five years. These trials are funded by the industry sponsor (usually the pharmaceutical company developing the new medicine under study), but they are typically conducted by independent contract research organizations (CROs). Sponsors engage CROs to conduct any or all trial-related duties and functions. These duties may include selecting study investigators and investigational sites, conducting study-specific training of study site personnel, and monitoring and reconciling study-generated data. Use of CROs is contracted with the understanding that the CRO is acting in lieu of the sponsor, but that the responsibility for the quality and integrity of the study data remains with the sponsor.[9,10]

Data from these trials are typically analyzed by independent data coordinating centers hired by the sponsor to conduct biostatistical analyses, and an additional level of periodic monitoring of study data is typically performed by a data safety monitoring board (DSMB). A DSMB is another entity independent from the sponsor of the trial and is charged with the evaluation of the safety of a study and the determination of whether a study should be continued or terminated based on benefit-to-risk ratio. DSMBs review study protocols and data collection methods, define safety parameters, review adverse events occurring during a clinical trial, and determine whether an interim analysis of a clinical trial is appropriate based on safety and efficacy findings, and if so, conduct the interim analysis. A DSMB may decide to terminate a trial exhibiting an unfavorable safety profile for the new medicine, may terminate a trial if the new medicine exhibits overwhelming benefits over the comparative treatment, or may decide to let the trial continue to its conclusion. DSMBs are composed of clinicians with expertise in relevant safety concerns. DSMBs have ethical responsibilities to the study subjects participating in the trial and scientific responsibilities to the investigators to ascertain that the study's objectives are being met.[9,10]

2.1.2. Product Placement Studies

When the FDA approves a new medicine, the clinical trials on which the approval is based are typically designed to evaluate the new medicine in study patients against placebo (a chemical entity with no medicinal effects) and against the accepted standard medicine prescribed for the condition under study. However, while the investigational medicine undergoes years of clinical testing, it is possible that the FDA will approve another new medicine developed for the same medical condition. Therefore, it is likely that the pharmaceutical company will pursue additional clinical trials after approval to compare the safety and effectiveness of its new medicinal product against other medicines to assure the new product's viability in the marketplace. These product placement studies are typically conducted against the following types of medicines:

- Current market leaders
- Expected future market leaders
- Medicines in the same chemical class
- Medicines in the same therapeutic class

2.1.3. Pharmacoeconomic Trials

Postmarketing clinical trials may also be conducted to ascertain cost-effectiveness of a new medicine. In these studies, costs and consequences of treatment are simultaneously measured to determine whether the benefits of a new medicine justify its costs. These trials may be conducted with the new medicine alone or in comparison with other medicines currently available or other modes of treatment (i.e., hospitalization and/or surgical intervention). The goal of a pharmacoeconomic study is to determine whether the expense incurred by the use of a new medication is justified in comparison with the cost of existing medication as well as potential savings resulting from a decrease in the number of physician visits, emergency room visits, length and number of hospitalizations, ancillary transportation costs, and the number of days of work lost by patients taking the new medication. The results of these pharmacoeconomic studies are analyzed by the large providers of prescription medicines (i.e., national, state, and local governments), health maintenance organizations, and pharmacy benefit management companies to determine the new medicine's place in their formularies (a compilation of medicines for which the providers will pay).[11]

2.1.4. Quality of Life Studies

Other postmarketing studies are conducted to determine the effect(s) of the new medication on patients' quality of life. These trials are designed to measure patients' reactions to a new medicine (including efficacy measures, safety measures, ease of use, convenience, and costs). Data are collected using quality of life questionnaires completed by patients addressing their current lifestyles, past experiences, and expectations for the future. The questionnaires are then assessed to determine whether patients are pleased or displeased with the new medication.[12]

2.1.5. Patient Subpopulation Studies

Other marketing studies are conducted to evaluate a new medicine in specific patient populations (i.e., elderly, pediatric, or immunosuppressed patients) or in patients taking specific concurrent medications that may not have been studied in-depth during the studies conducted for regulatory approval. These trials are designed to evaluate particular issues that might arise in these subpopulations that could not have been ascertained from earlier studies.

2.1.6. Postmarketing Surveillance of Medical Devices

Postmarketing surveillance of medical devices is conducted to evaluate the device in actual use by documenting the following parameters:

- Frequency of ongoing service and preventive maintenance of the devices
- Monitoring of performance
- Frequency of adverse events occurring during use of the device

Based on the postmarketing information, the manufacturer may either provide upgrades for the device or, more likely, develop an improved model based on the experiences of the first model and repeat the regulatory process.[2]

2.1.7. Ethical Considerations

As discussed in Chapter 2, there is an abundance of issues surrounding ethics of clinical trials and protection of human subjects participating in these trials.

One issue that impacts on pharmaceutical company-sponsored clinical trials as well as nonpharmaceutical company-sponsored trials (e.g., those sponsored by NIH, the Centers for Disease Control and Prevention, the World Health Organization, and the Bill and Melinda Gates Foundation) involves the ethical considerations of conducting placebo-controlled clinical trials in areas of socioeconomic depression.[13,14]

Historically, regulatory agencies have favored placebo-controlled studies for proving safety and efficacy of new medicines. Although this method is useful for the initial approval of a new medicine, in the case of long-term, costly therapies (e.g., antiretroviral medication for the treatment of HIV infection), there are often large-scale initiatives to study the new medication in areas where the disease under study has had devastating effects and the area is unable to afford the new medication (e.g., sub-Saharan Africa). The ethical issue in this case is whether placebo-controlled studies are ethically acceptable or whether an active therapy that is less expensive than the new medication (either a different medication or the new medication at a lesser dose and/or duration of therapy) is more appropriate. Proponents of placebo-controlled trials argue that placebo is essentially equivalent to the current "standard of care" in the region—that is, that no therapy is available to patients. They also make a statistical case that the number of patients required to compare an active medication to placebo is less than with an active control, which also leads to reduced costs to conduct the trial. Opponents of placebo-controlled trials state that equivalency studies—those conducted when a particular regimen that has already been proved effective is compared to a second regimen that is about as effective but less toxic or expensive[13]—are more acceptable.

2.2. Regulatory Issues

2.2.1. Biostatistical Analysis

Regulatory requirements for biostatistical analysis of industry-sponsored clinical trials have become increasingly rigorous. Statistical methodology for identifying primary end points and for analyzing all clinical trials data must be defined prior to beginning a clinical trial, and primary end points must exhibit statistical significance at the $p < 0.05$ level. Secondary end points may also be defined, but the FDA will not accept secondary end points alone to support labeling. Furthermore, even if clinical trials show clinical or medical significance, without statistical significance as defined previously, the FDA will not accept clinical or medical significance alone.[9]

2.2.2. Postmarketing Studies

In addition to the postmarketing studies performed voluntarily by the pharmaceutical company, regulatory agencies may also require that additional clinical trials be conducted after regulatory approval. One issue facing regulatory agencies is the dichotomy between the types of patients participating in preapproval clinical trials and those patients who will ultimately use the new medication postapproval. Patients selected for phase II and III clinical trials (phase I subjects are usually healthy normal volunteers) tend to be of young to middle age and relatively disease-free except for the condition under investigation and free of medications other than the investigational drug. By selecting these patients, physicians, biostatisticians, and regulators can evaluate the effects of an investigational drug without the confounding issues of drug–drug interactions and adverse events that might stem from concomitant illnesses rather than the condition being evaluated.

Although this methodology is useful and albeit necessary in evaluating investigational new drugs, it presents formidable issues when the investigational drug is approved for use by the general public. Upon approval, the drug will be used by different types of patients than those used in evaluating the drug for the approval process. In recent years, several medicines, including the following, were withdrawn from the U.S. market as a result of safety issues that were not apparent until the medicines were used by the general public under less stringent conditions than those under which clinical trials were conducted:

- Terfenadine, an antihistamine that exhibited drug interactions causing cardiotoxicity
- Ticrynafen, an antihypertensive medication that caused hepatotoxicity
- Flosequinan, a congestive heart failure medication shown to increase mortality[15]

In addition, rofecoxib (sold under the name Vioxx®), an anti-inflammatory medication, was voluntarily withdrawn from the U.S. market after the DSMB overseeing a long-term study of the drug in gastrointestinal disease recommended the study be halted because of an increased risk of serious cardiovascular events, including myocardial infarction and stroke.[16]

These issues have led the FDA to expand its requirements for postmarketing studies.[17] Postmarketing studies are required by the FDA of pharmaceutical companies not for initial approval of a drug but to provide the regulators additional information. Data from postmarketing studies typically address the following issues:

- Safety and efficacy in a wider patient population than that tested during phases I through III of the drug approval process
- Additional information on prescribing/use of a product
- Drug-interaction data
- Product quality information

The FDA is currently requesting postmarketing studies in 73% of approved new medications,[17] with a steady increase in the median number of patients from 30 in the 1970s to 123 in the 1980s and 920 in 2003.[18]

2.2.3. Patent Issues

2.2.3.1. Medicines and Biologics

A patent gives an inventor the right to be the only entity to manufacture and sell an invention for the life of the patent, typically 20 years. In the case of pharmaceutical companies, the invention is a new chemical entity, a new device, a new process, or a new biological product.

Pharmaceutical companies rely on government-granted patents to protect their huge research and development investments in new medicines they believe will exhibit medicinal efficacy. Without these patents to protect all of the inventions necessary to develop a drug for this period of time, other companies could copy the drugs immediately and offer their versions at prices that do not have to reflect the costs incurred to develop the drugs. This would seriously impact the pharmaceutical companies' abilities to recoup their expenses and reinvest in other research projects. Since the length of time to develop a new chemical entity into an approved medicine typically exceeds 10 years, the number of years remaining to recoup expenses and make a profit is reduced accordingly.[19]

In response to this issue, the FDA has developed a new initiative making it more attractive for pharmaceutical companies to conduct research leading to a second indication in a recently approved medication. Upon approval of the second indication, the FDA can extend a drug patent for an additional three years. This is attractive for the pharmaceutical company since it allows the company to recoup some of its development costs for the new indication prior to the medicine's patent expiration.

2.2.3.2. Medical Devices

Unlike new chemical entities that are patented early in their development, discoveries leading to the development of new medical devices are often not patented. These discoveries occur during the basic research stage of development, and either they are not patentable or the decision is made that the likelihood of requiring an early patent is outweighed by the likelihood that the usable patent life after full development of a device would be restricted. This extends the number of years that a device can be sold with patent protection.[2]

2.3. Financial Pressures

The R&D expenditures in the pharmaceutical and biotechnology industries continue to exceed R&D expenses (as a percentage of annual revenues) in any other area of the U.S. economy. The 2005 *Pharmaceutical Industry Profile* published by PhRMA[9] provides a spending figure for total biopharmaceutical R&D in 2004 of $49.3 billion, which represents approximately 17% of annual revenues. In comparison, the percentage of R&D compared with annual revenues for all U.S. industries in 2004 was 3.9%.

Table 28-2 contrasts published sales and R&D data for several major pharmaceutical companies and biotechnology companies for the years 2003–2005.[20]

TABLE 28-2 Sales and Research & Development Data for Selected Major Pharmaceutical and Biotechnology Companies

	2003			2004			2005		
	Sales	R&D	R&D as % of Sales	Sales	R&D	R&D as % of Sales	Sales	R&D	R&D as % of Sales
Pfizer	$45.19	$7.13	15.78%	$52.52	$7.68	14.62%	$51.3	$7.4	14.5%
GlaxoSmithKline	$38.27	$4.96	12.96%	$39.22	$5.47	13.95%	$37.3	$5.4	14.5%
Novartis	$24.86	$3.76	15.12%	$28.25	$4.21	14.90%	$32.5	$4.9	14.9%
Merck	$22.49	$3.18	14.14%	$22.94	$4.01	17.48%	$22.0	$3.9	17.5%
Bristol-Myers Squibb	$20.89	$2.28	10.91%	$19.38	$2.50	12.90%	$19.2	$2.8	14.3%
Eli Lilly	$12.58	$2.35	18.68%	$13.86	$2.69	19.41%	$14.7	$3.0	20.7%
Amgen	$8.36	$1.66	19.86%	$10.55	$2.03	19.24%	$12.4	$2.3	18.6%
Genetech	$3.30	$0.72	21.82% 16.16%	$4.62	$0.95	20.56% 16.63%	46.6	$1.3	19.0% 16.8%

2.3.1. Early Termination Strategies

Competitive economic forces have led to productivity and quality improvement mandates for all pharmaceutical companies, and as the cost of developing new medicines rises, decisions regarding continuing or terminating unpromising R&D programs have become increasingly critical. Reasons for terminating unpromising new drugs and the time to terminate are presented in Figure 28-3. Although safety and efficacy considerations have historically led to decisions to terminate clinical programs, economic factors—apart from safety and efficacy—are currently foremost in determining the viability of an investigational drug.[21]

2.3.2. Exportation of Clinical Trials

In an environment of increasing numbers of clinical trials every year, the pool of principal study investigators has dropped in the United States by 11% between 2001 and 2003. Not only has this pool decreased in size but also it has become significantly more male (even more male than the percentage of males in the population of board-certified physicians in the United States) and more regional. In other words, there has been a significant decline of principal study investigators practicing in regions with declining population, which potentially diminishes scientific and economic benefits to these areas. Conversely, the number of principal investigators outside the United States increased by 8% during the same period of time. In response to this shift in principal investigators and the concomitant availability of study subjects abroad, there has been a continued exportation of clinical trials from the United States to those areas with increasing populations of principal investigators and resulting patient populations.[22]

The impact of this shift on pharmaceutical companies is that, by placing studies offshore, their clinical trials can be conducted more efficiently with respect to both time and expenses to the detriment of the U.S. economy.

2.3.3. Cost Containment

The cost of drug development has increased 250% in the past 10 years. The price of medicines has reflected this increase despite the fact that only 3 approved new medicines in 10 return their development costs to the pharmaceutical company. Growing pressures to contain costs from government and private health benefits organizations in the United States and pricing and reimbursement authorities abroad will continue to levy pressure on pharmaceutical companies to move new drugs with clear advantages in safety, efficacy, or economic value to market quickly. In addition, the Medicare Prescription Drug, Improvement, and Modernization Act of 2003, launched in 2006, is causing increased pressure on pharmaceutical companies by health care plans and pharmacy benefit managers to exhibit advantageous cost and benefit profiles of their products in order to add them to their formularies.[21]

Financial pressures are generated not only by the pricing of new medicines but also by the robust generic medicine companies positioned to manufacture and sell innovator medicines as patents expire. In recent years, 30–40 innovative drugs with worldwide sales exceeding $10 billion lost patent protection each year and were subject to generic entry.[19]

The United States has a vital generic drug industry, largely as a result of the 1984 passage of the Waxman–Hatch Act, which facilitates entry of generic products as the patents for innovative products expire. Generic product entry drives down the sales of the nonpatent-protected innovator products and drives down the prices of the generic counterparts through market competition. Strong protection of intellectual property (patent protection) preserves the incentive to develop improved treatments, whereas the cost of new treatments declines rapidly after patents expire, leading to cost containment.[19]

2.3.4. Accelerated Approval of Medications (Fast Track)

Fast track programs became available under the FDA Modernization Act of 1997 and are designed to facilitate development and expedited review of new drugs intended to treat serious or life-threatening conditions and to demonstrate the potential to address unmet medical needs. Seriousness with regard to fast track designation is defined by the FDA as a disease that "impacts such factors as survival, day-to-day

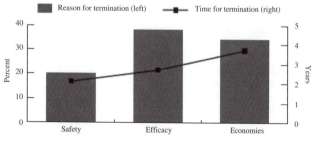

FIGURE 28-3 Reasons for terminating unpromising new drugs and time to termination. Source: Tufts Center for the Study of Drug Development.

functioning, or the likelihood that the disease, if left untreated, will progress from a less severe condition to a more serious one." Examples include AIDS and HIV, Alzheimer's disease, angina pectoris, heart failure, and cancer.[23]

In order for a product to be considered for the fast track regulatory approval process, "it must not only be used in patients with a serious condition, it must be intended to treat a serious aspect of that condition."[23] The following are examples of products that might meet these criteria:

• Therapeutic products directed at some aspect of a serious condition
• Diagnostic products used to improve diagnosis or detection of a condition with the presumption that the improvements in diagnosis or detection would lead to improved outcome
• Preventive products used for their ability to prevent serious manifestation(s) of a condition or to prevent a condition thereby preventing its serious consequences
• Products intended to ameliorate or prevent a serious side effect of another therapy treating a serious condition
• Therapeutic products with the ability to treat a condition while avoiding the serious sequelae of currently accepted treatments of the condition[23]

In order to qualify for the FDA fast track program, a product also must demonstrate the potential to address unmet medical needs in any of the following scenarios:

• There is no available therapy for the condition.
• There is available therapy for the condition, but
the new therapy exhibits superiority used alone or in combination with other therapies in morbidity end point controlled clinical trial(s);
the new therapy exhibits a positive effect on progressive disability that available therapy does not exhibit;
the new therapy provides benefit in patients who are unable to tolerate or are unresponsive to available therapy;
the new therapy provides benefit(s) similar to available therapy while avoiding serious toxicity present in available therapy or common, less serious toxicity that causes discontinuation of available therapy; or
the new therapy provides similar benefit to available therapy but exhibits improvement of some factor (e.g., compliance or convenience) that leads to improved effects on serious outcomes.

• The only available therapy had accelerated approval (either on the basis of an effect on a surrogate end point or for restricted distribution).[23]

In summary, the FDA's Fast Track Drug Development Program is an example of industry and regulators working together to solve a problem to the benefit of each party. As more investigational medicines are reviewed under the FDA's fast track designation, the resultant efficiencies should lead to a speedier, less expensive drug development process, which in turn should lead to more accessibility to new medicines.

3. INDUSTRY OUTLOOK

The pharmaceutical industry continues to be an exciting and innovative industry in which huge strides in medical advancement will continue to be realized in the years to come. Despite the innovations still to be made, the pharmaceutical industry faces serious challenges both from government regulators and the marketplace.

As traditional pharmaceutical companies continue to streamline the new drug development process and many incorporate biotechnological methodologies into at least part of their research effort, they will not only continue to develop new chemical entities for large portions of the population but also be able to develop drugs and biologicals used to treat small, unserved or underserved portions of the population, as biotechnology companies currently do, thus relieving, in part, some of the hurdles they have traditionally faced.

Biotechnology companies will likely prosper in the next few years as they continue to engage in innovative R&D strategies and investigate therapeutic and diagnostic products with high approval success rates. Their successes are likely to occur in the areas of serious and life-threatening diseases, due in part to the availability of fast track designation by the FDA.[18,21] Successes are likely to come in the areas of:

• diseases and conditions eligible for fast track designation;
• recombinant therapeutics currently in development (more than 30 are likely to be approved by the FDA in the short term);
• innovative and orphan therapeutics, if firms capitalize on scientific advice available from the FDA and its counterpart in Europe, the European Agency for Evaluation of Medicinal Products; and
• increasing numbers of oncology monoclonal antibodies will enter clinical trials due to recent successful launches.[18,21]

The outlook for biotechnology companies is strong. Biotechnology companies tend to have fewer problems amassing patients for their studies. These patients actively seek out promising medicines for their unmet needs through Web searches and patient advocacy groups. Since these patients are not already taking effective therapy for their medical conditions, they are much more willing to participate in clinical trials than are patients already being treated with available therapy.[24]

3.1. Public–Private Opportunities

Although some biopharmaceutical companies conduct their entire drug discovery and development programs in-house, many companies engage cooperatively with other organizations to share their expertise in drug discovery and development. These partnerships may take the form of licensing agreements between large pharmaceutical companies and biotechnology companies, in which the larger company typically conducts the large, expensive clinical trials of a promising investigational product discovered by a biotechnology firm. Other types of partnerships also exist between a pharmaceutical company and academic medical centers and/or government research agencies.[21]

Through these public–private partnerships, each entity brings its unique resources and strengths to the partnership (including intellectual property and other proprietary materials, experimental compounds, scientific expertise, and financial resources), which results in a more efficient development process and ultimately a better product than either partner could accomplish alone. Public–private partnerships have become a model for advancing science and communicating results of medical advances. As increasingly complex biomedical problems are addressed, strategic partnerships between pharmaceutical companies, government research agencies, academic medical centers, and other research centers will become increasingly important.[25]

4. SUMMARY

The successes and challenges the pharmaceutical industry faces in its conduct of clinical trials are in some ways similar and in many ways different from those faced by nonindustry participants in clinical research. Changes in the way clinical trials are conducted are beginning to reflect these issues and should continue to do so. Since the overwhelming percentage of the world's new medicines are developed by U.S.

pharmaceutical companies, it is in the regulators' and the public's best interest to facilitate pharmaceutical research and development in a manner that foremost protects patient safety while providing new medicines to promote patient health and well-being.

References

1. CenterWatch website. Available at CenterWatch.com. Accessed June 13, 2005.
2. Copyright 2005 Pfizer, Inc. Used with permission.
3. Tufts Center for the Study of Drug Development. *Backgrounder: How New Drugs Move through the Development and Approval Process.* Boston, Tufts Center for the Study of Drug Development, November 1, 2001.
4. Tufts Center for the Study of Drug Development. *Backgrounder: A Methodology for Counting Cost for Pharmaceutical R&D.* Boston, Tufts Center for the Study of Drug Development, November 1, 2001.
5. *Testing Drugs in People.* Available at www.fda.gov/cder/about/whatwedo/testtube-3a.pdf. Accessed May 13, 2005.
6. Bio.org website. Available at http://bio.org. Accessed May 31, 2005.
7. FDA.gov website. Available at www.fda.gov/cder/genomics/PGX_biomarkers.pdf. Accessed June 1, 2005.
8. Pharmaceutical Research and Manufacturers of America. *Pharmaceutical Industry Profile 2005.* Washington, DC, Pharmaceutical Research and Manufacturers of America, 2005.
9. Michael T. Gaffney, Ph.D., Senior Director, Statistical Science and Consulting, Pfizer, Inc. Personal communication, 2005.
10. FDA.gov website. Available at www.fda.gov/cder/livertox/Presentations/im7130/sld008.htm. Accessed June 2, 2005.
11. Spilker B. Pharmacoeconomic trials. In *Guide to Clinical Trials.* Philadelphia, Lippincott Williams & Wilkins, 1991.
12. Spilker B. Quality of life trials. In *Guide to Clinical Trials.* Philadelphia, Lippincott Williams & Wilkins, 1991.
13. Lurie P, Wolfe S. Unethical trials of interventions to reduce perinatal transmission of the human immunodeficiency virus in developing countries. *N Engl J Med* 1997;337:853–856.
14. Angell M. Investigators' responsibilities for human subjects in developing countries. *N Engl J Med* 2000;342:967–969.
15. Lasser K, Allen P, Woolhandler S, Himmelstein D, Wolfe S, Bor D. Timing of new black box warnings and withdrawals for prescription medications. *JAMA* 2002;287:2215–2220.
16. Food and Drug Administration. FDA issues public health advisory on VIOXX as its manufacturer voluntarily withdraws the product. *FDA News,* September 30, 2004.
17. Tufts Center for the Study of Drug Development. FDA requested postmarketing studies in 73% of recent new drug approvals, Impact Report Vol. 6, No. 4. Boston, Tufts Center for the Study of Drug Development, July/August 2004.
18. Tufts Center for the Study of Drug Development. *Outlook 2004.* Boston, Tufts Center for the Study of Drug Development, 2004.
19. Pfizer.com website. Available at Pfizer.com. Accessed June 1, 2005.
20. Paul J. Resnik, C.F.A., Principal, Resnik Asset Management Company, Inc. Personal communication, 2006.
21. Tufts Center for the Study of Drug Development. *Outlook 2005.* Boston, Tufts Center for the Study of Drug Development, 2005.
22. Tufts Center for the Study of Drug Development. Number of principal investigators in the U.S. is declining, according to Tufts CSDD [News release] Boston, Tufts Center for the Study of Drug Development, May 3, 2005.

23. *Guidance for Industry: Fast Track Drug Development Programs—Designation, Development, and Application Review.* Rockville, MD, U.S. Department of Health and Human Services, Food and Drug Administration, Center for Drug Evaluation and Research, Center for Biologics Evaluation and Research, July 2004.

24. Paul E. Freiman, President and CEO, Neurobiological Technologies, Inc. Personal communication, 2005.

25. NIH.gov website. Available at http://nihroadmap.nih.gov/publicprivate. Accessed June 13, 2005.

29

Human Genome Project, Genomics, and Clinical Research

HELEN N. LYON* AND BRUCE R. KORF**

*Division of Genetics, Program in Genomics, Children's Hospital Boston, Boston, Massachusetts
**Department of Genetics, University of Alabama, Birmingham, Alabama

1. OVERVIEW

During the past century, research on the role of genetics in health and disease has mostly focused on rare monogenic and chromosomal disorders. There have been major advances in our understanding of genetic mechanisms of disease and translation of this understanding to clinical practice. Examples of the latter include cytogenetic analysis, prenatal diagnosis, newborn screening for inborn errors of metabolism, and molecular diagnostic testing. The completion of the human genome sequence as well as a comprehensive sequence variation map (HapMap) has further empowered this research and also enabled attention to be directed toward elucidation of genetic contributions to common disorders. It is expected that this will lead to new insights into pathophysiology and to the development of new approaches to diagnosis, prevention, and treatment.

Analysis of complex disorders seeks to elucidate pathologic mechanisms by dissecting the many contributing genetic and environmental factors. There are several steps involved in this research. The first is to establish the degree to which genetics contributes to the disorder, and whether that contribution is best attributed to a single gene or multiple genes. This is done by determining whether a trait clusters in families and is seen among related individuals more often than in the general population. Heritability and variance components analysis estimate the relative strength of genetic and nongenetic risk factors. Possible inheritance models are delineated by segregation analysis.

Genes that contribute to disease susceptibility can be located with linkage mapping. Association studies can determine relative risk of disease in individuals with specific genetic variants and can help elucidate gene–environment interactions.[1]

The study of common disorders using genetic methods is often referred to as *genetic epidemiology*. This term reinforces the concept that such studies require careful design with epidemiologic methods, even when genetic factors are being investigated. Even the most thorough molecular and statistical analysis cannot overcome the detriments of poor study design. The study population needs to be clearly defined, recognizing that small isolated populations are the most genetically homogeneous but reduce the generalizability of results. For case–control studies, the cases need to be selected from the source population and matched for race. The phenotypic outcomes should be chosen carefully to represent the most narrow disease phenotype in order to decrease the impact of phenocopies and variable expression. All measurements should be done consistently for all study participants. Although DNA samples may be processed in several labs for multicenter studies, the same methods and quality control procedures should be used since differences in the quality of the genotyping can cause spurious results.[2] Optimal study design is necessary to dissect a complex trait because any single component may have a very modest effect on the outcome and therefore be indiscernible.

Studies to date have revealed few replicable gene associations for common disorders,[3,4] but new tools are available to enable more powerful studies. Improved technologies have provided high-throughput sequencing and genotyping with results available in public databases such as the human genome sequence, HapMap,[5,6] and the National Institutes of Health's Entrez. Chip technology has evolved to the extent that high-throughput whole genome association scans are now possible. These tools, combined with newly evolving statistical methods, provide the ability to assess large groups of people for variation across their entire genome for association with common diseases,[7] bringing such discoveries closer.[8,9] Modeling of the complex interactions resulting in disease creates very challenging statistical problems. This chapter reviews genetic models and describes the basic methods currently used to conduct complex trait inheritance research.

2. GENETIC MODELS

Medical genetics has typically focused on the diagnosis and treatment of disorders with recognizable patterns of genetic transmission. These patterns follow Mendelian inheritance in that they are determined largely by mutations in single genes, inherited in a dominant or recessive pattern. Specific disease manifestations and severity are modified by the effects of other genes and environmental factors. The pathogenesis of a more common disease is governed by complex interactions of multiple genes with one another and with the environment, so inheritance of these traits is referred to as multifactorial inheritance.

Each person has 22 pairs of nonsex chromosomes, one inherited from each parent, and either two X chromosomes in females or an X and a Y chromosome in males. An individual therefore has two copies of each nonsex linked gene. The term *allele* is used to refer to a variant form of a specific gene and can refer to a whole region or to a specific place in the sequence. A *locus* is a specific position in the DNA sequence and can refer to a single nucleotide or a whole region. At any given locus, a human has two alleles. A silent allele would be a variant that has no effect on phenotype, whereas a neutral allele has no effect on reproductive fitness but could alter the phenotype. In Mendelian inheritance, a recessive trait is one in which both copies of an altered allele need to be inherited for expression of the phenotype. A dominant trait only requires inheritance of one copy of an altered allele to cause the phenotype so that heterozygous individuals who have one altered and one normal copy express the phenotype. Recessively inherited traits are often associated with alleles that lead to loss of function of the corre-

sponding protein, often an enzyme. Residual function in heterozygotes is sufficient to prevent expression of the phenotype. A variety of molecular mechanisms underlie dominant inheritance. A dominant trait can result from mutations that lead to gain of function of the gene product, loss of function (haploinsufficiency), or interference with function of the normal allele (dominant negative effect).

The expected distribution of alleles in a population can be predicted by the population genetics theory of Hardy–Weinberg equilibrium. This concept is derived from the binomial equation, $p^2 + 2pq + q2 = 1$, and is based on assumptions that a population is stable, without selective pressure, migration, new mutation, or genetic drift. Assuming a two-allele system, A and a, with allele frequencies of p and q, the frequency of the homozygote for allele A is p^2, the frequency of heterozygous genotype is $2pq$, and the frequency of the genotype homozygous for allele a is q^2. Disruption of Hardy–Weinberg equilibrium could be due to selective pressure, mutation, or drift. In a small population, a new mutation might descend from a common ancestor and result in an anomalously high frequency of disease in a restricted population. This is called the *founder effect* and is the cause of the increased frequency of rare diseases such as Tay–Sachs in populations once isolated by culture or geography, such as the Ashkenazi Jews or the French–Canadians.

In contrast to rare monogenic disorders, common disorders are the result of epistatsis, or the interaction of multiple genes and environmental factors. In some cases, the trait is a binary one, such as the presence or absence of cleft lip. Other times, the trait is quantitative, such as height or blood pressure. Binary traits are explained by the threshold model of multifactorial inheritance, in which liability toward the trait is distributed in the population, but the trait is only expressed in individuals whose liability exceeds a threshold. Liability is determined both by genes and the environment, as well as by interactions among these factors. Other traits are expressed as a point along a continuum as a quantity, such as height or blood pressure. Such quantitative traits can be modeled by the additive effects of multiple genes (polygenic inheritance), along with interactions between genes and the environment.

The study of common disorders relies on the definition of disease factors. Since disease features tend to vary from one affected individual to another, it is often difficult to define traits that represent the disease completely. The term *phenotype* as used in this genetic analysis refers to the clinical presentation of a disease process. To discover the genetic interactions that result in a phenotype, the phenotype must be a distinct and homogeneous outcome. There are individuals who express similar clinical signs that are caused by different disease mechanisms, states referred to as phenocopies. Inclusion of a phenocopy in a genetic study will decrease the power to detect involved genes. In contrast, the same phenotype can be caused by two or more distinct genetic pathways, which is referred to as *genetic heterogeneity*. Furthermore, there are some individuals who inherit the genetic predisposition but do not develop the disease, which is referred to as *nonpenetrance*. The presence of phenocopies, genetic heterogeneity, and incomplete penetrance add to the complexity of the study of common disorders. One approach toward definition of a more homogeneous phenotype is to use intermediates in a disease pathway to provide distinct clinical measurements or disease outcomes. These intermediate phenotypes might be quantitative measurements, such as creatinine levels for renal failure or FEV_1 for asthma severity. The more narrow disease definition with intermediate phenotypes will lead to increased power to detect the specific genetic factors in at least part of the disease process.

3. COMPLEX TRAIT DISSECTION

A starting point in the genetic analysis of a common disease is to obtain evidence in support of a genetic contribution and to estimate the relative strength of genetic and environmental factors. One approach is to use a family study, in which individuals with a known amount of shared genetic material and similar environments are compared to unrelated people sharing a similar environment. This approach attempts to minimize or remove the impact of the environment and to isolate the genetic component. Familial aggregation is determined by comparing the frequency of affected relatives in cases compared to controls. If the cases and controls are classified correctly and genetic factors contribute significantly, the disease trait would be more frequent in the family members of cases. Affectation status as a binary or qualitative trait can be estimated through three different approaches, depending on the level of detail of the family history obtained. One approach involves reliance on the proband to give an accurate counting of affected and unaffected individuals in his or her family. Alternatively, to reduce information bias, a more detailed family history can be obtained. The most labor-intensive method is to undertake a full-fledged family study, in which all members of a pedigree are investigated for the trait.[10]

Aggregation can be quantified by a risk ratio, denoted by λ. The ratio of risk of disease in relatives of type R compared with the population risk is λ_R.[11]

$$\text{Autosomal dominant } \lambda = \frac{\frac{1}{2}}{x} = \frac{1}{2x}$$

$$\text{Autosomal recessive } \lambda = \frac{\frac{1}{4}}{x} = \frac{1}{4x}$$

$$\text{Multifactorial } \qquad \lambda = \frac{1}{\sqrt{x}}$$

FIGURE 29-1 For an autosomal dominant trait, risk in sibs will be one-half, so λ_S will be $1/2x$, where x is the population frequency of the trait. Similarly, λ_S will be $1/4x$ for an autosomal recessive trait. For a multifactorial trait, λ_S is estimated by $1/\sqrt{x}$.

The most widely used of the relatedness coefficients is the λ_S, the sibling risk ratio. Disorders with a genetic component have a λ_R greater than 1, whereas those with no genetic component have a λ_R of 1. A λ greater than 2 indicates a significant genetic component (Fig. 29-1).

Through family studies, we can also obtain an idea of the mechanism of inheritance by estimating λ decay in a pedigree. For a Mendelian trait, we would expect to see λ_{R-1} decrease by a factor of 2 with each degree of relationship. If more than one gene is involved in disease determination, λ_R would still decrease by a factor of 2 for an additive model, where each gene adds liability. If λ_R decays more rapidly, a multiplicative model is assumed. A multiplicative model represents the occurrence of epistasis as part of the disease process. We can therefore examine the family recurrence patterns to assess both the number of loci and the inheritance patterns.[12]

Once familial aggregation is identified and segregation estimated, variance components analysis can be used to quantify the amount of variation in a trait attributable to an environmental or a genetic component. These studies are often done with quantitatively measured traits; continuous variables that represent the disease phenotype, such as systolic blood pressure for hypertension or pulmonary function values for chronic obstructive pulmonary disease. These studies are complicated by the fact that family members sharing genetic components usually also share environmental exposures, so several approaches have been created, each adding more information about the relative strength of each component.

Monozygotic twins are identical in their nuclear genetic material because they result from the division of a zygote into two embryos. Dizygotic twins ("fraternal twins") are the result of two separate fertilized ova and are only as related as siblings. If raised together, each of the twins would be exposed to similar environmental factors and would therefore be at the same risk for disease development if identical. If both twins are affected with the disease, they are said to be concordant and the proportion of concordant twin pairs is the concordance rate. If the genetic component of a complex disorder only explains part of the disease phenotype, the concordance rate would be low in both mono- and dizygotic twins, but it would be higher in monozygotic twins. If environment were not a significant contributor to disease development, then monozygotic twins reared apart would be expected to have the same concordance rate as twins reared in the same environment. Twin studies may also be used to determine heritability of a trait.

Heritability can be thought of as the degree of similarity between related individuals that is due to shared genes. If a trait is heritable, individuals who share genes should have higher correlation of trait values than individuals who do not share genes. Heritability is estimated by the variable h^2, which is the proportion of genetic contribution to the variance of a trait, with 0 signifying no contribution and 1 signifying complete genetic control (Table 29-1). Heritability can be estimated from correlation of a quantitative trait among individuals with known degrees of relationship (Fig. 29-2).

These approaches estimate the genetic versus environmental contributions, but the possibilities of phenocopy, incomplete penetrance, differential age of onset, and ascertainment bias are all problems that can affect these estimates. The importance of precise phenotypic definition must once again be stressed as a critical prerequisite to successful genetic analysis.

$$V_P = \overbrace{V_A + V_D}^{\text{genetic variance}} + \overbrace{V_E + V_I}^{\text{environmental variance}} + Cov_{GE} + \overbrace{V_M}^{\text{measurement variance}}$$

V_A = additive genetic variance
V_D = deviation due to dominance and epistasis
V_E = environmental variance
V_I = interaction variance
Cov_{GE} = covariance of genetics and environment

Heritability in broad sense $\quad h^2 = \dfrac{V_A}{V_P}$

Heritability in narrow sense $\quad h^2 = \dfrac{V_G}{V_P}$

Relationship	Heritability
Monozygotic twins	$h^2 = r$
Sib-sib or Dizygotic twins	$h^2 = 2r$
One parent – One offspring	$h^2 = 2r$
Midparent – Offspring	$h^2 = r/\sqrt{0.5} = r/0.7071$
First cousins	$h^2 = 8r$

FIGURE 29-2 Heritability concept. Phenotypic variance can be partitioned into genetic variance, environmental variance, the covariance between the two, and measurement variance. Genetic variance can be further partitioned into additive affects and gene–gene interactions (epistasis); environmental variance can likewise be partitioned into additive and interactive effects. Heritability is defined as the proportion of phenotypic variance accounted for by genetic variance. If all genetic variance is considered, the value is referred to as heritability in the narrow sense; if only additive genetic variance is considered, the value is heritability in the broad sense. One can estimate heritability in the narrow sense from correlation of a quantitative trait between relatives, as shown in the table.

TABLE 29-1   Heritability Estimates for Common Diseases: Hypertension, Obesity,[40] Systemic Lupus Erythematosus (SLE),[41] Multiple Sclerosis,[42] and Asthma[43–45]

Disease	h^2 Overall (%)
Hypertension	15–35
Systolic	46
Diastolic	31
Obesity	45–60
SLE	66
Multiple sclerosis	40–50
Asthma (adult)	35–60

4. SAMPLE COLLECTION AND INFORMED CONSENT

Initiation of a genetic study requires recruitment of participants who are willing to provide phenotypic information and DNA samples for study. This section considers both the technical issues of isolation of DNA and some principles of the informed consent process that are specific to genetic research.

DNA can be extracted from fresh tissue, or cells can be grown in culture, providing a permanent source of DNA from the individual. Cell culture is usually done by Epstein–Barr virus transformation of lymphoblasts from peripheral blood, requiring access to a tissue culture facility. Although this can be costly, once a permanent cell line is established aliquots can be kept frozen indefinitely and used to expand a cell line whenever more DNA is required. Lymphoblast cultures can also be used as a source of RNA if the gene of interest is expressed in these cells. Skin fibroblasts can also be grown in long-term culture and frozen as a source of DNA and RNA, although it is usually more convenient to work with lymphoblast cultures if DNA isolation is the goal.

For most studies, however, DNA isolated from fresh tissues will provide adequate material. DNA can be obtained from blood, buccal swabs, saliva, and tissue samples. Commercial kits have high rates of success in DNA extraction from uncontaminated samples, using salt precipitation, silica gel membrane binding, or magnetic bead binding. All samples in a study ideally should be obtained by using the same technique to allow for homogeneity of biomarkers and quality of DNA. Whole genome amplification has been successfully used to increase the amount of DNA severalfold, providing sufficient DNA for hundreds to thousands of genotype assays. Blood samples can also provide biomarkers that may be used as intermediate phenotypes. Samples need to be labeled with unique identification codes and verification data, but without the participant's name. Samples should be aliquoted and stored in separate storage units to prevent loss of an individual's sample in the event of equipment failure. Laboratory information management systems have been developed to aid in processing, storage, and subsequent use of DNA samples. These systems ensure security of participant data by restricting access to the database. They can be used to manage, quality control, and export data. Adherence to sample management protocol and automation of the database are important for obtaining valid, replicable results.

There are probably no risks that are unique to genetic research, but there are a set of concerns that are more commonly raised by genetic studies or that may be viewed in a unique light. No single study is likely to raise all of these issues. Examples include the following:

1. Impact on families: Genetic traits are, for the most part (with the exception of new mutations), shared by multiple members of a family. This means that family members may be asked to participate in genetic studies. It also means that results may have significance for multiple family members, not just the proband. Unexpected family relationships, such as misattributed parentage, may be inadvertently discovered. Relatives may feel coerced to participate in research. The investigator may learn about previously unsuspected risk of disease in a family member who participates in a study. Some individuals may blame family members who transmitted a trait, whereas others may feel guilty either for transmitting the trait or for escaping its effects.

2. Vagaries of genetic testing: A genetic test result has the appearance of an ironclad objective finding and, indeed, may stand as a "fact" about an individual for a lifetime. The clinical relevance may be much less certain, however. Errors of genotyping are possible, particularly in research laboratories that are not set up as clinical testing facilities. Some mutations may not be pathogenic and may be erroneously assumed to predict disease. Conversely, failure to find a mutation may not rule out a particular disorder. Also, tests may be performed that do not guide clinical management but lead to anxiety and/or stigmatization. Note that these issues are not always encountered. Some genetic tests give clear-cut results that can be instrumental in providing counseling or guiding care for an individual. Each test needs to be evaluated on its merits before being clinically implemented.

3. Stigmatization: Genetic testing often has the power of estimating risk of disease well in advance of onset of signs or symptoms. This can be beneficial by permitting family planning, prenatal diagnosis, or medical surveillance. It can also expose individuals to risk of discrimination for employment; discrimination for health, disability, or life insurance; social stigmatization; and create emotional distress such as anxiety or guilt. The risk/benefit consideration is seldom clear-cut. It should be recognized, for example, that sometimes testing can resolve issues for an individual who is at risk based on family history, if the test shows that the individual does not carry a particular gene. Similarly, some individuals are vulnerable to stigmatization by virtue of existing signs or symptoms of disease, and genetic testing may have little additional impact. To further complicate matters, some states have laws that address discrimination for employment or insurance based on genetic testing; currently, there is no federal legislation on this issue.

4. Group stigmatization: Genetic traits may be shared by groups of individuals with a common geographic or ethnic origin. Genetic studies of such groups may lead to a perception that the group is "genetically inferior" and may lead to group stigmatization and discrimination.

The great interest in genetic research is fueled by the power of the approach in elucidating the basis of both rare and common disorders. To some extent, the promise may have been exaggerated by those who preach a tenet of "genetic determinism"—that a person's fate is sealed in his or her genes. This extreme view is inaccurate, but this does not diminish the potential power of genetic research. From the point of view of a research participant, there may be direct and/or indirect benefits of participation in research. These include the following:

Diagnosis: An individual may already be experiencing symptoms but may not know the specific diagnosis. Many people in this situation, or their parents or relatives, may wish to determine a precise diagnosis; genetic testing may make this feasible. Learning the diagnosis may bring peace of mind, information about risk of recurrence in the family, knowledge of expected health outcomes, and may guide further clinical management. For predictive tests, surveillance or other risk reduction approaches may be feasible.

Development of therapies: Discovery of genes that contribute to disease may reveal the cellular, tissue, and organ pathways that lead to the pathological process. These pathways become targets for new therapeutic approaches, including new pharmaceuticals. Genetic approaches are revealing pathogenic mechanisms that had previously been unknown and are opening doors to treatment of conditions that were previously considered to be intractable. Research participants may see hope of development of new treatments for themselves, their families, or future generations.

The process of educating a potential participant requires a careful explanation of potential risks, benefits, and alternatives to participation. It should be made clear whether there is a possibility of return of results that may be used for clinical decision making. If this is the case, arrangements need to be made for testing to be done in an appropriately equipped and licensed clinical laboratory and access provided to genetic counseling. If there is no plan to return results, that, too, must be made clear. In many cases, there is a mismatch of expectations, with research participants waiting for results that are never communicated. Providing a general update on results, such as using a newsletter format, can be helpful to engage the trust of participants. Studies of large populations may lead to risks of group stigmatization. Engagement of the community in providing advice about the design and conduct of the study and gaining their assent to participate can help to anticipate and address such concerns to earn the trust of the group.

5. APPROACH TO GENOTYPING

With evidence for a genetic contribution to a trait and access to DNA from individuals whose phenotype has been determined, the next step is to use genetic linkage or association approaches in an effort to more precisely define contributing genes. Either of these approaches relies on determination of genotypes at polymorphic loci. The term *polymorphism* derives from Greek *poly* (many) and *morphe* (shape), and in complex trait analysis it is used to describe genetic variation that occurs in > 1% of a population. This can be variation in the number of repeat elements in a repeated region or it may be a difference of a single nucleotide within a sequence. A polymorphism can include two or more alleles and can occur at coding or noncoding regions, splice sites, promoter and enhancer regions, and within introns. Allele frequencies of polymorphisms may differ in different populations. The genetic difference representing the polymorphism may not itself cause disease but may have been inherited down a lineage along with a disease-causing locus. The processes of mutation and recombination influence the inheritance and the linkage to a disease-causing locus.

Major types of polymorphism are listed in Table 29-2. The most useful polymorphisms are those in which there is a high frequency of heterozygosity and in which there are multiple alleles. Frequency of heterozygosity is calculated as 1 − sum of the frequency of homozygosity for each allele (Fig. 29-3). A highly heterozygous polymorphism will enhance the likelihood that an individual under study will have two distinct alleles to track. If there are only two alleles, it is possible that both members of a couple will have the

TABLE 29-2   Major Types of Polymorphisms, Methods of Detection, and Range of Heterozygosity

Type	Description	Detection	Heterozygosity
RFLP	Single base change	Southern analysis or PCR	< 0.4
VNTR	14–100 bp repeat unit with variable number of repeats	Southern analysis or PCR	> 0.6
STR	Di-, tri-, tetranucleotide repeats	PCR	> 0.7
SNP	Single base change	PCR	Variable

PCR, polymerase chain reaction; RFLP, restriction fragment length polymorphism; SNP, single nucleotide polymorphism; STR, short tandem repeat; VNTR, variable number tandem repeat.

$$HET = 1 - \sum_{i=1}^{x} P_i^2$$

consider four alleles, with frequencies 0.1, 0.2, 0.3, 0.4

$$HET = 1 - [(0.1)^2 + (0.2)^2 + (0.3)^2 + (0.4)^2] = 0.70$$

FIGURE 29-3 Calculation of heterozygosity. In the example, heterozygosity is calculated for a four-allele system, with the indicated allele frequencies.

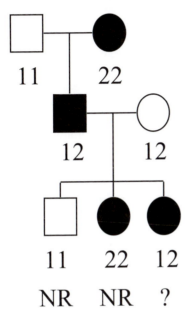

FIGURE 29-4 In this pedigree, the father in the second generation is heterozygous for a marker locus having two alleles, labeled 1 and 2. He inherited marker 2 from his affected mother along with the disease. He has passed allele 1 along with the nondisease allele to his first son and marker 2 along with disease to the middle daughter (both labeled "NR" in the figure, meaning nonrecombinant). The third daughter, however, is heterozygous for 12. Since his wife also has the 12 genotype, one cannot determine which parent transmitted which allele to the daughter, making it unclear whether or not she is recombinant. This mating is therefore uninformative for this child, despite both parents being heterozygous.

same genotype, creating ambiguity in interpretation of the segregation of alleles to offspring (Fig. 29-4). Use of polymorphisms that are both highly heterozygous and involve multiple alleles maximizes the information that can be obtained from a parent-to-offspring transmission.

Restriction fragment length polymorphism (RFLP) analysis was the earliest DNA-based method to be used for genotyping. This technique relies on the occurrence of sequence variation that disrupts or creates a restriction endonuclease cutting site. Enzyme digestion of the DNA from the variant allele then yields a fragment of different length than the wild-type allele. These size differences are detected by gel electrophoresis. This technique is limited to detection of polymorphic sites that alter a restriction endonuclease cutting site, but detection is simple and robust. Heterozygosity frequencies are variable but never higher than 0.50.

Variable number tandem repeat (VNTR) polymorphisms involve regions of the genome in which a run of nucleotides is repeated multiple times, with the repeat number being different in different individuals. Repeat number is determined by agarose gel electrophoresis and visualization of the allele size by Southern hybridization. These polymorphisms offer the advantage of a high rate of heterozygosity and multiple alleles. Short tandem repeat (STR; also called microsatellite) polymorphisms involve smaller repeated units, usually of two to four bases, in which the number of repeats is variable between individuals. Repeat number is usually detected by polyacrylamide gel electrophoresis following polymerase chain reaction (PCR) amplification. The technique is prone to artifacts, but the loci tend to be highly polymorphic and have multiple alleles.

Single nucleotide polymorphisms (SNPs) are variants at a particular base pair, usually with two different forms seen in the population. Although the rate of heterozygosity is not as high as for VNTR or STR polymorphisms, there is a greater likelihood that a SNP may correspond with a variant that is of pathological significance. SNPs in exons are described as coding SNPs if they change the resulting amino acid in the protein, and they are called synonymous SNPs if the variant codes for the same amino acid. SNPs are detected by sequencing the region of interest in a representative population sample and determining the frequency of variation. A variety of methods have been developed to enable high-throughput determination of SNP genotypes.

6. LINKAGE ANALYSIS (PARAMETRIC)

Linkage analysis provides an approach to mapping a single gene responsible for a phenotype. It makes use of the genetic relation of family members to track an inherited disease allele through a pedigree, assuming a particular genetic model. The model specifies parameters such as inheritance pattern, gene frequency, and penetrance of the disorder. To find the disease locus, linkage analysis determines if affected relatives share a segment of DNA more often than other segments. The shared segment would be more likely to contain the altered gene. The term *linkage* refers to alleles being inherited together. Two loci are said to be completely linked (or in complete linkage) if they are always inherited together. This can be true if the distance between them is very small. A block of alleles that are linked is known as a haplotype. The frequency of recombination between two loci is referred to as θ, or the recombination fraction (the number of recombinations/meiosis). A θ of 0 indicates that the loci are completely linked, whereas a θ of 0.5 indicates independent assortment of the two loci.

Linkage is measured in centimorgans (cM), with 1 cM representing one map unit and 1% recombination frequency. Although a cM is a genetic distance, it can be roughly related to physical distance of approximately 10^6 base pairs. Recombination occurs through crossover events during meiosis. To find linked regions, DNA markers are used to map an area of interest or the entire genome. These markers can be RFLPs, VNTRs, microsatellites (STRs), or SNPs.

To estimate linkage between markers, the number of recombinants is compared to the number of nonrecombinants (Fig. 29-5). The occurrence of few recombinants would indicate that the two loci are closely linked. Since it is not always possible to detect recombination due to an uninformative type of mating, the overall likelihood of the data represented in a pedigree (i.e., a specific number of recombinant and nonrecombinant offspring) is calculated for a set of different values for the recombination fraction. The likelihood of the pedigree at a particular recombination fraction is compared to the likelihood of random segregation (i.e., where $\theta = 0.5$). A likelihood ratio presuming linkage compared with no linkage is calculated for a set of values of θ. This odds ratio is expressed in $\log_{10}$ and called a lod score, or the log of the odds of the likelihood ratio. The marker loci with the highest lod scores are more likely to contain a segment segregating with disease. Lod scores are additive across families at each locus. A total score greater than 3.0, indicative of a 1000 : 1 odds favoring linkage, meets a threshold for significance with 5% chance of error.

With the availability of polymorphic markers that span the whole genome, it is possible to scan all chromosomes for a site of probable linkage. Lod scores > 3 are considered to indicate regions that merit further investigation (that meet genomewide significance).[13,14] Additional polymorphic markers within these regions are then tested for linkage using multipoint linkage analysis. In multipoint analysis, a putative disease locus is moved across a region spanned by multiple markers, searching for a positioning that produces a best fit to the observed frequency of recombination between the disease locus and any of the polymorphic

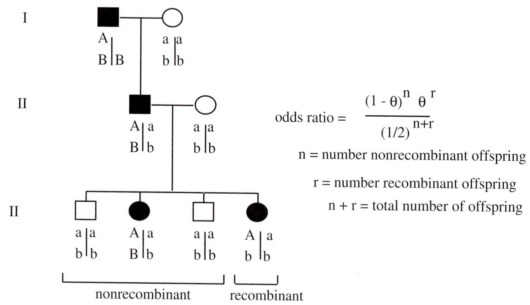

$$\text{odds ratio} = \frac{(1-\theta)^n \, \theta^r}{(1/2)^{n+r}}$$

n = number nonrecombinant offspring

r = number recombinant offspring

n + r = total number of offspring

FIGURE 29-5 Example of linkage analysis. In this family, the disease-causing allele is designated A, and the polymorphic alleles are designated B and b. The affected father in generation II is heterozygous for A and a, as well as for B and b. We know that he received A and B together from his affected father. Therefore, the first three children are nonrecombinant with respect to A and B, whereas the fourth child has inherited A along with b from the father and, therefore, is a recombinant. The formula for calculating the odds ratio is shown.

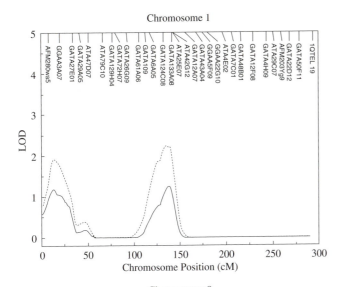

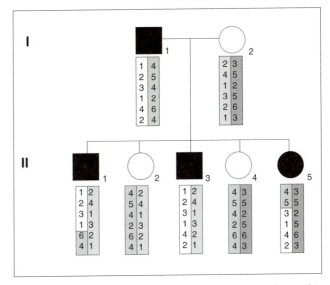

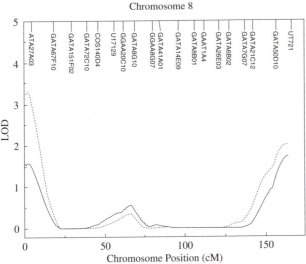

FIGURE 29-6 Example of multipoint linkage analysis. Whole genome scan at 9 cM intervals for linkage with FEV₁ (pre- and post-bronchodilator administration) in 72 pedigrees of patients with chronic obstructive pulmonary disease. *Solid line* is pre-bronchodilator and *dotted line* is post-bronchodilator.

FIGURE 29-7 Haplotype analysis for multiple polymorphic markers. Children II-1 and II-5 are recombinants, narrowing the region of interest to the third and fourth loci.

rately for linkage if they are of sufficient size. It is also possible to perform a statistical test of admixture. This assumes that a group of families consists of two subsets, a proportion α that is linked to the locus and a proportion $1 - \alpha$ that is not. Linkage data are analyzed for values of α between 0 and 1, searching for a value that produces a best fit with the data set.[15]

Parametric analysis is limited by the need to assume a genetic model and a set of parameters, but it is the most powerful mapping approach if the chosen model is correct. However, since complex trait disorders, by definition, are not inherited in a particular Mendelian pattern, this form of analysis is not as useful in such disorders. Alternative nonparametric approaches in which no assumptions are made about genetic mechanisms may be more useful.

loci (Fig. 29-6). Detailed study of the haplotypes in a region of interest can further narrow the region by identification of critical crossover events (Fig. 29-7).

Genetic heterogeneity constitutes a major pitfall in linkage analysis. Locus heterogeneity, in which the identical clinical phenotype is caused by defects at two or more different loci, can result in a subset of families having linkage between a marker and disease, whereas another set does not. Locus heterogeneity can therefore produce false-negative results. This can be overcome by using large pedigrees or large numbers of families. If there are distinct subphenotypes of a disorder, families can be stratified into subclasses and tested sepa-

7. NONPARAMETRIC ANALYSIS

To increase the power of linkage analysis in the face of the complex nature of common disease, an analysis method was devised that does not necessitate the definition of parameters or models—the so-called nonparametric or model-free method. This approach depends only on the assumption that two relatives affected with the same disease will share predisposing alleles, and it requires no further assumption of inheritance or recombination parameters. Although this is robust for complex trait analysis, it is also imprecise and less sensitive than parametric methods.

Nonparametric analysis is dependent on allele sharing. At any given locus, two people could share an allele for two different reasons: Either they have inherited the allele from a common ancestor, or the allele arose in two different populations of ancestors (Fig. 29-8). Two alleles are called identical by state (IBS) if they exist in the same detectable form but are only identical by descent (IBD) if they were inherited from the same ancestor.[16]

Several methods have evolved around allele-sharing nonparametric methods. In the affected pedigree member approach, members of multigenerational pedigrees are genotyped for multiple polymorphic loci. If affected relatives share alleles IBS more often than expected due to chance, this favors linkage. The most common application of this method is by testing affected siblings, who are easier to ascertain than more distant relatives in large pedigrees. These methods use only parts of pedigrees (i.e., affected relatives) and are not fully nonparametric in that, to extend to multipoint analysis, recombination fractions between markers need to be specified. A method called nonparametric linkage was developed to address the shortfalls of previous nonparametric approaches.[17]

The first step of nonparametric linkage is a multipoint analysis to extract the most data from a pedigree by establishing IBD based on a large number of markers. Since the inheritance pattern in these pedigrees is usually ambiguous, the nonparametric linkage method averages the test statistic over all possible patterns of transmission. The statistic is normalized and weighted across pedigrees. It establishes a probability of linkage for any point on the chromosome with the available marker information. A further extension of this method is the adaptation to utilization of discordant sibpairs to identify areas that are coinherited infrequently.

Nonparametric linkage analysis is a useful tool for detection of chromosomal regions that may contain disease loci, although there has been considerable debate about a lod score threshold that would represent significance in a genomewide study.[13] Another caution in the interpretation of the nonparametric lod score results is that the linkage calculations are influenced by the presence of a strong gene disease association with a particular allele. Linkage analysis searches for within-pedigree allele sharing, but if there is a significant allele association in the population, there will also be between-pedigree sharing and the linkage results will be distorted.

8. ASSOCIATION AND LINKAGE DISEQUILIBRIUM

The mapping methods described previously use linkage to search for a region of the genome that segregates in families with disease status. The specific alleles at a linked locus are not necessarily associated with the disease; in fact, the allele that is ultimately found to be linked with the disease mutation will vary from family to family. As the search for a disease gene narrows, however, alleles may be identified that are associated with disease because they are either etiologically involved or in very close proximity (linkage disequilibrium) to the causal variant. The concept of linkage disequilibrium can be conceptualized by imagining an isolated population in which an individual develops a new mutation. When inherited by subsequent generations, the allele with the mutation is accompanied by a large DNA segment. Through recombination at meiosis, the segment containing the mutation gradually gets shorter with each generation. This shortening of the shared segment is called linkage decay. Linkage decay is a relatively slow process because it occurs only through recombination of alleles transmitted to the offspring. In analysis of linkage data, a haplotype may be identified that appears intact in chromosomes in unrelated individuals, suggesting the occurrence of linkage disequilibrium in that population. If a haplotype is found more frequently in a population of people affected with a disease, the markers may be closely linked to the disease gene. Insufficient time has passed to achieve "linkage equilibrium" in which the frequency of any particular haplotype is the product of the frequency of the individual alleles.[18]

Association analyses are conducted on candidate genes or linkage areas using the following methods:

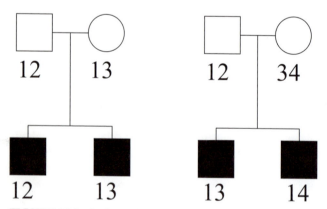

FIGURE 29-8 In the pedigree on the left, the two siblings share allele 1, which is identical by state (IBS) but not identical by descent (IBD) since the first child received the 1 allele from his mother, whereas the second received it from his father. In the pedigree on the right, the 1 alleles are IBS and IBD since both children must have received the allele from their father.

case–control studies, population-based studies, and transmission disequilibrium testing. In a case–control study, the frequencies of individual polymorphisms or haplotypes can be compared in affected and unaffected individuals (Table 29-3). Comparing the allele frequencies in cases and controls is usually easier than the collection of entire affected families, although determination of phenotype is complicated by temporal issues such as age of onset and penetrance.

Case–control studies can be confounded by ethnic background. Some alleles and diseases will be found more often in certain ethnic groups, which may result in spurious conclusion of an association of the allele and the disease (Table 29-4). Although a well-designed case–control study will match subjects by ethnicity, it is not always possible to detect population substructure. Undetected substructure could lead to undetected confounding by ethnicity, which is called population stratification. This has been proposed as an explanation of the poor replication of results of association studies; however, it is likely that many false positives are the result of tests in samples that are too small to have power to detect the modest effect of a single gene variant.[19,20]

TABLE 29-3 Hypothetical Case–Control Study Comparing the Presence or Absence of Allele 2 of a Polymorphism in Individuals with or without Asthma[a]

	Asthma	No Asthma
Allele 2 Present	300	100
Allele 2 Not Present	700	900

[a]In this example, allele 2 is found more often in asthmatics than controls.

TABLE 29-4 Same Study as Illustrated in Table 29-3, Except Now the Study Population Is Known to Consist of a Mixture of Two Groups, Designated A and B[a]

	Population	Asthma	No Asthma
Allele 2 present	A	0	0
	B	50	0
Allele 2 not present	A	0	100
	B	50	0

[a]In this example, allele 2 is never seen in population A, and no one in population A has asthma. Among people in population B, allele 2 is seen in 50%, and all of these individuals have asthma. When the two populations are mixed, it appears as though allele 2 is associated with asthma, but in fact this is due to the fact that both allele 2 and asthma occur uniquely in population B—there is no evidence presented that allele 2 and asthma are otherwise associated.

One solution that has been proposed is to estimate the substructure and adjust the analysis accordingly. With the technical advances in genotyping allowing for rapid typing of large numbers of samples, researchers have the ability to type random unlinked markers to test for substructure. If a panel of random markers shows no population substructure, it is unlikely that a positive association result is due to population stratification. If such a scan detects subgroups, the case–control chi-square can be calculated for each subgroup stratum or adjusted by a factor by rescaling the chi-square statistic.[21]

To obviate the effects of stratification, association studies can also be done with transmission disequilibrium tests. This method requires at least trios of two parents and an affected child. The parents' affectation status does not need to be determined because they are only genotyped to establish the possible inherited alleles. If an allele is not associated with a disease, one would expect it to be transmitted 50% of the time to affected individuals. If it is associated with a disease, this random segregation would be violated, which is called transmission disequilibrium (Fig. 29-9). Since this method requires nuclear families, it is better suited to the study of disorders with an early age of onset, where parents are still alive at the time of diagnosis of a child. There are statistical packages using more extensive pedigrees in family-based analysis, incorporating other first-degree relations. These methods require the use of highly polymorphic markers since transmission from homozygous parents cannot be detected.[22]

In instances in which continuous outcomes (quantitative traits) represent disease status or severity, population studies are also a powerful method of establishing disease association. This method involves linear regression of genotypes from affected people with intermediate phenotypic variables to establish an association, basically assessing the effect of the dose of polymorphism on the resulting intermediate phenotype. Failure to replicate results in these studies is more likely due to inadequate sample size and poor study design as well as undetected stratification.

Although many genetic associations with disease have been published, only a few have been consistently replicated. ApoE4 and cardiovascular disease, PPARG and type 2 diabetes, and CTLA4 and type 1 diabetes are examples of reports that have been confirmed in multiple large studies.[23–25] There are three explanations for any observed association: The association is due to artifact or chance, the allele is in linkage disequilibrium with an allele at another locus that causes disease, or the allele alters function and contributes to the phenotype. To reduce the error rate in

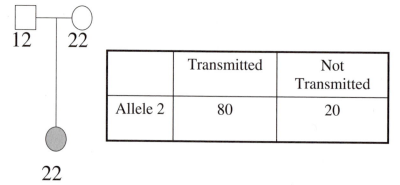

	Transmitted	Not Transmitted
Allele 2	80	20

FIGURE 29-9 Transmission disequilibrium test. In this data set, allele 2 is transmitted from a heterozygous parent more than the expected 50%, supportive of disequilibrium of the marker allele and the trait.

reporting false or chance associations, stringent significance levels and corrections for multiple testing need to be implemented. Previously published associations that are not replicable may have been underpowered to detect a true association. Larger sample sizes and careful consideration of the power in study design will help to reduce the number of false-positive studies, thereby reducing unnecessary follow-up studies.[3,8,26]

Since each individual gene studied may only contribute a modest amount to the variation in a trait, its effect may be too small to be detected by association analysis. Since common disease traits are likely due to interactions among multiple genes and with the environment, methods to detect these effects are being developed. With increased computing power, more complex models can be built. One method called combinatorial partitioning has been proposed as a way of detecting combinations of alleles whose variation explains a degree of trait variation. This is a model-free method designed to combine large numbers of loci and detect nonadditive locus effects. It partitions genotype data into different combinations and determines which sets have subjects with phenotypic similarities and differences. These sets are then validated with a multifold cross-validation.[27] Another recently developed method is called multifactor dimensionality reduction (MDR). MDR is a nonparametric and model-free method that reduces the dimensionality of genotype combinations to improve detection of disease associations. This method can be used in case–control analyses and discordant sib-pair analyses.[28,29]

The identification of large numbers of SNPs has opened the way toward unbiased screening across the genome for association of alleles with disease, rather than confining the focus to candidate loci.[30] The obvious obstacle, however, is that thousands of markers would need to be genotyped if the entire genome were to be tested with SNPs spaced 1000 base pairs apart. There is evidence that larger blocks of DNA, in the range of 10–20 kb, are transmitted more or less intact from generation to generation in human populations.[31] It is therefore possible to choose an SNP or combination of SNPs in a haplotype block that will serve as a proxy for the rest of the SNPs. These SNPs, so-called tag SNPs, reduce the number of variants to genotype by several orders of magnitude. A coalition of researchers recognized the potential value of having a map of common haplotypes in people to provide easy selection of tag SNPs. This large-scale effort involved genotyping SNPs every 0.5–2 kb in genomes of people of several different ethnicities. This HapMap is now complete, and it is available as a public resource (www.hapmap.org).[5] The linkage disequilibrium patterns have already been advantageous in the development of whole genome scans. In the design of two genotyping methods, 500,000 SNPs can be genotyped in an individual sample, and this density of SNPs has been estimated to capture the vast majority of common variation for that individual using association tests of the tag SNPs and combinations. The massive amount of genotype data poses a daunting analysis problem but is also likely to provide enough power to find variants that have a modest effect on the disease phenotype.[7,32]

Whole genome association studies have been used to identify causal common variants in unexpected genes. The example of age-related macular degeneration (AMD) highlights the power of whole genome association studies. A whole genome association analysis and subsequent fine mapping identified a common causal variant in complement factor H that increases the risk of AMD 2.5-fold for each additional allele. All published association studies (4/4) reproduced this finding with high statistical significance (p values ranging from 10^{-5} to 10^{-20}), documenting that this is a major risk allele for AMD.[33-36] This example demonstrates the power of comprehensive association analyses to unveil pathways not previously known to play a role in common diseases. A whole genome association test for obesity identified a variant in the gene *INSIG2* (insulin-induced gene 2) that replicated in four out of five follow-up studies. Careful replications of results from whole genome scans will be necessary to avoid a deluge of false-positive association results from the large number of tests done to capture variation across the genome.

9. GENE IDENTIFICATION

The ultimate goal of genetic analysis in clinical research is the identification of the gene or genes that contribute to a particular disorder. Gene identification opens the door to development of new approaches to diagnosis and therapy (see the next section). The path to gene identification can be an arduous one, but it has been made considerably smoother using the tools developed through the Human Genome Project.

The first genes to be identified were those for which the associated proteins were known. Examples include the alpha and beta globin genes involved in hemoglobinopathies, the gene for α-1-antitrypsin, or the gene for hexosaminidase A (Tay–Sachs disease). In each case, the abnormal protein was known in advance based on structural or biochemical studies. The corresponding gene could be identified from libraries of cDNA (copies of mRNA) or genomic DNA using information about the structure or function of the protein to identify a clone that included part of the sequence. The approach proved to be successful for understanding the molecular basis for a wide variety of single gene disorders.

Beginning in the mid-1980s, another approach, referred to as positional cloning, was introduced. This approach was ideally suited to the identification of disease genes where the underlying protein defect was not known. The concept was to find the gene by first mapping it and then isolating DNA from the mapped region and searching the region for the actual disease gene.

Gene mapping is done by linkage analysis. In the early days, this was accomplished using RFLPs, but throughout the years the advent of highly polymorphic markers such as STRs has simplified the process of finding linkage. In some cases, the identification of rare affected individuals who have a visible chromo-

somal abnormality accelerates the mapping process. Some of these individuals harbor chromosomal rearrangements such as deletions or translocations that disrupt the gene that is responsible for their disorder. Such rearrangements immediately focus attention on the location of the gene and also provide useful tools for cloning the disrupted region.

Once the gene is mapped, the next step is to clone a region of the chromosome near the mapped locus. Relatively large areas can be covered by starting from a closely linked marker and cloning overlapping segments around that marker, a process referred to as gene walking. The advent of cloning vectors that can accommodate large inserts, such as yeast or bacterial artificial chromosomes, significantly aided this effort. The availability of the human genome sequence makes the process even easier since one can now scan the map for candidate genes in a region in which closely linked markers are known to reside.[37]

Identification of the disease gene within a cloned region presents a significant challenge. The region spanned by closely linked markers might encompass 1 million or more base pairs, including several or perhaps dozens of genes. In the early days of positional cloning, there was no map to identify the genes along the cloned segment. Genes were identified by searching for regions of interspecies homology (indicative of conserved segments that might be expressed genes) or searching for structures, such as CpG islands, that tend to occur near promoter sequences. Candidate genes were screened for expression in tissues known to be affected by the disease process and then for the occurrence of mutations in affected individuals. The human genome sequence streamlines this process since the location of genes at the region of interest is now available in computer databases.

Proof that a candidate gene is indeed the gene of interest is the final challenge. As noted previously, characteristics include expression in affected tissues and the occurrence of mutations in affected individuals. These mutations must be distinguished from clinically insignificant sequence variants (e.g., polymorphisms). Pathogenic mutations will not be seen in unaffected individuals and will have a significant impact on the expression or function of the gene product. Mutations that lead to loss of function (e.g., frameshifts and stop mutations) are likely to be pathogenic. It can be more difficult to establish pathogenicity of amino acid substitutions. Pathological changes tend to occur at amino acids that are highly conserved in evolution and that involve substitution of amino acids with significantly different chemical properties. Replication of the phenotype in a model system, such as a mouse knockout, represents a further level of proof that a candidate gene is the actual disease gene.

The availability of the human genome sequence has vastly facilitated the process of finding genes for clinical disorders. Candidate genes can be readily identified within a region closely linked to a disease locus. In some cases, candidates are identified based on hypotheses about the physiology of the disease process. Gene association studies can also be used to implicate a candidate gene as being involved in the pathogenesis of disease. The latter will be particularly important in the identification of genes that contribute to common disorders.

10. INTEGRATION OF GENETICS INTO MEDICAL PRACTICE

A major goal of clinical research involving genetics is the development of new knowledge that will improve the ability to provide clinical care for both rare and common disorders. Although it has been widely suggested that genetics will transform the practice of medicine, it is likely that this transformation will occur gradually and at different rates for different medical conditions. Health providers will increasingly use genetics in daily practice, although in some cases it is likely that genetic approaches will be so intimately integrated into practice that physicians will not realize that they are making medical decisions using genetic information.[38] This section considers the integration of genetics into medical practice, focusing on genetic testing and the development of new treatments.

10.1. Genetic Testing

Genetic testing may be defined as analysis of changes in the genome, either inherited or acquired, to diagnosis disease or risk of disease. Genetic testing began in 1959 with the analysis of chromosomes, initially to diagnosis trisomy syndromes such as Down syndrome. Throughout the years, the resolution of chromosomal analysis has gradually improved. Molecular cytogenetic tests now permit the detection of submicroscopic deletions or very subtle chromosome rearrangements. DNA testing began in the late 1970s. Initially, it was confined to genetic linkage analysis, but as more genes responsible for disease have been identified, direct mutation testing is increasingly available.

Genetic testing for monogenic or chromosomal disorders is generally done to establish a diagnosis in a symptomatic or at-risk individual or to provide prenatal diagnosis. If the gene mutation in the family is known, direct testing of DNA from an individual or a fetus is relatively straightforward. Sensitivity and specificity of testing is very high, although incomplete penetrance can lead to uncertainty regarding the likelihood of symptoms developing in an individual found to have a mutation. In some instances, a genetic variant may be detected that is of unknown significance and may represent either a pathogenic mutation or benign variant. The clinician must be cautious in the interpretation of results of DNA testing, taking these ambiguities into account. Sensitivity to the psychological impact of testing, including decisions to terminate a pregnancy or identification of a mutation for an untreatable disorder in an individual at risk, is likewise critically important.

The discovery of genetic factors that contribute to common disorders will enable genetic testing for these as well. Here, though, the genetic tests will indicate risk but will not be highly predictive. Such tests will therefore not be diagnostic but can be of use in determining disease susceptibility, refining a diagnosis, and guiding treatment.[39]

Predispositional testing for common disorders offers the hope of early identification of individuals at risk, who can then be offered preventative measures. These may include alteration of life style or use of medications. Common disorders generally result from a complex interaction of multiple genes and environmental factors. Therefore, the predictive value of individual genetic tests may not be high, resulting only in modest changes in estimates of relative risk of disease. It is possible that multiple genetic variants may be tested simultaneously, and that these may result in more highly predictive tests. On the other hand, the specific combination of genetic variants in a given individual may be nearly unique to that individual, making it difficult to accumulate data on predictive value of any specific group of results. Predispositional testing also raises major ethical concerns, such as stigmatization and discrimination of asymptomatic individuals found to be at increased risk of disease. Quality control of testing will be another challenge. Predispositional testing is most likely to be used in instances in which results have substantial predictive value and when there are proven interventions to modify risk. Legislation to avoid the risks of discrimination will also be important.

Genetic testing may also find application in establishing a precise diagnosis to guide treatment decisions. If genetic factors are identified that reveal subphenotypes of common disorders, such as asthma or hypertension, it may emerge that different medications are more appropriate for specific subgroups of patients. This would result in use of treatments that are tailored to a particular individual's needs, more likely to be effective, and less likely to cause side effects. Drug dosage can also be tailored to an individual by testing variants in genes that encode enzymes involved in drug metabolism. This would reduce the risk of over- or undermedication. Finally, idiosyncratic drug reactions may, in some cases, be due to genetic susceptibilities. Identification of individuals at risk would permit the clinician to avoid use of offending drugs in these patients, reducing the rate of severe side effects while allowing the medication to be used in others.

10.2. Treatment

One of the most exciting prospects for the application of genetics in clinical research is the development of new approaches to treatment. Identification of genetic factors that contribute to risk of disease provides insights into pathophysiology that can, in turn, suggest new targets for treatment. Drug development can be directed toward these targets, resulting in treatments that are highly specific for particular disease mechanisms. Such drugs are likely to be more effective and less likely to cause side effects. As noted previously, their use may be coupled with genetic testing to identify patients whose disease is most likely to benefit from a particular therapeutic agent.

A number of approaches have been used to develop drugs that precisely target specific components of a pathway. One is to define the molecular structure of the target and characterize its biological pathway. A small molecule is then designed to interact with the target in a manner that will alter its behavior in a desired way. Alternatively, thousands of candidate molecules can be developed as libraries through combinatorial chemistry. Individual molecules can be tested for interaction with the target using robotic systems in the hopes of identifying one or a few that are candidates for further development. Identification of genes that modify a disease phenotype can also be used in drug development. The proteins encoded by modifying genes may interact with a target protein and reveal mechanisms of altering the function of that protein, suggesting new avenues of treatment.

Advances in genetics will also provide new approaches to test potential new drugs. Animal models of disease and tissue culture systems provide safe systems for preclinical testing. Use of expression arrays to identify patterns of gene expression resulting from exposure of a tissue to a drug may reveal patterns that are predictive of successful therapy, or of toxicity, before introduction to humans. As noted previously, genetic testing may also provide information to modify drug dosage or avoid use of a drug in individuals at risk of severe side effects.

Gene therapy has been a long-anticipated addition to the treatment armamentarium. There have been several areas of success and major setbacks in recent years, but the promise remains bright. Challenges include achieving expression in physiological quantities, targeting the delivery of a gene to the correct cells, maintaining expression over long periods, avoiding immune reactions, and the risk of causing pathogenic mutations by insertional mutagenesis. Overcoming these challenges will be a critically important prerequisite to the integration of gene therapy into routine practice.

11. SUMMARY

A full working knowledge of disease, genetic mechanisms, epidemiology, genotyping, and statistical analysis will be required for successful study of complex disease. The future discoveries will therefore be made by teams of researchers cooperating to integrate all parts, including clinicians, laboratory scientists, biostatisticians, bioinformaticians, and pharmaceutical designers. The thoughtful application of discoveries to cure or ameliorate diseases will be a great challenge to clinicians as we enter a new age of genomic medicine.

Further Reading

Guttmacher AE, Collins FS. Genomic medicine—A primer. *N Engl J Med* 2002;347:1512–1520.

Haines JL and Pericak-Vance MA (ed.). *Approaches to Gene Mapping in Complex Human Diseases.* New York, Wiley–Liss, 1998.

Olson JM, Witte JS, Elston RC. Genetic mapping of complex traits. *Stat Med* 1999;18:2961–2981.

Rao DC. Genetic Dissection of Complex Traits. Boston, *Adv Genet* 2001;42:13–34.

Wolfsberg TG, Wetterstrand KA, Guyer MS, Collins FS, Baxevanis AD. A user's guide to the human genome. *Nat Genet* 2002;32(Suppl.):1–79.

References

1. Haines JL and Pericak-Vance MA. *Approaches to Gene Mapping in Complex Human Diseases.* New York, Wiley–Liss, 1998.
2. Little J, *et al.* Reporting, appraising, and integrating data on genotype prevalence and gene–disease associations. *Am J Epidemiol* 2002;156:300–310.
3. Hirschhorn JN, Altshuler D. Once and again—Issues surrounding replication in genetic association studies. *J Clin Endocrinol Metab* 2002;87:4438–4441.
4. Lohmueller KE, Pearce CL, Pike M, Lander ES, Hirschhorn JN. Meta-analysis of genetic association studies supports a contribution of common variants to susceptibility to common disease. *Nat Genet* 2003;33:177–182.

5. Altshuler D, *et al.* A haplotype map of the human genome. *Nature* 2005;437:1299–1320.

6. Gibbs RA, *et al.* The International HapMap Project. *Nature* 2003;426:789–796.

7. Hirschhorn J, Daly M. Genome-wide association studies for common diseases and complex traits. *Nature Rev Genet* 2005;6:95–108.

8. Hirschhorn JN, Lohmueller K, Byrne E, Hirschhorn K. A comprehensive review of genetic association studies. *Genet Med* 2002;4:45–61.

9. Altmuller J, Palmer LJ, Fischer G, Scherb H, Wjst M. Genomewide scans of complex human diseases: True linkage is hard to find. *Am J Hum Genet* 2001;69:936–950.

10. Khoury MJ, James LM. Population and familial relative risks of disease associated with environmental factors in the presence of gene–environment interaction. *Am J Epidemiol* 1993;137:1241–1250.

11. Risch N. Linkage strategies for genetically complex traits: I. Multilocus models. *Am J Hum Genet* 1990;46:222–228.

12. Risch N. Linkage strategies for genetically complex traits: II. The power of affected relative pairs. *Am J Hum Genet* 1990;46:229–241.

13. Kruglyak L, Daly MJ, Reeve-Daly MP, Lander ES. Parametric and nonparametric linkage analysis: A unified multipoint approach. *Am J Hum Genet* 1996;58:1347–1363.

14. Lander E, Kruglyak L. Genetic dissection of complex traits: Guidelines for interpreting and reporting linkage results. *Nat Genet* 1995;11:241–247.

15. Bishop DT. Linkage analysis: Progress and problems. *Philos Trans R Soc London B Biol Sci* 1994;344:337–343.

16. Bishop DT, Williamson JA. The power of identity-by-state methods for linkage analysis. *Am J Hum Genet* 1990;46:254–265.

17. Kruglyak L. Nonparametric linkage tests are model free. *Am J Hum Genet* 1997;61:254–255.

18. Ardlie KG, Kruglyak L, Seielstad M. Patterns of linkage disequilibrium in the human genome. *Nat Rev Genet* 2002;3:299–309.

19. Campbell CD, *et al.* Demonstrating stratification in a European–American population. *Nat Genet* 2005.

20. Cardon LR, Palmer LJ. Population stratification and spurious allelic association. *Lancet* 2003;361:598–604.

21. Reich DE, Goldstein DB. Detecting association in a case–control study while correcting for population stratification. *Genet Epidemiol* 2001;20:4–16.

22. Horvath S, Xu X, Laird NM. The family based association test method: Strategies for studying general genotype–phenotype associations. *Eur J Hum Genet* 2001;9:301–306.

23. Elosua R, *et al.* Association of APOE genotype with carotid atherosclerosis in men and women: The Framingham Heart Study. 2004;45:1868–1875.

24. Ueda H, *et al.* Association of the T-cell regulatory gene CTLA4 with susceptibility to autoimmune disease. *Nature* 2003;423:506–511.

25. Altshuler D, *et al.* The common PPARgamma Pro12Ala polymorphism is associated with decreased risk of type 2 diabetes. *Nat Genet* 2000;26:76–80.

26. Silverman EK, Palmer LJ. Case–control association studies for the genetics of complex respiratory diseases. *Am J Respir Cell Mol Biol* 2000;22:645–648.

27. Nelson MR, Kardia SL, Ferrell RE, Sing CF. A combinatorial partitioning method to identify multilocus genotypic partitions that predict quantitative trait variation. *Genome Res* 2001;11:458–470.

28. Ritchie MD, *et al.* Multifactor-dimensionality reduction reveals high-order interactions among estrogen-metabolism genes in sporadic breast cancer. *Am J Hum Genet* 2001;69:138–147.

29. Ritchie MD, Hahn LW, Moore JH. Power of multifactor dimensionality reduction for detecting gene–gene interactions in the presence of genotyping error, missing data, phenocopy, and genetic heterogeneity. *Genet Epidemiol* 2003;24:150–157.

30. Sachidanandam R, *et al.* A map of human genome sequence variation containing 1.42 million single nucleotide polymorphisms. *Nature* 2001;409:928–933.

31. Gabriel SB, *et al.* The structure of haplotype blocks in the human genome. *Science* 2002;296:2225–2229.

32. Shephard N, John S, Cardon L, McCarthy MI, Zeggini E. Will the real disease gene please stand up? *BMC Genet* 2005;6(Suppl. 1):S66.

33. Abecasis GR, *et al.* Age-related macular degeneration: A high-resolution genome scan for susceptibility loci in a population enriched for late-stage disease. *Am J Hum Genet* 2004;74:482–494.

34. Edwards AO, *et al.* Complement factor H polymorphism and age-related macular degeneration. *Science* 2005;308:421–424.

35. Haines JL, *et al.* Complement factor H variant increases the risk of age-related macular degeneration. *Science* 2005;308:419–421.

36. Klein RJ, *et al.* Complement factor H polymorphism in age-related macular degeneration. *Science* 2005;308:385–389.

37. Wolfsberg TG, Wetterstrand KA, Guyer MS, Collins FS, Baxevanis AD. A user's guide to the human genome. *Nat Genet* 2002;32(Suppl.):1–79.

38. Guttmacher AE, Jenkins J, Uhlmann WR. Genomic medicine: Who will practice it? A call to open arms. *Am J Med Genet* 2001;106:216–222.

39. Burke W, *et al.* Genetic test evaluation: Information needs of clinicians, policy makers, and the public. *Am J Epidemiol* 2002;156:311–318.

40. Allison DB, *et al.* The heritability of body mass index among an international sample of monozygotic twins reared apart. *Int J Obesity Related Metab Disorders: J Int Assoc Study Obesity* 1996;20:501–506.

41. Lawrence JS, Martins CL, Drake GL. A family survey of lupus erythematosus. 1. Heritability. *J Rheumatol* 1987;14:913–921.

42. Feltkamp TE, Aarden LA, Lucas CJ, Verweij CL, de Vries RR. Genetic risk factors for autoimmune diseases. *Immunol Today* 1999;20:10–12.

43. Duffy DL, Martin NG, Battistutta D, Hopper JL, Mathews JD. Genetics of asthma and hay fever in Australian twins. *Am Rev Respir Dis* 1990;142:1351–1358.

44. Edfors-Lubs ML. Allergy in 7000 twin pairs. *Acta Allergol* 1971;26:249–285.

45. Nieminen MM, Kaprio J, Koskenvuo M. A population-based study of bronchial asthma in adult twin pairs. *Chest* 1991;100:70–75.

Index